Mouse Models in the Study of Genetic Neurological Disorders

Advances in Neurochemistry

Volumes 1–4	Edited by B. W. Agranoff and M. H. Aprison
Volume 5	OLIGODENDROGLIA Edited by William T. Norton
Volume 6	AXONAL TRANSPORT IN NEURONAL GROWTH AND REGENERATION Edited by John S. Elam and Paul Cancalon
Volume 7	NEUROCHEMICAL CORRELATES OF CEREBRAL ISCHEMIA Edited by Nicolas G. Bazan, Pierre Braquet, and Myron D. Ginsberg
Volume 8	MAGNETIC RESONANCE SPECTROSCOPY AND IMAGING IN NEUROCHEMISTRY Edited by Herman Bachelard
Volume 9	MOUSE MODELS IN THE STUDY OF GENETIC NEUROLOGICAL DISORDERS Edited by Brian Popko

A Continuation Order Plan is available for this series. A continuation order will bring delivery of each new volume immediately upon publication. Volumes are billed only upon actual shipment. For further information please contact the publisher.

Mouse Models in the Study of Genetic Neurological Disorders

Edited by

Brian Popko
University of North Carolina at Chapel Hill
Chapel Hill, North Carolina

Kluwer Academic / Plenum Publishers
New York, Boston, Dordrecht, London, Moscow

Cover illustration: Transgenic mouse that expresses a human spinocerebellar ataxia type 1 gene containing an expanded CAG trinucleotide repeat. The mouse displays evidence of ataxia and Purkinje cell degeneration. For details see Chapter 8 by Orr and Zoghbi in this volume. Reproduced with permission from Burright *et al.*, (1995) *Cell* **82**:937–948.

ISBN 0-306-45965-5

233 Spring Street, New York, N.Y. 10013

10 9 8 7 6 5 4 3 2 1

A C.I.P. record for this book is available from the Library of Congress.

Printed in the United States of America

CONTRIBUTORS

DAVID R. BORCHELT • *Department of Pathology and Division of Neuropathology, The Johns Hopkins University School of Medicine, Baltimore, Maryland 21205–2196*

DON W. CLEVELAND • *Ludwig Institute for Cancer Research and Departments of Medicine and Neuroscience, University of California, San Diego, La Jolla, California 92093*

ALBERTO C. S. COSTA • *The Jackson Laboratory, Bar Harbor, Maine 04609*

SÉBASTIEN COUILLARD-DESPRÉS • *Centre for Research in Neuroscience, McGill University, The Montréal General Hospital Research Institute, Montréal H3G 1A4 Canada*

MURIEL T. DAVISSON • *The Jackson Laboratory, Bar Harbor, Maine, 04609*

JOHN R. GILBERT • *Division of Neurology, Department of Medicine, Joseph and Kathleen Bryan Alzheimer's Disease Research Center, Durham, North Carolina 27705*

JEAN-PIERRE JULIEN • *Centre for Research in Neuroscience, McGill University, The Montréal General Hospital Research Institute, Montréal, H3G 1A4, Canada*

DOMNA KARAGOGEOS • *Institute of Molecular Biology and Biotechnology and Department of Basic Sciences, University of Crete Medical School, Heraklion 711 10, Crete, Greece*

CARMEN LAMPRON • *Centre for Research in Neuroscience, McGill University, The Montréal General Hospital Research Institute, Montréal, H3G 1A4, Canada*

J. DAVID MCDONALD • *Department of Biological Sciences, Wichita State University, Wichita, Kansas 67260*

KLAUS-ARMIN NAVE • *Center for Molecular Biology, University of Heidelberg, D-69120 Heidelberg, Germany*

HARRY T. ORR • *Departments of Laboratory Medicine and Pathology, and Biochemistry, and Institute of Human Genetics, University of Minnesota, Minneapolis, Minnesota 55455*
OURANIA PAVLOU • *Institute of Molecular Biology and Biotechnology, Heraklion 711 10, Crete, Greece*
MICHAEL PODERYCKI • *Department of Biology, Boston College, Chestnut Hill, Massachusetts 02167*
BRIAN POPKO • *Department of Biochemistry and Biophysics, Program in Molecular Biology and Biotechnology, UNC Neuroscience Center, University of North Carolina at Chapel Hill, Chapel Hill, North Carolina 27599–7250*
DONALD L. PRICE • *Departments of Pathology, Neurology, and Neuroscience and Division of Neuropathology, The Johns Hopkins University School of Medicine, Baltimore, Maryland 21205–2196*
ALLEN D. ROSES • *Division of Neurology, Department of Medicine, Joseph and Kathleen Bryan Alzheimer's Disease Research Center, Durham, North Carolina 27705*
DONALD E. SCHMECHEL • *Division of Neurology, Department of Medicine, Joseph and Kathleen Bryan Alzheimer's Disease Research Center, Durham, North Carolina 27705; Department of Neurobiology, Duke University Medical Center, Durham, North Carolina; and Durham Veterans Administration Medical Center, Durham, North Carolina*
THOMAS N. SEYFRIED • *Department of Biology, Boston College, Chestnut Hill, Massachusetts 02167*
INMACULADA SILOS-SANTIAGO • *Laboratory of Developmental Neurogenetics, Department of Molecular Oncology, Bristol-Myers Squibb, PR1, Princeton, New Jersey 08543*
SANGRAM S. SISODIA • *Department of Pharmacological and Physiological Sciences, University of Chicago, Chicago, Illinois 60637*
WILLIAM D. SNIDER • *Center for the Study of Nervous System Injury, Department of Neurology, Washington University School of Medicine, St. Louis, Missouri 63110*
UELI SUTER • *Institute of Cell Biology, ETH-Hönggerberg, CH-8093 Zürich, Switzerland*
KUNIHIKO SUZUKI • *Neuroscience Center and Departments of Neurology and Psychiatry, University of North Carolina School of Medicine, Chapel Hill, North Carolina 27599*
KINUKO SUZUKI • *Neuroscience Center and Department of Pathology & Laboratory Medicine, University of North Carolina School of Medicine, Chapel Hill, North Carolina 27599*
MARIANA TODOROVA • *Department of Biology, Boston College, Chestnut Hill, Massachusetts 02167*

MARIE T. VANIER • *INSERM-CNRS U189, Lyon-Sud School of Medicine and Fondation Gillet-Mérieux, Lyon-Sud Hospital, F-69921 Oullins Cedex, France*
TONI L. WILLIAMSON • *Ludwig Institute for Cancer Research, University of California, San Diego, La Jolla, California 92093*
PHILIP C. WONG • *Department of Pathology and Division of Neuropathology, The Johns Hopkins University School of Medicine, Baltimore, Maryland 21205–2196*
PU-TING XU • *Division of Neurology, Department of Medicine, Joseph and Kathleen Bryan Alzheimer's Disease Research Center, Durham, North Carolina 27705*
HUDA Y. ZOGHBI • *Departments of Pediatrics, and Molecular and Human Genetics, and Howard Hughes Medical Institute, Baylor College of Medicine, Houston, Texas 77030*

PREFACE

The number of mouse models that are available for the study of human genetic neurological disorders is large and growing rapidly. Therefore, it was difficult to select the models that were reviewed in this volume. Clearly, there are important models that are not discussed, and perhaps a volume twice this size would have been more appropriate. Moreover, the pace at which new models are being developed and analyzed is rapid. As this volume goes to press, I am sure that additional mouse genes responsible for naturally occurring neurological disorders are being discovered and that many new transgenic and mutant mouse strains are being developed. Therefore, this volume should not be viewed as a comprehensive compendium, but rather as an update of work in progress. It is exhilarating to witness the fast pace at which these models are being established as important tools in the study of basic neuroscience and neurological disorders. It will be even more exciting to see their utilization in the development and testing of therapeutic interventions for these diseases.

I would like to thank each of the authors who have contributed to this volume for their time and their expertise. I would also like to thank Drs. Timothy Coetzee and Joshua Corbin for their advice in the selection of the topics covered. I am deeply indebted to Dr. Kunihiko Suzuki, who first approached me with the idea for this volume, for his guidance throughout its preparation. I would also like to express my utmost gratitude to Kathy Toews for her untiring assistance in all aspects of the preparation of my chapter and of the completed volume. I would also like to thank Michael Hennelly at Kluwer Academic/Plenum Publishers for his patience and counsel in the preparation of the manuscript and for facilitating the rapid publication of the final product.

Brian Popko

CONTENTS

CHAPTER 1

AN OVERVIEW OF MOUSE MODELS IN NEUROSCIENCE RESEARCH

BRIAN POPKO

CHAPTER 2

X-LINKED DYSMYELINATION: MOUSE MODELS OF PELIZAEUS–MERZBACHER DISEASE

KLAUS-ARMIN NAVE

CHAPTER 3

CHARCOT–MARIE–TOOTH DISEASE: PATHOLOGY, GENETICS, AND ANIMAL MODELS

UELI SUTER

CHAPTER 4

MOUSE MUTATIONS IN THE STUDY OF CEREBELLAR DEVELOPMENT

DOMNA KARAGOGEOS AND OURANIA PAVLOU

CHAPTER 5

THE ROLE OF NEUROTROPHIC FACTORS IN DEVELOPMENT AND NEURODEGENERATIVE DISORDERS

WILLIAM D. SNIDER AND INMACULADA SILOS-SANTIAGO

CHAPTER 6

TRANSGENIC MICE WITH NEUROFILAMENT ABNORMALITIES

CARMEN LAMPRON, SÉBASTIEN COUILLARD-DESPRÉS AND JEAN-PIERRE JULIEN

CHAPTER 7

MOUSE MODELS OF AMYOTROPHIC LATERAL SCLEROSIS

TONI L. WILLIAMSON AND DON W. CLEVELAND

CHAPTER 8

TRANSGENIC MOUSE MODELS OF CAG TRINUCLEOTIDE REPEAT NEUROLOGIC DISEASES

HARRY T. ORR AND HUDA Y. ZOGHBI

CHAPTER 9

ALZHEIMER'S DISEASE AND GENETICALLY ENGINEERED ANIMAL MODELS

DAVID R. BORCHELT, PHILIP C. WONG, SANGRAM S. SISODIA, AND DONALD L. PRICE

CHAPTER 10

MODEL OF GENETIC SUSCEPTIBILITY TO LATE-ONSET ALZHEIMER'S DISEASE: MICE TRANSGENIC FOR HUMAN APOLIPOPROTEIN E ALLELES

DONALD E. SCHMECHEL, PU-TING XU, JOHN R. GILBERT, AND ALLEN D. ROSES

Chapter 11

LYSOSOMAL DISORDERS

Kunihiko Suzuki, Marie T. Vanier, and Kinuko Suzuki

CHAPTER 12

NEUROLOGICAL IMPLICATIONS OF THE GENETIC MOUSE MODELS FOR HUMAN PHENYLKETONURIA AND HYPERPHENYLALANINEMIA

J. DAVID MCDONALD

CHAPTER 13

MOUSE MODELS OF DOWN SYNDROME

MURIEL T. DAVISSON AND ALBERTO C. S. COSTA

CHAPTER 14

MODELING EPILEPTIC DISORDERS IN MICE

THOMAS N. SEYFRIED, MICHAEL PODERYCKI, AND MARIANA TODOROVA

CHAPTER 1

AN OVERVIEW OF MOUSE MODELS IN NEUROSCIENCE RESEARCH

BRIAN POPKO

1. INTRODUCTION

In 1965, Sidman *et al.* published the *Catalog of the Neurological Mutants of the Mouse.* This now classic treatise cataloged and described the naturally occurring mouse mutants that were available and relevant to neuroscientists at that time. This work represented the first—and last—comprehensive review of the neurological mouse mutants and provided the framework and vision for the exploitation of these important models for neural development and disease. In the 30 years plus that have passed since this publication, not only have the molecular lesions responsible for many of the naturally occurring mutations been identified, but also the number of mouse models available that are of interest to neuroscientists has increased remarkably. The techniques that allow for the manipulation of the mouse genome have resulted in the generation of a multitude of new models

---

BRIAN POPKO • *Department of Biochemistry and Biophysics, Program in Molecular Biology and Biotechnology, UNC Neuroscience Center, University of North Carolina at Chapel Hill, Chapel Hill, North Carolina 27599-7250.*

Mouse Models in the Study of Genetic Neurological Disorders, Volume 9 of *Advances in Neurochemistry,* edited by Popko, Kluwer Academic/Plenum Publishers, New York, 1999.

of neurobiological interest. Importantly, these techniques have also allowed for the modeling of human diseases, or of predetermined aspects of these disorders. Consequently, as predicted by Sidman *et al.* (1965), it is no longer feasible to produce a comprehensive compendium describing in detail the available genetic models of interest to the neuroscientist. This current volume, therefore, does not aspire to represent itself as the updated version of the inclusive catalog of Sidman *et al.* (1965). If, however, the following pages educate with regard to the potential utility of the mouse models described or inspire the design and generation of similar models, we will consider this effort a success.

2. NATURALLY OCCURRING MOUSE MODELS

Although the naturally occurring mouse models provide a great resource for neuroscientists, the full potential of these mutants is realized only once the disrupted gene has been identified. The first of the neurological mutant loci to be identified were uncovered in the mid-1980s primarily through reverse genetics approaches. For example, the cDNAs for the myelin base protein (MBP) and proteolipid protein (PLP) were isolated and examined as logical candidates for the naturally occurring dysmyelinating mouse mutants (Roach *et al.,* 1983; Dautigny *et al.,* 1986; Nave *et al.,* 1986). These studies demonstrated that the MBP gene is altered in shiverer mice and that the PLP gene is mutated in jimpy animals. As the techniques used for the analysis of the mouse genome have become increasingly sophisticated and our knowledge of the genome has become increasingly detailed, the frequency of mutant loci identification has risen dramatically. The use of genetic linkage analyses in combination with innovative cloning strategies has recently aided in the isolation of a number of mutant loci. As the density of mouse chromosomal markers (genetic and physical) increases and, ultimately, as the entire sequence of the mouse genome becomes known, the ease of mutant allele identification through such forward genetic approaches will undoubtedly be greatly facilitated. In the next few years, the molecular lesions responsible for most of the mouse models discussed in Sidman *et al.* (1965) will likely be known.

Table 1 provides a list of the naturally occurring neurological mutants for which the molecular lesions have been characterized at this writing. The consequences of these gene identifications have ranged from the confirmation of the expected role of a particular protein to the completely unanticipated uncovering of novel pathways. For example, biochemical evidence had suggested MBP as a potential mediator in the compaction of the myelin sheath and the maintenance of the tight opposition of the cytoplasmic membrane faces of the myelinating cell (reviewed in Mendz, 1992). Therefore, it was not an entirely unexpected discovery that the hypomyelinating mouse mutation shiverer, in which com-

TABLE 1. Naturally Ocurring Mouse Mutants that Have Been Molecularly Characterized

Mutation	Gene/Protein	Reference
arrested development (*adr*)	Muscle chloride channel (*Clc-l*)	Steinmeyer *et al.* (1991)
dilute (*dv*)	myosin heavy chain	Mercer *et al.* (1991)
dystonia musculorum (*dt*)	dystonin (bullous pemphigoid antigen 1)	Brown *et al.* (1995)
dystrophic (*dy*)	merosin	Sunada *et al.* (1994)
extra-toes (*Xt*)	zinc (*Gli3*) finger gene	Schimmang *et al.* (1992); Hui and Joyner (1993)
jimpy (*jp*)	proteolipid protein (*PLP*)	Dautigny *et al.* (1986); Nave *et al.* (1986)
lethargic (*lh*)	Ca^{2+} channel (*Cchβ-4*)	Burgess *et al.* (1997)
leaner (Tg^{la})	neuronal calcium channel (*Cacnl1a4*)	Doyle *et al.* (1997)
lurcher (*Lc*)	glutamate receptor (*GluRdelta2*)	Zuo *et al.* (1997)
mdx (*mdx*)	dystrophin	Hoffman *et al.* (1987); Sicinski *et al.* (1989)
motor endplate disease (*med*)	sodium channel (*Scn8a*)	Burgess *et al.* (1995)
mottled (*Mo*)	copper-transporting ATPase	Levinson *et al.* (1994)
mucopolysaccharidedosis (gus^{mps})	β-glucuronidase	Sands and Birkenmeier (1993)
muscle dysgenesis (*mdg*)	muscle-calcium channel	Chaudhari (1992)
oscillator (*ot*)	α_1 glycine receptor (*Glral*)	Buckwalter *et al.* (1994)
pink-eye cleft-palate (p^{cp})	$GABA_A$ receptor	Nakatsu *et al.* (1993)
quaking (*qk*)	quaking (*qkI*)	Ebersole *et al.* (1996)

continues

TABLE 1. Continued

Mutation	Gene/Protein	Reference
reeler (*rl*)	reelin	D'Arcangelo *et al.* (1995)
retinal degeneration (*rd*)	cGMP-phosphodiesterase	Bowes *et al.* (1990)
retinal degenerative-slow (*rds*)	*rds*/peripherin	Travis *et al.* (1989)
rostral cerebellar malformation (*rcm*)	*UNC-5*	Ackerman *et al.* (1997)
scrambler (*scm*)	disabled (*mdabl*)	Ware *et al.* (1997); Sheldon *et al.* (1997)
shiverer (shi^{mbp})	myelin basic protein (*MBP*)	Roach *et al.* (1983)
slow-wave epilepsy (*swe*)	sodium/hydrogen exchanger (*Nhe1*)	Cox *et al.* (1997)
spastic (*spa*)	glycine receptor (*Glyrb*)	Mulhardt *et al.* (1994); Kingsmore *et al.* (1994)
splotch (*Sp*)	*Pax-3*	Epstein *et al.* (1991)
staggerer (*sg*)	orphan (*ROR* α) nuclear hormone receptor	Hamilton *et al.* (1996)
swaying (*sw*)	*wnt-1*	Thomas *et al.* (1991)
tottering (*tg*)	voltage-sensitive calcium channel (*alpha1A*)	Fletcher *et al.* (1996)
trembler (*Tr*)	peripheral myelin protein 22 (*PMP22*)	Suter *et al.* (1992)
twitcher (*twi*)	galactocerebrosidase	Sakai *et al.* (1996)
vibrator (*vb*)	phosphatidylinositol transfer protein α	Hamilton *et al.* (1997)
weaver (*wv*)	potassium channel (*GIRK*)	Patil *et al.* (1995)
yotari (*yot*)	disabled (*mdab1*)	Sheldon *et al.* (1997)

pacted myelin is completely lacking, was due to a partial deletion of the MBP gene (Roach *et al.*, 1983). On the other hand, the recent discovery that the weaver mutation is due to a point mutation in a gene encoding a G-protein coupled, inward rectifying potassium channel was unexpected (Patil *et al.*, 1995). In weaver mice, cerebellar granule cells fail to differentiate prior to migration and synaptogenesis (reviewed in Hatten and Heintz, 1995), such that ion channel genes were not high on the list of candidate loci. The weaver mutation was identified based on its position within the mouse genome (Karagogeos and Pavlou, *this volume*). One of the great strengths of this so-called "new genetics" is that preconceived notions of the likely function of a mutated gene do not serve as the sole basis of disease gene identification.

3. TRANSGENIC MODELS

In 1980, it was demonstrated that DNA could be introduced into the mouse germline through the direct injection of single-celled zygotes (Gordon *et al.*, 1980). Soon afterward, it was shown that properly designed DNA constructs have the potential to be consistently expressed *in vivo* in defined cell types and developmental periods (Brinster *et al.*, 1981). In addition to providing critical information concerning molecular aspects of the development and function of the nervous system, this technology has provided the opportunity to model aspects of human neurological diseases through the targeted expression of chimeric genes to neural tissue. These transgenic studies have been particularly useful in modeling the so-called "gain of function" mutations. Transgenic studies have also been very helpful in distinguishing the causes of neurological disorders from their phenotypic consequences.

A general outline of the steps involved in the introduction of DNA constructs into the mouse germline is shown in Fig. 1. The DNA of interest is injected into one of the pronuclei of single-celled zygotes, which are isolated from the oviduct of pregnant females approximately 12 hr after fertilization. A female that has mated with a vasectomized male serves as the surrogate mother to the embryos that survive the injection. At about 3 weeks of age, pups born to the surrogate mothers are analyzed for the presence of the transgene, and mice identified as transgenic are used to generate a line of animals. Because the DNA constructs integrate into the host's cells genome randomly, each line of animals generated with a particular DNA construct is distinct. The expression of a transgene is influenced by the position at which it integrates, such that multiple lines need to be examined to ensure that a particular phenotype is due solely to the transgene and not to a position effect.

It can be argued strongly that some of the best examples of the use of the transgenic technology to decipher the etiology of particular diseases have in-

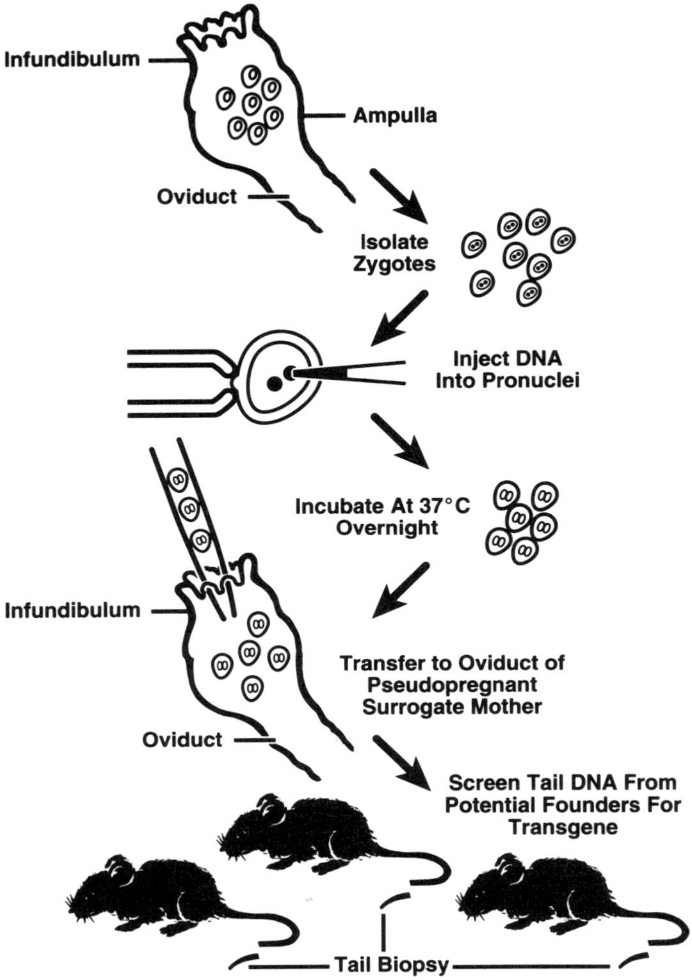

FIGURE 1. Transgenic mice. Schematic diagram of the steps involved in the generation of transgenic mice through the microinjection of single-celled mouse embryos.

cluded disorders of the nervous system. For example, the use of the transgenic approach has been crucial in our current understanding of the prion disorders, particularly with regard to species specificity and incubation time (Prusiner, 1996). Similarly, the demonstration of the neurodegenerative consequences of neurofilament overexpression (Lampron *et al., this volume*) and the neuronal expression of a mutated form of superoxide dismutase (Williamson and Cleveland, *this volume*) have significantly furthered our understanding of the etiology of amyotrophic lateral sclerosis (ALS). Moreover, recent exciting advances in the efforts to model various aspects of Alzheimer's disease (Borchelt *et al., this volume;* Schmechel *et al., this volume*) and the trinucleotide repeat disorders (Orr and Zoghbi, *this volume*) will also likely have profound consequences on our understanding of these complex diseases. The ability to model particular aspects of neurological disorders in mice also provides a unique opportunity to test new therapeutic approaches. Table 2 provides a brief overview of the efforts to model various human neurological disorders in transgenic animals.

4. GENE TARGETING IN EMBRYONIC STEM CELLS

The ability to introduce or select for specific mutations in prokaryotes and lower eukaryotes has made these organisms very attractive experimental systems. Over the past decade, technology has been developed that allows for the similar manipulation of the mouse germline (reviewed in Capecchi, 1989; Smithies, 1993). The acquisition of this capability has in short order revolutionized biology. In addition to providing critical *in vivo* information concerning the function of a particular mammalian protein, this technology has allowed for the modeling of human genetic disorders, especially those caused by loss of gene function. The availability of small animal models of human diseases is critically important in furthering our understanding of the pathogenesis of these disorders, as well as providing experimental models for testing rational approaches to therapeutic intervention.

The most common approach toward the generation of predetermined mouse mutants is outlined in Figs. 2 and 3. This approach requires the generation of a DNA construct in which a positive selectable marker, usually the gene that confers resistance to neomycin, is incorporated into an exon of the gene of interest (Fig. 2). The neo-gene acts as an insertable mutagen as well as a means to select for embryonic stem (ES) cells that take up the construct following electroporation. Also included in the construct flanking the gene of interest is a negative selectable marker, typically the thymidine kinase (TK) gene of the herpes simplex virus. This marker does not incorporate when the DNA construct recombines in a homologous manner with the target gene, such that its presence is indicative of a random integration event. The construct is electroporated into ES cells that have been established from the inner cell mass of a blastocyst and that are cultured under conditions that suppress differentiation. ES cells in which the DNA construct has successfully recombined with the endogenous sequences are selected for and reintroduced into a preimplantation mouse embryo (Fig. 3). These "chimeric" mouse embryos are then placed into the uterus of a pseudopregnant female that acts as the surrogate mother of the embryos. The ES cells

TABLE 2. Examples of Transgenic Studies Relevant to Neurological Disorders

Model (transgene)	Phenotype	Reference
Motor neuron disease		
Neurofilament overexpressors	Motor neuron degeneration	Cote *et al.* (1993); Lampron *et al.* (*this volume*)
Mutant Cu/ZN superoxide dismutase	Vacuolar degeneration of mitchondria	Wong *et al.* (1995); Williamson *et al.* (*this volume*)
Trinucleotide repeat disorders		
Spinocerebellar ataxia type 1 (SCA1)	Ataxia and Purkinje cell degeneration	Burright *et al.* (1995); Orr and Zoghbi (*this volume*)
Exon 1 of Huntington's disease gene	Involuntary stereotypic movements, seizures	Mangiarini *et al.* (1996); Orr and Zoghbi (*this volume*)
Myelin models		
PLP/DM-20 overexpression	Hypo-, dys-, and demyelination	Readhead *et al.* (1994); Johnson *et al.* (1995)
PMP-22 overexpression	Peripheral hypomyelination	Sereda *et al.* (1996); Suter (*this volume*)
PMP-22 underexpression	Thickened myelin sheaths and tomacula	Maycox *et al.* (1997); Suter (*this volume*)
CNP overexpression	Dysmyelination	Gravel *et al.* (1996)
Alzheimer's disease		
β-Amyloid precursor protein	Various "Alzheimer-like" abnormalities	Hsiao *et al.* (1996); Borchelt *et al.* (*this volume*)
Mutant presenilins	Increased β-amyloid deposition	Citron *et al.* (1997); Borchelt *et al.* (*this volume*)
Viral Disorders		
Creutzfeldt–Jakob disease	Various "CJD-like" abnormalities	Prusiner (1996)
PML (JC virus early region)	CNS hypomyelination	Small *et al.* (1986)
HIV genome in oligodendrocytes	Vacuolar myelopathy	Goudreau *et al.* (1996)
HIV gp120 in astrocytes	Glial and neuronal abnormalities	Toggas *et al.* (1994)
Neuromuscular disorders		
Mutant acetylcholine receptor	Neuromuscular disorders (slow channel syndrome)	Gomez *et al.* (1996)
Enervated insertional mutation	Muscle degeneration, PNS abnormalities	Kelly *et al.* (1994); Rath *et al.* (1995)
Epileptic disorders		
Jerky insertional mutation	Generalized clonic seizures; epileptic brain activity	Toth *et al.* (1995); Seyfried *et al.* (*this volume*)

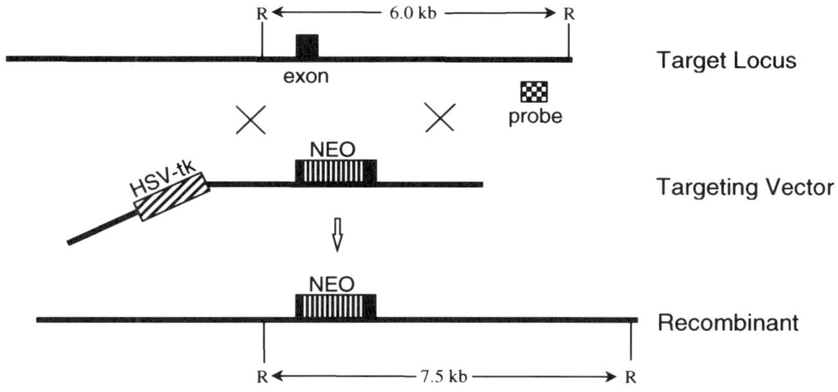

FIGURE 2. Homologous recombination of a gene targeting vector. Schematic diagram of a gene targeting event.

that have remained totipotent while in culture have the potential to contribute to all cell types of the animals that result from the chimeric embryos, including germinal cells. Animals homozygous for the genetic alteration are produced by interbreeding progeny animals with the ES cells genotype.

Such efforts to manipulate the mouse germline in a predetermined manner have been exploited exceedingly well by neuroscientists. As outlined in Table 3, these studies have shed considerable light on the molecules critically involved in the early development of the nervous system, as well as the growth factors that play an essential role in the modeling of neural circuits. Such studies have also been particularly useful in helping to elucidate the function of proteins and lipids critical to the myelination process. Importantly, several human genetic neurological diseases have also been modeled using the gene targeting approach. For example, many of the lysosomal storage disorders that lead to neurological dysfunction have been modeled in mice (Suzuki *et al., this volume*). The availability of these authentic models of human disease provides a wonderful opportunity to test approaches toward therapeutic intervention.

5. FUTURE DIRECTIONS

A number of recent developments suggest that the techniques associated with the manipulation of the mouse genome are becoming increasingly sophisticated, such that the models produced for the further analysis of human neurological disorders should become similarly improved. In the following sections sev-

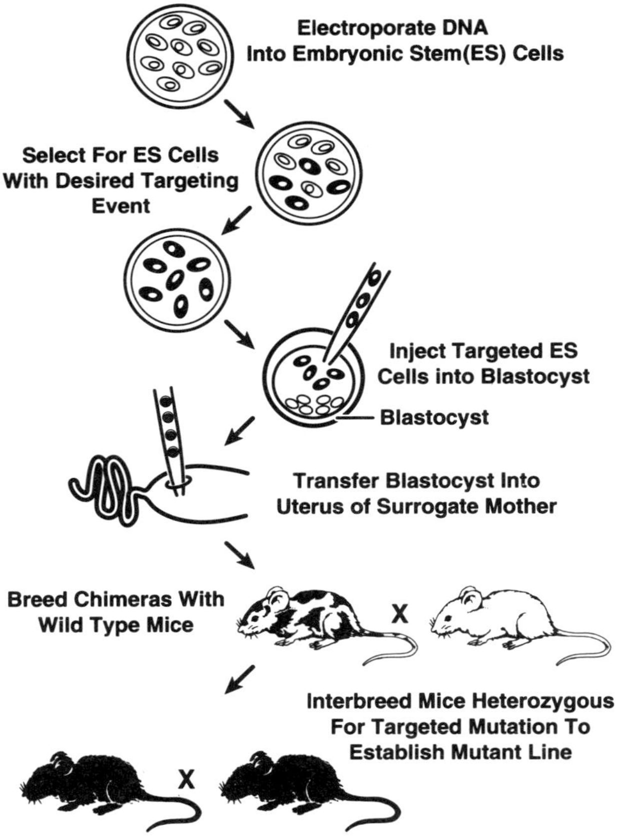

FIGURE 3. Gene targeting in embryonic stem cells. Schematic diagram of the steps involved in the generation of mutant animals through the use of the gene targeting approach in embryonic stem cells.

eral of the recent developments that should allow for the establishment of even better animal models in the near future are outlined.

5.1. Single-Copy Transgenic Mice with Chosen Site of Integration

A problem that is encountered when transgenic mice are generated by the microinjection of DNA into single-celled embryos is that the injected DNA integrates into the host cells genome randomly. The expression of the transgene is usually influenced to a large degree by the host sequences surrounding it in the genome. Consequently, multiple transgenic lines need to be examined to distinguish the influence of the surrounding DNA from that contained in the transgene.

TABLE 3. Examples of Knockout Models Relevant to Human Neurological Disorders

System	Gene	References
Neural development		
	Bcl-x	Motoyama *et al.* (1995)
	Cdk5	Ohshima *et al.* (1996)
	engrailed-1	Wurst *et al.* (1994)
	engrailed-2	Joyner *et al.* (1991)
	Fyn	Umerori *et al.* (1994)
	HNF-3	Weinstein *et al.* (1994); Ang and Rossant (1994)
	Hoxa-1	Chisaka *et al.* (1992); Lufkin *et al.* (1992)
	ISL1	Pfaff *et al.* (1996)
	Krox20	Schneider-Maunoury *et al.* (1993); Swiatek and Gridley (1993)
	Mash-1	Guillemot *et al.* (1993)
	Otx1	Acampora *et al.* (1996)
	Otx2	Ang *et al.* (1996); Matsuo *et al.* (1995)
	pax2	Torres *et al.* (1996)
	pax5	Urbanek *et al.* (1994)
	Sonic hedgehog	Chiang *et al.* (1996)
	Twist	Chen and Behringer (1995)
	wnt-1	Thomas and Capecchi (1990); McMahon and Bradley (1990)
Cell adhesion		
	L1	Furley *et al.* (1997); Dahme *et al.* (1997)
	NCAM	Cremer *et al.* (1994)
	TAG-1	Furley *et al.* (1997)
Neurotrophins		Snider and Silos-Santiago (*this volume*)
	BDNF	Conover *et al.* (1995); Jones *et al.* (1994)
	CNTF	Masu *et al.* (1993)
	GDNF	Moore *et al.* (1996); Pichel *et al.* (1996); Sanchez *et al.* (1996)
	GGF	Meyer and Birchmeier (1995)
	NGF	Crowley *et al.* (1994)
	NT3	Ernfors *et al.* (1994)
	NT4	Liu *et al.* (1995)
Neurotrophin receptors		
	c-ret	Schuchardt *et al.* (1994)
	CTNF receptor	DeChiara *et al.* (1995)
	p75	Lee *et al.* (1992)
	trkA	Smeyne *et al.* (1994)
	trkB	Klein *et al.* (1993)
	trkC	Klein *et al.* (1994)

continues

TABLE 3. Continued

System	Gene	References
Myelination		
	CGT	Coetzee *et al.* (1996); Bosio *et al.* (1996)
	MAG	Li *et al.* (1994); Montag *et al.* (1994)
	P0	Giese *et al.* (1992)
	PLP	Klugman *et al.* (1997); Nave (*this volume*)
	PMP22	Adlkofer *et al.* (1995)
	SCIP	Jaegle *et al.* (1996); Bermingham *et al.* (1996)
Neurotransmission		
	Dopamine D1 receptor	Xu *et al.* (1994)
	Dopamine D2 receptor	Baik *et al.* (1995)
	Dopamine D3 receptor	Accili *et al.* (1996)
	Dopamine transporter	Giros *et al.* (1996)
	Dopamine	Zhou and Palmiter (1995)
	glutamate receptor-B	Brusa *et al.* (1995)
	mglutamate receptor 1	Aiba *et al.* (1994)
	mglutamate receptor 2	Yokoi *et al.* (1996)
	glutamate transporter	Tanaka *et al.* (1997)
	neuronal NO synthase	Nelson *et al.* (1995)
	NMDA receptor 1	Forrest *et al.* (1994); Li *et al.* (1994)
	synapsin 1	Rosahl *et al.* (1993)
	synapsin 2	Rosahl *et al.* (1995)
Miscellaneous		
	CaMKII	Silva *et al.* (1992); Bach *et al.* (1995)
	GAP-43	Strittmatter *et al.* (1995)
	GFAP	Gomi *et al.* (1995); Pekny *et al.* (1995)
	Na^+, K^+*-ATPase*	Magyar *et al.* (1994)
	netrim-1	Serafini *et al.* (1996)
	Vimentin	Galou *et al.* (1996)

Moreover, transgenes can (5–10%) act as insertional mutagens such that when animals from a particular line are bred to homozygosity they often display an unexpected phenotype due to the inactivation of an endogenous gene. Another problem encountered with transgenic studies is that the transgenes usually integrate at variable copy numbers. Bronson *et al.* (1996) have addressed these issues with a technique that uses homologous recombination in ES cells to target the integration of a single copy of the transgene to a specific predetermined location in the mouse genome. Although this technique is a bit more cumbersome than the standard microinjection approach, the uniformity of transgene expression in the animals generated should make this a useful means of establishing lines that express the transgene at consistent, predictable levels.

5.2. Inducible Transgene Expression

Another major shortcoming of the transgenic approach has been the lack of a reliable means of targeting the expression of the transgene to specific cell types in an inducible manner. Although expression from the metalithione transcriptional control region is strongly stimulated by heavy metals in liver cells, this regulatory region is active in only a few cell types and there is considerable basal expression in the absence of inducer (Palmiter *et al.,* 1993). Therefore, this system cannot be easily adapted to provide an inducible system of interest to neuroscientists. Recently, however, two approaches toward inducible transgene expression have been developed that appear suitable for the desired, targeted expression to the CNS. In one, the repressor of the prokaryotic tetracycline (*tet*) operon has been fused to the eukaryotic transcriptional activating domain of VP16 (Furth *et al.,* 1994; Kistner *et al.,* 1996). This fusion gene is placed under the transcriptional control of a specific regulatory region that is active in the desired cells. The gene of interest is placed under the transcriptional control of the *tet* operator. In one version of this system, the *tet* fusion protein binds to the *tet* operator in the absence of *tet,* thereby activating transcription of the transgene. When *tet* is added to the system it binds to the *tet* repressor, which is released from the *tet* operator, thus shutting down transcription of the transgene. In the other version of this system, a mutant form of the *tet* repressor is used that only binds to the *tet* operator when associated with *tet* (Gossen *et al.,* 1995). In this version, transcription of the transgene is off in the absence of the *tet,* but is induced by up to five orders of magnitude in its presence. Preliminary evidence suggests that the *tet* systems will be useful for the controlled induction of genes in the nervous system (Mayford *et al.,* 1996).

In the other recently developed inducible system, the *Drosophila* hormone ecdysone is exploited as the activator of transcription (No *et al.,* 1996). In this system, the ecdysone receptor is placed under the transcriptional control of the cell-specific regulatory region. Transgenic mice expressing the ecdysone receptor are capable of activating an ecdysone responsive transcriptional control region, to which the gene of interest is linked, upon the administration of hormone. Ecdysone administration alone reportedly has no effect in mice. In the only published report of the use of this system in mice, inductions of up to four orders of magnitude were achieved with a lower level of basal activity than that seen in the *tet* system (No *et al.,* 1996). The use of these inducible systems should provide a valuable means for the targeted expression of neurotoxic substance to the CNS of adult animals.

5.3. *Cre/loxP*

Gene targeting in embryonic stem cells provides an extremely powerful means by which to assess the function of a particular protein. Nevertheless, mutations in genes that are expressed in multiple cell types often result in pleotropic phenotypes, such that it is difficult to implicate a particular cell type in a specific phenotypic consequence. Moreover, certain mutations result in embryonic lethality due to the developmental expression of a gene that is also important in the mature nervous system. For these reasons, considerable effort has been exerted in an effort to develop a means by which to introduce mutations in restricted cell types or at defined developmental stages, or both. One means by which to achieve regional or cell specific gene inactivation exploits the *Cre/loxP* site-specific recombination system of the P1 bacteriophage (review in Sauer, 1994). In this approach, the expression of the *Cre* recombinase is targeted to the specific cells of interest in transgenic animals using an appropriate transcriptional control element. The *loxP* sites, which are the targets of the *Cre* recombinase, are inserted into the mouse genome flanking an exon of the gene of interest, exploiting embryonic stem cell technology. When mice are generated that are homozygous for the targeted gene tagged with the *loxP* sites and that contain the *Cre* transgene, the targeted gene will be inactivated only in those cells in which the transgene is expressed. This approach has recently been used in a particularly exciting set of studies in which the *N*-methyl-*D*-aspartate (NMDA) receptor gene was specifically disrupted in the CA1 region of the hippocampus (McHugh *et al.*, 1996; Tsien *et al.*, 1996a,b). It is likely that the *Cre/loxP* recombination approach will also be useful in combination with the inducible systems described previously such that the developmental period in which the mutation is induced can be controlled (St. Onge *et al.*, 1996).

5.4. Point Mutations

The standard approach to gene targeting results in null, or complete loss of function, mutations. Nevertheless, to carry out detailed structure–function analyses of a particular protein *in vivo,* or to create precise animal models of a specific human genetic disease, it is necessary to introduce subtle mutations into the mouse germline. Over the past few years, several methods have been developed that allow for such an approach. Deng *et al.* (1993) describe a process in which the subtle mutation is linked to the positive selectable marker in the targeting vector. This approach has the advantage that only a single targeting event is required, but it has the disadvantage that the positive selectable marker remains in the targeted locus. The presence of the marker within the transcriptional unit might adversely effect the expression of the gene of interest. In response to such concerns, the "in–out" (Valancius and Smithies, 1991) and "hit and run" (Hasty *et al.,* 1991) procedures were developed. These approaches are more cumbersome than the previously described approach in that they require two steps of selection—one for the properly targeted locus and one for the precise excision of the positive selectable marker. Nevertheless, they have the considerable advantage that mutant loci are produced that are devoid of DNA sequences outside of the introduced mutation. The further refinement of these approaches, perhaps by flanking the positive selectable marker with *lox* sites to increase the efficiency of excision, will make the introduction of subtle mutations into the mouse genome a more common route to gene alteration.

6. SUMMARY

Clearly, the number of animal models available to neuroscientists for study is increasing at a rapid pace. The sophisticated techniques that have been developed for the efficient and detailed analysis of large genomes have allowed for the identification of the genes responsible for a number of naturally occurring mutations that result in phenotypes of interest to neuroscientists. The identification of the naturally occurring disease loci is particularly useful in that many of these models have already been examined in considerable detail by neuroscientists at the morphological, physiological, and biochemical levels. Moreover, the various approaches toward the manipulation of the mouse genome have resulted in the establishment of a multitude of animal models that express neurological phenotypes of predetermined design. The development of new variations of the approaches to the alteration of the mouse genome, as well as other innovative approaches toward the generation of mouse models (McDonald, *this volume;* Davisson and Costa, *this volume*), will allow for the development of even more sophisticated models in the near future.

REFERENCES

Acampora, D., Mazan, S., Avantaggiato, V., Barone, P., Tuorto, F., Lallemand, Y., Brulet, P., and Simeone, A., 1996, Epilepsy and brain abnormalities in mice lacking the *Otx1* gene, *Nat. Genet.* **14:**218–222.

Accili, D., Fishburn, C. S., Drago, J., Steiner, H., Lachowicz, J. E., Park, B. H., Gauda, E. B., Lee, E. J., Cool, M. H., Sibley, D. R., Gerfen, C. R., Westphal, H., and Fuchs, S., 1996, A targeted mutation of the D3 dopamine receptor gene is associated with hyperactivity in mice, *Proc. Natl. Acad. Sci. USA* **93:**1945–1949.

Ackerman, S. L., Kozak, I. P., Przyborski, S. A., Rund, L. A., Boyer, B. B., and Knowles, B. B., 1997, The mouse rostral cerebellar malformation gene encodes an UNC-5-like protein, *Nature* **386:**838–842.

Adkofer, K., Martin, R., Aguzzi, A., Zielasek, J., Toyka, K. V., and Suter, U., 1995, Hypermyelination and demyelinating peripheral neuropathy in *Pmp22*-deficient mice, *Nat. Genet.* **11:**274–280.

Aiba, A., Chen, C., Herrup, K., Rosenmund, C., Stevens, C. F., and Tonegawa, S., 1994, Reduced hippocampal long-term potentiation and context-specific deficit in associative learning in *mGluR1* mutant mice, *Cell* **79:**365–375.

Ang, S. L., and Rossant, J., 1994, HNF-3 beta is essential for node and notochord formation in mouse development, *Cell* **78:**561–574.

Ang, S. L., Jin, O., Rhinn, M., Daigle, N., Stevenson, L., and Rossant, J., 1996, A targeted mouse *Otx2* mutation leads to severe defects in gastrulation and formation of axial mesoderm and to deletion of rostral brain, *Development* **122:**243–252.

Bach, M. E., Hawkins, R. D., Osman, M., Kandel, E. R., and Myaford, M., 1995, Impairment of spatial but not contextual memory in *CaMKII* mutant mice with a selective loss of hippocampal LTP in the range of the theta frequency, *Cell* **81:**905–915.

Baik, J. H., Picetti, R., Saiardi, A., Thiriet, G., Dierich, A., Depaulis, A., Le Meur, M., and Borrelli, E., 1995, Parkinsonian-like locomotor impairment in mice lacking dopamine D2 receptors, *Nature* **377:**424–428.

Bermingham, J. R., Jr., Scherer, S. S., O'Connell, S., Arroyo, E., Kalla, K. A., Powell, F. L., and Rosenfeld, M. G., 1996, *Tst-1/Oct-6/SCIP* regulates a unique step in peripheral myelination and is required for normal respiration, *Genes Dev.* **10:**1751–1762.

Bosio, A., Binczek, E., and Stoffel, W., 1996, Functional breakdown of the lipid bilayer of the myelin membrane in central and peripheral nervous system by disrupted galactocerebroside synthesis, *Proc. Natl. Acad. Sci. USA* **93:**13280–13285.

Bowes, C., Li, T., Daneiger, M., Baxter, L. C., Applebury, M. L., and Farber, D. B., 1990, Retinal degeneration in the rd mouse is caused by a defect in the beta subunit of rod cGMP-phosphodiesterase, *Nature* **347:**677–680.

Brinster, R. L., Chen, H. Y., Trumbauer, M., Senear, A. W., Warren, R., and Palmiter, R. D., 1981, Somatic expression of herpes thymidine kinase in mice following injection of a fusion gene into eggs, *Cell* **27:**223–231.

Bronson, S. K., Plaehn, E. G., Kluckman, K. D., Hagaman, J. R., Maeda, N., and Smithies, O., 1996, Single-copy transgenic mice with chosen-site integration, *Proc. Natl. Acad. Sci. USA* **93:**9067–9072.

Brown, A., Bernier, G., Mathieu, M., Rossant, J., and Kothary, R., 1995, The mouse *dystonia musculorum* gene is a neural isoform of bullous pemphigoid antigen I, *Nat. Genet.* **10:**301–306.

Brusa, R., Zimmermann, F., Koh, D. S., Feldmeyer, D., Gass, P., Seeburg, P. H., and Sprengel, R., 1995, Early-onset epilepsy and postnatal lethality associated with an editing-deficient *GluR-B* allele in mice, *Science* **270:**1677–1680.

Buckwalter, M. S., Cook, S. A., Davisson, M. T., White, W. F., and Camper, S. A., 1994, A frameshift mutation in the mouse alpha 1 glycine receptor gene (*Glral*) results in progressive neurological symptoms and juvenile death, *Hum. Mol. Genet.* **3:**2025–2030.

Burgess, D. L., Kohrman, D. C., Galt, J., Plummer, N. W., Jones, J. M., Spear, B., and Meisler, M. H., 1995, Mutation of a new sodium channel gene, *Scn8a,* in the mouse mutant 'motor endplate disease', *Nat. Genet.* **10:**461–465.

Burgess, D. L., Jones, J. M., Meisler, M. H., and Noebels, J. L., 1997, Mutation of the Ca^{2+} channel beta subunit gene *Cchb4* is associated with ataxia and seizures in the lethargic (lh) mouse, *Cell* **88:**385–392.

Burright, E. N., Clark, H. B., Servadio, A., Matilla, T., Feddersen, R. M., Yunis, W. S., Duvick, L. A., Zoghbi, H. Y., and Orr, H. T., 1995, SCA1 transgenic mice: A model for neurodegeneration caused by an expanded CAG trinucleotide repeat, *Cell* **82:**937–948.

Capecchi, M. R., 1989, Altering the genome by homologous recombination, *Science* **244:**1288–1292.

Chaudhari, N., 1992, A single nucleotide deletion in the skeletal muscle-specific calcium channel transcript of muscular dysgenesis (*mdg*) mice, *J. Biol. Chem.* **267:**25636–25639.

Chen, Z. F., and Behringer, R. R., 1995, Twist is required in head mesenchyme for cranial neural tube morphogenesis, *Genes Dev.* **9:**686–699.

Chiang, C., Litingtun, Y., Lee, E., Young, K. E., Corden, J. L., Westphal, H., and Beachy, P. A., 1996, Cyclopia and defective axial patterning in mice lacking Sonic hedgehog gene function, *Nature* **383:**407–413.

Chisaka, O., Musci, T. S., and Capecchi, M. R., 1992, Developmental defects of the ear, cranial nerves and hindbrain resulting from targeted disruption of the mouse homeobox gene *Hax-1.6, Nature* **355:**516–520.

Citron, M., Westaway, D., Xia, W., Carlson, G., Diehl, T., Levesque, G., Johnson-Wood, K., Lee, M., Seubert, P., Davis, A., Kholodenko, D., Motter, R., Sherrington, R., Perry, B. Yao, H., Strome, R., Lieberburg, I., Rommens, J., Kim, S., Schenk, D., Fraser, P., St. George Hyslop, P., and Selkoe, D. J., 1997, Mutant presenilins of Alzheimer's disease increase production of 42-residue amyloid beta-protein in both transfected cells and transgenic mice, *Nature Med.* **3:**67–72.

Coetzee, T., Fujita, N., Dupree, J., Shi, R., Blight, A., Suzuki, K., Suzuki, K., and Popko, B., 1996, Myelination in the absence of galactocerebroside and sulfatide: Normal structure with abnormal functional and regional instability, *Cell* **86:**209–219.

Conover, J. C., Erickson, J. T., Katz, D. M., Blanchi, L. M., Poueymirou, W. T., McClain, J., Pan, L., Helgren, M., Ip, N. Y., Boland, P., Friedman, B., Wiegand, S., Vejsada, R., Kato, A. C., DeChiara, T. M., and Yancoupoulos, G. D., 1995, Neuronal deficits, not involving motor neurons, in mice lacking BDNF and/or NT4, *Nature* **375:**235–238.

Cote, F., Collard, J. F., and Julien, J. P., 1993, Progressive neuronopathy in transgenic mice expressing the human neurofilament heavy gene: A mouse model of amyotrophic lateral sclerosis, *Cell* **73:**35–46.

Cox, G. A., Lutz, C. M., Yang, C.-L., Biemesderfer, D., Bronson, R. T., Fu, A., Aronson, P. S., Noebels, J. L., and Frankel, W. N., 1997, Sodium/hydrogen exchanger gene defect in slow-wave epilepsy mutant mice, *Cell* **91:**139–148.

Cremer, H., Lange, R., Christoph, A., Plomann, M., Vopper, G., Roes, J., Brown, R., Baldwin, S., Kraemer, P., Scheff, S., Barthels, D., Rajewsky, K., and Wille, W., 1994, Inactivation of the N-*CAM* gene in mice results in size reduction of the olfactory bulb and deficits in spatial learning, *Nature* **367:**455–459.

Crowley, C., Spencer, S. D., Nishimura, M. C., Chen, K. S., Pitts-Meek, S., Armanini, M. P., Ling, L. H., MacMahon, S. B., Shelton, D. L., and Levinson, A. D., 1994, Mice lacking nerve growth factor display perinatal loss of sensory and sympathetic neurons yet develop basal forebrain cholinergic neurons, *Cell* **76:**1001–1011.

Dahme, M., Bartsch, U., Martini, R., Anliker, B., Schachner, M., and Mantei, N., 1997, Disruption of the mouse *L1* gene leads to malformations of the nervous system, *Nat. Genet.* **17:**346–349.

D'Arcangelo, G., Miao, G. G., Chen, S. C., Soares, H. D., Morgan, J. I., and Curran, T., 1995, A protein related to extracellular matrix proteins deleted in the mouse mutant reeler, *Nature* **374:**719–723.

Dautigny, A., Mattei, M. G., Morello, D., Alliel, P. M., Pham-Dinh, D., Amar, L., Arnaud, D., Simon, D., Mattei, J. F., Guenet, J. L., Jolles, P., and Avners, P., 1986, The structural gene coding for myelin-associated proteolipid protein is mutated in jimpy mice, *Nature* **321:**867–869.

DeChiara, T. M., Vejsada, R., Poueymirou, W. T., Acheson, A., Suri, C., Conover, J. C., Friedman, B., McClain, J., Pan, L., Stahl, N., Ip, N. Y., Kato, A., and Yancopoulos, G. D., 1995, Mice lacking the CNTF receptor, unlike mice lacking CNTF, exhibit profound motor neuron deficits at birth, *Cell* **83:**313–322.

Deng, C., Thomas, K. R., and Capecchi, M. R., 1993, Location of crossovers during gene targeting with insertion and replacement vectors, *Mol. Cell. Biol.* **13:**2134–2140.

Doyle, J., Ren, X., Lennon, G., and Stubbs, L., 1997, Mutations in the *Cacnl1a4* calcium channel gene are associated with seizures, cerebellar degeneration, and ataxia in tottering and leaner mutant mice, *Mamm. Genome* **8:**113–120.

Ebersole, T. A., Chen, Q., Justice, M. J., and Artzt, K., 1996, The quaking gene product necessary in embryogenesis and myelination combines features of RNA binding and signal transduction proteins, *Nat. Genet.* **12:**260–265.

Epstein, D. J. Vekemans, M., and Gros, P., 1991, *Splotch (Sp2H),* a mutation affecting development of the mouse neural tube, shows a deletion within the paired homeodomain of *Pax-3, Cell* **67:**767–774.

Ernfors, P., Lee, K. F., Kucera, J., and Jaenisch, R., 1994, Lack of neurotrophin-3 leads to deficiencies in the peripheral nervous system and loss of limb proprioceptive afferents, *Cell* **77:**503–512.

Fletcher, C. F., Lutz, C. M., O'Sullifan, T. N., Shaughnessy, J. D., Jr., Hawkes, R., Frankel, W. N., Copeland, N. G., and Jenkins, N. A., 1996, Absence epilepsy in tottering mutant mice is associated with calcium channel defects, *Cell* **87:**607–617.

Forrest, D., Yuzaki, M., Soares, H. D., Ng, L., Luk, D. C., Sheng, M., Stewart, C. L., Morgan, J. L., Connor, J. A., and Curran, T., 1994, Targeted disruption of *NMDA receptor 1* gene abolishes NMDA response and results in neonatal death, *Neuron* **13:**325–338.

Furley, A. J., Stewart, C. J., Jessel, T. M., and Soriano, P., 1997, Analysis of mice carrying mutations in neural cell adhesion molecules, *in* "Keystone Symposia on Molecular and Cellular Biology," Keystone, Colorado (March 31–April 6, 1997), p. 95.

Furth, P. A.,St. Onge, L., Boger, H., Gruss, P., Gossen, M., Kistner, A., Bujard, H., and Hennighausen, L., 1994, Temporal control of gene expression in transgenic mice by a tetracycline-responsive promoter, *Proc. Natl. Acad. Sci. USA* **91:**9302–9306.

Galou, M., Colucci-Guyon, E., Ensergueix, D., Ridet, J. L., Gimenez y Ribotta, M., Privat, A., Babinet, C., and Dupouey, P., 1996, Disrupted glial fibrillary acidic protein network in astrocytes from vimentin knockout mice, *J. Cell Biol.* **133:**853–863.

Giese, K. P., Martini, R., Lemke, G., Soriano, P., and Schachner, M., 1992, Mouse P_0 gene disruption leads to hypomyelination, abnormal expression of recognition molecules, and degeneration of myelin and axons, *Cell* **71:**565–576.

Giros, B., Jaber, M., Jones, S. R., Wightman, R. M., and Caron, M. G., 1996, Hyperlocomotion and indifference to cocaine and amphetamine in mice lacking the dopamine transporter, *Nature* **379:**606–612.

Gomez, C. M., Bhattacharyya, B. B., Charnet, P., Day, J. W., Labarca, C., Wollmann, R. L., and Lambert, E. H., 1996, A transgenic mouse model of the slow-channel syndrome, *Muscle Nerve* **19:**79–87.

Gomi, H., Yokoyama, T., Fujimoto, K., Ikeda, T., Katoh, A., Itoh, T., and Itohara, S., 1995, Mice devoid of the glial fibrillary acidic protein develop normally and are susceptible to scrapie prions, *Nature* **14:**29–41.

Gordon, J. W., Scangos, G. A., Plotkin, D. J., Barbosa, J. A., and Ruddle, F. H., 1980, Genetic transformation of mouse embryos by microinjection of purified DNA, *Proc. Natl. Acad. Sci. USA* **77:**7380–7384.

Gossen, M., Freundlieb, S., Bender, G., Muller, G., Hillen, W., and Bujard, H., 1995, Transcriptional activation by tetracyclines in mammalian cells, *Science* **268:**1766–1769.

Goudreau, G., Carpenter, S., Beaulieu, N., and Jolicoeur, P., 1996, Vacuolar myelopathy in transgenic mice expressing human immunodeficiency virus type 1 proteins under the regulation of the myelin basic protein gene promoter, *Nat. Med.* **2:**655–661.

Gravel, M., Peterson, J., Yong, V. W., Kottis, V., Trapp, B., and Braun, P. E., 1996, Overexpression of 2′,3′-nucleotide 3′-phosphodiesterase in transgenic mice alters oligodendrocyte development and produces aberrant myelination, *Mol. Cell. Neurosci.* **7:**453–466.

Guillemot, F., Lo, L.-C., Johnson, J. E., Auerbach, A. B., Anderson, D. J., and Joyner, A. L., 1993,

Mammalian *achaete-scute* homolog-1 is required for the early development of olfactory and autonomic neurons, *Cell* **75**:463–476.
Hamilton, B. A., Frankel, W. N., Kerrebrock, A. W., Hawkins, T. L., FitzHugh, W., Kusumi, K., Russell, L. B., Mueller, K. L., van Berkel, V., Birren, B. W., Kruglyak, L., and Lander, E. S., 1996, Disruption of the nuclear hormone receptor RORalpha in staggerer mice, *Nature* **379**:736–739.
Hamilton, B. A., Smith, D. J., Mueller, K. L., Kerrebrock, A. W., Bronson, R. T., van Berkel, V., Daly, M. J., Kruglyak, L., Reeve, M. P., Nemhauser, J. L., Hawkins, T. L., Rubin, E. M., and Lander, E. S., 1997, The vibrator mutation causes neurodegeneration via reduced expression of PITPα: Positional complementation cloning and extragenic suppression, *Neuron* **18**:711–722.
Hasty, P., Ramirez-Solis, R., Krumlauf, R., and Bradley, A., 1991, Introduction of a subtle mutation into the *Hox-2.6* locus in embryonic stem cells, *Nature* **350**:243–246.
Hatten, M. E., and Heintz, N., 1995, Mechanisms of neural patterning and specification in the developing cerebellum, *Annu. Rev. Neurosci.* **18**:385–408.
Hoffman, E. P., Brown, R. H., Jr., and Kunkel, L. M., 1987, Dystrophin: The protein product of the Dechenne muscular dystrophy locus, *Cell* **51**:919–928.
Hsiao, K., Chapman, P., Nilsen, S., Eckman, C., Harigaya, Y., Younkin, S., Yang, F., and Cole, G., 1996, Correlative memory deficits, Aβ elevation, and amyloid plaques in transgenic mice, *Science* **274**:99–102.
Hui, C. C., and Joyner, A. L., 1993, A mouse model of greig cephalopolysyndactyly syndrome: The *extra-toes* mutation contains an intragenic deletion of the *Gli3* gene, *Nat. Genet.* **3**:241–246.
Jaegle, M., Mandemakers, W., Broos, L., Zwart, R., Karis, A., Visser, P., Grosveld, F., and Meijer, D., 1996, The POU factor Oct-6 and Schwann cell differentiation, *Science* **273**:507–510.
Johnson, R. S., Roder, J. C., and Riordan, J. R., 1995, Over-expression of the DM-20 myelin proteolipid causes central nervous system demyelination in transgenic mice, *J. Neurochem.* **64**:967–976.
Jones, K. R., Farinas, I., Backus, C., and Reichardt, L. F., 1994, Targeted disruption of the BDNF gene perturbs brain and sensory neuron development but not motor neuron development, *Cell* **76**:989–999.
Joyner, A. L., Herrup, K., Auerbach, A., Davis, C. A., and Rossant, J., 1991, Subtle cerebellar phenotype in mice homozygous for a targeted deletion of the *En-2* homeobox, *Science* **251**:1239–1243.
Kelly, D., Chancellor, K., Milatovich, A., Francke, U., Suzuki, K., and Popko, B., 1994, Autosomal recessive neuromuscular disorder in a transgenic line of mice, *J. Neurosci.* **14**:198–207.
Kingsmore, S. F., Giros, B., Suh, D., Bieniarz, M., Caron, M. G., and Seldin, M. F., 1994, Glycine receptor beta-subunit gene mutation in spastic mouse associated with LINE-1 element insertion, *Nat. Genet.* **7**:136–141.
Kistner, A., Gossen, M., Zimmermann, F., Jerecic, J., Ullmer, C., Lübbert, H., and Bujard, H., 1996, Doxycycline-mediated quantitative and tissue-specific control of gene expression in transgenic mice, *Proc. Natl. Acad. Sci. USA* **93**:10933–10938.
Klein, R., Smeyne, R. J., Wurst, W., Long, L. K., Auerbach, B. A., Joyner, A. L., and Barbacid, M., 1993, Targeted disruption of the *trkB* neurotrophin receptor gene results in nervous system lesions and neonatal death, *Cell* **75**:113–122.
Klein, R., Silos-Santiago, I., Smeyne, R. J., Lira, S. A., Brambilla, R., Bryant, S., Zhang, L., Snider, W. D., and Barbacid, M., 1994, Disruption of the neurotrophin-3 receptor gene *trkC* eliminates Ia muscle afferents and results in abnormal movements, *Nature* **368**:249–251.
Klugmann, M., Schwab, M. H., Pühlhofer, A., Schneider, A., Zimmermann, F., Griffiths, I. R., and Nave, K.-A., 1997, Assembly of CNS myelin in the absence of proteolipid protein, *Neuron* **18**:59–70.
Lee, K.-F., Li, E., Huber, J., Landis, S. C., Sharpe, A. H., Chao, M. V., and Jaenisch, R., 1992,

Targeted mutation of the gene encoding the low affinity NGF receptor p75 leads to deficits in the peripheral sensory nervous system, *Cell* **69:**737–749.

Levinson, B., Vulpe, C., Elder, B., Martin, C., Verley, F., Packman, S., and Gitschier, J., 1994, The *mottled* gene is the mouse homologue of the Menkes disease gene, *Nat. Genet.* **6:**369–373.

Li, Y., Erzurumlu, R. S., Chen, C., Jhaveri, S., and Tonegawa, S., 1994, Whisker-related neuronal patterns fail to develop in the trigeminal brainstem nuclei of NMDAR1 knockout mice, *Cell* **76:**427–437.

Liu, X., Ernfors, P., Wu, H., and Jaenisch, R., 1995, Sensory but not motor neuron deficits in mice lacking NT4 and BDNF, *Nature* **375:**238–241.

Lufkin, T., Dierich, A., LeMeur, M., Mark, M., and Chambon, P., 1992, Disruption of the *Hox-1.6* homeobox gene results in defects in a region corresponding to its rostral domain of expression, *Cell* **68:**1105–1119.

Magyar, J. P., Bartsch, U., Wang, Z. Q., Howells, N., Aguzzi, A., Wagner, E. F., and Schachner, M., 1994, Degeneration of neural cells in the central nervous system of mice deficient in the gene for the adhesion molecule on Glia, the beta 2 subunit of murine Na,K-ATPase, *J. Cell Biol.* **127:**835–845.

Mangiarini, L., Sathasivam, K., Seller, M., Cozens, B., Harper, A., Hetherington, C., Lawton, M., Trottier, Y., Lehrach, H., Daview, S. W., and Bates, G. P., 1996, Exon 1 of the *HD* gene with an expanded CAG repeat is sufficient to cause a progressive neurological phenotype in transgenic mice, *Cell* **87:**493–506.

Masu, Y., Wolf, E., Holtman, B., Sendtner, M., Brem, G., and Thoenen, H., 1993, Disruption of the *CNTF* gene results in motor neuron degeneration, *Nature* **365:**27–32.

Matsuo, I., Kuratani, S., Kimura, C., Takeda, N., and Aizawa, S., 1995, Mouse *Otx2* functions in the formation and patterning of rostral head, *Genes Dev.* **9:**2646–2658.

Maycox, P. R., Ortuño, D., Burrola, P., Kuhn, R., Bieri, P. L., Arrezo, J. C., and Lemke, G., 1997, A transgenic mouse model for human hereditary neuropathy with liability to pressure palsies, *Mol. Cell. Neurosci.* **8:**405–416.

Mayford, M., Bach, M. E., Huang, Y. Y., Wang, L., Hawkins, R. D., and Kandel, E. R., 1996, Control of memory formation through regulated expression of a *CaMKII* transgene, *Science* **274:**1678–1683.

McHugh, T. J., Blum, K. I., Tsien, J. Z., Tonegawa, S., and Wilson, M. A., 1996, Impaired hippocampal representation of space in CA1-specific NMDAR1 knockout mice, *Cell* **87:**1339–1349.

McMahon, A. P., and Bradley, A., 1990, The *wnt-1 (int-1)* proto-oncogene is required for development of a large region of the mouse brain, *Cell* **62:**1073–1085.

Mendz, G. L., 1992, Structure and molecular interactions of myelin basic protein and its antigenic peptides, *in* "Myelin: Biology and Chemistry" (R. E. Martenson, ed.), pp. 277–366, CRC Press, Boca Raton, FL.

Mercer, J. A., Seperack, P. K., Strobel, M. C., Copeland, N. G., and Jenkins, N. A., 1991, Novel myosin heavy chain encoded by murine dilute coat colour locus, *Nature* **349:**709–713.

Meyer, D., and Birchmeier, C., 1995, Multiple essential functions of neuregulin in development, *Nature* **378:**386–390.

Montag, D., Giese, K. P., Bartsch, U., Martini, R., Lang, Y., Blüthmann, H., Harthigasan, J., Kirschner, D. A., Wintergerst, E. S., Nave, K., Zielasek, J., Toyka, K. V., Lipp, H., and Schachner, M., 1994, Mice deficient for the myelin-associated glycoprotein show subtle abnormalities in myelin, *Neuron* **13:**229–246.

Moore, M. W., Klein, R. D., Farinas, J., Sauer, H., Armanini, M., Phillips, H., Reichardt, L. F., Ryan, A. M., Carver-Moore, K., and Rosenthal, A., 1996, Renal and neuronal abnormalities in mice lacking GDNF, *Nature* **382:**76–79.

Motoyama, N., Wang, F., Roth, K. A., Sawa, H., Nakayama, K., Nakayama, K., Negishi, I., Senju, S.,

Zhang, Q., Fujii, S., and Loh, D. Y., 1995, Massive cell death of immature hematopoietic cells and neurons in *Bcl-x*-deficient mice, *Science* **267:**1506–1510.

Mulhardt, C., Fischer, M., Gass, P., Simon-Chazottes, D., Guenet, J. L., Betz, H., and Becker, C. M., 1994, The spastic mouse: Aberrant splicing of glycine receptor beta subunit mRNA caused by intronic insertion of L1 element, *Neuron* **13:**1003–1015.

Nakatsu, Y., Tyndale, R. F., DeLorey, T. M., Durham-Pierre, D., Gardner, J. M., McDanel, H. J., Nguyen, Q., Wagstaff, J., Lalande, M., Sikela, J. M., Olsen, R. W., Tobin, A. J., and Brilliant, M. H., 1993, A cluster of three $GABA_A$ receptor subunit genes is deleted in a neurological mutant of the mouse *p* locus, *Nature* **364:**448–450.

Nave, K.-A., Lai, C., Bloom, F. E., and Milner, R. J., 1986, Jimpy mutant mouse: A 74-base deletion in the mRNA for myelin proteolipid protein and evidence for a primary defect in RNA splicing, *Proc. Natl. Acad. Sci. USA* **83:**9264–9268.

Nelson, R. J., Demas, G. E., Huang, P. L., Fishman, M. C., Dawson, V. L., Dawson, T. M., and Snyder, S. H., 1995, Behavioural abnormalities in male mice lacking neuronal nitric oxide synthase, *Nature* **378:**383–386.

No, D., Yao, T.-P., and Evans, R. M., 1996, Ecdysone-inducible gene expression in mammalian cells and transgenic mice, *Proc. Natl. Acad. Sci. USA* **93:**3346–3351.

Ohshima, T., Ward, J. M., Huh, C. G., Longenecker, G., Veeranna, Pant, H. C., Brady, R. O., Martin, L. J., and Kulkarni, A. B., 1996, Targeted disruption of the cyclin-dependent kinase 5 gene results in abnormal corticogenesis, neuronal pathology and perinatal death, *Proc. Natl. Acad. Sci. USA* **93:**11173–11178.

Palmiter, R. D., Sandgren, E. P., Koeller, D. M., and Brinster, R. L., 1993, Distal regulatory elements from the mouse metallothionein locus stimulate gene expression in transgenic mice, *Mol. Cell. Biol.* **13:**5266–5275.

Patil, N., Cox, D. R., Bhat, D., Faham, M., Myers, R. M., and Peterson, A. S., 1995, A potassium channel mutation in weaver mice implicates membrane excitability in granule cell differentiation, *Nat. Genet.* **11:**126–129.

Pekny, M., Leveen, P., Pekna, M., Eliasson, C., Berthold, C. H., Westermark, B., and Betsholtz, C., 1995, Mice lacking glial fibrillary acidic protein display astrocytes devoid of intermediate filaments but develop and reproduce normally, *EMBO J.* **14:**1590–1598.

Pfaff, S. L., Mendelsohn, M., Stewart, C. L., Edlund, T., and Jessell, T. M., 1996, Requirement for LIM homeobox gene *Isl1* in motor neuron generation reveals a motor neuron-dependent step in interneuron differentiation, *Cell* **84:**309–320.

Pichel, J. G., Shen, H. Z., Granholm, A. C., Drago, J., Grinberg, A., Lee, E. J., Huang, S. P., Saarma, M., Hoffer, B. J., Sariola, H., and Westphal, H., 1996, Defects in enteric innervation and kidney development in mice lacking GDNF, *Nature* **382:**73–76.

Prusiner, S. B., 1996, Transgenetics of prion diseases, *Curr. Top. Microbiol. Immunol.* **206:**275–304.

Rath, E., Kelly, D., Bouldin, T., and Popko, B., 1995, Impaired peripheral nerve regeneration in a mutant strain of mice (*Enr*) with a Schwann cell defect, *J. Neurosci.* **15:**7226–7237.

Readhead, C., Schneider, A., Griffiths, I., and Nave, K. A., 1994, Premature arrest of myelin formation in transgenic mice with increased proteolipid protein gene dosage, *Neuron* **12:**583–595.

Roach, A., Boylan, K., Horvath, S., Prusiner, S. B., and Hood, L. E., 1983, Characterization of cloned cDNA representing rat myelin basic protein: Absence of expression in brain of shiverer mutant mice, *Cell* **34:**799–806.

Rosahl, T. W., Geppert, M., Spillane, D., Herz, J., Hammer, R. E., Malenka, R. C., and Sudhof, T. C., 1993, Short-term synaptic plasticity is altered in mice lacking synapsin I, *Cell* **75:**661–670.

Rosahl, T. W., Spillane, D., Missler, M., Herz, J., Selig, D. K., Wolff, J. R., Hammer, R. E., Malenka,

R. C., and Sudhof, T. C., 1995, Essential functions of synapsins I and II in synaptic vesicle regulation, *Nature* **375**:488–493.

Sakai, N., Inui, K., Tatsumi, N., Fukushima, H., Nishigaki, T., Taniike, M., Nishimoto, J., Tsukamoto, H., Yanagihara, I., Ozono, K., and Okada, S., 1996, Molecular cloning and expression of cDNA for murine galactocerebrosidase and mutation analysis of the twitcher mouse, a model of Krabbe's disease, *J. Neurochem.* **66**:1118–1124.

Sanchez, M. P., Silos-Santiago, I., Frisen, J., He, B., Lira, S. A., and Barbacid, M., 1996, Renal agenesis and the absence of enteric neurons in mice lacking GDNF, *Nature* **382**:70–73.

Sands, M. S., and Birkenmeier, E. H., 1993, A single-base-pair deletion in the beta-glucuronidase gene accounts for the phenotype of murine mucopolysaccharidosis type VII, *Proc. Natl. Acad. Sci. USA* **90**:6567–6571.

Sauer, B., 1994, Site-specific recombination: Developments and applications, *Curr. Opin. Biotechnol.* **5**:521–527.

Schimmang, T., Lemaistre, M., Vortkamp, A., and Ruther, U., 1992, Expression of the zinc finger gene *Gli3* is affected in the morphogenetic mouse mutant extra-toes (*Xt*), *Development* **116**:799–804.

Schneider-Maunoury, S., Topilko, P., Seitandou, T., Levi, G., Cohen-Tannoudji, M., Pournin, S., Babinet, C., and Charnay, P., 1993, Disruption of *Krox-20* results in alteration of rhombomeres 3 and 5 in the developing hindbrain, *Cell* **75**:1199–1214.

Schuchardt, A., D'Agati, V., Larsson-Blomberg, L., Costantini, F., and Pachnis, V., 1994, Defects in the kidney and enteric nervous system of mice lacking the tyrosine kinase receptor Ret, *Nature* **367**:380–383.

Serafini, T., Colamarino, S. A., Leonardo, E. D., Wang, H., Beddington, R., Skarnes, W. C., and Tessier-Lavigne, M., 1996, Netrin-1 is required for commissural axon guidance in the developing vertebrate nervous system, *Cell* **87**:1001–1014.

Sereda, M., Griffiths, I., Puhlhofer, A., Steward, H., Rossner, M. J., Zimmerman, F., Magyar, J. P., Schneider, A., Hund, E., Meinck, H. M., Suter, U., and Nave, K. A., 1996, A transgenic rat model of Charcot–Marie–Tooth disease, *Neuron* **16**:1049–1060.

Sheldon, M., Rice, D. S., Arcangelo, D., Yoneshima, H., Nakajima, K., Mikoshiba, K., Howell, B. W., Cooper, J. A., Goldowitz, D., and Curran, T., 1997, *Scrambler* and *yotari* disrupt the disabled gene and produce a reeler-like phenotype in mice, *Nature* **389**:730.

Sicinski, P., Geng, Y., Ryder-Cook, A. S., Barnard, E. A., Darlison, M. G., and Barnard, P. J., 1989, The molecular basis of muscular dystrophy in the mdx mouse: A point mutation, *Science* **244**:1578–1580.

Sidman, R. L., Green, M. C., and Appel, S. H., 1965, "Catalog of the Neurological Mutants of the Mouse," Harvard University Press, Cambridge, MA.

Silva, A. J., Stevens, C. F., Tonegawa, S., and Wang, Y., 1992, Deficient hippocampal long-term potentiation in alpha-calcium-calmodulin kinase II mutant mice, *Science* **257**:201–206.

Small, J. A., Scangos, G. A., Cork, L., Jay, G., and Khoury, G., 1986, The early region of human papovavirus JC induces dysmyelination in transgenic mice, *Cell* **46**:13–18.

Smeyne, R. J., Klein, R., Schnapp, A., Long, L. K., Bryant, S., Lewin, A., Lira, S. A., and Barbacid, M., 1994, Severe sensory and sympathetic neuropathies in mice carrying a disrupted *TRK/NGF* receptor gene, *Nature* **368**:246–249.

Smithies, O., 1993, Animal models of human genetic diseases, *Trends Genet.* **9**:112–116.

Steinmeyer, K., Klocke, R., Ortland, C., Gronemeier, M., Jockusch, H., Grunder, S., and Jentsch, T. J., 1991, Inactivation of muscle chloride channel by transposon insertion in myotonic mice, *Nature* **354**:304–308.

St. Onge, L., Furth, P. A., and Gruss, P., 1996, Temporal control of the *Cre* recombinase in transgenic mice by a tetracycline responsive promoter, *Nucleic Acids Res.* **24**:3875–3877.

Strittmatter, S. M., Fankhauser, C., Huang, P. L., Mashimo, H., and Fishman, M. C., 1995, Neuronal pathfinding is abnormal in mice lacking the neuronal growth cone protein GAP-43, *Cell* **80:**445–452.

Sunada, Y., Bernier, S. M., Kozak, C. A., Yamada, Y., and Campbell, K. P., 1994, Deficiency of merosin in dystrophic *dy* mice and genetic linkage of laminin M chain gene to dy locus, *J. Biol. Chem.* **269:**13729–13732.

Suter, U., Welcher, A. A., Ozcelik, T., Snipes, G. J., Kosaras, B., Francke, U., Billings-Gagliardi, S., Sidman, R. L., and Shooter, E. M., 1992, Trembler mouse carries a point mutation in a myelin gene, *Nature* **356:**241–244.

Swiatek, P. J., and Gridley, T., 1993, Perinatal lethality and defects in hindbrain development in mice homozygous for a targeted mutation of the zinc finger gene *Krox20, Genes Dev.* **7:**2071–2084.

Tanaka, K., Watase, K., Manabe, T., Yamada, K., Watanabe, M., Takahashi, K., Iwama, H., Nishikawa, T., Ichihara, N., Kikuchi, T., Okuyama, S., Kawashima, N., Hori, S., Takimoto, M., and Wada, K., 1997, Epilepsy and exacerbation of brain injury in mice lacking the glutamate transporter GLT-1, *Science* **276:**1699–1702.

Thomas, K. R., and Capecchi, M. R., 1990, Targeted disruption of the murine *int-1* proto-oncogene resulting in severe abnormalities in midbrain and cerebellar development, *Nature* **346:**847–850.

Thomas, K. R., Musci, T. S., Neumann, P. E., and Capecchi, M. R., 1991, Swaying is a mutant allele of the proto-oncogene *wnt-1, Cell* **67:**969–976.

Toggas, S. M., Masliah, E., Rockenstein, E. M., Rall, G. F., Abraham, C. R., and Mucke, L., 1994, Central nervous system damage produced by expression of the HIV-1 coat protein gp120 in transgenic mice, *Nature* **367:**188–193.

Torres, M., Gomez-Pardo, E., and Gruss, P., 1996, *Pax2* contributes to inner ear patterning and optic nerve trajectory, *Development* **122:**3381–3391.

Toth, M., Grimsby, J., Guzsaki, G., and Donovan, G. P., 1995, Inactivation of a novel gene related to *CENP-B* is associated with neuronal hyperexcitability and seizure in transgenic mice, *Nat. Genet.* **11:**71–75.

Travis, G. H., Brennan, M. B., Danielson, P. E., Kozak, C. A., and Sutcliffe, J. G., 1989, Identification of a photoreceptor-specific mRNA encoded by the gene responsible for retinal degeneration slow (*rds*), *Nature* **338:**70–73.

Tsien, J. Z., Chen, D. F., Gerber, D., Tom, C., Mercer, E. H., Anderson, D. J., Mayford, M., Kandel, E. R., and Tonegawa, S., 1996a, Subregion- and cell type-restricted gene knockout in mouse brain, *Cell* **87:**1317–1326.

Tsien, J. Z., Huerta, P. T., and Tonegawa, S., 1996b, The essential role of hippocampal CA1 NMDA receptor-dependent synaptic plasticity in spatial memory, *Cell* **87:**1327–1338.

Umemori, H., Sata, S., Yagi, T., Aizawa, S., and Yamamoto, T., 1994, Initial events of myelination involve Fyn tyrosine kinase signalling, *Nature* **367:**572–576.

Urbanek, P., Wang, Z. Q., Fetka, I., Wagner, E. F., and Busslinger, M., 1994, Complete block of early B cell differentiation and altered patterning of the posterior midbrain in mice lacking *Pax5/BSAP, Cell* **79:**901–912.

Valancius, V., and Smithies, O., 1991, Testing an "in–out" targeting procedure for making subtle genomic modifications in mouse embryonic stem cells, *Mol. Cell. Biol.* **11:**1402–1408.

Ware, M. L., Fox, J. W., Gonzalez, J. L., Davis, N. M., Lambert de Rouvroit, C., Russo, C. J., Chua, S. C., Jr., Goffinet, A. M., and Walsh, C. A., 1997, Aberrant splicing of a mouse disabled homolog, *mdab1,* in the scrambler mouse, *Neuron* **19:**239–249.

Weinstein, D. C., Ruiz I., Altaba, A., Chen, W. S., Hoodless, P. Prezioso, V. R., Jessell, T. M., and Darnell, J. E., Jr., 1994, The winged-helix transcription factor HNF-3 beta is required for notochord development in the mouse embryo, *Cell* **78:**575–588

Wong, P. C., Pardo, C. A., Borchelt, D. R., Lee, M. K., Copeland, N. G., Jenkins, N. A., Sisodia, S. S., Cleveland, D. W., and Price, D. L., 1995, An adverse property of a familial ALS-linked *SOD1* mutation causes motor neuron disease characterized by vacuolar degeneration of mitochondria, *Neuron* **14:**1105–1116.

Wurst, W., Auerbach, A. B., and Joyner, A. L., 1994, Multiple developmental defects in *Engrailed-1* mutant mice: An early mid-hindbrain deletion and patterning defects in forelimbs and sternum, *Development* **120:**2065–2075.

Xu, M., Moratalla, R., Gold, L. H., Hiroi, N., Koob, G. F., Graybiel, A. M., and Tonegawa, S., 1994, Dopamine D1 receptor mutant mice are deficient in striatal expression of dynorphin and in dopamine-mediated behavioral responses, *Cell* **79:**729–742.

Yokoi, M., Kobayashi, K., Manabe, T., Takahashi, T., Sakaguchi, I., Katsuura, G., Shigemoto, R., Ohishi, H., Nomura, S., Nakamura, K., Nakao, K., Katsuki, M., and Nakanishi, S., 1996, Impairment of hippocampal mossy fiber LTD in mice lacking *mGluR2, Science* **273:**645–647.

Zhou, Q. Y., and Palmiter, R. D., 1995, Dopamine-deficient mice are severely hypoactive, adipsic, and aphagic, *Cell* **83:**1197–1209.

Zuo, J., De Jager, P. L., Takahashi, K. A., Jiang, W., Linden, D. J., Heintz, N., 1997, Neurodegeneration in Lurcher mice caused by mutation in delta2 glutamate receptor gene, *Nature* **388:**769–773.

CHAPTER 2

X-LINKED DYSMYELINATION: MOUSE MODELS OF PELIZAEUS–MERZBACHER DISEASE

KLAUS-ARMIN NAVE

The analysis of naturally occurring mouse mutants with genetic defects in myelin-specific genes has helped to increase understanding of corresponding human neurological diseases. This chapter focuses on one example, the leukodystrophy Pelizaeus–Merzbacher disease (PMD) and a less severe disorder, X-linked spastic paraplegia type 2 (SPG-2). These two diseases are clinically distinct with respect to the degree of hypomyelination and motor dysfunction but involve the same gene encoding myelin proteolipid protein (PLP). Dysmyelination in a natural mouse mutant, termed *jimpy,* is a model for the connate form of PMD and largely the result of abnormal oligodendrocyte death. In the allelic *rumpshaker* mouse, a model for SPG-2, oligodendrocytes appear normal in number but fail to assemble myelin correctly. PLP-overexpressing transgenic mice provide experi-

---

DR. KLAUS-ARMIN NAVE • Center for Molecular Biology, University of Heidelberg, D-69120 Heidelberg, Germany.

Mouse Models in the Study of Genetic Neurological Disorders, Volume 9 of *Advances in Neurochemistry,* edited by Popko, Kluwer Academic/Plenum Publishers, New York, 1999.

mental evidence that even duplications of the normal PLP gene are not tolerated. A comparison of PLP point mutations with a PLP null allele, generated by gene targeting in mouse embryonic stem cells, demonstrates that dysmyelination and glial cell death are largely abnormal gain-of-function effects caused by aberrant protein expression. The various mutations of the PLP gene illustrate some of the strengths and weaknesses of mouse models for human neurological diseases.

Inherited defects of myelination are frequently divided into "dysmyelinating" diseases, to describe the abnormal development of the myelin sheath, and "demyelinating" diseases, which are myelin degenerations occurring after an apparently normal development. As we will see later such a distinction is useful from a pathophysiological point of view, but is a frequent cause of confusion because demyelinating processes can start early and thus overlap in time with developmental processes. More than 100 years ago, Pelizaeus (1885) and later Merzbacher (1910) described a human dysmeylinating disease that constitutes the prototype of the first group of inherited myelin disorders. PMD is characterized by impaired motor development, an early onset, and X-linked mode of transmission. The associated neuropathological findings are a characteristic and severe hypomyelination that is limited to the CNS with no obvious neuronal involvement or inflammation. Islets of preserved myelin around blood vessels result in a sometimes patchy appearance of the white matter. The demonstration of neutral lipid material in white matter areas by Sudan black staining led to the classification as a "sudanophilic leukodystrophy." Using these neuropathological characteristics, Seitelberger (1970) subdivided PMD into several forms, taking into account its mode of transmission, age of onset, and more recently (Seitelberger *et al.*, 1996) molecular genetic data: PMD type I corresponds to the slow progressive "classical" form, described by Pelizaeus; the more severe PMD type II with rapid progression and death within the first 10 years of life is the "connate" form. As discussed later, SPG-2 is clinically distinct from either form of PMD but, by molecular genetic criteria, constitutes a milder, allelic form (Saugier-Veber *et al.,* 1994). Today PMD and SPG-2 are genetically defined as mutations of the X chromosome-linked gene for PLP, the major membrane protein of CNS myelin. To understand the PLP disease mechanism from the level of the affected gene to the systems level has been greatly helped by analyzing various PLP gene mutations in corresponding murine models.

1. PELIZAEUS–MERZBACHER DISEASE AND X-LINKED SPASTIC PARAPLEGIA

The current diagnostic criteria of human PMD were redefined by Boulloche and Aicardi (1986) and involve, in addition to the X-linkage, electrophysiological and magnetic resonance imaging (MRI) diagnostics to demonstrate abnormalities of CNS white matter formation. These criteria can be applied at any time during the life of a patient. As first noticed by Pelizaeus, the clinical characteristic of impaired motor development presents very early, usually before 1 year of age. Motor handicaps are, in general, greater than the impairments of psychointellectual development, and a rapid deterioration of CNS functions is not observed. However, in the type II ("connate") form of PMD, affected males show virtually no psychomotor development and death occurs in the first decade. In the more frequent type I ("classical") form, PMD patients show some motor improvements for up to 10–12 years of age, but after a plateau a slow deterioration begins after ages 15–20, and death occurs, on average, in the fourth decade. The level of performance reached varies from one family to another but, interestingly, also between the affected males of the same family. This suggests that yet unidentified modifier genes play a role, and that a clinical continuum exists between the connate and classical form of PMD.

By histological, electrophysiological, and neuroradiological criteria, PMD is a dysmyelination that is restricted to the CNS. Peripheral nerve conduction velocities are relatively normal, although a peripheral involvement has been reported in a few cases (Garbern *et al.,* 1997). That hereditary spastic paraplegia (SPG) can be genetically allelic to PMD was an unexpected finding. Clinically, SPG is a heterogeneous group of diseases, divided into "pure" and "complicated" forms according to the presence or absence of other associated neurological signs. Linkage studies have demonstrated at least two SPG loci on the X chromosome. SPG-2, which is allelic to PMD in the Xq22 region, is characterized by progressive gait abnormalities after almost normal motor development in the first 2 years of life. This late-onset disease shares several clinical features with PMD, including nystagmus, cerebellar ataxia, and a pyramidal syndrome. However, hypotonia, choreoathetosis, or bobbing movements of head and trunk, which are frequently seen in PMD patients, are rarely observed. SPG-2 patients can walk unaided for many years, and most of them are able to attend school.

2. MYELIN PROTEOLIPID PROTEIN (PLP/DM20)

Human PMD was associated with mutations of the X-linked PLP gene, following the discovery that the dysmyelinated *jimpy* mouse, a natural model of PMD, expresses abnormally spliced PLP mRNA. PLPs are major structural components of CNS myelin, comprising more than 50% of total myelin protein by weight (Lees and Brostoff, 1984). Full-length PLP, also termed proteolipid apoprotein or lipophilin, has a molecular mass of 20 kDa and is an integral membrane protein of 276 amino acids (Fig. 1). A smaller isoform, termed DM20 (molecular weight 26.5 kDa) is derived by alternative mRNA splicing (Milner *et al.,* 1985; Nave *et al.,* 1987). The primary structure of both proteins, which is

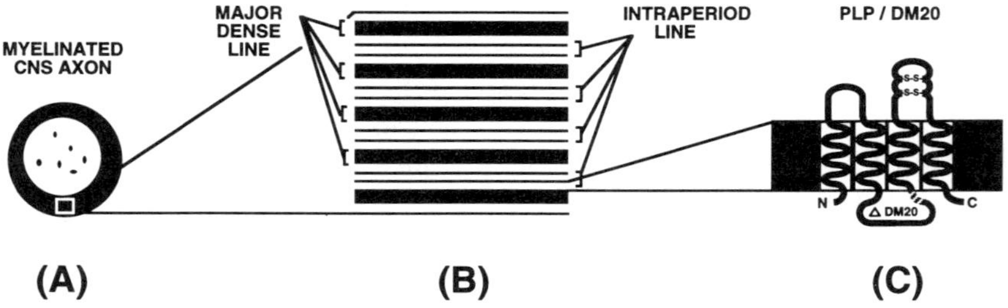

FIGURE 1. Schematic depiction of (A, B) compact myelin and (C) PLP, a 276-amino-acid membrane protein with four transmembrane domains, as a structural component of compact myelin. Indirect evidence suggests that PLP assembles into an oligomeric structure prior to incorporation into the myelin membrane. Amino acid substitutions, caused by point mutations in the human and mouse PLP gene, cluster in the membrane-embedded domains and in the second extracellular loop, leaving the mutant protein misfolded. The stippled part of the molecule (residues 136–151) is absent from the smaller DM20 isoform that is derived by alternative RNA splicing.

100% conserved between mouse and human, predicts the presence of four transmembrane domains and places PLP/DM20 into a diverse family of "four-helix bundle" proteins (Popot *et al.,* 1991). A topological model, based on the intracellular location of the acylated cystein residues, assumes that both the amino- and carboxyl-terminus of PLP/DM20 protrude at the cytoplasmic surface (Weimbs and Stoffel, 1992). In this model, DM20 differs by the absence of a 35-amino-acid peptide from the intracellular loop region. A functional difference between PLP and DM20 is not known, but preliminary data suggest that the PLP-specific loop region specifies a differential intracellular sorting of PLP and DM20 (Anderson *et al.,* 1997). Proteolipids are posttranslationally acylated proteins and disulfide bonds stabilize the extracellular loop regions and thus the overall topology of PLP/DM20. Overexpression experiments in heterologous cells have failed to reveal properties of PLP or DM20 as a cell adhesion molecule (in contrast to P0, the major membrane protein of PNS myelin; Filbin *et al.,* 1990). Transgenic experiments suggest, however, that PLP contributes to a stable association of the adjacent membranes that form the "intraperiod line" of compacted myelin.

PLP is a CNS-specific myelin component, whereas the transcript of this gene can also be detected, at a lower level, in peripheral nerve (Pucket *et al.,* 1987). Interestingly, it is the alternative DM20 mRNA that predominates in Schwann cells, and this proteolipid is subsequently unable to reach the myelin compartment. The reason for this apparent intracellular retention of DM20 in Schwann cells is unknown. In PLP-overexpressing transgenic mice (see later), PLP/DM20 becomes a major component of peripheral myelin, colocalized with P0 (Anderson *et al.,* 1997).

It is now recognized that PLP is the prototype of a small gene family including two additional neuronal proteins, termed M6A and M6B (Yan *et al.*, 1993), which correspond to the DM20 isoform. There is more than 50% sequence identity at the amino acid level between these proteins which allows identification of those amino acid positions that have been more strictly conserved in evolution than others. Mutations that alter these strictly conserved residues in PLP/DM20 turn out to cause the most severe dysmyelination phenotype in mouse and man (Nave and Boespflug-Tanguy, 1996).

2.1. Myelin Defects in *jimpy* and *rumpshaker* Mice

The genetic defects of myelination comprise one of the large subgroups of neurological mutations in the mouse (Nave, 1994). Following the cloning of the PLP gene and its assignment to the X chromosome in mouse and human (Milner *et al.*, 1985; Willard and Riordan, 1985) it was soon recognized that mutations in this gene underlie the most severe dysmyelination in the mouse, the spontaneous *jimpy* mutation.

The *jimpy* phenotype is marked by a CNS-specific dysmyelination and has provided an authentic animal model for PMD (Sidman *et al.*, 1964). Typically, 50% of the male progeny of *jimpy* carrier females are affected, whereas the heterozygous carrier female itself is phenotypically normal. Thus, by formal genetics *jimpy* is X-linked recessive. The behavioral symptoms of *jimpy* begin in the second postnatal week, coinciding with the onset of CNS myelination in wild-type mice. Ataxis and a generalized body tremor (which usually precede locomotor activity) are quite similar in this and other myelin-deficient mutants (Fig. 2). After 2–3 weeks, affected mice develop seizures that vary in extent and duration. Prolonged convulsions result in premature death, generally before 4 weeks of age.

Histologically, *jimpy* mice are almost completely myelin deficient in the CNS (only 1–2% of the axons are myelinated), whereas the PNS appears unaffected. This reflects the CNS-specific assembly, if not expression, of PLP. The allelic mouse mutant *jimpy–msd* (for "myelin-synthesis-deficient") is phenotypically undistinguishable from *jimpy,* although more axons are ensheathed (3–5% of normal). Both *jimpy* and *jimpy–msd* mice are good models for the connate form of PMD. Other well-studied mutants from this group include *myelin-deficient* (*md*) in the rat (Boison and Stoffel, 1989) and *shaking pup* in the dog (Nadon *et al.*, 1990). As larger and longer-lived animals they are well suited for physiological experiments in which the consequences of myelin deficiency are studied (Young *et al.*, 1989; Utzschneider *et al.*, 1992).

Another PLP mutant allele in mice, *rumpshaker* (*rsh*), has a much milder phenotype (Fanarraga *et al.*, 1992; Schneider *et al.*, 1992). The preservation of oligodendrocytes that escape cell death results in a substantially improved my-

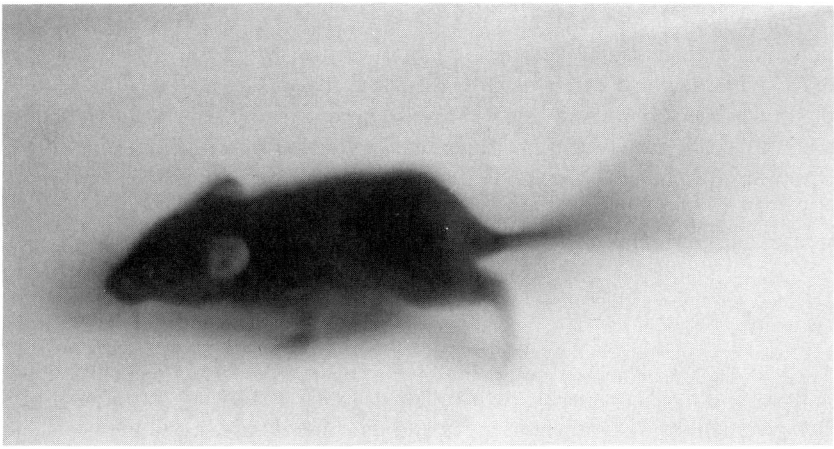

FIGURE 2. Behavioral consequences of CNS dysmyelination in the mouse include tremors and severe ataxia, similar to motor dysfunctions associated with PMD. With increasing age, lethal seizures and convulsions are more frequent in dysmyelinated rodents than in human patients (hemizygous *jimpy* mouse at 3 weeks).

elination compared to *jimpy. Rumpshaker* mice, although visibly handicapped, are long lived and reproduce well. They show less seizure activity, but the degree of their tremoring phenotype is markedly influenced by modifier genes that are specific to the background strain (unpublished observation). *Rumpshaker* mice are a good animal model for SPG-2; in fact the same point mutation (I186T) was identified later in the index family of the human disease (Kobayashi *et al.,* 1994, Johnston and MacKusick, 1962).

2.2. Cellular Consequences: From Genotype to Phenotype

Spontaneous mutations in the PLP/DM-20 gene are usually point mutations that alter the primary structure of the encoded protein (for review see Hodes *et al.,* 1994; Nave and Boespflug-Tanguy, 1996). However, to trace the link from genotype to phenotype has proven to be complicated, even in the mouse models available. In the "classical" *jimpy* mutation, a loss of the strictly conserved RNA acceptor splice site (exon 5) results in the deletion of that exon from the mature PLP mRNA, a shift in the open-reading frame, and consequently an aberrant protein that is predicted to lack most of the second extracellular loop-region and the entire fourth transmembrane (TM) domain (Nave *et al.,* 1986; Hudson *et al.,* 1987). In *jimpy–msd,* on the other hand, a conservative amino acid substitution (A242V) is located within that TM domain, at first glance a minor change to the overall structure (Gencic and Hudson, 1990). It is remarkable that the cellular consequences of these mutations are nearly the same; oligodendrocytes fail to mature and the majority of them die with morphological features of apoptosis even before myelination has begun (Knapp *et al.*, 1986; Vermeesch *et al.*, 1990). At the same time, the number of immature oligodendrocytes is increased severalfold, and mitotic precursor cells have an abnormally short cell cycle time (Skoff, 1982; Knapp and Skoff, 1987). Thus, abnormal cell death and a shift in the oligodendrocyte population toward an immature state is the main cause of myelin deficiency in *jimpy* and presumably in the connate form of human PMD. The cause of cell death, however, is not fully understood. Dying oligodendrocytes have morphological features of apoptotic cells and are laden with (sudanophilic) lipid vesicles (Meier and Bischoff, 1974; Knapp *et al.*, 1986). Interestingly, some mutant oligodendrocytes survive and give rise to scattered "islands" of myelin. These cells may correspond to patchy ("tigroid") areas of myelination in PMD (Seitelberger, 1970). Some *jimpy* oligodendrocytes are also able to survive for several months after transplantation into a non-*jimpy* host brain (Lachapelle *et al.*, 1994). It is remarkable that *jimpy* oligodendrocytes that degenerate also when taken into culture survive and differentiate to a much higher degree in the presence of an astrocyte-conditioned medium (Bartlett *et al.*, 1988), as visualized by strong galactocerebroside staining. The lethal cellular consequences of PLP mutations are thus not strictly genetically determined but can, at least in part, be influenced by "environmental" factors. The identity of these factors has not been determined. It is possible that the expression and accumulation of mutant PLP (see later) interferes specifically with a growth-factor-dependent survival mechanism of oligodendrocytes.

In *jimpy* mice, the surviving oligodendrocytes assemble myelin that is thinner than in wild-type and ultrastructurally abnormal. Specifically, the "intraperiod line," normally a double line, is collapsed and no longer distinguishable from the alternating "major dense line" (Duncan *et al.*, 1987, 1989; Schneider *et al.*, 1995). Since *jimpy* myelin is devoid of functional PLP, this abnormality illustrates most likely the loss of the major structural component of myelin.

The first demonstration of a defective PLP biosynthesis in PMD, a biochemical study, was published by Koeppen *et al.* (1987). The identification of a mutation in the PLP gene of this patient by Hudson *et al.* (1989) may have been one of the first examples that a neurological disease could be defined at the nucleotide level. As of today, more than 20 different base substitutions in the coding sequence of the PLP gene have been reported. In two thirds of these cases the nucleotide changes, identified by cloning and sequencing PCR fragments, predicted amino acid substitutions for the second extracellular loop region of PLP. A good correlation was observed between the severity of the disease and the fact that mutations led to the substitution of a highly conserved amino acid, as judged by aligning normal PLP/DM-20 with other members of the "DM family" of membrane proteins (Kitagawa *et al.*, 1993; Yan *et al.*, 1993). In addition, frameshift mutations leading to truncated forms of PLP/DM20 have been associated with PMD of different clinical severity. Unfortunately, the relationship between the molecular defect and the resulting clinical phenotype remains complicated. Perhaps most surprising, the complete deletion of the PLP gene has been reported in a family in which the two affected males expressed a very mild form of PMD (Raskind *et al.*, 1991; and see later). All this is best explained with a disease model in which the severe cellular consequences arise from expression of abnormal PLP-like polypeptides rather than the loss of PLP-function.

2.3. Studying PLP Mutations *In Vitro*

An important approach to analyze the cellular effect of PLP gene mutations has been the experimental overexpression of the cloned cDNA for wild-type and mutant PLP isoforms in transfected COS-7 cells (Gow *et al.*, 1994; Gow and Lazzarini, 1996). These studies have demonstrated two important facts: first, even nonglial cells recognize mutant PLP as an improperly folded protein that is retained in intracellular compartments, whereas the wild-type protein reaches the cell surface. Secondly, the degree of retainment correlates well with the clinical severity of the corresponding mutation *in vivo*. Thus, abnormal oligodendrocyte death could be secondary to retaining misfolded PLP that is toxic. On the other hand, transfected COS-7 cells as well as other permanent cell lines tolerate the expression of misfolded PLP surprisingly well, suggesting that any toxicity must be specific to oligodendroglial cells (also Schwann cells in *jimpy* mice express the mutant gene without signs of premature death). A monoclonal antibody, termed O10, has become a useful tool with which to distinguish wild-type PLP/DM20 (O10-positive) from any of the mutant isoforms (PLP-negative) when expressed in COS-7 cells. Obviously, the O10 epitope is a folding-sensitive domain of PLP that is not generated when the polypeptide misfolds (Jung *et al.*, 1996). It is possible that the O10 epitope emerges as PLP and DM20 molecules form oligomeric complexes in the plane of the membrane.

2.4. PLP Transgenic Mice

In an attempt to complement the defect of *jimpy* mice genetically, several lines of transgenic mice have been generated that transmit autosomal extra copies of the entire wildtype PLP gene (Kagawa *et al.*, 1994; Readhead *et al.*, 1994). A quite unexpected observation in the course of these studies was a severely dysmyelinated phenotype that marked all those transgenic mice that overexpressed the wild-type PLP gene at least twofold (as quantified at the RNA level)—in the absence of any mutation of the endogenous X-lined PLP gene. In the study of Readhead *et al.* (1994), these mice were homozygous for either one of two cosmid-derived PLP transgenes or double-heterozygous. The novel phenotype was similar to that of *jimpy* mice, but PLP-overepxressing transgenics were less severely affected and lived several weeks longer. Abnormal oligodendrocyte death was less obvious than *jimpy* but a reactive astrocytosis was more pronounced (Readhead *et al.,* 1994). It was concluded that the mere increase in PLP gene dosage is not tolerated and turns PLP into a disease gene. The molecular mechanisms that cause dysmyelination upon PLP overexpression are not understood. It is possible that PLP interacts stoichiometrically with proteins that are required for proper PLP export. If such a stoichiometry were disturbed, "excess" PLP could accumulate and mimic the retention of misfolded PLP in mutant mice. Transgenic mice that overexpress PLP/DM20 at a lower level (i.e., hemizygous transgenics) develop normally but, after 12 months of age, show also ataxia and behavioral abnormalities. Surprisingly, they are caused by a noninflammatory demyelination with clear signs of axonal involvement (Anderson *et al.,* 1998). The nature of this late-onset degenerative disease is unclear and its relationship to the developmental disorder remains to be determined. Demyelination has also been observed in adult transgenic mice overexpressing multiple copies of a DM20 cDNA construct driven by a myelin-specific promoter (Simons-Johnson *et al.,* 1995). Owing to the duration of human neural development, it is possible that any demyelination that begins 1 or 2 years after myelin deposition would still be considered an "early onset" childhood disease.

A small interstitial duplication of the chromosomal region Xq21–122, encompassing the PLP locus and numerous other marker genes, has been associated with a complex syndrome that includes PMD in a boy who was clinically not further investigated (Cremers *et al.,* 1987). The PLP/DM20 transgenic mice have provided strong evidence that duplications of the human PLP gene, and potentially other forms of dysregulated PLP gene expression, turn a normal PLP gene into a disease gene. This led to a systematic search for PLP gene duplications in PMD. A large duplication of the entire PLP gene has been reported in two classical forms of PMD (Ellis and Malcolm, 1994) and one other has been mentioned by Hodes *et al.* (1994). When a large number of PMD families was analyzed by Boespflug-Tanguy and colleagues, among 64 PMD index cases without obvious PLP gene mutation, a quantitative Southern analysis suggested PLP gene duplications in 21 cases: 11 were familial and 13 sporadic cases, and all had a typical classical form of PMD (Boespflug-Tanguy *et al.,* unpublished data; Nave and Boespflug-Tanguy, 1996). Thus, PLP gene duplications are a major cause of PMD (Table 1) that had escaped detection by standard DNA sequence analysis.

PLP gene duplications as a disease mechanism underlying PMD bear a striking resemblance to the genetics of Charcot–Marie–Tooth disease type 1A (CMT1A), a peripheral neuropathy most frequently caused by a duplication of the gene for peripheral myelin protein-22, or less frequently by point mutations in the PMP22 gene. PMP-22 is an integral membrane protein of the myelin

TABLE 1. Classification of PLP Gene Mutations

Mutations of the PLP/DM20 gene	Human PMD/SPG-2 (examples)	Rodent models
Amino acid substitutions	Connate PMD (P14L)	*md* rat (T74P)
	Connate PMD (W162R)	*jimpy*–msd (A242V)
	Connate PMD (P215S)	
	Classical PMD (V218F)	
	SPG-2 (I186T)	*rumpshaker* (I186T)
	SPG-2 (H139T)	
Frame shift mutations	Connate PMD (M257 →fs)	*jimpy* (G207 →fs)
PLP/DM20 overexpression	Classical PMD: PLP gene duplication (Xq22)	PLP-transgenic mice
PLP *null* allele	SPG-2: genomic deletion	PLP "knockout" mice

Point mutations, deletions, and duplication of the human PLP gene result in dys- and demyelination and a variable degree of motor dysfunction and intellectual impairment. They define a clinical continuum of disease expression, from pure forms of spastic paraplegia (SPG-2) to severe forms of Pelizaeus–Merzbacher disease (connate PMD). The human disease is modeled in mouse mutants that can differ in the relative time course of disease expression.

sheath, like PLP/DM20, but specific to the peripheral nervous system (see U. Suter, Chapter 3, *this volume*).

Random X chromosome inactivation in early embryonic development results in a mosaic distribution of PLP-mutant and wild-type oligodendrocytes in carrier females. With respect to PLP gene expression, there is no heterozygosity at the cellular level: females are phenotypically normal, because oligodendrocytes that die prematurely are functionally replaced by wild-type cells. If both mutant and wild-type alleles of the PLP gene were coexpressed in oligodendrocytes, would heterozygous females be dysmyelinated? All attempts to rescue *jimpy* mice by expressing an additional autosomal PLP transgene have failed. In these experiments, wild-type PLP transgenes were expressed in the correct cell type and with the same efficiency as the endogenous (mutant) PLP gene. Moreover, wild-type PLP assembled into the few compacted *jimpy*-myelin sheath where it restored the "intraperiod line." However, overall myelination and the motor impairment of *jimpy* mice were not improved. Thus, in an experimental "heterozygous" situation, the *jimpy* allele acted in a dominant-negative fashion: glial cell death and dysmyelination appeared to result from the expression of *jimpy* PLP in oligodendrocytes (abnormal gain-of-function) rather than from the absence of the wild-type protein (loss-of-function) (Schneider *et al.,* 1995). On close inspection we noted a small but significant increase of cell survival and myelination in all PLP-transgenic *jimpy* mice. This further suggests that the dominant-negative effect of misfolded *jimpy*-PLP (see "toxicity" earlier) is partly shielded by the presence of wild-type PLP. The implications of these findings for PMD is that, even if one could reintroduce the wild-type PLP gene to mutant oligodendrocytes, myelination could not be restored unless expression of the mutant PLP allele has stopped.

2.5. PLP "Knockout" Mice

A missing link in demonstrating the role of PLP and DM20 in myelin disease directly have been mutant mice that do not express any gene product, that is, PLP/DM20-*null* mutant mice. Attempts to generate such a loss-of-function allele by homologous recombination in mouse embryonic stem cells have been made in two laboratories, following different experimental strategies, and giving slightly different results. One approach was to remove the alternative RNA splice site in exon 3B of the PLP gene and selectively express one of two PLP isoforms (Boison and Stoffel, 1994). However, this manipulation, which left the neomycine resistance gene within intron 3, resulted in transgenic mice, termed "PLP-neo" and "DM20-neo," that showed an unexpected defect of mRNA splicing and a truncation of the encoded polypeptide. Moreover the rate of transcription was reduced and the combined effect resulted in the apparent absence of PLP and DM20 from myelin. The second line of PLP-deficient mice was created directly as a PLP/DM20 *null* allele by removing exon 1 of the PLP/DM20 gene, including the translation start codon (Klugmann *et al.,* 1997; Griffiths, *et al.,* 1998).

Unexpectedly, mice in both lines lacked severe motor problems and revealed a rather normal development of CNS white matter areas at the macroscopic level. At the cellular level, substantial myelin synthesis was obvious and oligodendrocytes were differentiated without the signs of abnormal cell death that mark *jimpy* mice. At the ultrastructural level, Boison *et al.* (1995) reported that myelin membranes are uncompacted whorles of loose membranes, whereas Klugmann *et al.* (1997) demonstrated compacted myelin of reduced physical stability (Figure 3). A reinvestigation of DM20-neo mice (Rosenbluth *et al.,* 1996) confirmed that delamination is most likely secondary to fixation for electron microscopy and thus an informative "artefact." Taken together, PLP and DM20 are not necessary for spiral membrane wrapping and subsequent myelin compaction, but they contribute to the overall adhesive forces that stabilize the extracellular "intraperiod line" of compacted CNS myelin.

Interestingly, in DM20-neo mice (but not in the PLP *null* allele) a significant fraction of small-caliber axons remained unmyelinated, as illustrated by Boison *et al.* (1995) for the optic nerve. This suggests that even small amounts of a truncated PLP-like polypeptide can inhibit myelinating oligodendrocytes (Klugmann *et al.,* 1997), and that these myelin deficits could underlie the subtle behavioral phenotype detected in DM20-neo mice (Boison *et al.,* 1995). Another unexpected finding in PLP-deficient mice were clear signs of axonal involvement. From the age of 6 to 8 weeks focal axonal swellings containing organelles

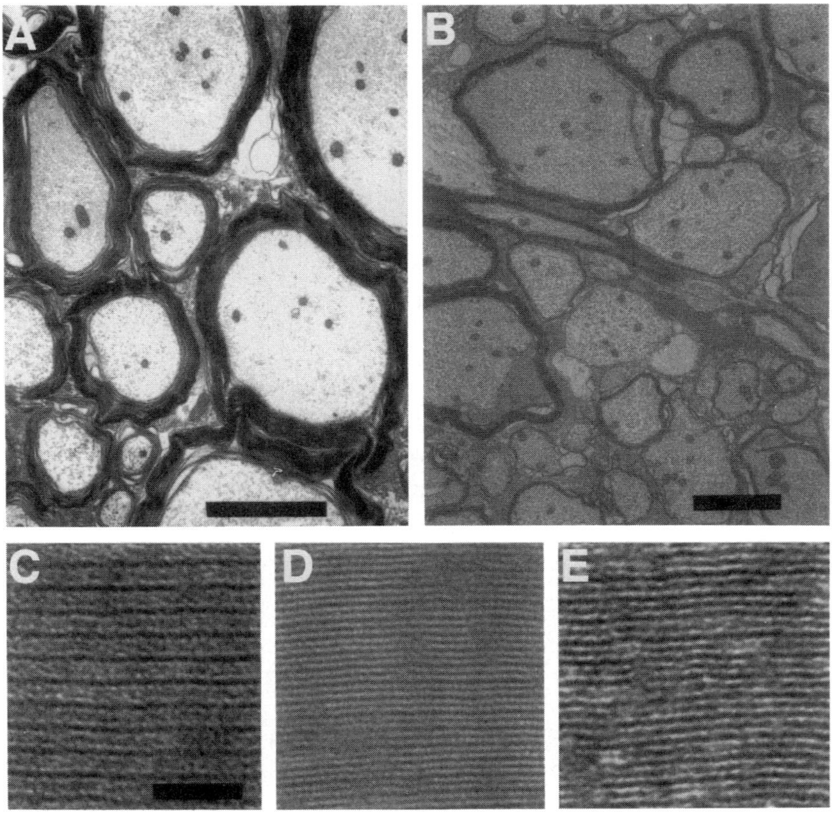

FIGURE 3. Ultrastructure of CNS myelin in PLP mutant mice. (A) Electron micrograph of spinal cord from adult *plp*^{null}/*Y* mice showing axons surrounded by compacted myelin sheaths of appropriate thickness. Bar = 2 mm. (B) In comparison with the well-myelinated tissue shown in A, the spinal cord of an adult *rumpshaker* mouse is severely dysmyelinated. Many axons are naked or thinly myelinated. Note that *plp*^{rsh} represents the least severe phenotype of the natural mutant mice. Bar = 2 μm. (C–E) Higher magnification reveals differences in the periodicity of myelin from wild-type (C), PLP-deficient (D), and *jimpy* mice (E). Note that the "intraperiod line" is condensed in the absence of PLP/DM20, resembling the MDL. Bar = 50 nm. (Adapted from Klugmann *et al.,* 1997)

were detected throughout the white and gray matter in all regions of the CNS, particularly in areas where small-diameter myelinated axons predominate. In animals more than 1 year of age axonal degeneration was so prominent that also the animals' motor performance deteriorated compared to age-matched controls (Griffiths *et al.,* 1998). The molecular basis of this remarkable interaction between oligodendrocytes and axons is under investigation.

The comparison of *jimpy* mice (or any allele with a point mutation in the PLP gene) and PLP-deficient mice leads to the firm conclusion that abnormal PLP itself is responsible for dysmyelination, whereas the loss of PLP function is not related to the disease phenotype, at least in mice. This interpretation is in agreement with the mentioned inability to rescue *jimpy* mutants using a wild-type PLP transgene (Nadon *et al.*, 1994; Schneider *et al.*, 1995), and with the intracellular accumulation of misfolded PLP when expressed in heterologous cells (Gow *et al.*, 1994; Jung *et al.*, 1996). In contrast, these experimental data are difficult to reconcile with the findings of human geneticists. Independent reports have shown that some PMD patients diagnosed clinically with hypomyelination have an altered PLP gene that must be considered a *null* allele, such as a large genomic deletion (Raskind *et al.*, 1991) or a point mutation destroying the AUG initiation codon (Sistermans *et al.*, 1996). Is PLP required for myelination in humans but not in mice? The biochemical composition of myelin is nearly identical and it is very unlikely that structural proteins, as highly conserved as PLP, serve different functions in mouse and human. More likely, the loss of PLP/DM20 is so obvious in humans because they are long-lived species in which myelin assembly continues for several years after birth. In fact, the absence of PLP causes a late-onset and mild form of PMD that has many features of SPG-2. The proband in the family studied by Sistermans *et al.* (1996) had a slowly progressive deterioration of mental functions at age 33, whereas his spastic paraplegia was not noted before 4 years of age (note that all other forms of PMD are detected in the first year of life). It is most plausible that in the absence of proteolipids, myelin is assembled but remains physically unstable and at a high risk of premature breakdown. If such a degeneration occurs within a few years, the resulting disease would have the features of a developmental disorder similar to other forms of PMD because it interferes with the acquisition of mature motor functions. In contrast, myelin degeneration that begins years after myelin synthesis may have minor consequences in short-lived mice. Follow-up examinations of PLP-deficient mice are in progress, but there is no convincing evidence that compact myelin that lacks PLP/DM20 delaminates *in vivo* for at least 22 months of age (unpublished results).

The principal issue that has emerged again with PLP K.O mice concerns the onset and progression of a disease in relation to the species-specific life span. As summarized in this chapter, it is the experience with alterations of the PLP gene that an abnormality of myelin assembly (dysmyelination) can be adequately modeled in the corresponding stage of mouse development. In contrast, the degenerative changes of the same structure (demyelination) appear to follow their own, possibly more species-independent time course. Consequently, the same genetic defect may have a different outcome for a short-lived mouse and a long-lived human being. Whether a "true" mouse model of human disease can be generated is therefore also a function of the underlying disease mechanism, an issue likely to be relevant for a number of genetic defects.

ACKNOWLEDGMENTS

I thank all members of my laboratory in Heidelberg, Drs. Odile Boespflug-Tanguy (Clermont-Ferrand), and Ian Griffiths (Glasgow) for many helpful discussions. The author's work is supported by grants from the Deutsche Forschungsgemeinschaft/SFB317 and the EC Biomed-2 program.

REFERENCES

Anderson, T. J., Montague, P., Nadon, N., Nave, K.-A., and Griffiths, I. R., 1997, Modification of Schwann cell phenotype with PLP transgenes: Evidence that the PLP and DM20 isoproteins are targeted to different cellular domains, *J. Neurosci. Res.* **50:**13–22.

Anderson, T. J., Schneider, A., Barrie, J.-A., Klugmann, M., McCulloch, M. C., Kirkham, D., Kyriakides, E., Nave, K.-A., and Griffiths, I. R., 1988, Increased dosage of the proteolipid protein gene causes late-onset neurodegeneration, *J. Comp. Neurol.* **394:**506–519.

Bartlett, W. P., Knapp, P. E., and Skoff, R. P., 1988, Glial conditioned medium enables *jimpy* oligodendrocytes to express properties of normal oligodendrocytes: Production of myelin antigens and membranes, *Glial* **1:**253–259.

Boison, D., and Stoffel, W., 1989, Myelin-deficient rat: A point mutation in exon III (A→C, Thr75→Pro) of the myelin proteolipid protein causes dysmyelination and oligodendrocyte death, *EMBO J.* **8:**3295–3302.

Boison, D., and Stoffel, W., 1994, Disruption of the compacted myelin sheath of axons of the central nervous system in proteolipid protein-deficient mice, *Proc. Natl. Acad. Sci. USA* **91:**11709–11713.

Boison, D., Büssow, H., D'Urso, D., Müller, H. W., and Stoffel, W., 1995, Adhesive properties of proteolipid protein are responsible for the compaction of CNS myelin sheaths, *J. Neurosci.* **15:**5502–5513.

Boulloche, J., and Aicardi, J., 1986, Pelizaeus–Merzbacher disease: Clinical and nosological study, *J. Child Neurol.* **1:**233–239.

Cremers, F. P. M., Pfeiffer, R. A., van de Pol, T. J. R., Hofker, M. H., Kruse, T. A., Wieringa, B., and Ropers, H. H., 1987, An interstitial duplication of the X chromosome in a male allows physical fine mapping of probes from the Xq13–q22 region, *Hum. Genet.* **77:**23–27.

Duncan, I. D., Hammang, J. P., and Trapp, B. D., 1987, Abnormal compact myelin in the myelin-deficient rat: Absence of proteolipid protein correlates with a defect in the intraperiod line, *Proc. Natl. Acad. Sci. USA* **84:**6287–6291.

Duncan, I. D., Hammang, J. P., Goda, S., and Quarles, R. H., 1989, Myelination in the *jimpy* mouse in the absence of proteolipid protein, *Glia* **2:**148–154.

Ellis, D., and Malcolm, S., 1994, Proteolipid protein gene dosage effect in Pelizaeus–Merzbacher disease, *Nat. Genet.* **6:**333–334.

Fanarraga, M. L., Griffiths, I. R., McCulloch, M. C., Barrie, J. A., Kennedy, P. G. E., and Brophy, P. J., 1992, *Rumpshaker*: An X-linked mutation causing hypomyelination: Developmental differences in myelination and glial cells between the optic nerve and spinal cord, *Glia* **5:**161–170.

Filbin, M. T., Walsh, F. S., Trapp, B. D., Pizzey, J. A., and Tennekoon, G. I., 1990, Role of myelin P0 protein as a homophilic adhesion molecule, *Nature* **344:**871–872.

Garbern, J. Y., Cambi, F., Tang, X. M., Sima, A. A., Vallat, J. M., Bosch, E. P., Lewis, R., Shy, M., Sohi, J., Kraft, G., Chen, K. L., Joshi, I., Leonard, D. G., Johnson, W., Raskind, W., Dlouhy, S. R., Pratt, V., Hodes, M. E., Bird, T., and Kamholz, J., 1997, Proteolipid protein is necessary in peripheral as well as central myelin, *Neuron* **19:**205–218.

Gencic, S., and Hudson, L., 1990, Conservative amino acid substitution in the myelin proteolipid protein of *jimpy*^{msd} mice, *J. Neurosci.* **10:**117–124.

Gow, A., and Lazzarini, R. A., 1996, A cellular mechanism governing the severity of Pelizaeus–Merzbacher disease, *Nat. Genet.* **13:**422–428.

Gow, A., Friedrich, V. L., and Lazzarini, R. A., 1994, Many naturally occurring mutations of myelin proteolipid protein impair its intracellular transport, *J. Neurosci. Res.* **37:**574–583.

Griffiths, I. R., Klugmann, M., Anderson, T., Yool, D., Thomson, C., Schwab, M. H., Schneider, A., Zimmermann, F. McCulloch, M., Nadon, N., and Nave, K.-A., 1998, Axonal swellings and degeneration in mice lacking myelin proteolipid protein. *Science* **280:**1610–1613.

Hodes, M. E., Pratt, V. M., and Dlouhy, S. R., 1994, Genetics of Pelizaeus–Merzbacher disease, *Dev. Neurosci.* **15:**383–394.

Hudson, L. D., Berndt, J. A., Puckett, C., Kozak, C. A., and Lazzarini, R. A., 1987, Aberrant splicing of proteolipid protein mRNA in the dysmyelinating *jimpy* mutant mouse, *Proc. Natl. Acad. Sci. USA* **84:**1454–1458.

Hudson, L. D., Puckett, C., Berndt, J., Chan, J., and Gencic, S., 1989, Mutation of the proteolipid protein gene (PLP) in a human X chromosome-linked disorder, *Proc. Natl. Acad. Sci. USA* **86:**8128–8131.

Johnston, A. W., and MacKusick, V. A., 1962, A sex-linked recessive form of spastic paraplegia, *Am. J. Hum. Genet.* **14:**83–94.

Jung, M., Sommer, I., Schachner, M., and Nave, K.-A., 1996, Monoclonal antibody O10 defines a conformationally sensitive cell-surface epitope of proteolipid protein (PLP): Evidence that PLP misfolding underlies dysmyelination in mutant mice, *J. Neurosci.* **24:**7920–7929.

Kagawa, T., Ikenaka, K., Inoue, Y., Kuriyama, S., Tsujii, T., Nakajima, J., Nakao, J., Aruga, J., Okano, H., and Mikoshiba, K., 1994, Glial cell degeneration and hypomyelination caused by overexpression of the myelin proteolipid protein gene, *Neuron* **13:**427–442.

Kitagawa, K., Sinoway, M. P., Yang, C., Gould, R. M., and Colman, D. R., 1993, A proteolipid protein gene family: Expression in sharks and rays and possible evolution from an ancestral gene encoding a pore-forming polypeptide, *Neuron* **11:**433–448.

Klugmann, M., Schwab, M., Pühlhofer, A., Schneider, A., Zimmermann, F., Griffiths, I., and Nave, K.-A., 1997, Assembly of CNS myelin in the absence proteolipid protein, *Neuron* **18:**59–70.

Knapp, P. E., and Skoff, R. P., 1987, A defect in the cell cycle of neuroglia in the myelin deficient *jimpy* mouse, *Dev. Brain Res.* **35:**301–306.

Knapp, P. E., Skoff, R. P., Redstone, D. W., 1986, Oligodendroglial cell death in *jimpy* mice: An explanation for the myelin deficit, *J. Neurosci.* **6:**2813–2822.

Kobayashi, H., Hoffman, E., and Marks, H., 1994, The *rumpshaker* mutation in spastic paraplegia, *Nat. Genet.* **7:**351–352.

Koeppen, A. H., Ronca, N. A., Greenfield, E. A., and Hans, M. B., 1987, Defective biosynthesis of proteolipid protein in Pelizaeus–Merzbacher disease, *Ann. Neurol.* **21:**159–170.

Lachapelle, F., Gumpel, M., and Baumann, N., 1994, Contribution of transplantations to the understanding of the role of the PLP gene, *Neurochem. Res.* **19:**1083–1090.

Lees, M., and Brostoff, S. L., 1984, Proteins of myelin, in "Myelin," 2nd edit. (P. Morell, ed.), pp. 197–224, Plenum Press, New York.

Meier, C., and Bischoff, A., 1974, Dysmyelination in "*jimpy*" mouse, *J. Neuropathol. Exp. Neurol.* **33:**343–353.

Merzbacher, L., 1910, Eine eigenartige familiäre Erkrankungsform (Aplasia axialis extracorticolis congenita), *Z. Ges. Neurol. Psychiatr.* **3:**1–138.

Milner, R. J., Lai, C., Nave, K.-A., Lenoir, D., Ogata, J., and Sutcliffe, J. G., 1985, Nucleotide sequences of two mRNAs for rat brain myelin proteolipid protein, *Cell* **42:**931–939.

Nadon, N. L., Duncan, I. D., and Hudson, L. D., 1990, A point mutation in the proteolipid protein gene of the '*shaking pup*' interrupts oligodendrocyte development, *Development* **110:**529–537.

Nadon, N. L., Arnheiter, H., and Hudson, L. D., 1994, A combination of PLP and DM20 transgenes promotes partial myelination in the *jimpy* mouse, *J. Neurochem.* **63:**822–833.

Nave, K.-A., 1994, Neurological mouse mutants and genes of myelin, *J. Neurosci. Res.* **38:**607–612.

Nave, K.-A., and Boespflug-Tanguy, O., 1996, Developmental defects of myelin formation: From X-linked mutations to human dysmyelinating diseases, *The Neuroscientist* **2:**33–43.

Nave, K.-A., Lai, C., Bloom, F. E., and Milner, R. J., 1986, *Jimpy* mutant mouse: A 74-base deletion in the mRNA for myelin proteolipid protein and evidence for a primary defect in RNA splicing, *Proc. Natl. Acad. Sci. USA* **83:**9264–9268.

Nave, K.-A., Lai, C., Bloom, F. E., and Milner, R. J., 1987, Splice site selection in the proteolipid protein (PLP) gene transcript and primary structure of the DM-20 protein of central nervous system myelin, *Proc. Natl. Acad. Sci. USA* **84:**5665–5669.

Pelizaeus, F., 1885, Über eine eigenthümliche Form spastischer Lähmung mit Cerebral erscheinungen auf hereditärer Grundlage (Multiple Sklerose), *Arch. Psychiatr.* **16:**698–710.

Popot, J. L., Pham-Dinh, D., and Dautigny, A., 1991, Major myelin proteolipid: The 4-α-helix topology, *J. Membr. Biol.* **120:**233–246.

Pucket, C., Hudson, L., Ono, K., Friedrich, V., Benecke, J., Dubois-Dalcq, M., and Lazzarini, R. A., 1987, Myelin specific proteolipid protein is expressed in myelinating Schwann cells but not incorporated into myelin sheaths, *J. Neurosci. Res.* **18:**511–518.

Raskind, W. H., Williams, C. A., Hudson, L. D., and Bird, T. D., 1991, Complete deletion of the proteolipid protein gene (PLP) in a family with X-linked Pelizaeus–Merzbacher disease, *Am. J. Med. Genet.* **49:**1355–1360.

Readhead, C., Schneider, A., Griffiths, I., and Nave, K.-A., 1994, Premature arrest of myelin formation in transgenic mice with increased proteolipid protein gene dosage, *Neuron* **12:**583–595.

Rosenbluth, J., Stoffel, W., and Schiff, R., 1996, Myelin structure in proteolipid protein (PLP)-*null* mouse spinal cord, *J. Comp. Neurol.* **371:**336–344.

Saugier-Veber, P., Munnich, A., Bonneau, D., Rozet, J. M., Le Merrer, M., Gil, R., and Boespflug-Tanguy, O., 1994, X-linked spastic paraplegia and Pelizaeus–Merzbacher disease are allelic disorders at the proteolipid protein locus, *Nat. Genet.* **6:**257–261.

Schneider, A., Montague, P., Griffiths, I., Fanarraga, M., Kennedy, P., Brophy, P., and Nave, K.-A., 1992, Uncoupling of hypomyelination and glial cell death by a mutation in the proteolipid protein gene, *Nature* **358:**758–761.

Schneider, A., Griffiths, I. R., Readhead, C., and Nave, K.-A., 1995, Dominant-negative action of the *jimpy* mutation in mice complemented with an autosomal transgene for myelin proteolipid protein, *Proc. Natl. Acad. Sci. USA* **92:**4447–4451.

Seitelberger, F., 1970, Pelizaeus-Merzbacher disease, in "Handbook of Clinical Neurology, Vol. 10" (P. J. Vinken, G. W. Bruyn, eds.), pp. 150–202, North Holland, Amsterdam.

Seitelberger, F., Urbanitz, I., and Nave, K.-A., 1996, Pelizaeus–Merzbacher Disease, in "Handbook of Clinical Neurology, Vol. 22" (P. J. Vinken and G. W. Bruyn, eds.), pp. 559–579, Elsevier Science B. V. Amsterdam.

Sidman, R. L., Dickie, M. M., and Appel, S. H., 1964, Mutant mice (*quaking* and *jimpy*) with deficient myelination in the central nervous system, *Science* **144:**309–311.

Simons-Johnson, R., Roder, J. C., and Riordan, J. R., 1995, Over-expression of the DM-20 myelin proteolipid causes central nervous system demyelination in transgenic mice, *J. Neurochem.* **64:**967–976.

Sistermans, E. A., de Wijs, I. J., de Coo, R. F., Smit, L. M., Menko, F. H., and van Oost, B. A., 1996,

A (G-to-A) mutation in the initiation codon of the proteolipid protein gene causing a relatively mild form of Pelizaeus-Merzbacher disease in a Dutch family, *Hum. Genet.* **97**:337–339.

Skoff, R. P., 1982, Increased proliferation of oligodendrocytes in the hypomyelinated mouse mutant-*jimpy*, *Brain Res.* **248**:19–31.

Utzschneider, D., Black, J. A., and Kocsis, J. D., 1992, Conduction properties of spinal cord axons in the myelin-deficient rat mutant, *Neuroscience* **49**:221–228.

Vermeesch, M. K., Knapp, P. E., Skoff, R. P., Studzinski, D. M., and Benjamins, J. A., 1990, Death of individual oligodendrocytes in *jimpy* brain precedes expression of proteolipid protein, *Dev. Neurosci.* **12**:303–315.

Weimbs, T., and Stoffel, W., 1992, Proteolipid protein (PLP) of CNS myelin: Positions of free, disulfide-bonded, and fatty acid thioester-linked cysteine residues and implications for the membrane topology of PLP, *Biochemistry* **31**:12289–12296.

Willard, H. F., Riordan, J. R., 1985, Assignment of the gene for myelin proteolipid protein to the X chromosome: Implications for X-linked myelin disorders, *Science* **230**:940–942.

Yan, Y., Lagenaur, C., and Narayanan, V., 1993, Molecular cloning of M6: Identification of a PLP/DM-20 gene family, *Neuron* **11**:423–431.

Young, W., Rosenbluth, J., Wojak, J. C., and Sakatani, K., 1989, Extracellular potassium activity and axonal conduction in spinal cord of the myelin-deficient mutant rat, *Exp. Neurol.* **106**:41–51.

CHAPTER 3

CHARCOT–MARIE–TOOTH DISEASE: PATHOLOGY, GENETICS, AND ANIMAL MODELS

UELI SUTER

1. INTRODUCTION

This chapter focuses on a particular group of hereditary disorders of the peripheral nervous system (PNS) known as the hereditary motor and sensory neuropathies (HMSNs) or Charcot–Marie–Tooth disease (CMT). The first descriptions of the clinical symptoms associated with these disorders were reported more than a century ago by Charcot and Marie (Charcot and Marie, 1886), and independently by Tooth (Tooth, 1886). Subsequent epidemiological studies have revealed that CMT is probably the most commonly inherited disorder of the nervous system, with an estimated frequency as high as one in 2500 individuals (Skre, 1974). In the early 1990s, progress in human genetics in combination with basic research using the spontaneous mouse mutant *Trembler* (Falconer, 1951)

UELI SUTER • *Institute of Cell Biology, ETH-Hönggerberg, CH-8093 Zürich, Switzerland.*
Mouse Models in the Study of Genetic Neurological Disorders, Volume 9 of *Advances in Neurochemistry,* edited by Popko, Kluwer Academic/Plenum Publishers, New York, 1999.

led to a major breakthrough in our understanding of this heterogeneous group of disorders (reviewed by Suter *et al.,* 1993). Today, many facets of the molecular basis of CMT have been unravelled and several fascinating novel animal models have been developed by genetic engineering (reviewed by Suter, 1997b). These animals now provide an excellent basis for future studies aimed at the elucidation of the disease mechanisms and as models for the development of potential treatment strategies.

In this chapter, I first briefly summarize our current knowledge of the clinical, electrophysiological, neuropathological, and genetic aspects of CMT, followed by a comparative survey of CMT animal models and a discussion of their potential benefits for CMT research.

2. CHARCOT–MARIE–TOOTH DISEASE

2.1. Clinical Evaluation, Electrophysiology, and Neuropathology

Clinical, electrophysiological, and neuropathological evaluations as well as the mode of inheritance have been traditionally used for the differential diagnosis of this heterogeneous group of diseases (reviewed by Ouvrier *et al.,* 1996). Based on these criteria, CMT has been classified into at least seven subgroups (Dyck *et al.,* 1993). This chapter focuses mainly on the most common forms—CMT type 1 (CMT1), CMT type 2 (CMT2), and X-linked CMT (CMTX). In addition, Dejerine–Sottas syndrome (DSS; Dejerine and Sottas, 1893) and hereditary neuropathy with liability to pressure palsy (HNPP; Windebank, 1993) are included because of their genetic relationship to CMT1.

2.1.1. Charcot–Marie–Tooth Disease Type 1, Type 2, and X-linked

Clinically, CMT1, CMT2, and CMTX are polyneuropathy syndromes with onset usually in the second decade of life manifested as a slowly progressive weakness of the distal limb muscles (reviewed by Lupski *et al.,* 1991). Classical features include prominent atrophy of the muscles of the leg and feet and, less frequently, of the hands. Pes cavus deformities develop often with age whereas prominent sensory symptoms are relatively rare. In general, the severity of the clinical phenotype is highly variable not only between affected, unrelated individuals but also within the same family, or even in identical twins (Garcia *et al.,* 1995; Thomas *et al.,* 1997).

CMT1 and CMT2 can be distinguished based on electrophysiological and pathological criteria. Whereas CMT1 displays significantly reduced motor nerve conduction velocities (NCV) and decreased signal amplitudes, CMT2 is characterized by normal or only mildly reduced NCV accompanied by reduced amplitudes of the evoked motor and sensory responses (Harding and Thomas, 1980). In agreement with the electrophysiological findings, the main pathology observed in CMT1 biopsies are signs of demyelination and remyelination including thinly myelinated large-caliber axons and Schwann cell proliferation. This abnormal proliferation results in concentric layers of supernumerary Schwann cell processes around axons forming onion bulb structures (Dyck *et al.,* 1993). In contrast, the neuropathological hallmarks of CMT2 are axonal degeneration with little, if any, myelin defects detectable (Harding and Thomas, 1980).

Male CMTX patients have mildly reduced NCV whereas female carriers are variably affected by intermediate NCV (Nicholson and Nash, 1993), resulting in some overlap of the electrophysiological features with CMT2 (Timmerman *et al.,* 1996). The CMTX pathology is similar to CMT1 but includes also prominent axonal loss and specific features such as abnormal swelling of a subset of periaxonal collars (Hahn *et al.,* 1990).

2.1.2. Dejerine–Sottas Syndrome

The criteria for DSS are historically not well defined (Dejerine and Sottas, 1893) and, as an unfortunate consequence, the term DSS has been used lately as a bare synonym for a particular class of CMT1 with severe symptoms and early disease onset (reviewed by Tyson *et al.,* 1997). These patients show very slow nerve conduction velocities and are characterized by abundant Schwann cell and basal lamina onion bulb formation in nerve biopsies (Dyck *et al.,* 1993). Occasionally, severe amyelination can also be observed in the subtype congenital hypomyelination (Harati and Butler, 1985; Warner *et al.,* 1996).

2.1.3. Hereditary Neuropathy with Liability to Pressure Palsies

Until the advent of molecular genetics, HNPP was recognized as a separate disease entity different from HMSN (Snipes and Suter, 1995). HNPP patients experience first symptoms usually in adolescence and the main symptoms are characterized by recurrent pressure palsies precipitated by minor trauma to peripheral nerves (Davies, 1954). In particular, entrapment neuropathies such as carpal tunnel syndrome have been frequently found to be associated with the genetic defect linked to HNPP. Furthermore, affected individuals may also develop a chronic disease reminiscent of CMT at older age (Windebank, 1993). Electrophysiological findings show a widely variable spectrum of abnormalities that are usually mild: however, conduction block has been observed in some affected patients (Gouider *et al.,* 1995). Prominent focal hypermyelinated structures called *tomacula* are the pathological hallmark of HNPP (Meier and Moll, 1982). Segmental demyelination is occasionally observed and some partial loss of myelin compaction has been detected, suggesting impaired adhesion between myelin leaflets (Yoshikawa and Dyck, 1991).

2.2. Genetics

CMT is genetically heterogeneous and includes autosomal dominant, autosomal recessive, and X-linked disorders (reviewed by Suter and Snipes, 1995a). The genes defective in the most common forms of CMT have now been identified and mutational analysis has become an important tool for the diagnosis and genetic counseling of CMT patients.

2.2.1. Charcot–Marie–Tooth Disease Type 1 and Hereditary Neuropathy with Liability to Pressure Palsies

Autosomal dominant CMT1 is the most frequently found form of this group of diseases in the general population (approx. 70% of CMT cases; Nelis *et al.*, 1996), and the genetic defect in the vast majority of these cases has been linked to chromosome 17p11.2–p12 (reviewed by Suter and Patel, 1994). This specific CMT1 subtype has been designated CMT1A and human molecular genetics revealed tight linkage to an intrachromosomal duplication of 1.5 mb (Lupski *et al.*, 1991; Raeymaekers *et al.*, 1991). Interestingly, a deletion of the same chromosomal segment was subsequently found to be associated with HNPP (Chance *et al.*, 1993). Detailed studies of the affected chromosomal region revealed that CMT1A and HNPP can be viewed as reciprocal genetic diseases due to an unequal crossover that occurs predominantly in male meiosis and is mediated by recombination between repeat sequences that flank the monomeric segment and show sequence similarities to insect transposons (Reiter *et al.*, 1996).

The natural mouse mutant *Trembler* (*Tr*) had been viewed as a potential animal model for CMT1 owing to the similarities in pathology. The finding of point mutations in the peripheral myelin protein 22 (*PMP22*) gene in *Tr* (Suter *et al.*, 1992a) as well as in the related spontaneous *Trembler–J* (*Tr–J*) mouse, which also exhibits an autosomal dominant demyelinating neuropathy (Suter *et al.*, 1992b), as well as the mapping of *PMP22* on mouse chromosome 11 to a segment syntenic to human chromosome 17p11.2 (Suter *et al.*, 1992b), identified *PMP22* as the prime CMT1A candidate gene. In agreement with this hypothesis, the entire *PMP22* gene was found to be located in the CMT1A duplication/HNPP deletion (Patel *et al.*, 1992), and subsequent studies demonstrated that *PMP22* is indeed overexpressed in CMT1A and underexpressed in HNPP patients (Hanemann *et al.*, 1994; Yoshikawa *et al.*, 1994; Vallat *et al.*, 1996). Additional support for the causative role of *PMP22* in these diseases was provided when several point mutations in the human *PMP22* gene were detected in nonduplication CMT1A and DSS patients (reviewed by Suter and Snipes,

1995b). Furthermore, a frame-shifting mutation near the amino-terminus of *PMP22* (likely generating a null allele) was discovered in a HNPP family not harboring the usual large deletion on chromosome 17 (Nicholson *et al.,* 1994).

2.2.2. Charcot–Marie–Tooth Disease Type 1B

Like CMT1A, CMT1B is autosomal dominantly inherited and the corresponding disease locus was originally mapped near the Duffy locus on chromosome 1 and with a later refinement to 1q22–q23 (Bird *et al.,* 1982, 1997). The affected gene encodes the peripheral myelin protein zero (PO), the major protein component of PNS myelin. To date, more than two dozen different point mutations in this gene have been linked to CMT1B (reviewed by Shapiro *et al.,* 1996; Warner *et al.,* 1996) and it has been estimated that CMT1B accounts for less than 5% of CMT (Nelis *et al.,* 1996).

2.2.3. X-Linked Charcot–Marie–Tooth Disease

CMTX is characterized by the lack of male-to-male transmission. The gene responsible for the most common form of CMTX has been identified as *connexin32* (*Cx32*) by a combined positional cloning and candidate gene approach (Bergoffen *et al.,* 1993). More than 80 different *Cx32* mutations have been found that result in CMTX, accounting for approx. 10–15% of all CMT patients (reviewed by Scherer *et al.,* 1997).

2.2.4. Dejerine-Sottas Syndrome

DSS is genetically heterogeneous and mutations in the *PMP22* and *P0* genes have been reported in DSS patients (reviewed by De Jonghe *et al.,* 1997). Because of the nagging problems mentioned earlier concerning the classification of DSS, it is anticipated that this clinical concept will be redefined soon based on the individual genetic defects that are responsible for the phenotype.

2.2.5. Charcot–Marie–Tooth Disease Type 2

Although several CMT2 disease loci have been mapped on chromosomes 1p35–p36 (CMT2A), 3q13–q22 (CMT2B), and 7p14 (CMT2D), none of the affected genes has yet been described (reviewed by De Jonghe *et al.,* 1997).

3. ANIMAL MODELS FOR CHARCOT–MARIE–TOOTH DISEASE

Animal models have played a major role in the progress of CMT research. In particular, the natural mouse mutants *Tr* and *Tr–J* have provided the key to the identification of *PMP22* as the disease gene in CMT1A. These findings also indicated that genes encoding myelin-associated proteins may be mutated in other CMT1 subtypes, a hypothesis that proved to be correct (reviewed by Suter and Snipes, 1995b). In addition, human genetics and basic research on the regulation, structure, and function of CMT genes have contributed novel clues toward our current knowledge of these disorders. However, a thorough understanding of the molecular mechanisms that underlie disease development and progression depends on animal models that accurately reflect the biology of CMT subtypes. Using genetic approaches, several rat and mouse transgenic lines carrying appropriate mutations homologous to pathological CMT alleles have been developed. This approach has led to the characterization of a number of phenotypes that quite faithfully reflect the symptoms of the corresponding human disorders.

3.1. Animal Models for Charcot–Marie–Tooth Disease Type 1A and Hereditary Neuropathy with Liability to Pressure Palsies

3.1.1. *Trembler, Trembler–J, and Trembler–Ncnp*

As noted earlier, the autosomal dominant natural mouse mutants *Tr* and *Tr–J* carry different point mutations in the *PMP22* gene that encodes a protein component of compact myelin (Aguayo *et al.*, 1977; Snipes *et al.*, 1992; Haney *et al.*, 1996; reviewed by Suter and Snipes, 1995b). Clinically, both *Tr* and *Tr–J* are affected by an unsteady gait, tremor, and quadriparesis (Falconer *et al.*, 1951; Henry *et al.*, 1983). Pathological examinations in *Tr–J* (associated with the substitution of a leucine by a proline residue in the first membrane-associated domain of *PMP22;* Suter *et al.*, 1992b) revealed a demyelinating neuropathy including thinly myelinated axons, aberrant Schwann cell proliferation, and Schwann cell onion bulb formation (Henry *et al.*, 1983; Notterpek *et al.*, 1997). In addition, morphological evidence indicative of abnormal Schwann cell–axon interactions has been described (Robertson *et al.*, 1997). Immunohistochemical and biochemical analysis showed an upregulation of the endosomal–lysosomal pathway consistent with the hypothesis that the *Tr–J* mutation alters myelin stability (Notterpek *et al.*, 1997). Most interestingly, the identical *Tr–J* point mutation has also been found in a severely affected CMT1 family (Valentijn *et al.*, 1992) verifying the value of this mouse strain as animal model for non-duplication CMT1A.

Tr mice [associated with the replacement of a glycine by an aspartate residue in the last putative membrane-associated *PMP22* domain (Suter *et al.*, 1992a)] display prominent dysmyelination early in development, with only a few thinly myelinated fibres visible on peripheral nerve sections (Henry *et al.*, 1983). Some myelin sheaths show focal abnormalities in compaction (Heath *et al.*, 1991) and empty basal lamina onion bulbs, most probably remnants of degenerated supernumerary Schwann cells, are frequently found in older *Tr* animals (Low, 1977). These pathological characteristics are similar to DSS (Ayers and Anderson, 1973) and indeed, the *Tr* point mutation has also been found in a human pedigree of a family that had been diagnosed with DSS based on the severe clinical symptoms (Ionasescu *et al.*, 1997). Recent immunohistochemical analysis in *Tr* animals and studies in transfected cells indicate that the molecular basis of the *Tr* phenotype involves impaired intracellular trafficking of the Tr protein (Naef *et al.*, 1997). Furthermore, Tr protein exhibits a potential dominant-negative effect on the transport of wild-type *PMP22,* suggesting a disease mechanism for some CMT1A-causing *PMP22* point mutations (Naef *et al.*, 1997). Additional evidence indicates that *Tr* can also act as a gain-of-function allele depending on the genetic context (Adlkofer *et al.*, 1997b) and as a consequence of the primary genetic defect in *PMP22,* the expression of other myelin proteins and components of the lipid metabolism is also heavily affected in *Tr* (Bascles *et al.*, 1992; Garbay and Cassagne, 1994; Garbay *et al.*, 1995; Boiron-Sargueil *et al.*, 1995; Heape *et al.*, 1995; Bartoszewics *et al.*, 1996).

A third spontaneous autosomal dominant mutation affecting the *PMP22* gene is associated with the *Trembler–Ncnp* (*Tr–Ncnp*) mouse strain (Suh *et al.*, 1997). These animals carry an in-frame deletion of the entire exon IV of the *PMP22* gene (Suter *et al.*, 1994) and show behavioral abnormalities similar to those of *Tr* and *Tr–J* with initially normal myelination but signs of segmental demyelination with progressing age (Suh *et al.*, 1997). Homozygous *Tr–Ncnp* mice (similar to homozygous *Tr* mice) remain completely amyelinated (Henry and Sidman, 1988). However, the presence of large vacuolor structures, which likely represent swollen endoplasmic reticulum, are unique to *Tr–Ncnp* (most prominent in homozygous animals; Suh *et al.*, 1997). It is tempting to speculate that this intriguing feature may reflect the consequences of a defect in protein transport similar to that observed in *Tr.* Interestingly, the sciatic nerves of homozygous *Tr–Ncnp* contain a significant number of nuclei with signs of apoptosis but the identity of the affected cells remains to be established (Suh *et al.*, 1997). These findings are of particular relevance because overexpression of wild-type *PMP22* in NIH3T3 fibroblasts has been reported to induce apoptosis (Fabbretti *et al.*, 1995; Zoidl *et al.*, 1997). Furthermore, a modulatory role of *PMP22* on the cell cycle has been suggested by experimental alterations of *PMP22* levels in cultured Schwann cells (Zoidl *et al.*, 1995) whereas no effects on early steps of myelination were observed in an *in vitro* myelination system (D'Urso *et al.*, 1997).

Beside the considerable benefits that have been gained from the various *Trembler* animals as models to study disease mechanisms of nonduplication CMT1A, these mutant mouse strains have also been very instructive in basic science. The intriguing effects of impaired axon–glia interactions on axon diameter, phosphorylation of neurofilament proteins, velocity of slow axonal transport, density of microfilaments, microtubule stability, and growth factor biosynthesis (De Waegh and Brady, 1990, 1991; De Waegh *et al.,* 1992; Kirkpatrick and Brady, 1994; Friedman *et al.,* 1996) have been discovered first in these animals. Although at first glance those studies seem not directly related to CMT, they are still of major importance for our understanding of the intense bidirectional dialogue between axons and Schwann cells in processes that are likely to be disturbed in many genetic forms, as well as in acquired peripheral neuropathies (Snipes and Suter, 1994; Figure 1).

3.1.2. Animal Models with Increased PMP22 Gene Dosage

Because the most common form of CMT is associated with an intrachromosomal duplication containing the *PMP22* gene (CMT1A), transgenic rats and mice were generated to obtain accurate animal models for this subtype and to prove that increased *PMP22* gene dosage is sufficient to elicit the hallmarks of the disease. Rats carrying three additional copies of a cosmid containing the *PMP22* gene controlled by its own regulatory elements show many features that are strikingly similar to CMT1A patients (Sereda *et al.,* 1996; Thomas *et al.,* 1996). Behaviorally, these animals display an unsteady gait and loss of muscle strength leading to deficits in motor performance. However, in agreement with the findings in CMT1A families, quite a large variation in the severity of the symptoms is observed between individual transgenic littermates. Electrophysiological examinations reveal reduced NCV and dispersed muscle action potentials comparable to CMT1A patients. On morphological analysis, adult animals show normally myelinated fibers intermingled with amyelinated large-caliber axons and too thinly myelinated axons associated with Schwann cell onion bulbs indicative of demyelination and remyelination events. In addition, some hypermyelinated fibers are present. Interestingly, motor nerves appear to be more heavily affected than sensory fibers. However, whether this peculiarity, which has also been observed in various other rodent myelin mutants, is reflected in human patients remains to be clarified (Adlkofer *et al.,* 1995, 1997a; Martini *et al.,* 1995b; Anzini *et al.,* 1997).

Further increase of *PMP22* gene dosage by breeding the transgenic rats to homozygosity led to almost complete PNS amyelination (Sereda *et al.,* 1996), comparable to the findings in transgenic mice carrying approx. 15 or 30 copies of the same *PMP22*-containing cosmid (Magyar *et al.,* 1996). In these animals with highly increased *PMP22* gene dosage, Schwann cells and larger caliber axons segregate correctly into the 1:1 relationship. However, the differentiation into myelinating units is impaired as manifested by continued Schwann cell proliferation and the maintained expression of proteins specific for nonmyelinating Schwann cells (Magyar *et al.,* 1996). A pronounced demyelinating phenotype has also been observed in transgenic mice with eight copies of a large yeast

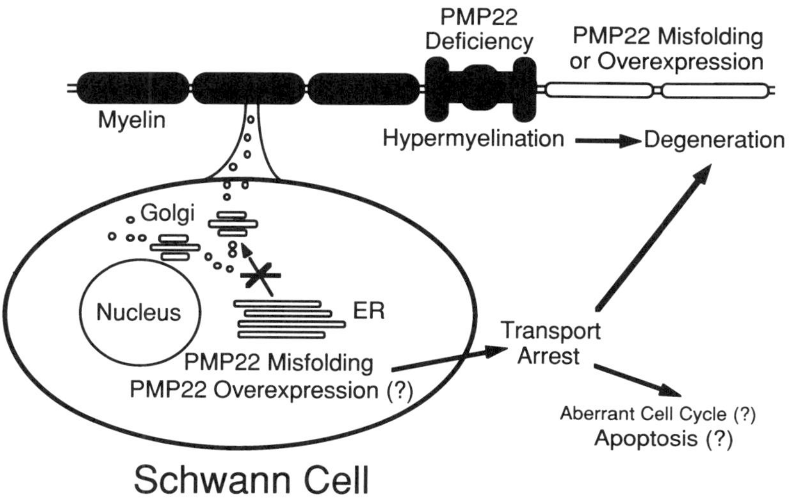

FIGURE 1. Molecular mechanisms of PMP22-mediated neuropathies.

artificial chromosome (YAC) containing the *PMP22* gene (Huxley *et al.*, 1996). These findings are in agreement with the generally more severe phenotype of homozygous versus heterozygous CMT1A duplication patients although there is considerable clinical variability (Kilian and Kloepfer, 1979; Le Guern *et al.*, 1997).

3.1.3. Mice with Decreased PMP22 Gene Dosage

Based on genetic analysis, the recurrent peripheral neuropathy HNPP has been suggested to be due to haploinsufficiency of the *PMP22* gene (Chance *et al.*, 1993). In agreement with this hypothesis, transgenic mice generated by gene targeting in embryonic stem cells and retaining only one functional copy of the *PMP22* gene (*PMP22*$^{0/+}$) display the classical pathological hallmarks of HNPP (Adlkofer *et al.*, 1997b). Tomacula develop progressively with age albeit with high variability between individual animals. Analysis of *PMP22*$^{0/+}$ mice of various ages reveals that tomacula are unstable structures that tend to degenerate, leading to the classical sign of a CMT1A-type demyelinating neuropathy. These findings offer a plausible explanation for the sometimes overlapping clinical and pathological features of HNPP and CMT1 (Windebank *et al.*, 1993). Furthermore, it is tantalizing to speculate that the increased vulnerability of tomaculous myelin to degeneration may also be involved in the disease mechanism that is responsible for the transient symptoms experienced by HNPP patients after minor trauma to peripheral nerves.

Recently, another transgenic mouse model for HNPP has been generated using the P0 promoter to drive the expression of *PMP22*-antisense RNA (Maycox *et al.,* 1997). The resulting reduction of *PMP22* expression levels in these mice was estimated to only approx. 15% (compared to 50% in *PMP22*$^{0/+}$ mice), but some pathological signs such as tomacula-like myelin structures accompanied by electrophysiological abnormalities and a pronounced high-stepping gait are already prominent in the sciatic nerves of these animals.

Despite the large amount of information that has been gained from CMT1A- and HNPP-related studies in human molecular genetics, electrophysiology, and pathology, the molecular function of *PMP22* remains largely unknown. This question is particularly relevant because *PMP22* is also expressed in various nonneural tissues during development and in adulthood, but no consistent abnormalities have been described outside of the PNS in CMT1A or HNPP patients (Welcher *et al.,* 1991; Baechner *et al.,* 1995; Parmatier *et al.,* 1995, 1997). Significant progress toward the elucidation of the function was expected from the analysis of mice completely devoid of *PMP22* (*PMP22*$^{0/0}$). Surprisingly, these animals are born and develop normally until approximately postnatal day 14 when signs of walking difficulties due to progressive weakness of the hindlimbs become obvious (Adlkofer *et al.,* 1995). Pathological examinations revealed that *PMP22*$^{0/0}$ mice are delayed in PNS myelination and that at postnatal day 24, nearly every internodal segment is affected by tomacula. In contrast, at 10 weeks of age, tomacula are rarely observed but characteristics of demyelination and remyelination including thinly myelinated axons and Schwann cell onion bulb formations become abundant. Furthermore, older animals show definitive signs of axonal loss. Based on these findings, *PMP22* is likely to be involved in the initiation of myelination, the maintenance of myelin and axons, and the regulation of myelin thickness. Interestingly, no phenotype was detected outside of the nervous system which could not be recognized as the consequences of the myelination defects and axonal loss, despite the widespread expression of *PMP22* in different tissues and the suggestions from *in vitro* studies that the molecule may play an important role in the regulation of the cell cycle and apoptosis (Fabbretti *et al.,* 1995; Zoidl *et al.,* 1995, 1997). The recent discovery that *PMP22* belongs to an extended gene family with overlapping expression patterns in different tissues may hint toward potential compensatory mechanisms in *PMP22*$^{0/0}$ mice (Marvin *et al.,* 1995; Taylor *et al.,* 1995; Ben-Porath and Benvenisty, 1996; Lobsiger *et al.,* 1996; Ruegg *et al.,* 1996; Taylor and Suter, 1996; Bolin *et al.,* 1997).

The described studies on rodents with altered *PMP22* gene dosage show that these animals provide suitable animal models for hereditary peripheral neuropathies with different severities. Interestingly, however, increased *PMP22* gene dosage and the total lack thereof in *PMP22*$^{0/0}$ mice are both associated with neuromyotonia in older animals (Toyka *et al.,* 1997). This disorder is characterized by muscle stiffness and cramping caused by hyperactivity of the motor units (Isaacs, 1961) and may be due to autoimmunity against peripheral nerve components (Vasilescu *et al.,* 1984). Whether there is a causal relationship between *PMP22* expression, hereditary peripheral neuropathies, and certain forms of neuromyotonia remains to be clarified.

3.2. Animal Models for Charcot–Marie–Tooth Disease Type 1B

3.2.1. P0-Deficient Mice

The major myelin protein gene *P0* is exclusively expressed by Schwann cells as a potential lineage-specific marker and the most prominent component of compact PNS myelin (reviewed by Lee *et al.,* 1997; Suter, 1997a). As a prototypic member of the immunoglobulin (Ig)-superfamily of cell–cell recognition proteins, *P0* carries a single extracellular Ig domain. The extracellular portion of *P0* has been suggested to function in the compaction of the extracellular aspects of myelin membranes and the formation of the intraperiod line (D'Urso *et al.,* 1990; Filbin *et al.,* 1990; Schneider-Schaulies *et al.,* 1990). This hypothesis is strongly supported by the ultrastructural analysis of young adult *P0*-deficient ($Po^{0/0}$) mice which reveals axon Schwann cell units with widened intraperiod lines (Giese *et al.,* 1992). Furthermore, some large-caliber axons are associated with spiraling loops of completely loose myelin membranes containing cytoplasm. This observation is consistent with earlier suggestions that the basic intracellular domain of *P0* interacts with negatively charged phospholipid head groups on the opposing myelin membrane to stabilize the major dense line (Lemke, 1988; Martini *et al.,* 1995a). With progressing age, severe hypomyelination, reduced nerve conduction velocities, and Schwann cell onion bulb formation become predominant, indicating that *P0* is involved not only in the correct formation of the myelin structure but also in its maintenance.

In the vast majority of cases, CMT1B is associated with dominant point mutations in the *P0* gene that lead to rather divergent severities of the disease with distinct pathological hallmarks ranging from mainly partially uncompacted myelin and hypomyelination to focal hypermyelination similar to the tomacula seen in HNPP (Thomas *et al.,* 1994; Gabreels-Festen *et al.,* 1996; Warner *et al.,* 1996). Thus, a careful evaluation of the phenotype of $P0^{0/+}$ mice was of particular interest (Martini *et al.,* 1995b). These mutant animals, which have only one functional *P0* allele, develop normally but acquire a CMT1-like phenotype at older age including myelin degeneration and Schwann cell onion bulb formation. Thus, it is anticipated that a heterozygous *P0* null allele is also likely to cause CMT1 in humans. The different severities of $Po^{0/0}$ and $P0^{0/+}$ mice probably reflect the observed varying disease phenotypes in a family whose affected members carry a frameshift mutation at position 74 in the P0 protein that possibly results in a loss of function. Whereas the heterozygous parents are affected by mild CMT1B, the homozygous children show a severe DSS phenotype (Warner *et al.,* 1996).

In most of the patients affected by *P0* point mutations, however, dominant-negative and toxic gain-of-function mechanisms are likely to contribute to the observed pathologies. Based on the recently solved three-dimensional structure of the extracellular domain of the P0 protein and in conjunction with the known functions of the P0 protein, it is now possible to model theoretically the effect of particular mutations and to test the resulting hypotheses using appropriate *in vitro* systems (Wong and Filbin, 1996; Shapiro *et al.,* 1996). In addition, deleterious effects of a subset of *P0* mutant alleles on protein biosynthesis and intracellular trafficking must also be considered as potential disease mechanisms as has been described for *PMP22* (Naef *et al.,* 1997).

3.3. Animal Models for X-Linked Charcot–Marie–Tooth Disease

3.3.1. Connexin32-Deficient Mice

The gap junction protein Cx32 is expressed mainly in the liver and in myelinating Schwann cells. Within PNS myelin, Cx32 is found in uncompacted regions of myelin including the Smidt-Lanterman incisures and the paranodal loops (Bergoffen *et al.,* 1993; Scherer *et al.,* 1995). Based on this localization, it has been speculated that Cx32 may be involved in the formation of reflexive gap junctions between adjacent Schwann cell myelin membranes to provide a short-cut pathway through the myelin sheath (Paul, 1995; Scherer *et al.,* 1995, 1997). Thus, a disturbance of such gap-junction-mediated channels may underlie the demyelinating phenotype observed in CMTX patients carrying *Cx32* mutations (Scherer *et al.,* 1997). This hypothesis is supported by *in vitro* studies suggesting disturbed gap junction functions of CMTX-associated *Cx32* mutations (Bruzzone *et al.,* 1994; Omori *et al.,* 1996).

To analyze the function of *Cx32* in the living organism and to produce an animal model of CMTX, *Cx32*-deficient animals have been generated (Nelles *et al.,* 1996). Animals devoid of *Cx32* show a large reduction of morphologically detectable gap junction plaques in the liver, and stimulation of sympathetic nerves leads to reduced glucose release from mutant hepatic cells compared to wild-type. Examinations of peripheral nerves in *Cx32*-deficient animals reveal no morphologically detectable effect on the development and formation of myelin in 1-month-old mutants. However, classical signs of myelin degeneration including too thin myelin sheaths and Schwann cell onion bulb formation were progressively observed in older animals (Anzini *et al.,* 1997). Detailed morphological analysis of *Cx32*-deficient nerves revealed additional specific features of this mutant. Consistent with the proposed functional role of *Cx32* in the uncompacted domains of myelin, the structure of some Schmidt–Lanterman incisures and paranodal loops is abnormal in *Cx32*$^{0/0}$ animals. These abnormalities are characterized by expanded cytoplasm containing many vesicular structures. In particular, lysosomes and complex membranous foldings are frequently found. In addition, abnormally thick periaxonal collars and occasional features reminiscent of axonal regeneration can also be observed (Anzini *et al.*, 1997). Similar features are also frequently seen in biopsies of CMTX patients, supporting the view that many CMTX-associated *Cx32* mutations may generate null alleles (Martini, 1997).

In contrast to CMTX patients, little evidence for axonal loss was observed in *Cx32*-deficient mice (Hahn *et al.*, 1990; Anzini *et al.*, 1997). Although a decrease in the muscle compound action potential in *Cx32*-deficient mice and signs of regeneration may indicate some axonal degeneration, the striking difference may also be due to differences in the body plan and life span of mice and human. Because the longest nerves in CMT1 patients tend to develop the most severe clinical and pathological signs and the course of the disease is clinically progressive, the comparably small-sized nerves of the mouse model may not be able to develop all the prominent disease features that are found in the human.

4. SUMMARY AND OUTLOOK

The recent progress in the understanding of Charcot–Marie–Tooth disease and related neuropathies has provided an instructive example of how molecular genetics can contribute to a better understanding of an important group of neurological diseases. Human molecular genetics and the analysis of spontaneous as well as genetically engineered animal models have contributed largely to this success. We now have a comprehensive array of animal models available for the most common forms of CMT, and the correlations between these animal models and the corresponding human diseases are surprisingly accurate. However, there are also substantial differences. Animal models for most of the disease-causing null alleles have been constructed. However, given the complex pathological picture, many CMT alleles may act by dominant-negative and/or gain-of-function mechanisms (Adlkofer *et al.*, 1997b; Naef *et al.*, 1997). Thus, to obtain a detailed understanding of the phenotype/genotype correlation will require the introduction of more subtle mutations that correctly reflect those found in human disease. Furthermore, detailed studies using inbred mouse and rat lines are required to assess the basis of the variability observed between CMT patients with identical disease-causing mutations and to assess the influence of epigenetic factors (Garcia *et al.*, 1995). Such approaches will be instrumental in the design of any therapeutic intervention with the aim to avoid aggravation of the disease, or even to ameliorate the symptoms of patients.

ACKNOWLEDGMENTS

I thank Dr. Verdon Taylor for critically reading the manuscript. The work in the laboratory of the author is supported by the Swiss National Science Foundation, the Swiss National Research Program for Neurodegenerative Disorders, the European Union project: "Clinical, Genetic and Functional Analysis of Peripheral Neuropathies: An Integrated Approach," and the Swiss Bundesamt for Science.

REFERENCES

Adlkofer, K., Martini, R., Aguzzi, A., Zielasek, J., Toyka, K. V., and Suter, U., 1995, Regional hypermyelination and chronic demyelinating peripheral neuropathy in PMP22-deficient mice, *Nat. Genet.* **11:**274–280.

Adlkofer, K., Frei, R., Neuberg, D. H.-H., Zielasek, J., Toyka, K. V., and Suter, U., 1997a, Heterozygous peripheral myelin protein 22-deficient mice are affected by a progressive demyelinating neuropathy, *J. Neurosci.* **17:**4662–4671.

Adlkofer, K., Naef, R., and Suter, U., 1997b, Analysis of compound heterozygous mice reveals that the Trembler mutation can behave as a gain-of-function allele, *J. Neurosci. Res.* **49:**671–680.

Aguayo, A. J., Attiwell, M., Trecarten, J., Perkins, S., and Bray, G. M., 1977, Abnormal myelination in transplanted Trembler mouse Schwann cells, *Nature* **265:**73–74.

Anzini, P., Neuberg, D. H.-H., Schachner, M., Nelles, E., Willecke, K., Zielasek, J., Toyka, K. V., Suter, U., and Martini, R., 1997, Structural abnormalities and deficient maintenance of peripheral nerve myelin in mice lacking the gap junction protein connexin32, *J. Neurosci.* **17:**4545–4551.

Ayers, M. M., and Anderson, R., 1973, Onion bulb neuropathy in the Trembler mouse: A model of hypertrophic interstitial neuropathy (Dejerine–Sottas) in man, *Acta Neuropathol.* **25:**54–70.

Baechner, D., Liehr, T., Hameister, H., Altenberger, H., Grehl, H., Suter, U., and Rautenstrauss, B., 1995, Widespread expression of the peripheral myelin protein-22 gene (PMP22) in the neural and non-neural tissues during murine development, *J. Neurosci. Res.* **42:**733–741.

Bartoszewics, Z. P., Lauter, C. J., and Quarles, R. H., 1996, The myelin-associated glycoprotein of the peripheral nervous system in Trembler mutants contains increased alpha2-3 sialic acid and galactose, *J. Neurosci. Res.* **43:**587–593.

Bascles, L., Bonnet, J., and Garbay, B., 1992, Expression of the PMP-22 gene in Trembler mutant mice: Comparison with the other myelin protein genes, *Dev. Neurosci.* **14:**336–341.

Ben-Porath, I., and Benvenisty, N., 1996, Characterization of a tumor-associated gene, a member of a novel family of genes encoding membrane glycoproteins, *Gene* **183:**69–75.

Bergoffen, J., Scherer, S. S., Wang, S., Oronzi Scott, M., Bone, L. J., Paul, D. L., Chen, K., Lensch, M. W., Chance, P. F., and Fischbeck, K. H., 1993, Connexin mutations in X-linked Charcot–Marie–Tooth disease, *Science* **262:**2039–2042.

Bird, T. D., Ott, J., and Giblett, E. R., 1982, Evidence for linkage of Charcot–Marie–Tooth neuropathy to the Duffy locus on chromosome 1, *Am. J. Hum. Genet.* **34:**388–394.

Bird, T. D., Kraft, G. H., Lipe, H. P., Kenney, K. L., and Sumi, S. M., 1997, Clinical and pathological phenotype of the original family with Charcot–Marie–Tooth disease type 1B. A 20 year study, *Ann. Neurol.* **41:**463–469.

Boiron-Sargueil, F., Heape, A., Garbay, B., and Cassagne, C., 1995, Cerebroside formation in the peripheral nervous system of normal and Trembler mice, *Neurosci. Lett.* **186:**21–24.

Bolin, L., McNeil, T., Lucian, L. A., DeVaux, B., Franz-Bacon, K., Gorman, D. M., Zurawski, S., Murray, R., and McClanahan, T. K., 1997, HNMP-1: A novel hematopoietic and neural membrane protein differentially regulated in neural development and injury, *J. Neurosci.* **17:**5493–5502.

Bruzzone, R., White, T. W., Scherer, S. S., Fischbeck, K. H., and Paul, D. L., 1994, Null mutations of connexin32 in patients with X-linked Charcot–Marie–Tooth disease, *Neuron* **13:**1253–1260.

Chance, P. F., Alderson, M. K., Leppig, K. A., Lensch, M. W., Matsunami, N., Smith, B., Swanson, P. D., Odelberg, S. J., Disteche, C. M., and Bird, T. D., 1993, DNA deletion associated with hereditary neuropathy with liability to pressure palsies, *Cell* **72:**143–151.

Charcot, J.-M., and Marie, P., 1886, Sur une forme particulière d'atrophie musculaire progressive souvent familiale debutant par les pieds et les jambes et atteignant plus dard les mains, *Rev. Méd.* **6:**97–138.

Davies, D. M., 1954, Recurrent peripheral nerve palsies in a family, *Lancet ii:*266–268.

Dejerine, J., and Sottas, J., 1893, Sur la névrite interstitielle, hypertrophique et pregressive de l'enfance, *C. R. Soc. Biol.* **45:**63–96.

De Jonghe, P., Timmerman, V., Nelis, E., Martin, J.-J., and Van Broeckhoven, C., 1997, Charcot–Marie–Tooth disease and related peripheral neuropathies, *J. Periph. Nerv. Syst.* **2:**370–387.

De Waegh, S., and Brady, S. T., 1990, Altered slow axonal transport and regeneration in a myelin-deficient mutant mouse: The Trembler as an in vivo model for Schwann cell–axon interactions, *J. Neurosci.* **10:**1855–1865.

De Waegh, S. M., and Brady, S. T., 1991, Local control of axonal properties by Schwann cells: Neurofilaments and axonal transport in homologous and heterologous nerve grafts, *J. Neurosci. Res.* **30:**201–212.

De Waegh, S. M., Lee, V. M., and Brady, S. T., 1992, Local modulation of neurofilament phosphorylation, axonal caliber, and slow axonal transport by myelinating Schwann cells, *Cell* **68:**451–463.

D'Urso, D., Brophy, P. J., Staugaitis, S. M., Gillespie, C. S., Frey, A. B., Stempak, J. G., and Colman, D. R., 1990, Protein zero of peripheral nerve myelin: Biosynthesis: membrane insertion, and evidence for homotypic interaction, *Neuron* **4:**449–460.

D'Urso, D., Schmalenbach, C., Zoidl, G., Prior, R., and Mueller, H. W., 1997, Studies on the effects of altered PMP22 expression during myelination in vitro, *J. Neurosci. Res.* **48:**31–42.

Dyck, P. J., Chance, P. F., Lebo, R. V., and Carney, J. A., 1993, Hereditary motor and sensory neuropathies, *in* "Peripheral Neuropathy" (P. J. Dyck, P. K. Thomas, J. W. Griffin, P. A. Low, and J. F. Poduslo, eds.), pp. 1094–1136, W. B. Saunders, Philadelphia.

Fabbretti, E., Edomi, P., Brancolini, C., and Schneider, C., 1995, Apoptotic phenotype induced by overexpression of wild-type gas3/PMP22: Its relation to the demyelinating peripheral neuropathhy CMT1A, *Genes Dev.* **9:**1846–1856.

Falconer, D. S., 1951, Two new mutants, '*Trembler*' and '*Reeler*' with neurological actions in the house mouse (*Mus musculus L.*), *J. Genet.* **50:**192–201.

Filbin, M. T., Walsh, F. S., Trapp, B. D., Pizzey, J. A., and Tennekoon, G. I., 1990, Role of myelin P_0 protein as a homophilic adhesion molecule, *Nature* **344:**871–872.

Friedman, H. C., Jelsma, T. N., Bray, G. M., and Aguayo, A. J., 1996, A distinct pattern of trophic factor expression in myelin-deficient nerves of Trembler mice: Implications for trophic support by Schwann cells, *J. Neurosci.* **16:**5344–5350.

Gabreels-Festen, A. A. W. M., Hoogendijk, J. E., Meijerink, P. H. S., Gabreels, F. J. M., Bolhuis, P. A., Van Beersum, S., Kulkens, T., Nelis, E., Jennekens, F. G. I., De Visser, M., Van Engelen, B. G. M., Van Broeckhoven, C., and Mariman, E. C. M., 1996, Two divergent types of nerve pathology in patients with different P0 mutations in Charcot–Marie–Tooth disease, *Neurology* **47:**761–765.

Garbay, B., and Cassagne, C., 1994, Expression of the ceramide galactosyltransferase gene during myelination of the mouse nervous system. Comparison with the genes encoding myelin basic proteins, choline kinase and CTP:phosphocholine cytidylyltransferase, *Dev. Brain Res.* **83:**119–124.

Garbay, B., Boiron-Sargueil, F., and Cassagne, C., 1995, Expression of the exon 1A-containing PMP22 transcript is altered in the Trembler mouse, *Neurosci. Lett.* **198:**157–160.

Garcia, C. A., Malamut, R. E., England, J. D., Parry, G. S., Liu, P., and Lupsky, J. R., 1995, Clinical variability in two pairs of identical twins with the Charcot–Marie–Tooth disease type 1A duplication, *Neurology* **45:**2090–2093.

Giese, K. P., Martini, R., Lemke, G., Soriano, P., and Schachner, M., 1992, Mouse *P0* gene disruption leads to hypomyelination, abnormal expression of recognition molecules, and degeneration of myelin and axons, *Cell* **71:**565–576.

Gouider, R., LeGuern, E., Gugenheim, M., Tardieu, S., Maisonobe, T., Leger, J. M., Vallat, J. M., Agid, Y., Bouche, P., and Brive, A., 1995, Clinical, electrophysiological, and molecular correlations in 13 families with hereditary neuropathy with liability to pressure palsies and a chromosome 17p11.2 deletion, *Neurology* **45:**2018–2023.

Hahn, A. F., Brown, W. F., Koopman, W. J., and Feasby, T. E., 1990, X-linked dominant hereditary motor and sensory neuropathy, *Brain* **113:**1511–1525.

Hanemann, C. O., Stoll, G., D'Urso, D., Fricke, W., Martin, J. J., Van Broeckhoven, C., Mancardi, G. L., Bartke, I., and Müller, H. W. 1994, Peripheral myelin protein 22 expression in Charcot–Marie–Tooth disease type 1a sural biopsies, *J. Neurosci. Res.* **37:**654–659.

Haney, C., Snipes, G. J., Shooter, E. M., Suter, U., Garcia, C., Griffin, J. W., and Trapp, B., 1996, Ultrastructural distribution of PMP22 in Charcot–Marie–Tooth disease type 1A, *J. Neuropathol. Exp. Neurol.* **55:**290–299.

Harati, Y., and Butler, I. J., 1985, Congenital hypomyelinating neuropathy, *J. Neurol. Neurosurg. Psychiatry* **48:**1269–1276.

Harding, A. E., and Thomas, P. K., 1980, The clinical features of hereditary motor and sensory neuropathy types I and II, *Brain* **103:**259–280.

Heape, A. M., Bessoule, J.-J., Boiron-Sargueil, F., Garbay, B., and Cassagne, C., 1995, Spingolipid metabolic disorders in Trembler mouse peripheral nerves in vivo result from an abnormal substrate supply, *J. Neurochem.* **65:**1665–1673.

Heath, J. W., Inuzuka, T., Quarles, R. H., and Trapp, B. D., 1991, Distribution of P0 protein and the myelin-associated glycoprotein in peripheral nerves from Trembler mice, *J. Neurocytol.* **20:**439–449.

Henry, E. W., and Sidman, R. L., 1988, Long lives for homozygous Trembler mutant mice despite virtual absence of peripheral myelin, *Science* **241:**344–346.

Henry, E. W., Cowen, J. S., and Sidman, R. L., 1983, Comparison of Trembler and Trembler–J phenotypes: Varying severity of peripheral hypomyelination, *J. Neuropathol. Exp. Neurol.* **42:**688–706.

Huxley, C., Passage, E., Manson, A., Putzu, G., Figarella-Branger, D., Pellisier, J. F., and Fontes, M., 1996, Construction of a mouse model of Charcot–Marie–Tooth disease type 1A by pronuclear injection of human YAC DNA, *Hum. Mol. Genet.* **5:**563–569.

Ionasescu, V. V., Searby, C. C., Ionasescu, R., Chatkupt, S., Patel, N., and Koenigsberger, R., 1997, Dejerine–Sottas neuropathy in mother and son with the same point mutation of PMP22 gene, *Muscle Nerve* **20:**97–99.

Isaacs, H., 1961, A syndrome of continuous muscle-fibre activity, *J. Neurol. Neurosurg. Psychiatry* **24:**319–325.

Kilian, J. M., and Kloepfer, H. W., 1979, Homozygous expression of a dominant gene for Charcot–Marie–Tooth neuropathy, *Ann. Neurol.* **5:**515–522.

Kirkpatrick, L. L., and Brady, S. T., 1994, Modulation of the axonal microtubule cytoskeleton by myelinating Schwann cells, *J. Neurosci.* **14:**7440–7450.

Lee, M.-J., Brennan, A., Blanchard, A., Zoidl, G., Dong, Z., Tabernero, A., Zoidl, C., Dent, M. A. R., Jessen, K. R., and Mirsky, R., 1997, P0 is constitutively expressed in the rat neural crest and embryonic nerves and is negatively and positively regulated by axons to generate non-myelin-forming and myelin-forming Schwann cells, respectively, *Mol. Cell. Neurosci.* **8:**336–350.

Le Guern, E., Gouider, R., Mabin, D., Tardieu, S., Birouk, N., Parent, P., Bouche, E., and Brice, A., 1997, Patients homozygous for the 17p11.2 duplication in Charcot–Marie–Tooth type 1A disease, *Ann. Neurol.* **41:**104–108.

Lemke, G., 1988, Unwrapping the genes of myelin, *Neuron* **1:**535–543.

Lobsiger, C. S., Magyar, J. P., Taylor, V., Wulf, P., Welcher, A. A., Amgen EST Program, and Suter, U., 1996, Identification and characterization of a cDNA and the structural gene encoding the mouse epithelial membrane protein-1, *Genomics* **36:**379–387.

Low, P. A., 1977, The evolution of "onion bulbs" in the hereditary hypertrophic neuropathy of the Trembler mouse, *Neuropathol. Appl. Neurobiol.* **3:**81–92.

Lupski, J. R., Garcia, C. A., Parry, G. J., and Patel, P. I., 1991, Charcot–Marie–Tooth polyneuropathy syndrome: Clinical, electrophysiological, and genetic aspects, *in* "Current Neurology" (S. Appel, ed.), pp. 1–25, Mosby-Yearbook, Chicago.

Lupski, J. R., Montes de Oca-Luna, R., Slaugenhaupt, S., Pentao, L., Guzzetta, V., Trask, B. J., Saucedo, C. O., Barker, D. F., Killian, J. M., Garcia, C. A., Chakravarti, A., and Patel, P. I., 1991, DNA duplication associated with Charcot–Marie–Tooth disease type 1A, *Cell* **66:**219–232.

Magyar, J. P., Martini, R., Ruelicke, T., Aguzzi, A., Adlkofer, K., Dembic, Z., Zielasek, J., Toyka, K. V., and Suter, U., 1996, Impaired differentiation of Schwann cells in transgenic mice with increased PMP22 gene dosage, *J. Neurosci.* **16:**5351–5360.

Martini, R., 1997, Animal models for inherited peripheral neuropathies, *J. Anat.* **191:**321–336.

Martini, R., Mohajeri, M. H., Kasper, S., Giese, K. P., and Schachner, M., 1995a, Mice doubly deficient in the genes for P0 and myelin basic protein show that both proteins contribute to the formation of the major dense line in peripheral nerve myelin, *J. Neurosci.* **15:**4488–4495.

Martini, R., Zielasek, J., Toyka, K. V., Giese, K. P., and Schachner, M., 1995b, Protein zero (P0)-deficient mice show myelin degeneration in peripheral nerves characteristic of inherited human neuropathies, *Nat. Genet.* **11:**281–286.

Marvin, K. W., Fujimoto, W., and Jetten, A. M., 1995, Identification and characterization of a novel squamous cell-associated gene related to PMP22, *J. Biol. Chem.* **270:**28910–18916.

Maycox, P. R., Ortuno, D., Burrola, P., Kuhn, R., Bieri, P. L., Arrezo, J. C., and Lemke, G., 1997, A transgenic mouse model for hereditary neuropathy with liability to pressure palsies, *Mol. Cell. Neurosci.* **8:**405–416.

Meier, C., and Moll, C., 1982, Hereditary neuropathy with liability to pressure palsies. Report of two families and review of the literature, *J. Neurol.* **228:**73–95.

Naef, R., Adlkofer, K., Lescher, B., and Suter, U., 1997, Aberrant protein trafficking in Trembler suggests a disease mechanism for hereditary motor and sensory neuropathies, *Mol. Cell. Neurosci.* **9:**13–25.

Nelis, E., Van-Broeckhoven, C., and the HMSN Consortium, 1996, Estimation of the mutation frequencies in Charcot–Marie–Tooth disease type 1 and hereditary neuropathy with liability to pressure palsies: A European collaborative study, *Eur. J. Hum. Genet.* **4:**25–33.

Nelles, E., Bützler, C., Jung, D., Temme, A., Gabriel, H.-D., Dahl, U., Traub, O., Stümpel, F., Jungermann, K., Zielasek, J., Toyka, K. V., Dermietzel, R., and Willecke, K., 1996, Defective propagation of signals generated by sympathetic nerve stimulation in the liver of connexin32-deficient mice, *Proc. Natl. Acad. Sci. USA* **93:**9565–9570.

Nicholson, G. A., and Nash, J., 1993, Intermediate nerve conduction velocities define X-linked Charcot–Marie–Tooth neuropathy families, *Neurology* **43:**2558–2564.

Nicholson, G. A., Valentijn, L. J., Cherryson, A. K., Kennerson, M. L., Bragg, T. L., DeKroon, R. M., Ross, D. A., Pollard, J. D., McLeod, J. G., Bolhuis, P. A., and Baas, F., 1994, A frame shift mutation in the PMP22 gene in hereditary neuropathy with liability to pressure palsies, *Nat. Genet.* **6**:263–266.

Notterpek, L., Shooter, E. M., and Snipes, G. J., 1997, Upregulation of the endosomal–lysosomal pathway in the Trembler–J neuropathy, *J. Neurosci.* **17**:4190–4200.

Omori, Y., Mesnil, M., and Yamasaki, H., 1996, *Connexin32* mutations from X-linked Charcot–Marie–Tooth disease patients: Functional defects and dominant-negative effects, *Mol. Biol. Cell* **7**:907–916.

Ouvrier, R., 1996, Correlation between the histopathologic, genotypic, and phenotypic features of hereditary peripheral neuropathies in childhood, *J. Child Neurol.* **11**:133–146.

Parmantier, E., Cabon, F., Braun, C., Urso, D. D., Muller, H. W., and Zalc, B., 1995, Peripheral myelin protein-22 is expressed in rat and mouse brain and spinal cord motorneurons, *Eur. J. Neurosci.* **7**:1080–1088.

Parmatier, E., Braun, C., Thomas, J.-L., Peyron, F., Martinez, S., and Zalc, B., 1997, PMP22 expression in the central nervous system of the embryonic mouse defines potential transverse segments and longitudinal columns, *J. Comp. Neurol.* **378**:159–172.

Patel, P. I., Roa, B. B., Welcher, A. A., Schoener-Scott, R., Trask, B. J., Pentao, L., Snipes, G. J., Garcia, C. A., Francke, U., Shooter, E. M., Lupski, J. R., and Suter, U., 1992, The gene for the peripheral myelin protein PMP-22 is a candidate for Charcot–Marie–Tooth disease type 1A, *Nat. Genet.* **1**:159–165.

Paul, D. L., 1995, New functions for gap junctions, *Curr. Opin. Cell Biol.* **7**:665–672.

Raeymaekers, P., Timmerman, V., Nelis, E., De Jonghe, J. P., Hoogendijk, J. E., Baas, F., Barker, D. F., Martin, J. J., De Visser, M., and Bolhuis, P. A., 1991, Duplication in chromosome 17p11.2 in Charcot–Marie–Tooth neuropathy type 1a (CMT 1a). The HMSN Collaborative Research Group, *Neuromuscul. Disord.* **1**:93–97.

Reiter, L. T., Murakami, T., Koeuth, T., Pentao, L., Muzny, D. M., Gibbs, R. A., and Lupski, J. R., 1996, A recombination hotspot responsible for two inherited peripheral neuropathies is located near a mariner transposon-like element, *Nat. Genet.* **12**:288–297.

Robertson, A. M., King, R. H. M., Muddle, J. R., and Thomas, P. K., 1997, Abnormal Schwann cell–axon interactions in the Trembler–J mouse, *J. Anat.* **190**:423–432.

Ruegg, C. L., Wu, H.-Y., Fagnoni, F. F., Engleman, E. G., and Laus, R., 1996, B4B, a novel growth-arrest gene, is expressed by a subset of progenitor/pre-B lymphocytes negative for cytoplasmic u-chain, *J. Immunol.* **157**:72–80.

Scherer, S. S., Deschenes, S. M., Xu, Y.-I., Grinspan, J. B., Fischbeck, K., and Paul, D. L., 1995, Connexin32 is a myelin-related protein in the PNS and CNS, *J. Neurosci.* **15**:8281–8294.

Scherer, S. S., Bone, L. J., Deschenes, S. M., Fischbeck, K., and Balice-Gordon, L. J., 1997, The role of the gap junction protein connexin32 in the myelin sheath, *in* "Cell Biology and Pathology of Myelin: Evolving Biological Concepts and Therapeutic Approaches" (M. Devon, R. Doucette, B. H. Juurlink, A. J. Nazarali, D. J. Schreyer, and V. M. K. Verge, eds.), pp. 83–102, Plenum Press, New York.

Schneider-Schaulies, J., von Brunn, A., and Schachner, M., 1990, Recombinant peripheral myelin protein P0 confers both adhesion and neurite outgrowth-promoting properties, *J. Neurosci. Res.* **27**:286–297.

Sereda, M., Griffiths, I., Puhlhofer, A., Stewart, H., Rossner, M. J., Zimmermann, F., Magyar, J. P., Schneider, A., Hund, E., Meinck, H.-M., Suter, U., and Nave, K. A., 1996, A rat transgenic model for Charcot–Marie–Tooth disease, *Neuron* **16**:1049–1060.

Shapiro, L., Doyle, J. P., Hensley, P., Colman, D. R., and Henderson, W. A., 1996, Crystal structure of the extracellular domain from P0, the major structural protein of peripheral nerve myelin, *Neuron* **17**:435–449.

Skre, H., 1974, Genetic and clinical aspects of Charcot–Marie–Tooth disease, *Clin. Genet.* **6**:98–118.

Snipes, G. J., and Suter, U., 1994, Signaling pathways mediating axon–Schwann cell interactions, *Trends Neurosci.* **17**:399–401.

Snipes, G. J., and Suter, U., 1995, Molecular basis of common hereditary motor and sensory neuropathhies in humans and in mouse models, *Brain Pathol.* **5**:233–247.

Snipes, G. J., Suter, U., Welcher, A. A., and Shooter, E. M., 1992, Characterization of a novel peripheral nervous system myelin protein (PMP-22/SR13), *J. Cell Biol.* **117**:225–238.

Suh, J. G., Ichihara, N., Saigoh, K., Nakabayashi, O., Yamanishi, T., Tanaka, K., Wada, K., and Kikuchi, T., 1997, An in-frame deletion in peripheral myelin protein-22 gene causes hypomyelination and cell death of the Schwann cells in the new Trembler mutant mice, *Neuroscience* **79**:735–744.

Suter, U., 1997a, Myelin: Keeping nerves well wrapped up, *Curr. Biol.* **7**:R21–R24.

Suter, U., 1997b, Molecular biology of hereditary motor and sensory neuropathies, *in* "Cell Biology and Pathology of Myelin: Evolving Biological Concepts and Therapeutic Approaches" (R. M. Devon, R. Doucette, B. H. J. Juurlink, A. J. Nazarali, D. J. Schreyer, and V. M. K. Verge, eds.), pp. 51–59, Plenum Press, New York.

Suter, U., and Patel, P. I., 1994, Genetic basis of inherited peripheral neuropathies, *Hum. Mut.* **3**:95–102.

Suter, U., and Snipes, G. J., 1995a, Biology and genetics of hereditary motor and sensory neuropathies, *Annu. Rev. Neurosci.* **18**:45–75.

Suter, U., and Snipes, G. J., 1995b, Peripheral myelin protein PMP22: Facts and hypotheses, *J. Neurosci. Res.* **40**:145–151.

Suter, U., Welcher, A. A., Oezcelik, T., Snipes, G. J., Kosaras, B., Francke, U., Billings-Gagliardi, G. S., Sidman, R. L., and Shooter, E. M., 1992a, Trembler mouse carries a point mutation in a myelin gene, *Nature* **356**:241–244.

Suter, U., Moskow, J. J., Welcher, A. A., Snipes, G. J., Kosaras, B., Sidman, R. L., Buchberg, A. M., and Shooter, E. M., 1992b, A leucine-to-proline mutation in the putative first transmembrane domain of the 22-kDa peripheral myelin protein in the Trembler–J mouse, *Proc. Natl. Acad. Sci. USA* **89**:4382–4386.

Suter, U., Welcher, A. A., and Snipes, G. J., 1993, Progress in the molecular understanding of hereditary peripheral neuropathies reveals new insights into the biology of the peripheral nervous system, *Trends Neurosci.* **16**:50–56.

Suter, U., Snipes, G. J., Schoener-Scott, R., Welcher, A. A., Pareek, S., Lupski, J. R., Murphy, R. A., Shooter, E. M., and Patel, P. I., 1994, Regulation of tissue-specific expression of alternative peripheral myelin protein-22 (*PMP22*) gene transcripts by two promoters, *J. Biol. Chem.* **269**:25795–25808.

Taylor, V., and Suter, U., 1996, Epithelial membrane protein-2 and epithelial membrane protein-3: Two novel members of peripheral myelin protein 22 gene family, *Gene* **175**:115–120.

Taylor, V., Welcher, A. A., AMGEN Est Programm, and Suter, U., 1995, Epithelial membrane protein-1, peripheral myelin protein 22 and lens membrane protein 20 define a novel gene family, *J. Biol. Chem.* **270**:28824–28833.

Thomas, F. P., Lebo, R. V., Rosoklija, G., Ding, X.-S., Lovelance, R. E., Latov, N., and Hays, A. P., 1994, Tomaculous neuropathy in chromosome 1 Charcot–Marie–Tooth syndrome, *Acta Neuropathol.* **87**:91–97.

Thomas, P. K., King, R. H. M., Small, J. R., and Robertson, A. M., 1996, The pathology of Charcot–Marie–Tooth disease and related disorders, *Neuropathol. Appl. Neurobiol.* **22**:269–284.

Thomas, P. K., Marques, W., Davis, M. B., Sweeney, M. G., King, R. H. M., Bradley, J. L., Muddle, J. R., Tyson, J., Malcolm, S., and Harding, A. E., 1997, The phenotypic manifestations of chromosome 17p11.2 duplication, *Brain* **120**:465–478.

Timmerman, V., De Jonghe, P., Spoelders, P., Simokovic, S., Lofgren, A., Nelis, E., Vance, J., Martin, J. J., and Van-Broeckhoven, C., 1996, Linkage and mutation analysis of Charcot–Marie–Tooth neuropathy type 2 families with chromosome 1p35–p36 and Xp13, *Neurology* **46**:1311–1318.

Tooth, H. H., 1886, "The Peroneal Type of Progressive Muscular Atrophy," Lewis, London.

Toyka, K. V., Zielasek, J., Ricker, K., Adlkofer, K., and Suter, U., 1997, Hereditary neuromyotonia: A mouse model associated with deficiency or overexpression of PMP22, *J. Neurol. Neurosurg. Psychiatry* **63**:812–813.

Tyson, J., Ellis, D., Fairbrother, U., King, R. H. M., Muntoni, F., Jacobs, J., Malcolm, S., Harding, A. E., and Thomas, P. K., 1997, Hereditary demyelinating neuropathy of infancy: A genetically complex syndrome, *Brain* **120:**47–63.

Valentijn, L. J., Baas, F., Wolterman, R. A., Hoogendijk, J. E., van der Bosch, N. H., Zorn, I., Gabreels-Festen, A. W., de Visser, M., Bolhuis, P. A., 1992, Identical point mutations of PMP-22 in Trembler–J mouse and Charcot–Marie–Tooth disease type 1A. *Nat. Genet.* **2:**288–291.

Vallat, J. M., Sindou, P., Preux, P. M., Tabaraud, F., Milor, A. M., Couratier, P., Le Guern, E., and Brice, A., 1996, Ultrastructural PMP22 expression in inherited demyelinating neuropathies, *Ann. Neurol.* **39:**813–817.

Vasilescu, C., Alexianu, M., and Dan, A., 1984, Muscle hypertrophy and syndrome of continuous motor activity in prednisone-responsive Guillain–Barré polyneuropathy, *J. Neurol.* **231:**276–279.

Warner, L. E., Hilz, M. J., Appel, S. H., Killian, J. M., Kolodny, E. H., Karpati, G., Carpenter, S., Watters, G. V., Wheeler, C., Witt, D., Bodell, A., Nelis, E., van Broeckhoven, C., and Lupski, J. R., 1996, Clinical phenotypes of different MPZ (P0) mutations may include Charcot–Marie–Tooth type 1B, Dejerine–Sottas, and congenital hypomyelination, *Neuron* **17:**451–460.

Welcher, A. A., Suter, U., De Leon, M., Snipes, G. J., and Shooter, E. M., 1991, A myelin protein is encoded by the homologue of a growth arrest-specific gene, *Proc. Natl. Acad. Sci. USA* **88:**7195–7199.

Windebank, A. J., 1993, Inherited recurrent neuropathy, *in* "Peripheral Neuropathy" (P. J. Dyck, P. K. Thomas, J. W. Griffin, P. A. Low, and J. F. Poduslo, eds.), pp. 1137–1148, W. B. Saunders, Philadelphia.

Wong, M.-H., and Filbin, M. T., 1996, Dominant-negative effects on adhesion by myelin P0 protein truncated in its cytoplasmic domain, *J. Cell Biol.* **134:**1531–1541.

Yoshikawa, H., and Dyck, P. J., 1991, Uncompacted inner lamellae in inherited tendency to pressure palsy, *J. Neuropathol. Exp. Neurol.* **50:**649–657.

Yoshikawa, H., Nishimura, T., Nakatsuji, Y., Fujimura, H., Himoro, M., Hayasaka, K., Sakoda, S., and Yanagihara, T., 1994, Elevated expression of messenger RNA for peripheral myelin protein 22 in biopsied peripheral nerves of patients with Charcot–Marie–Tooth disease type 1A, *Ann. Neurol.* **35:**445–450.

Zoidl, G., Blass-Kampmann, S., D'Urso, D., Schmalenbach, C., and Müller, H. W., 1995, Retroviral-mediated gene transfer of the peripheral myelin protein PMP22 in Schwann cells: Modulation of cell growth, *EMBO J.* **14:**1122–1128.

Zoidl, G., D'Urso, D., Blass-Kampmann, S., Schmalenbach, C., Kuhn, R., and Muller, H. W., 1997, Influence of elevated expression of rat wild type PMP22 and its mutant PMP22 (Trembler) on cell growth of NIH3T3 fibroblasts, *Cell Tissue Res.* **287:**459–470.

CHAPTER 4

MOUSE MUTATIONS IN THE STUDY OF CEREBELLAR DEVELOPMENT

DOMNA KARAGOGEOS and OURANIA PAVLOU

1. CEREBELLAR DEVELOPMENT

1.1. Cerebellar Cytoarchitecture

The cerebellar cortex has been a favorite system for neuroscientists to study both developmental events and adult neural circuitry. A characteristic feature of the adult cerebellum that is, at least in part, under genetic control (see Section 2) is its unique pattern of lobules and folia that provide it with its highly convoluted structure. The main part of the cerebellum is separated by the primary fissure into the anterior and posterior lobes. Smaller fissures divide the anterior and posterior lobes into nine lobules, the tenth being the flocculonodular lobe. The cerebellum

DOMNA KARAGOGEOS • Institute of Molecular Biology and Biotechnology and Department of Basic Sciences, University of Crete Medical School, Heraklion 711 10, Crete, Greece. OURANIA PAVLOU • Institute of Molecular Biology and Biotechnology, Heraklion 711 10, Crete, Greece.
Mouse Models in the Study of Genetic Neurological Disorders, Volume 9 of *Advances in Neurochemistry,* edited by Popko, Kluwer Academic/Plenum Publishers, New York, 1999.

consists of three gross functional regions: the central (vermis) and the two lateral ones (hemispheres).

From the time of Ramon y Cajal and continuing up to the present, the cerebellar cortex has been utilized as a model system for migration and cell differentiation studies, in part because of its simple cytoarchitecture. The adult cerebellar cortex consists of three layers: on the outside the molecular layer, in the middle the Purkinje cell layer, and on the inside the granule cell layer (Fig. 1).

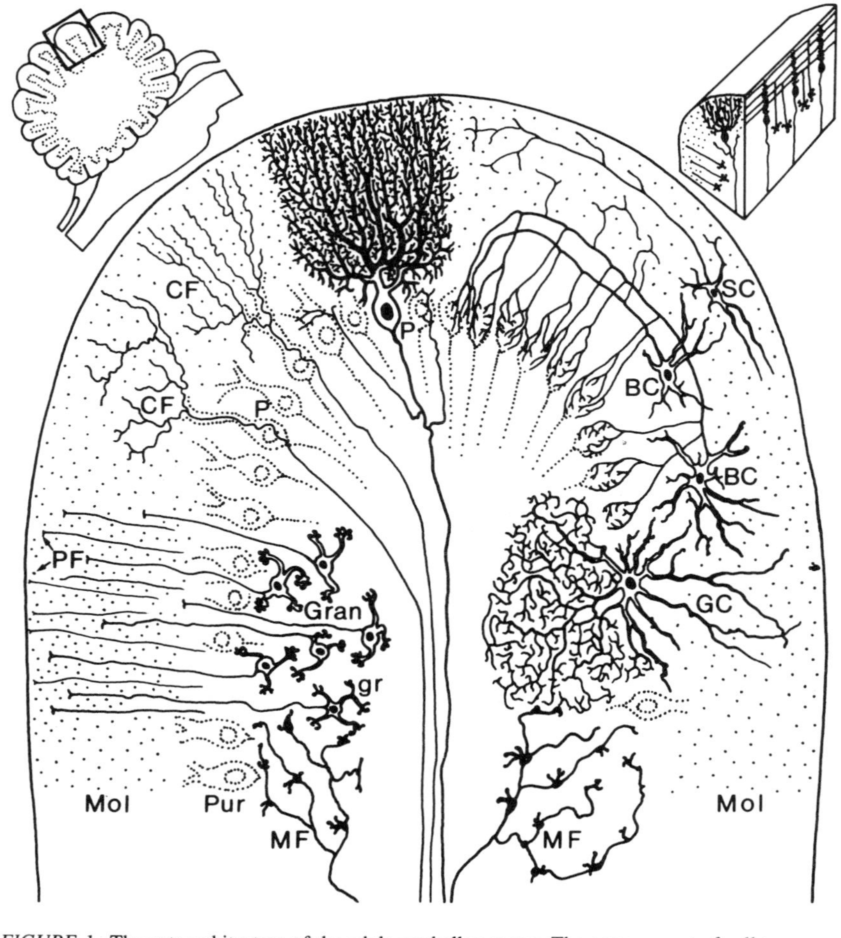

FIGURE 1. The cytoarchitecture of the adult cerebellar cortex. The arrangement of cell types comprising each cerebellar folium is depicted. BC = basket cell; CF = climbing fiber; Gran = granule cell layer; GC = Golgi cell; gr = granule cell; MF = mossy fiber; Mol = molecular layer; P = Purkinje cell layer; Pur = Purkinje cell; PF = parallel fibers; SC = stellate cell. [From the 3rd edition of "Developmental Neurobiology," (M. Jacobson), with permission.]

The five types of neurons and their connections are known in detail (Ramon y Cajal, 1889, 1911, 1960; Llinas and Hillman, 1969; Palay and Chan-Palay, 1974, 1982; for a recent review see Hatten and Heintz, 1995). The molecular layer consists mainly of the granule cell axons, called the parallel fibers because they are arranged parallel to the long axis of the folium. In this layer, the stellate and basket inhibitory interneurons are also found. The parallel fibers form excitatory synaptic connections on dendrites of Purkinje cells but also on the dendrites of stellate, basket, and Golgi interneurons.

The Purkinje cells are the unique output of the cerebellar cortex. They are inhibitory neurons that send their dendrites up to the molecular layer and their axons down to the white matter. They receive input from climbing fibers, originating in the inferior olive; the Purkinje-climbing fiber synaptic relationship has been one of the most well studied paradigms of connections in the nervous system.

The granule cell layer (IGL, see later) contains billions of granule cells (the most abundant cell type in the nervous system), which are excitatory neurons. A few Golgi cells are also found in this layer. The granule cells receive input from mossy fibers, the biggest afferent system in the cerebellum, originating from a number of brainstem nuclei. Golgi cells synapse with granule cell dendrites and receive input from parallel fibers.

The white matter consists of mossy and climbing fibers, Purkinje cell axons, as well as the deep cerebellar nuclei. These nuclei are the principal source of efferent fibers from the cerebellum, projecting mainly to the thalamus and the vestibular nuclei in the brainstem.

1.2. Early Developmental Events

The cerebellar anlage (Fig. 2) is derived from both the mesencenphalic and metencenphalic regions as was shown by chick-quail grafts and not solely from the metencephalon as was believed previously (Alvarado-Mallart and Sotelo, 1982; Martinez and Alvarado-Mallart, 1989; Alvarado-Mallart *et al.,* 1990; Hallonet *et al.,* 1990; Martinez *et al.,* 1991; Hallonet and Le Douarin, 1993; Alvarez Otero *et al.,* 1993). The appearance of two bilateral thickenings of the roof of the

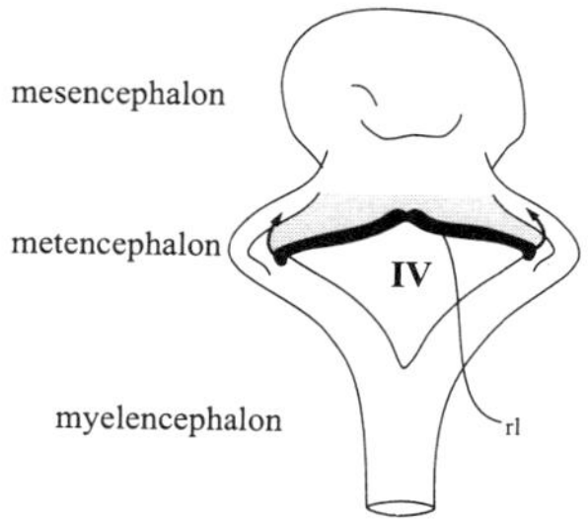

FIGURE 2. Cell movements during EGL formation in the embryo. The cerebellar primordium occupies the roof of the IVth ventricle. At E14, granule cell precursors, originating only from the metencephalic brain vesicle, are generated at a subdomain of the rhombic lip (rl, *darkly shaded area*), and move over and up the rhombic lip (*arrow*), covering the roof of the anlage at the metencephalic surface area.

anterior metencephalon signals the first events in cerebellar development. These thickenings expand and finally fuse on the midline. Within the cerebellar anlage, the first cerebellar neurons are born (in this order: deep nuclei, Purkinje, Golgi cells) (Miale and Sidman, 1961; Altman and Bayer, 1985a–c). They undergo extensive migrations to occupy their position in the regularly organized cerebellar architecture.

Precursors of the deep cerebellar nuclei are born first in the cerebellar plate (Pierce, 1975), followed shortly afterwards by Purkinje cell precursors. Both types of cells originate from the ventricular germinal zone of the rhombic lip. Then the Purkinje cell precursors migrate on radial glia and settle beneath the proliferating granule cells to form an irregular cell layer that undergoes a period of alignment, during which it becomes a single cell layer. The precursors of the deep cerebellar nuclei migrate also on radial glia but occupy a more rostral position in the anlage.

Granule cell precursors are derived from a subdomain of the rhombic lip at the posterior limit of the cerebellar anlage in contact with the choroid plexus. They spread across the surface of the metencephalon anteriorly and medially. These movements are taking place at the end of the deep nuclei and Purkinje cell precursor movements and establish the external granular layer (EGL), a germinal zone that gives rise to all granule cells. After birth, in rodents, the EGL precursors undergo extensive proliferation in the most superficial part of the EGL (EGLa) (Fugita *et al.,* 1966; Fugita, 1967). Subsequently, the postmitotic cells occupy the most internal part of the EGL (EGLb) and from there start their inward migration over radial glia and the molecular and Purkinje cell layers to generate the IGL (internal granule cell layer) (Rakic, 1971; Sidman and Rakic, 1973; Edmondson and Hatten, 1987; Gregory *et al.,* 1988; Fishman and Hatten, 1993). Except for this well studied, inward migration on radial glia, experiments with replication-incompetent retroviruses in the chick demonstrated the existence of transverse migrations in the EGL as well (Ryder and Cepko, 1994). Thus, in the EGLb, postmitotic granule cell progenitors actively migrate long distances in the transverse plane in both medial and lateral directions.

Stage-specific expression of genes has revealed the existence of at least four steps of granule cell development: genesis, initiation of differentiation, axon outgrowth and migration, and formation of synaptic contacts (Kuhar *et al.,* 1993). Of these stages, the migration on Bergmann radial glia has been extensively studied both *in vivo* and *in vitro* (Hatten and Mason, 1990; Hatten, 1993). In these studies, a key player has been the mouse mutation *weaver* (see Section 2.2.2 of this chapter). The neural glycoprotein astrotactin has been implicated as a receptor system for the radial migration process (Edmondson *et al.,* 1988; Fishman and Hatten, 1993; Zheng *et al.,* 1996). More recent studies implicate the brain-specific member of the lipid-binding protein family (BLBP), expressed by the radial glia, in the signaling pathway involved in establishing the glial fiber system (Feng *et al.,* 1994). At a more mature stage, as the granule cell enters the IGL, the parallel fibers form connections with the Purkinje cell dendrites and the granule cells with the mossy fibers.

The Purkinje cells form their axons in the absence of other cell types and send them to synapse with neurons of the deep cerebellar nuclei. The axon outgrowth phase is succeeded by the growth of a single dendrite followed by extensive branching. The synaptic partners of Purkinje cells are the climbing fibers and the parallel fibers. Ramon y Cajal first postulated that the initial steps of Purkinje axon extension take place without "neurotrophic" influences but that the growth of Purkinje cell dendrites is stimulated by the growth of parallel fibers (Ramon y Cajal, 1911). When climbing fibers start growing from the contralateral inferior olive toward the cerebellum, they segregate in a manner reflecting the topography of the brainstem (Wassef *et al.,* 1992). Initially each Purkinje cell is innervated by multiple developing climbing fibers; these transient synapses regress later in postnatal development to generate the one-to-one relationship between climbing fibers and Purkinje cells (synapse elimination; Crepel, 1982; for a recent review, see Lohof *et al.,* 1996). Finally, Purkinje cells elaborate their dendritic trees and form synaptic contacts on the dendritic spines. These processes, undefined as yet in terms of molecular detail, are known to require granule cell–Purkinje cell interactions *in vivo* and *in vitro* (Baptista *et al.,* 1994). The Purkinje cells are molecularly heterogeneous and their parcellation into identifiable, putatively functional subclasses of Purkinje cells has been described in detail (Sotelo and Wasset, 1991; Hawkes *et al.,* 1993).

In summary, as with most regions of the developing brain, early events in cerebellar development involve patterning of the cerebellar territory while later events involve the differentiation of particular cell types. Earlier events may (granule cells) or may not (Purkinje cells) involve local signaling but later events are influenced by axon–target interactions.

2. CEREBELLAR MOUSE MUTANTS

There is a still expanding list of mouse mutants that affect the development of the cerebellum (Sidman *et al.,* 1965; Sidman, 1968; Sotelo, 1980; Mullen, 1982; Sidman, 1983; Heintz *et al.,* 1993). The resulting phenotypes cover the whole range of developmental events and represent a continuum of mutants, each presumably blocking a distinct step in the neuronal differentiation process. In this chapter we focus on a few of these mutants (Table 1), emphasizing what is being uncovered about the molecular mechanisms shaping the mature neuronal phenotype. The development of the cerebellum has been studied in detail by traditional neuroanatomical methods and has been reviewed extensively in the past. Over the last few years, however, there has been an explosion in the number of

TABLE 1. Spontaneous Mouse Mutants that Affect Cerebellar Development

Name and allele symbol(s)	Inheritance and chromosome	Gene mutated	Protein	Primary cell target of the mutation
Patterning mutants				
swaying (sw)	Recessive 15	*Wnt-1*	Proto-oncogene	Met–mes region
meander tail (mea)	Recessive 4	?	?	Granule cells
leaner/tottering (tg^{la}/tg)	Recessive 8	*Cacnl/1a4*	Calcium channel	Multiple
Migration mutants				
rostral cerebellar malformation (rcm)	Recessive 3	*rcm*	Netrin receptor (chemorepulsive)	Granule cells Purkinje cells
weaver (wv)	Recessive[a] 16	*GIRK2*	Potassium channel	Granule cells
reeler (rl)	Recessive 5	*reelin*	extracellular matrix protein	Purkinje cells
scrambler (sc)	Recessive 4	*mdab1*	Phosphoprotein	Purkinje cells
Differentiation mutants				
Lurcher (Lc)	Semidominant 6	*Gluδ2*	Receptor/channel	Purkinje cells
staggerer (sg)	Recessive 9	*RORα*	Receptor	Purkinje cells
purkinje cell degeneration (pcd)	Recessive 13	?	?	Purkinje cells
motor endplate disease/jolting ($med/med^{J}/med^{jo}$)	Recessive 15	*Scn8a*	Sodium channel	Purkinje cells

Note: This table includes only the spontaneous mutants discussed in the text. It does not include the targeted knockouts or the transgenics with cerebellar phenotypes included in this chapter. For detailed references see text.

[a]Or semidominant given the migration defect observed in heterozygotes (details in section 2.2.2.).

mutant loci that have been cloned and characterized (Table 1). The identification of the responsible genes and the realization that some of these genes were previously known and their cellular functions already under scrutiny promises new adventures in developmental neurobiology, where molecular data confirm, reinforce, and complement knowledge obtained during the last 50 years by diligent neuroanatomists. In addition to linking gene function to terminally differentiated neurons, the newly identified loci pave the way for uncovering the defects underlying human neurodegenerative disorders.

2.1. Patterning Mutations

2.1.1. Wnt-1 Null Mutants

The *Wnt-1 (int-1)* proto-oncogene is the mouse homolog of the *Drosophila* segment polarity gene, *wingless* (Nusse and Varmus, 1982). The *Wnt-1* proto-oncogene has been shown to be necessary for the formation or stabilization of the boundary between the mesencephalic–metencephalic region (Bally-Cuif *et al.*, 1995). Embryos homozygous for null alleles of the *Wnt-1* gene show severe defects in the mesencephalon and metencephalon while the cerebellum is missing (among other severe defects, McMahon and Bradley, 1990; Thomas and Capecchi, 1990; McMahon *et al.*, 1992). However, in the sole mouse that survived until adulthood, the posterior portion of the cerebellum was present. The authors hypothesize that *Wnt-1* inductive influences form a gradient starting at the mesencephalon and spreading to the metencephalon. Thus, there may be a differential penetrance of the mutation so that some mice lack all cerebellar territory whereas others lack only the anterior part. The *Wnt-1* gene is localized to mouse chromosome 15. Although the *Wnt-1*^{neo} is a loss-of-function mutation, the phenotype is expressed in quite a variable fashion, in terms of life span and the amount of midbrain and cerebellar tissue present.

2.1.2. Swaying (sw)

This mutation results from a deletion of a single base pair in the protein coding region (third exon) of the *Wnt-1* gene (Lane, 1967; Thomas *et al.*, 1991). A large part of the COOH-terminal portion of the protein, essential for its biological activity, is deleted. The phenotype of *swaying* is similar to that of the *Wnt-1* null mutants. *Swaying* mutants are characterized by ataxia and lack of the anterior part of their cerebellum. The mutant brains are asymmetric and *Wnt-1* expression is maintained on the dorsal and ventral midlines of mesencephalon and diencephalon.

2.1.3. *Engrailed* -½ *(En-1, En-2)* Null Mutants

The mouse *Engrailed (En)* genes were identified as homologs of the Drosophila *engrailed,* a segment polarity gene (Joyner and Martin, 1987; Davis and Joyner, 1988). In the nervous system, *En-1* is expressed in a band of cells that will later give rise to the mesencephalic–metencephalic junction, the region from which the cerebellum is derived, while *En-2* expression begins a little later and is also found in a band around the mesencephalic–metencephalic junction. *En-2* is broadly expressed in the embryonic cerebellum whereas postnatally it becomes restricted to granule cells and cells in the molecular layer.

Mice homozygous for a targeted deletion of the *En-1* homeobox die shortly after birth and display, among others, a severe cerebellar phenotype. A deletion of the mid-hindbrain junction is observed early in development and the phenotype is reminiscent of the *Wnt-1* phenotype described previously (Wurst *et al.,* 1994).

Mice homozygous for a targeted deletion of the *En-2* homeobox ($En\text{-}2^{hd}$), as well as mice homozygous for a null mutation of the *En-2* gene ($En\text{-}2^{ntd}$), display abnormal cerebellar development (Joyner *et al.,* 1991; Millen *et al.,* 1994). In addition, the mutants reveal an abnormal foliation pattern. In more detail, the mutant phenotype appears at an embryonic stage (E15.5 and E16.5) and consists of a decrease in the extent of fusion of cerebellar rudiments at the midline and an apparent size reduction of the cerebellum and the colliculi. Postnatally, the developmental profile of foliation in the hemispheres revealed the absence of the ansoparamedial fissure. In the vermis, the mutant cerebella exhibited a complex phenotype of general delay and abnormal fissure formation, evident primarily in the posterior region. As the mutant phenotype was evident in the embryo, before the initiation of the process of foliation, the defect in *En-2* mutants may lay in the process of generating precursor cells rather than in foliation.

2.1.4. *Math1* Null Mutants

Math1, the mouse homolog of the Drosophila gene *atonal,* encodes a basic helix–loop–helix (bHLH) transcription factor expressed in the precursors of the EGL and their derivatives, as well as in the dorsal spinal cord by E9 in the mouse (Ben-Arie *et al.,* 1996). Postnatal EGL cells continue to express *math1* (Ben-Arie *et al.,* 1996). Targeted mutants in the *math1* gene fail to form granule cells and are born with a cerebellum that is devoid of an external germinal layer (Ben-Arie *et al.,* 1997). This is the only gene known so far to be required *in vivo* for granule cell genesis.

Several other mutants with patterning phenotypes in the cerebellum exist. *Meander tail (mea), leaner* (tg^{la}) and *rostral cerebellar malformation (rcm,* described in the section on migration mutants) have primary effects on the anterior portion of the cerebellum. In some of them, the cerebellar foliation pattern is also affected.

2.1.5. Meander tail (mea)

In the *meander tail,* an autosomal recessive mutation that maps to chromosome 4 (Fletcher *et al.,* 1991), the mice exhibit mild ataxia and a skeletal deformity (kinked tail) (Hollander and Waggie, 1977). In the anterior lobes the Purkinje cells are positioned randomly throughout the cortex, the IGL and other layers do not form, and the Bergmann glia are absent. A discrete boundary between the abnormal anterior lobes and normal posterior lobes is observed at the junction of folia VI and VII (Ross *et al.,* 1990). This very precise boundary is probably the most impressive element in the *mea/mea* cerebellum. On the basis of this observation and of the phenotype of the *leaner* (see later) and the *swaying* mutants (see earlier), the hypothesis has been put forward that the anterior part of the cerebellum may represent a distinct compartment, in the developmental sense of the term.

Whether granule, Purkinje, or radial glia are the primary targets of this mutation was unknown until recently. Chimeric mice were examined to determine that the *mea/mea* granule cells are the primary cellular targets of the *mea* gene (Hamre and Goldowitz, 1997). Genotypically *mea/mea* Purkinje and glial cells were morphologically normal while *mea/mea* granule cells were few. In addition, the wild-type granule cells seem to be upregulated in numbers, thus possibly competing out the mutant granule cells (Hamre and Goldowitz, 1997). In a study published simultaneously, it was demonstrated that pluripotent neural progenitors, when transplanted into the anterior cerebellum of *mea/mea* neonates, replenished the granule cells absent from this part of the cerebellum (Rosario *et al.,* 1997). Although the nature of the *mea* gene product is still unknown, it is proposed to be involved in the developmental progression of granule cell precursors, either by playing a role in the migration toward the anterior part of the cerebellum or in the process of programmed cell death in the EGL (Hamre and Goldowitz, 1997).

2.1.6. Leaner (tg^{la}) and Tottering (tg)

The *leaner* mutant, another mouse mutation that shows anterior patterning defects, displays degeneration of granule, Purkinje, and Golgi cells in the anterior lobes (Sidman *et al.,* 1965; Herrup and Wilczynski, 1982). However, although the same region is affected in the two mutants (*meander tail* and *leaner*), the developmental defects are quite distinct. In the *leaner,* the cytoarchitecture is normal and the degeneration occurs postnatally, while in the *meander tail* the classical trilaminar organization of the cerebellar cortex is not observed. *Leaner*

(tg^{la}) is a severe allele of the *tottering (tg)* locus on chromosome 8, a model of absence epilepsy (Noebels and Sidman, 1979; Herrup and Wilczynski, 1982; Heller *et al.*, 1983; Green, 1989). *Tottering* does not display a cerebellar phenotype (Green and Sidman, 1962; Meier and MacPike, 1971; Hess and Wilson, 1991). By positional cloning, a voltage-sensitive calcium channel of the α_{1A} subtype that is mutated in *tottering* and *leaner* mice was found (Fletcher *et al.*, 1996). The tg^{la} allele contains a mutation of a splice donor consensus sequence that results in the truncation of the normal open reading frame, whereas the *tg* allele contains a single base pair change that results in a nonconservative amino acid change (Fletcher *et al.*, 1996). This type of calcium channel may be a major component of the P- and Q-type calcium channels (Fletcher *et al.*, 1996; Doyle *et al.*, 1997). The gene is expressed in several areas of the brain (olfactory bulb, cerebral cortex, hippocampus, inferior colliculus) and in the cerebellum, where prominent and uniform expression was detected in Purkinje and granule cells as well as in the deep cerebellar nuclei (Fletcher *et al.*, 1996).

The human homolog of this channel (CACNL1A4) has recently been identified as the gene affected in SCA6, a glutamine-repeat neurodegenerative disease (Matsuyama *et al.*, 1997; Riess *et al.*, 1997), opening up new avenues for the investigation of pathogenesis in this and related cerebellar disorders. The nature of the mutation in SCA6 (the expanded glutamine repeat) differs from that of the mouse mutation. However, the availability of animal mutants facilitates the investigation of the neurodegenerative mechanisms. This case demonstrates, once again, how independent avenues of research may complement each other and converge to the identification of the same gene.

2.1.7. Foliation Mutations

Two foliation pattern mutants have been identified, *cerebellar folial pattern (Cfp-1)* (Neumann *et al.*, 1990; Cooper *et al.*, 1991) and *declival sulcus of cerebellum (dsc)* (Wahlsten and Andison, 1991), each controlling the formation of specific fissures and sulci in the cerebellum. *Clp-1* maps on chromosome 4 and was identified on the basis of the folial patterns between C57BL/6J and DBA/2J crosses. DBA/2J mice had an intraculminate fissure (between lobules IV and V) that was absent in C57BL/6J mice (Neumann *et al.*, 1990). *dsc* has not been well characterized yet.

2.2. Migration Mutations

2.2.1. Rostral Cerebellar Malformation (rcm)

Mice homozygous for the spontaneous *rostral cerebellar malformation* mutation (rcm^{S}) display severe ataxia (Lane *et al.*, 1992). The *rcm* mutation is autosomal recessive, localized to mouse chromosome 3. Histological sections revealed a number of cerebellar abnormalities, among them a reduction in size and number of folia as well as the formation of five or six lobes instead of eight. A characteristic feature of the *rcm* mutant is the presence of ectopic granule and Purkinje cells in midbrain regions (inferior colliculus, tegmentum), as early as the beginning of granule cell migration. Aberrant cohorts of granule cells (indicating additional migration defects) are also observed, often associated with gaps in the Purkinje cell layer. Owing to these putative migration abnormalities, the laminar structure of the cerebellum is perturbed (Lane *et al.,* 1992; Ackerman *et al.,* 1997).

The same phenotype was observed by a transgenic insertion allele *(rcm^{tg}).* The mitochondrial uncoupling protein (Ucp) minigene had disrupted the *rcm* locus (Ackerman *et al.,* 1997). Thus, a genomic DNA fragment flanking the transgene insertion site was used in molecular mapping and polymerase chain reaction (PCR) studies indicating that approximately 8.5 kb of DNA adjacent to the transgene insertion site was deleted. cDNA clones isolated by RACE revealed that the *rcm* gene is the mouse homolog of *UNC-5,* the *C. elegans* receptor mediating the repulsive effect of the netrin ligand (Ackerman *et al.,* 1997), supporting the hypothesis that the *rcm* gene product is involved in cerebellar neuronal migration. The *rcm* transcript is expressed, among other structures in the brain, in the cells (granule, Purkinje) that undergo aberrant migration. The Rcm protein was shown to bind to Netrin-1 *in vitro.* The hypothesis has been put forward that *rcm* might act within the migrating precursors guiding these cells over the cerebellum by repulsion from the floor plate netrin (Hedgecock and Norris, 1997).

2.2.2. Weaver (wv)

The mouse mutant *weaver* was first described more than 30 years ago (Lane, 1964) and the extensive study of its phenotype has provided important information on the processes of neuronal development, migration, and cell death. It displays several neurological and reproductive defects and is characterized by severe ataxia, hypotonia, and a tremor (Sidman *et al.,* 1965; Sidman, 1968). Homozygous mice exhibit ataxia by about the second postnatal week and rarely survive more than 1 month after birth. These behavioral manifestations reflect the selective loss of neurons in two regions of the *wv/wv* brain: the cerebellum and the substantia nigra. In the cerebellum, precursor cells in the EGL proliferate normally but fail to undergo early steps in differentiation and extension of neurites (Rezai and Yoon, 1972; Rakic and Sidman, 1973a,b; Goldowitz and Mullen, 1982; Hatten *et al.,* 1984). The cells subsequently die during the first 2 weeks of postnatal life. Heterozygous mice are not ataxic and their cerebellar development is only partially affected. In heterozygotes, external granule cells proliferate normally but their migration through the molecular layer is abnormal, with some of them never reaching the granular layer (Rezai and Yoon, 1972). Despite the migration defect observed in heterozygotes, the mutation has been historically classified as recessive based on behavioral criteria and the manifestation of ataxia only in homozygotes.

It was established quite early that the major defect in *weaver* is a migration impairment that produces several secondary effects, including cell death (Rakic and Sidman, 1973a). However, it was originally thought that the primary site of the *weaver* gene action was the Bergmann glial cell (Rakic and Sidman, 1973a). Subsequent studies with *weaver*/wild-type chimeras suggested that the defect acts in a cell-autonomous manner and is intrinsic to granule cells (Goldowitz and Mullen, 1982; Goldowitz, 1989). The development of techniques to purify granule cells and astroglia from normal and mutant mice allowed the performance of cell mixing experiments. Granule neurons isolated from *weaver* cerebella failed to migrate on wild-type astroglia (Hatten *et al.,* 1986). In contrast, wild-type neurons, when mixed with *weaver* astroglia, formed tight appositions with them, characteristic of migrating neurons (Hatten *et al.,* 1986). These experiments further confirmed that the granule cell is the primary site of action of the *weaver* defect. However, other cell mixing experiments *in vitro* suggested that the weaver defect is indeed localized to the granule cells but acts in a nonautonomous manner (Gao *et al.,* 1992), thus contradicting earlier studies that implied a cell-autonomous *weaver* gene action (Goldowitz and Mullen, 1982; Goldowitz, 1989). The explanation given for this discrepancy was the difference in experimental design, namely genetic mosaics versus *in vitro* cell mixing experiments. Implantation experiments of *weaver* granule cell precursors into wild-type cerebellum further confirmed that the action of the *weaver* gene requires local interactions among progenitor cells in order to modulate neuronal differentiation (Gao and Hatten, 1993).

The *weaver* gene eluded identification until recently. The locus had been mapped in a region of chromosome 16 that is syntenic with the Down's syndrome region on human chromosome 21 (Reeves *et al.,* 1989; Mjaatvedt *et al.,* 1993). Patil and colleagues (1995) took advantage of the great number of human transcripts mapped in this region to identify the mouse homolog mutated in the *weaver* mutation. The gene responsible for the *weaver* defect is coding for one of the G protein-gated inwardly rectifying K^+ channels, namely GIRK2 (Patil *et al.,* 1995). GIRK2 is one of three such channels expressed in the brain. It is localized in the proliferating rhombic lip at 14.5 dpc and in both the outer proliferating and inner differentiating zones of the EGL from 17.5 dpc to P6 (Kofuji *et al.,* 1996).

The gene defect is a missense point mutation (G to A) that results in a glycine to serine substitution in the pore-forming region of the GIRK2 protein (Patil *et al.,* 1995). The mutant GIRK2 channel forms homomultimers that lose their selectivity for K^+ ions, producing instead constitutive Na^+ currents

(Slesinger *et al.,* 1996). In addition, *weaver* GIRK2, forming heteromultimers with wild-type GIRK1 channels, have reduced inwardly rectifying K^+ current (Slesinger et al., 1996). The pronounced reduction of K^+ currents in cerebellar granule cells from *weaver* homozygotes leads to cell death (Surmeier *et al.,* 1996). Further functional studies of the mutated channel have shown that the reason for EGL precursor damage in *weaver* is the Na^+ influx (Kofuji *et al.,* 1996). The GIRK2 protein does not appear to be required for granule cell differentiation, thus leaving open the issue for other locally acting factors that signal the entry of EGL cells into differentiation (Kofuji *et al.,* 1996).

The mutation affects several cell types within the CNS. Besides the granule neurons and the dopaminergic neurons of the substantia nigra, Purkinje neurons fail to migrate and die in the postnatal cerebellum (Smeyne and Goldowitz, 1990; Maricich *et al.,* 1997), while deep cerebellar nuclei are anatomically abnormal and decreased in cell numbers (Maricich *et al.,* 1997). It is noteworthy that a mutation in a potassium channel that is expressed in several areas within the CNS has multiple yet cell-specific effects on neuronal degeneration (Hess, 1996). The involvement of the *weaver* gene (*girk2*) in the survival of such diverse cell types hints at a consolidated model for cerebellar development where migration impairment affects cell survival (Maricich *et al.,* 1997). The investigators suggest that *girk2* may act in a cell-autonomous manner in each of the neuronal populations. However, the unifying feature is deficit in cell movement that leads to ectopic placement in the cortex and, subsequently, to cell death. One possible mechanism for cell loss due to ectopia is the target-related cell death, an event that has been suggested for *Lurcher* Purkinje neurons (see later). In this context, the alteration or loss of appropriate afferent and efferent synaptic contacts triggers a sequence of events that result in neuronal degeneration. Both granule cells in *weaver* and Purkinje cells in *Lurcher* undergo apoptotic death (Gillardon *et al.,* 1995; Wullner *et al.,* 1995). Perhaps, then, the common mechanism underlying degeneration in both mutants is indeed target-related cell death. Furthermore, it is possible that a single gene defect having different effects, depending on the neuronal cell type, arises from the interaction of the gene product with separate neurotransmitter receptor systems (Hess, 1996).

The *weaver* phenotype in heterozygous versus homozygous animals at the level of cerebellar morphology predicts a gene dosage effect (Rakic and Sidman, 1973a). This can presumably be explained by the levels of the mutant GIRK2 channels in the heterozygous versus the homozygous animals. However, detailed analysis of GIRK2 content in normal and mutant cerebella showed that differential amounts of the channel protein do not correlate directly with cell death (Lauritzen *et al.,* 1997).

When the *girk2* gene was inactivated by homologous recombination the null mutant mouse did not phenocopy the *weaver* mutation. The *girk2 -/-* mice are morphologically indistinguishable from wild-type mice (Signorini *et al.,* 1997).

This led to the hypothesis that the *weaver* mutation arises not from a loss-of-function but rather from an abnormal or gain-of-function effect on the GIRK2 mutated channel. This is in agreement with the studies of Lauritzen and colleagues (1997) mentioned earlier, thus suggesting that cell death is probably the result of both the abnormal function of GIRK2 along with other locally interacting factors. This observation supports the results of earlier studies that suggested a nonautonomous action for the *weaver* gene (Gao *et al.,* 1992; Gao and Hatten, 1993).

The human homolog of the mouse *girk2* gene has been cloned (Ferrer *et al.,* 1995). Its involvement in the etiology of Parkinson's disease (a disorder in which dopaminergic neurons in the substantia nigra are affected) has been ruled out (Bandmann *et al.,* 1996). However, several human disorders are caused by mutations in potassium channels (Sanguinetti and Spector, 1997). Solving the puzzle of how a single gene product affects several cell types in different ways might shed some light on the complex nature of human neurodegenerative disorders. Furthermore, deciding exactly how this channel triggers signaling cascades that direct cell movement will provide valuable information on the more basic aspect of neuronal maturation during cortical development.

The emerging picture on the action of the *weaver* gene during cerebellar development is, of course, much more complicated than the case of a single gene affecting a single cell type. However, this is to be expected in a complex environment such as the CNS, where the unique identity of a particular cell type is the result of the interplay between several gene products and local, spatially and temporally restricted factors. The most significant conclusion developmental neurobiologists draw from such studies is that the distribution of a protein does not necessarily predict the direct targets of the encoding gene and, thus, the areas affected in the mutant.

2.2.3. Reeler (rl)

The *reeler* mouse mutant has been around even longer than *weaver.* It was first described by Falconer (1951) and has been used as a model for the study of corticogenesis and the analysis of migration and assembly of neurons. Mutant mice display impaired motor coordination, tremors, and ataxia (Caviness and Rakic, 1978). In *reeler* brains the cortical layering is inverted. Neurons that are born at earlier time points end up in the most superficial, instead of the deeper layers, while neurons born later settle in deeper instead of more superficial layers. In contrast to *weaver,* in *reeler* mice both neuronal proliferation and migration are normal. At the end of the migratory events, however, the cellular architecture is perturbed. Although virtually all cortical structures display malformations, there is no cell death or degeneration associated with the abnormal patterns (Goffinet, 1984; Rakic and Caviness, 1995). All the major morphological classes of cells are represented and in correct cell numbers, but they are ectopically placed (Sidman, 1983; Rakic and Caviness, 1995). The mutant cerebellum is much smaller than normal, lacking foliation and a Purkinje cell layer. Besides the cerebellum, abnormal cortical development is evident in the cerebrum and the hippocampus as well. The *reeler* model is used successfully to study pattern formation in the developing cortex as this depends on the correct alignment and orientation of terminally differentiated neurons rather than on cell form.

The early neuroanatomical experiments described the defect in *reeler* as the inability of the migrating neurons to find their correct address in the cortex (Caviness and Rakic, 1978; Goffinet, 1984; Rakic and Caviness, 1995), the mechanism unidentified nevertheless. However, it took the application of molecular biology to give the field a real boost. We focus on the most recent efforts to identify the gene mutated in *reeler* and to examine its role in cortical development. In an attempt to identify the protein missing from mutant brains the team of Ogawa and colleagues (1995) performed the following experiment. They immunized *reeler* mice with brain homogenates from wild-type mice. The antibody obtained (CR-50) was used on wild-type brains to track the cells normally producing the protein mutated in *reeler.* The structures with the highest levels of CR-50 immunoreactivity are the cerebrum and the hippocampus, with lower levels in the cerebellum (Ogawa *et al.,* 1995). The cells most strongly staining for CR-50 are the Cajal–Retzius (C–R) neurons that are found in the part of the preplate that becomes the marginal zone (Ogawa *et al.,* 1995; Schiffmann *et al.,* 1997). This predicts the C–R neurons as the cellular site of action of the *reeler* mutant gene.

Concurrently with this work, a transgenic mouse for *fos* was found to carry an insertional inactivation of the *reeler* locus. The cloning of the DNA surrounding the *fos* insertion revealed the gene mutated in *reeler,* which they named *reelin* (D'Arcangelo *et al.,* 1995). The gene maps to chromosome 5 and codes for Reelin, a very large protein of approximately 390 kDa that is homologous to F-spondin and to tenascin. Reelin shares features with extracellular matrix proteins and is expressed in embryonic and postnatal neurons during periods of neuronal migration, an indication that it may be involved in the guidance of postmitotic cells to their final destination (D'Arcangelo *et al.,* 1995). The predicted cell–cell interaction impaired in *reeler* cortex (Ogawa *et al.,* 1995) is supported further by the structure of the Reelin protein (D'Arcangelo *et al.,* 1995). Reelin has a signal peptide but lacks any apparent transmembrane domains, suggesting that it might be secreted and function outside the cells (Goffinet, 1995). It is now known that Reelin is indeed secreted into the extracellular matrix and the defect in a particular allele of *reeler* is due to the lack of this secretion (De Bergeyck *et al.,* 1997). The search is on for the identification of the receptor binding Reelin and, thus, modulating the transport of the intracellular signal guiding migration. The rejuvenation of the research in *reeler* by molecular biology approaches is demonstrated by the fact that, in addition to the above experimental efforts, two independent groups, using positional cloning strategies, produced partial sequence data at about the same time (Bar *et al.*, 1995; Hirotsune *et al.*, 1995).

Subsequently to the cloning of *reelin* it was demonstrated that the CR-50 antibody indeed recognizes a Reelin epitope (D'Arcangelo *et al.*, 1997). The CR-50 antibody disrupts neural cell aggregation *in vitro* (D'Arcangelo *et al.*, 1997) and converts wild-type explants into *reeler*-like clusters (Miyata *et al.*, 1997). In addition, one of the earliest structural abnormalities observed in *reeler* brains is the failure of cortical plate neurons to aggregate within the preplate and the extracellular matrix (Sheppard and Pearlman, 1997).

The expression pattern of Reelin does not correlate with the *reeler* phenotype (Schiffman *et al.*, 1997). This was a surprising finding because the original prediction was that Reelin would be produced by the migrating cortical neurons or by the glia supporting the migration. One hypothesis suggests a cell-repulsive role for Reelin that prevents migrating neurons from entering Reelin-rich regions (Schiffmann *et al.*, 1997). On the other hand, grafting and explant experiments have suggested that C–R cells might play an important role in neuronal migration by regulating the identity of radial glia and the movement of migrating neurons (Soriano *et al.*, 1997). C–R cells were shown to alter dramatically the migration pattern of granule cells, perhaps through a chemoattractive influence (Soriano *et al.*, 1997). This does not contradict the above hypothesis, because it is possible that neuronal migration is under the influence of both chemoattractive and chemorepulsive signals, as is the case for axon guidance (Tessier-Lavigne and Goodman, 1996). The timing of Reelin expression in the cerebellum is consistent with the appearance of the external granule layer just as the granule neurons begin their migration across the molecular layer (Ogawa *et al.*, 1995). The distribution of CR-50 immunostaining in the premigratory zone, the inner half of the EGL and the IGL, with Purkinje neurons nonstaining, agrees with the possibility of Reelin involved in cell–cell interactions that modulate the assembly of the Purkinje cell layer (Miyata *et al.*, 1996).

In an attempt to account for the defect in *reeler* and the recent observations about Reelin, two models have been proposed for the function of the wild-type protein (Sheppard and Pearlman, 1997). The first suggests that Reelin regulates the adhesive ability of preplate neurons, while the second hypothesizes that Reelin acts as a stop signal for migrating neurons. This second model might be more consistent with data demonstrating that *reeler* neurons have increased adhesiveness compared to wild-type (Hoffarth *et al.*, 1995), implying a gain-of-function effect for the mutated gene. It is likely that both models hold true at different time points during development. Initially, increased adhesion of precursor cells is required to form the preplate aggregates. Later on, however, the adhesive bonds need to be broken for migration to stop, once neurons have reached their correct position in the cortex.

The human homolog of the *reelin* gene (*RELN*) has been cloned and mapped to chromosome 7 (De Silva *et al.*, 1997), thus providing the potential for investigating its role in human genetic diseases and, certainly, in normal human brain development.

2.2.4. Scrambler (scm)

Scrambler is a recently described neurological mutation in the mouse that displays cortical abnormalities that are very similar to *reeler* (Sweet *et al.*, 1996; Gonzalez *et al.*, 1997). Owing to this similarity in cerebellar phenotype, it was originally thought that perhaps *reelin* and the gene affected in *scrambler* are part of the same genetic cascade (Gonzalez *et al.*, 1997). It was even proposed that the *scrambler* gene product represents the missing receptor for Reelin (Schiffmann *et al.*, 1997). Subsequent studies, however, showed that Reelin expression is normal in *scrambler* mutants both spatially and temporally (Goldowitz *et al.*, 1997; Gonzalez *et al.*, 1997; Ware *et al.*, 1997). The suggestion was made that the *scrambler* gene product acts downstream of the Reelin product (Gonzalez *et al.*, 1997). The identification of the defect common in *scrambler* and *yotari*, another newly identified, autosomal, recessive mutation, revealed that the gene responsible is actually the mouse homologue of the *Drosophila disabled* gene (*mdab1*) (Sheldon *et al.*, 1997; Ware *et al.*, 1997). The *mdab1* gene codes for a phosphoprotein that binds nonreceptor tyrosine kinases. Based on the most recent data, the current working hypothesis is that Reelin, present in the extracellular matrix, signals to neighboring neurons by activating a kinase cascade (Goffinet, 1997). *Mdab1* knockout mice display abnormal neuronal layering in the cerebral cortex, cerebellum, and hippocampus (Howell *et al.*, 1997). Reelin expression is normal in these null mutants (Howell *et al.*, 1997), in keeping with data from *scrambler* (Goldowitz *et al.*, 1997; Gonzalez *et al.*, 1997). Mdab1 is disqualified as the Reelin receptor. However, Mdab1 itself might bind that receptor as part of the signal transduction system that regulates cell migration and final positioning of neurons.

2.3. Differentiation Mutations

2.3.1. Lurcher (Lc)

The *Lurcher* mutant was first described by Phillips (1960), and is categorized as an autosomal semidominant mutation, primarily affecting the Purkinje neurons of the cerebellum (Heintz *et al.*, 1993; Sidman, 1983). Homozygous mutant mice die shortly after birth because a large number of mid- and hindbrain neurons are lost during embryogenesis (Cheng and Heintz, 1997). Heterozygous carriers develop ataxia due to a selective and cell-autonomous death of Purkinje neurons during postnatal development (Caddy and Biscoe, 1979) that is apoptotic in nature (Herrup and Busser, 1995; Norman *et al.*, 1995; Wullner *et al.*, 1995).

Earlier studies with *Lurcher*/wild type chimeras showed that the defect might be intrinsic to Purkinje cells and extrinsic to others (Wetts and Herrup, 1982a, 1982b). The original hypothesis stated that the primary target of action of the *Lc* gene is the Purkinje neuron and degeneration of the other cell types is a secondary effect, presumably due to lack of appropriate connections with Purkinje cells. Later experiments with both *Lurcher* and *staggerer* mutants showed that indeed granule cells and inferior olive neurons die because of an absence of trophic support from their postsynaptic targets, the Purkinje cells (Herrup and Busser, 1995). In *Lurcher* mice, Purkinje cells retain their immature phenotype by receiving multiple innervations by climbing fibers (Rabacchi *et al.*, 1992). The *Lurcher* mouse provides an excellent model for cell-autonomous, preprogrammed death of the Purkinje neurons (Herrup and Busser, 1995).

The gene mutated in *Lurcher* was identified by positional cloning (Zuo *et al.*, 1997) after the chromosomal region around the *Lurcher* was saturated with DNA markers (Zuo *et al.*, 1995; De Jager *et al.*, 1997). It was revealed that the gene codes for one of the glutamate receptors, specifically δ2 (GluRδ2). The mutation causes a G to A transition, producing an alanine to threonine substitution in a highly conserved region of transmembrane domain III in the receptor (Zuo *et al.*, 1997). The result is a gain-of-function mutation in which the mutant receptor leads to increased excitatory activity and thus to apoptotic cell death. Expression of the mutant channel in *Xenopus* oocytes produced a constant depolarization of the resting potential, an indication that *in vivo* the mutation leads to cell death via excitotoxicity (Zuo *et al.*, 1997). The nature of the mutation is reminiscent of the gene defect described for *weaver,* where again a single amino acid substitution in a highly conserved portion of an ion channel causes cell damage due to disturbed ion permeability. As previously discussed in the section on *weaver,* both types of mutants display apoptotic cell death of Purkinje neurons (Wullner *et al.*, 1995). It is suggested that excitotoxicity triggers a loss of cell cycle control that leads to apoptosis.

The interesting feature of the *Lurcher* defect is that the *GluR δ2* gene was cloned and functionally studied before it was recognized as the *Lurcher* mutation (Yamazaki *et al.*, 1992; Araki *et al.*, 1993; Lomeli *et al.*, 1993). Investigators studying the physiology of the glutamate receptors have produced *GluRδ2* knockout mice that display ataxia and are defective in Purkinje cell synapse formation (Kashiwabuchi *et al.*, 1995). The *GluRδ2* null mutants parallel the *Lurcher* mice in the Purkinje cell selectivity of the defect. Furthermore, the *GluRδ2* expression pattern (Araki *et al.*, 1993) explains, in part, this cell specificity as well as the timing of the gene action in the mutants.

Studies with homozygous *Lurcher* mice, published shortly before the mutation became known (Resibois *et al.*, 1997), have suggested a model in which the mutation is semidominant in Purkinje cells and recessive in other cell types. Detailed analysis of homozygous and heterozygous *Lurcher* brains showed that although the defect is not restricted to Purkinje cells, the other cell types are affected only in the homozygotes. The action of the gene on Purkinje cells is characterized as semidominant because Purkinje neurons in heterozygotes present the same defects as in homozygotes but only later in development and less severely. Furthermore, the Purkinje cells of homozygous mice die at the end of migration whereas in heterozygotes they die at the beginning of the dendritic expansion (Resibois *et al.*, 1997). This makes the cell death event the result of a gene dosage effect. The selective action of the *Lurcher* defect on Purkinje cells could be due to the timing of its expression pattern, as already suggested, or to its interaction with local cell factors, as suggested for other gene products, such as the GIRK2 channel in *weaver* mice.

Double mutant analysis with *Lurcher–staggerer* chimeras (Messer *et al.*, 1991) was used to investigate the mechanisms of Purkinje cell death. The double mutants were characterized only by *sg* defects, suggesting that the *sg* gene acts earlier than the *lc* gene in the Purkinje cell developmental scheme. Subsequent studies have suggested that the two mutations block discrete steps in the maturation process of Purkinje neurons (Soha and Herrup, 1995; see also section on *staggerer*). Detailed dissection of the mechanisms governing Purkinje death in each of the mutants provides complementary information on the developmental processes controlling cell maturation. The description of a developmental continuum is of particular interest in light of the recent cloning of both of these loci. The precise function of each protein in the cell is currently under investigation in the context of dissecting the molecular mechanisms governing cell differentiation. In that perspective, the techniques used previously to investigate the behavior of the mutant cell types can be complemented by the newly revealed molecular details.

The *Lurcher* mutation falls into the category of cerebellar mutations that resemble human neurodegenerative diseases most and exhibit degeneration of a primary neuronal type. The cloning of the gene paves the way for further study of the role of excitotoxic cell death in human neurodegeneration.

2.3.2. Staggerer (sg)

This autosomal recessive mutant displays ataxia with tremors with a characteristic staggering phenotype (Sidman *et al.*, 1962). The *staggerer* homozygote has a small cerebellum in which granule cells are greatly diminished and Purkinje cell numbers show a 25% reduction. The Purkinje cells generate very few tertiary spines, and thus cannot form synapses with parallel fibers (Sidman, 1968, 1972;

Sotelo and Changeux, 1974; Hirano and Dembitzer, 1975). Analysis with mouse chimeras has demonstrated that the direct target of the gene is the Purkinje cell (Herrup and Mullen, 1979; Herrup, 1983). The granule cell loss appears to be a secondary effect of the mutation, probably due to the lack of synaptic activity with the Purkinje cell (Sidman, 1972, 1974; Landis and Sidman, 1978). Additional features of the mutant involve altered electrophysiological properties and multiple innervation of Purkinje cells by climbing fibers (Crepel and Mariani, 1975; Crepel *et al.,* 1980) as well as cell death in the inferior olive resulting from the indirect action of the *staggerer* gene (Zanjani *et al.,* 1990, 1994).

Staggerer granule cells multiply in the external granule cell layer and migrate inward. However, during the second half of the first postnatal week and in the next 2 weeks pycnotic nuclei appear and degeneration is obvious by the third postnatal week (Yoon, 1972, 1976; Landis and Sidman, 1978). Before granule cells die, several abnormalities are noticed. Cell division in the EGL occurs at a slower rate resulting in a thinner layer which, however, persists longer (Yoon, 1972, 1976).

In *staggerer* the Purkinje cells are generated normally during embryogenesis but their maturation is delayed during postnatal development (Sidman, 1968). It is evident right after birth that the *sg/sg* cerebellum is reduced in size before granule cells appear. By the second postnatal week, the reduction in Purkinje cell numbers, their failure to form the characteristic Purkinje monolayer, and their atrophic dendrites are noticeable (Sidman, 1968; Landis and Sidman, 1978). More recent morphological studies on *Lurcher* and *staggerer* mutants (Soha and Herrup, 1995) reveal that the stunted morphology of *staggerer* Purkinje cells is not rescued in mosaic animals, thus rendering support to the hypothesis that the dendritic abnormalities reflect cell autonomous genetic blocks.

Genetic and morphological studies provide evidence that the *staggerer* as well as the *Lurcher* mutations act by blocking Purkinje cell development. In *staggerer,* the developmental block seems to be acting earlier than in *Lurcher,* an observation supported by the fact that in *staggerer–Lurcher* double mutants, the *staggerer* characteristic persists and not the *Lurcher* (Messer *et al.,* 1991). Certain early markers of Purkinje cells such as calmodulin and zebrin are not expressed in *staggerer* Purkinje cells (Messer *et al.,* 1990; Sotelo and Wassef, 1991) while others, such as calbindin, are detected in the mutant. The current hypothesis is that normal Purkinje dendrite formation requires the action of the wild-type alleles at both the *sg* and *Lc* loci (Soha and Herrup, 1995).

The *sg* locus was mapped to mouse chromosome 9 (Green and Lane, 1967). Recently, a genomic deletion in the *ROR*α gene, a member of the nuclear hormone receptor gene superfamily, has been described as the cause of the *staggerer* mutation (Hamilton *et al.,* 1996; Matysiak-Scholze and Nehls, 1997). RORα is one of the three subtypes of retinoic acid receptors (Carlberg and Wiesenberg, 1995) and *staggerer* mice were found to carry a deletion within the *ROR*α gene that prevents translation of the ligand-binding homology domain (Hamilton *et al.*, 1996; Matysiak-Scholze and Nehls, 1997). *RORα* is expressed at the highest level in Purkinje cells from their early development (Matsui *et al.*, 1995; Hamilton *et al.*, 1996; Nakagawa *et al.*, 1997). Other than in Purkinje cells, *RORα* mRNA is present in the embryonic midbrain and more weakly in the thymus, whisker follicles, kidney, lung, eye, and gut. Hamilton and colleagues (1996), prompted by the role of thyroid hormone in cerebellar development, suggest that *RORα* interacts with a thyroid-hormone signaling pathway, necessary for Purkinje cell maturation.

2.3.3. Purkinje Cell Degeneration (pcd)

The *pcd* mutation can be classified into the same class of mutations that includes *Lurcher, staggerer,* and *tottering,* all of which are characterized by the postnatal degeneration of posmitotic neurons. The mutation, first described more than 20 years ago (Mullen *et al.*, 1976), is autosomal recessive and maps to chromosome 13 (Krulewski *et al.*, 1989; Campbell and Hess, 1996). The Purkinje cell is the primary neuronal type affected. However, other cell types, such as retinal photoreceptor cells and olfactory bulb mitral cells (Sidman, 1983), also degenerate. Spermatogenesis is also affected. The degeneration of Purkinje cells begins after the neurons have matured, positioned properly, and made synaptic contacts (Landis and Mullen, 1978) and the defect appears to act intrinsically on the particular cerebellar cell type (Sidman, 1983). The Purkinje cell death, virtually complete by the fourth postnatal week (Landis and Mullen, 1978), is apoptotic in nature (Gillardon *et al.*, 1995), similar to what is observed in other mouse neurological mutations previously discussed.

The *pcd* mutation has been extensively used as a model for Purkinje cell loss because it resembles the degenerative phenotype of human ataxias. Despite the fact that, several years ago, there was a report of a transgene insertion that serendipitously targeted the *pcd* locus (Krulewski *et al.*, 1989), the gene has not been identified as yet. It is expected that uncovering the defect in *pcd* will provide some insight into the common mechanism underlying degeneration of disparate cell types in this mutant. However, this is not always the case, as evident in the *weaver* mutant and its corresponding gene.

2.3.4. Motor Endplate Disease (med) and Jolting (med^{jo})

The *jolting* mutant arose as a spontaneous mutation (Dickie, 1965) and is characterized by body tremor and a poorly coordinated gait. By genetic mapping and complementation tests it was demonstrated that *jolting* (med^{jo}) is actually a milder allele of the more severe mutation *motor endplate disease (med)* which maps to chromosome 15 (Duchen and Stefani, 1971; Harris and Pollard, 1986).

The ataxic gait in both alleles is cerebellar in origin. Electrophysiological studies of both mutants have shown an impairment of spontaneous, regular discharges of the Purkinje neurons (Dick *et al.,* 1985, 1986; Harris *et al.,* 1992). The *med* locus was first identified by the random insertion of a transgene (Kohrman *et al.,* 1995), which was shown to have disrupted the gene coding for a novel voltage-gated sodium channel α subunit (Scn8a) (Burgess *et al.,* 1995). Complete loss of function of Scn8a in these insertional transgenics resulted in severe muscle atrophy, progressive paralysis, and juvenile death. The newly identified channel, Scn8a, was found to be expressed in the brain and spinal cord (Burgess *et al.,* 1995).

The two spontaneous alleles of *motor endplate disease, med* and *med^{J},* that produce nonfunctional Scn8a channels, also result in progressive paralysis and early death (Kohrman *et al.,* 1996a; Raman *et al.,* 1997). The milder allele *jolting (med^{jo})* is the result of a missense mutation that produces a shift in the voltage-dependence threshold of the channel (Kohrman *et al.,* 1996b; Raman *et al.,* 1997). The phenotypic effect in the *jolting* mice is the reduced spontaneous activity of Purkinje cells, reduced inhibitory input from the cerebellum, and, thus, a loss of motor control.

The Scn8a sodium channel adds to the growing list of ion channels identified as culprits in mouse neurological mutations, such as the potassium channel GIRK2 in *weaver,* the voltage-sensitive calcium channel of the α_{1A} subtype in *leaner* and *tottering,* and the glutamate receptor δ2 in *Lurcher.* The importance of membrane excitability during the development of mammalian central neurons and for their proper maintenance is thus highlighted. The human homolog of the sodium channel gene (*SCN8A*) was mapped to 12q13 and is a candidate for inherited neurodegenerative disease (Burgess *et al.,* 1995).

2.4. Transgenic or Targeted Deletion Lines with Cerebellar Phenotypes

In addition to the numerous spontaneous mouse mutations with neurological phenotypes available, the transgenic and knockout technology has produced several mouse lines that display perturbed cerebellar development. A very characteristic example, reviewed in another chapter, is the group of transgenics modeling human triple repeat neurodegenerative disorders. A few knockout mutants have already been described in the patterning mutants section. Some others are briefly discussed below but the list is by no means complete.

2.4.1. Purkinje Cell Targeting of SV40 Large T Antigen Protein

The targeting of the SV40 large T antigen (Tag) protein to Purkinje cells produced cell-specific neurodegeneration and a mouse model for perturbed cerebellar development (Feddersen *et al.,* 1992). The time of onset of the ataxic phenotype, in these transgenic mice, correlated with the amount of Tag protein produced in Purkinje cells. The higher the copy number of the transgene, the earlier Purkinje cell development is affected. In the transgenic lines with low copy numbers, Tag protein accumulates in amounts ample to overcome the threshold effect after the maturation of Purkinje cells. Thus, in an event paralleling the neurological mouse mutants *purkinje cell degeneration, nervous,* and *Lurcher,* cell death occurs in mature Purkinje cells. In contrast, a transgenic line with a high copy number displays ataxia earlier owing to the disruption of Purkinje cell development during their migration stage.

Subsequent experiments by the same investigators with mutant forms of the Tag protein provided a model for neurodegeneration by linking it to cell cycle regulation (Feddersen *et al.,* 1995), and further dissected the temporal pattern of Purkinje cell development as it relates to death susceptibility (Feddersen *et al.,* 1997).

2.4.2. BDNF Null Mutants

Brain-derived neurotrophic factor (BDNF) is highly expressed in granule cells while the BDNF receptor, TrkB, is synthesized in both granule and Purkinje cells (Wetmore *et al.,* 1990; Klein *et al.,* 1993; Rocamora *et al.,* 1993; Alvarez-Dolado *et al.,* 1994; Gao *et al.,* 1995; Segal *et al.,* 1995). In mice with targeted gene deletions of *BDNF* (Ernfors *et al.,* 1994) Trk receptors are not activated in granule and Purkinje cells and the mice exhibit abnormal gait (Schwartz *et al.,* 1997). In addition, there is increased death of granule cells; Purkinje cells lack a primary dendrite, appearing stunted; and the middle and posterior foliation pattern is altered in $BDNF^{-/-}$ mice. This work demonstrates that BDNF is required *in vivo* for the normal development of granule and Purkinje cells. BDNF seems to act in an autocrine–paracrine function, rather than as a target-derived factor to regulate granule and Purkinje cell survival and differentiation.

2.4.3. PKCγ Null Mutants

Protein kinase C (PKC) has been implicated in a variety of neuronal functions, including modulation of synaptic transmission. Of all its isoforms found in neural tissue, the γ isoform (PKCγ) is expressed only in central neurons (Saito *et al.,* 1988) with high levels in cerebellar Purkinje cells (Kose *et al.,* 1988; Saito *et al.,* 1988). Mice deficient in PKCγ, in addition to learning deficits and modified long-term potentiation, exhibited mild ataxia (Chen *et al.,* 1995; Kano *et al.,* 1995). Closer examination of the cerebella of these mice did not reveal any anatomical abnormalities. However, there was an increase in the number of Purkinje cells that remained innervated by multiple climbing fibers (CFs) (Kano *et al.,* 1995). Furthermore, it appeared that regression of multiple CF innervation is normal in the PKCγ mutant mice during the second postnatal week but is impaired afterwards, suggesting discrete steps in the elimination process, with PKCγ modulating only the subsequent step (Kano *et al.,* 1995). Thus, the investigators propose that PKCγ transduces a signal for the elimination of surplus CF synapses during cerebellar development. The corollary of this elimination deficit appears to be the motor coordination impairment (Chen *et al.,* 1995).

Indications about the role of proper synapse elimination on motor coordination come from two other types of mutant mice that parallel the PKCγ null mutants. Both the *GluRδ2* null mutants, previously discussed with respect to the *Lurcher* mutation (Kashiwabuchi *et al.,* 1995), and mice null for the metabotropic glutamate receptor mGluR1 (Aiba *et al.,* 1994; Conquet *et al.,* 1994) have multiple CF innervation that correlates with impaired motor coordination. However, in contrast to the PKCγ mutants, these mice also have impaired long-term depression that might be the primary reason for the ataxic behavior.

3. CONCLUSIONS

The mutant cerebellum has served as an important model in the last 30 years for the study of development of distinct cell types and the formation of complex cytoarchitectures. The availability of several neurological mutations that arose spontaneously, as well as the numerous transgenic mice produced with cerebellar defects, facilitate the precise dissection of the mechanisms governing cell birth, maturation, and differentiation. There are cerebellar mutants that at a gross morphological level exhibit phenotypic similarities. The pairs of *reeler/scrambler* and *Lurcher/staggerer,* discussed in the corresponding sections, serve as characteristic examples. In general, however, there is a continuum of mutant phenotypes reflecting the underlying genetic cascade of developmental events and the complex interplay of the gene products.

An interesting aspect of several of the mutations altering cerebellar development is the fact that usually the defect is intrinsic to one cell type (such as the Purkinje cell), which is also the one primarily degenerating, with additional cell types affected only as a secondary effect. This points to the importance of cell–cell interactions, accurate synaptic formation, and local architecture for the proper maturation of postmitotic neurons. In the majority of the mutants the gene defect targets several, apparently unrelated, cell types. The common mechanism governing different neuronal degeneration events is not obvious at first, save for the expression of the targeted gene in all these cell types. This is often the case with many human disorders as well. Finally, there is an additional aspect to the issue of how a gene defect leads to a particular phenotype, highlighted by the *reeler* mutant. In this mouse, the primary cell types expressing Reelin are not the ones affected by the mutation. Thus, it is not obvious at first how the mutant gene exerts its action during neuronal development. It is clearly not the straightforward case of a single gene affecting a single cell type directly. It becomes apparent, then, that identifying the affected gene is only the beginning in understanding how it fits into the grand scheme of neuronal differentiation.

A surprising finding has been the identification of so many ion channel genes affected in the various mutants (Table 1). Even in the case of *weaver*, where the original prediction had been for a surface receptor molecule involved in radial migration, the significance of membrane physiology emerged. Membrane excitability proved to be essential for cell differentiation and migration, pointing again to the existence of multiple mechanisms responsible for a given phenotype.

Finally, it should be particularly noted that the current molecular biology technology has enriched the field of mouse neurogenetics in two ways. First, it supplied traditional neuroanatomical approaches with new tools to isolate and analyze the mutant genes. Second, it provided us with a number of targeted mouse mutants or transgenic mice that exhibit cerebellar phenotypes and add to the growing body of knowledge on cerebellar development.

A significant contribution of mouse neurogenetics is that cerebellar mutants provide correlates for human ataxias in several ways. At a more basic level, the study of cortical development and shaping of neuronal identity in mutant versus wild-type animals reveals a complex sequence of events that includes precisely timed gene expression and local signaling mechanisms. Furthermore, the identification of several of the genes affected in the mouse mutants paves the way for the recognition of their *in vivo* function in normal development and, subsequently, for the role of the human homologs in neurodegenerative disorders. Finally, it is now possible to either specifically target a particular locus, creating a null mutant, or to produce a transgenic mouse overexpressing a mutant protein, thus studying gain-of-function effects, mechanisms that presumably act at the basis of human disorders. Therefore, the ability to manipulate the mouse genome can provide us with useful animal models for a variety of human neurodegenerative diseases.

Acknowledgments

We thank Nektaria Kelaidi for Fig. 2. We are grateful to Drs. Marion Wassef and Andre Goffinet for their critical comments on an earlier version of the manuscript. We acknowledge the Greek Secretariat for Research and Technology and the EU (BMH4-CT95-0524) for funding.

REFERENCES

Ackerman, S. L., Kozak, L. P., Przyborski, S. A., Rund, L. A., Boyer, B. B., and Knowles, B. B., 1997, The mouse rostral cerebellar malformation gene encodes an UNC-5-like protein, *Nature* **386:**838–842.

Aiba, A., Kano, M., Chen, C., Stanton, M. E., Fox, G. D., Herrup, K., Zwingman, T. A., and Tonegawa, S., 1994, Deficient cerebellar long-term depression and impaired motor learning in *mGluR1* mutant mice, *Cell* **79:**377–388.

Altman, J., and Bayer, S. A., 1985a, Embryonic development of the rat cerebellum. I. Delineation of the cerebellar primordium and early cell movements, *J. Comp. Neurol.* **231:**1–26.

Altman, J., and Bayer, S. A., 1985b, Embryonic development of the rat cerebellum. II. Translocation and regional distribution of the deep neurons, *J. Comp. Neurol.* **231:**27–41.

Altman, J., and Bayer, S. A., 1985c, Embryonic development of the rat cerebellum. III. Regional differences in the time of origin, migration and settling of Purkinje cells, *J. Comp. Neurol.* **231:**42–65.

Alvarado-Mallart, R.-M., and Sotelo, C., 1982, Differentiation of cerebellar anlage heterotypically transplanted to adult rat brain: A light and electron microscopic study, *J. Comp. Neurol.* **212:**247–267.

Alvarado-Mallart R.-M., Martinez, S., and Lance-Jones, C. C., 1990, Pluripotentiality of the 2-day-old avian germinative neuroepithelium, *Dev. Biol.* **139:**75–88.

Alvarez-Dolado, M., Iglesias, T., Rodriguez-Pena, Bernal, J., and Munoz, A., 1994, Expression of neurotrophins and the trk family of neurotrophin receptors in normal and hypothyroid rat brain, *Brain Res. Mol. Brain. Res.* **27:**249–257.

Alvarez Otero, R. M., Sotelo, C. M., Alvarado-Mallart, R.-M., 1993, Chick/quail chimeras with partial cerebellar grafts: An analysis of the origin and migration of cerebellar cells, *J. Comp. Neurol.* **333:**597–615.

Araki, K., Meguro, H., Kushiya, E., Takayama, C., Inoue, Y., and Mishina, M., 1993, Selective expression of the glutamate receptor channel δ2 subunit in cerebellar Purkinje cells, *Biochem. Biophys. Res. Commun.* **197:**1267–1276.

Bally-Cuif, L., Cholley, B., and Wassef, M., 1995, Involvement of *Wnt-1* in the formation of the mes/metencephalic boundary, *Mech. Dev.* **53:**23–34.

Bandmann, O., Davis, M. B., Marsden, C. D., and Wood, N. W., 1996, The human homologue of the *weaver* mouse gene in familial and sporadic Parkinson's disease, *Neuroscience* **72:**877–879.

Baptista, C. A., Hatten, M. E., Blazeski, R., and Mason, C. A., 1994, Cell–cell interactions influence survival and differentiation of purified Purkinje cells in vitro, *Neuron* **12:**243–260.

Bar, I., Lambert de Rouvroit, C., Royaux, I., Krizman, D. B., Dernoncourt, C., Ruelle, D., Beckers, M. C., and Goffinet, A. M., 1995, A YAC contig containing the *reeler* locus with preliminary characterization of candidate gene fragments, *Genomics* **26:**543–549.

Ben-Arie, N., McCall, A. E., Berkman, S., Eichele, G., Bellen, H. J., and Zoghbi, H. Y., 1996, Evolutionary conservation of sequence and expression of the bHLH protein Atonal suggests a conserved role in neurogenesis, *Hum. Mol. Genet.* **5:**1207–1216.

Ben-Arie, N., Bellen, H. J., Armstrong, D. L., McCall, A. E., Gordadze, P. R., Guo, Q., Matzuk, M. M., and Zoghbi, H. Y., 1997, *Math1* is essential for genesis of cerebellar granule neurons, *Nature* **390:**169–172.

Burgess, D. L., Kohrman, D. C., Galt, J., Plummer, N. W., Jones, J. M., Spear, B., and Meisler, M. H., 1995, Mutation of a new sodium channel gene, *Scn8a,* in the mouse mutant "motor endplate disease," *Nat. Genet.* **10:**461–465.

Caddy, K. W., and Biscoe, T. J., 1979, Structural and quantitative studies on the normal C3H and *Lurcher* mutant mouse, *Philos. Trans. R. Soc. (Lond.) Ser. B* **287:**167–201.

Campbell, D. B., and Hess, E. J., 1996, Chromosomal localization of the neurological mouse mutations *tottering (tg), Purkinje cell degeneration (pcd),* and *nervous (nr), Brain Res. Mol. Brain Res.* **37:**79–84.

Carlberg, C., and Wiesenberg, I., 1995, The orphan receptor family RZR/ROR, melatonin and 5-lipoxygenase: An unexpected relationship, *J. Pineal Res.* **18:**171–178.

Caviness, V. S., Jr., and Rakic, P., 1978, Mechanisms of cortical development: A view from mutations in mice, *Annu. Rev. Neurosci.* **1**:297–326.

Chen, C., Kano, M., Abeliovich, A., Chen, L., Bao, S., Kim, J. J., Hashimoto, K., Thompson, R. F., and Tonegawa, S., 1995, Impaired motor coordination correlates with persistent multiple climbing fiber innervation in PKCγ mutant mice, *Cell* **83**:1233–1242.

Cheng, S. S.-W., and Heintz, N., 1997, Massive loss of mid- and hindbrain neurons during embryonic development of homozygous *Lurcher* mice, *J. Neurosci.* **17**:2400–2407.

Conquet, F., Bashir, Z. I., Davies, C. H., Daniel, H., Ferraguti, F., Bordi, F., Franz-Bacon, K., Reggiani, A., Matarese, V., Conde, F., Collingridge, G. L., and Crepel, F., 1994, Motor deficit and impairment of synaptic plasticity in mice lacking *mGluR1, Nature* **372**:237–243.

Cooper, P. A., Benno, R. H., Hahn, M. F., and Hewitt, J. K., 1991, Genetic analysis of cerebellar foliation patterns in mice *(Mus musculus), Behav. Genet.* **21**:405–419.

Crepel, F., 1982, Regression of functional synapses in the immature mammalian cerebellum, *Trends Neurosci.* **5**:266–269.

Crepel, F., and Mariani, J., 1975, Anatomical, physiological and biochemical studies of the cerebellum from mutant mice. I. Electrophysiological analysis of cerebellar cortical neurons in the *staggerer* mouse, *Brain Res.* **98**:135–147.

Crepel, F., Delhaye-Bouchaud, N., Guastavino, J. M., and Sampaio, I., 1980, Multiple innervation of cerebellar Purkinje cells by climbing fibers in *staggerer* mutant mouse, *Nature* **283**:483–484.

D'Arcangelo, G., Miao, G. G., Chen, S.-C., Soares, H. D., Morgan, J. I., and Curran, T., 1995, A protein related to extracellular matrix proteins deleted in the mouse mutant *reeler, Nature* **374**:719–723.

D'Arcangelo, G., Nakajima, K., Miyata, T., Ogawa, M., Mikoshiba, K., and Curran, T., 1997, Reelin is a secreted glycoprotein recognized by the CR-50 monoclonal antibody, *J. Neurosci.* **17**:23–31.

Davis, C. A., and Joyner, A. L., 1988, Expression patterns of the homeobox-containing genes *En-1* and *En-2* and the proto-oncogene *int-1* diverge during mouse development, *Genes Dev.* **2**:1736–1744.

De Bergeyck, V., Nakajima, K., Lambert d Rouvroit, C., Naerhuyzen, B., Goffinet, A. M., Miyata, T., Ogawa, M., and Mikoshiba, K., 1997, A truncated Reelin protein is produced but not secreted in the "Orleans" *reeler* mutation *(*$Reln^{rl\text{-}Orl}$*), Brain Res. Mol. Brain Res.* **50**:85–90.

De Jager, P. L., Zuo, J., and Heintz, N., 1997, An ~1.2 Mb bacterial artificial chromosome contig refines the genetic and physical maps of the *Lurcher* locus on mouse chromosome 6, *Genome Res.* **7**:736–746.

De Silva, U., D'Arcangelo, G., Braden, V. V., Chen, J., Miao, G. G., Curren, T., and Green, E. D., 1997, The human *reelin* gene: Isolation, sequencing, and mapping on chromosome 7, *Genome Res.* **7**:157–164.

Dick, D. J., Boakes, R. J., and Harris, J. B., 1985, A cerebellar abnormality in the mouse with motor end-plate disease, *Neuropathol. Appl. Neurobiol.* **11**:141–147.

Dick, D. J., Boakes, R. J., Candy, J. M., Harris, J. B., and Cullen, M. J., 1986, Cerebellar structure and function in the murine mutant *"jolting," J. Neurol. Sci.* **76**:255–267.

Dickie, M. M., 1965, *Jolting, Mouse News Lett.* **32**:44.

Doyle, J., Ren, X., Lennon, G., and Stubbs, L., 1997, Mutations in the *Cacnl1a4* calcium channel gene are associated with seizures, cerebellar degeneration, and ataxia in *tottering* and *leaner* mutant mice, *Mamm. Genome* **8**:113–120.

Duchen, L. W., and Stefani, E., 1971, Electrophysiological studies of neuromuscular transmission in hereditary "motor end-plate disease" of the mouse, *J. Physiol.* **212**:535–548.

Edmondson, J. C., and Hatten, M. E., 1987, Glial-guided granule neuron migration in vitro: A high-resolution time-lapse video microscopic study, *J. Neurosci.* **7**:1928–1934.

Edmondson, J. C., Liem, R. K. H., Kuster, J. E., and Hatten, M. E., 1988, Astrotactin: A novel neuronal cell surface antigen that mediates neuron–astroglial interactions in cerebellar microcultures, *J. Cell Biol.* **106:**505–517.

Ernfors, P., Lee, K. F., and Jaenisch, R., 1994, Mice lacking brain-derived neurotrophic factor develop with sensory deficits, *Nature* **368:**147–150.

Falconer, D. S., 1951, Two new mutants, *"trembler"* and *"reeler,"* with neurological actions in the house mouse (*Mus musculus*), *J. Genet.* **50:**192–201.

Feddersen, R. M., Ehlenfeldt, R., Yunis, W. S., Clark, H. B., and Orr, H. T., 1992, Disrupted cerebellar cortical development and progressive degeneration of Purkinje cells in SV40 T antigen transgenic mice, *Neuron* **9:**955–966.

Feddersen, R. M., Clark, H. B., Yunis, W. S., and Orr, H. T., 1995, *In vivo* viability of postmitotic Purkinje neurons requires pRb family member function, *Mol. Cell. Neurosci.* **6:**153–167.

Feddersen, R. M., Yunis, W. S., O'Donnell, M. A., Ebner, T. J., Shen, L., Iadecola, C., Orr, H. T., and Clark, H. B., 1997, Susceptibility to cell death induced by mutant SV40 T-antigen correlates with Purkinje neuron functional development, *Mol. Cell. Neurosci.* **9:**42–62.

Feng, L., Hatten, M. E., Heintz, N., 1994, Brain lipid-binding protein (BLBP): A novel signaling system in the developing mammalian CNS, *Neuron* **12:**895–908.

Ferrer, J., Nichols, C. G., Makhina, E. N., Salkoff, L., Bernstein, J., Gerhard, D., Wasson, J., Ramanadham, S., and Permutt, A., 1995, Pancreatic islet cells express a family of inwardly rectifying K^+ channel subunits which interact to form G-protein-activated channels, *J. Biol. Chem.* **270:**26086–26091.

Fishman, R. B., and Hatten, M. E., 1993, Multiple receptor systems promote CNS neuronal migration, *J. Neurosci.* **13:**3485–3495.

Fletcher, C., Norman, D. J., and Heintz, N., 1991, Genetic mapping of *meander tail,* a mouse mutation affecting cerebellar development, *Genomics* **9:**647–655.

Fletcher, C. F., Lutz, M. C., O'Sullivan, T. N., Shaughnessy, J. D., Jr., Hawkes, R., Frankel, W. N., Copeland, N. G., and Jenkins, N. A., 1996, Absence epilepsy in *tottering* mutant mice is associated with calcium channel defects, *Cell* **87:**607–617.

Fugita, S., 1967, Quantitative analysis of cell proliferation and differentiation in the cortex of the postnatal mouse cerebellum, *J. Cell Biol.* **32:**277–287.

Fugita, S., Shimada, M., and Nakanuna, T., 1966, ^{3}H-thymidine autoradiographic studies on the cell proliferation and differentiation in the external and internal granular layers of the mouse cerebellum, *J. Comp. Neurol.* **128:**191–209.

Gao, W.-Q., and Hatten, M. E., 1993, Neuronal differentiation rescued by implantation of *weaver* granule cell precursors into wild-type cerebellar cortex, *Science* **260:**367–369.

Gao, W.-Q., Liu, X.-L., and Hatten, M. E., 1992, The *weaver* gene encodes a nonautonomous signal for CNS neuronal differentiation, *Cell* **68:**841–854.

Gao, W.-Q., Zheng, Z. L., and Karihaloo, M., 1995, Neurotrophin-4/5 (NT-4/5) and brain-derived neurotrophic factor (BDNF) act at later stages of cerebellar granule cell differentiation, *J. Neurosci.* **15:**2656–2667.

Gillardon, F., Baurle, J., Grusser-Cornehls, U., and Zimmermann, M., 1995, DNA fragmentation and activation of c-*Jun* in the cerebellum of mutant mice (*weaver, Purkinje cell degeneration*), *Neuroreport* **6:**1766–1768.

Goffinet, A. M., 1984, Events governing organization of postmigratory neurons: Studies on brain development in normal and *reeler* mice, *Brain Res.* **319:**261–296.

Goffinet, A. M., 1995, A real gene for *reeler, Nature* **374:**675–676.

Goffinet, A. M., 1997, Unscrambling a disabled brain, *Nature* **389:**668–669.

Goldowitz, D., 1989, The *weaver* granuloprival phenotype is due to intrinsic action of the mutant locus in granule cells: Evidence from homozygous *weaver* chimeras, *Neuron* **2:**1565–1575.

Goldowitz, D., and Mullen, R. J., 1982, Granule cells as a site of gene action in the *weaver* mouse cerebellum: Evidence from heterozygous mutant chimeras, *J. Neurosci.* **2:**1474–1485.

Goldowitz, D., Cushing, R. C., Laywell, E., D'Arcangelo, G., Sheldon, M. Sweet, H. O., Davisson, M., Steindler, D., and Curran, T., 1997, Cerebellar disorganization characteristic of *reeler* in *scrambler* mutant mice despite presence of Reelin, *J. Neurosci.* **17:**8767–8777.

Gonzalez, J. L., Russo, C. J., Goldowitz, D., Sweet, H. O., Davisson, M. T., and Walsh, C. A., 1997, Birthdate and cell marker analysis of *scrambler:* A novel mutation affecting cortical development with a *reeler*-like phenotype, *J. Neurosci.* **17:**9204–9211.

Green, M. C., 1989, Catalog of mutant genes and polymorphic loci, *in* "Genetic Variants and Strains of the Laboratory Mouse" (M. F. Lyon and A. G. Searle, eds.), Oxford University Press, Oxford.

Green, M. C., and Lane, P. W., 1967, Linkage group II of the house mouse, *J. Hered.* **58:**225–228.

Green, M. C., and Sidman, R. L., 1962, *Tottering,* a neuromuscular mutation in the mouse and its linkage with oligosyndactylism, *J. Hered.* **53:**233–237.

Gregory, W. A., Edmondson, J. C., Hatten, M. E., and Mason, C. A., 1988, Cytology and neuron–glial apposition of migrating cerebellar granule cells *in vitro, J. Neurosci.* **8:**1728–1738.

Hallonet, M. E. R., and Le Douarin, N. M., 1993, Tracing neuroepithelial cells of the mesencephalic and metencephalic alar plates during cerebellar ontogeny in quail-chick chimaeras, *Eur. J. Neurosci.* **5:**1145–1155.

Hallonet, M. E. R., Teillet, M.-A., and Le Douarin, N. M., 1990, A new approach to the development of the cerebellum provided by the quail-chick marker system, *Development* **108:**19–31.

Hamilton, B. A., Frankel, W. N., Kerrebrock, A. W., Hawkins, T. L., FitzHugh, W., Kusumi, K., Russell, L. B., Mueller, K. L., van Berkel V., Birren, B. W., Kruglyak, L., and Lander, E. S., 1996, Disruption of the nuclear hormone receptor RORα in *staggerer* mice, *Nature* **379:**736–739.

Hamre, K. M., and Goldowitz, D., 1997, *Meander tail* acts intrinsic to granule cell precursors to disrupt cerebellar development: Analysis of *meander tail* chimeric mice, *Development* **124:**4201–4212.

Harris, J. B., and Pollard, S. L., 1986, Neuromuscular transmission in the murine mutants "*motor end-plate disease*" and "*jolting,*" *J. Neurol. Sci.* **76:**239–253.

Harris, J. B., Boakes, R. J., and Court, J. A., 1992, Physiological and biochemical studies on the cerebellar cortex of the murine mutants "*jolting*" and "*motor end-plate disease,*" *J. Neurol. Sci.* **110:**186–194.

Hatten, M. E., 1993, The role of migration in central nervous system neuronal development, *Curr. Opin. Neurobiol.* **3:**38–44.

Hatten, M. E., and Heintz, N., 1995, Mechanisms of neural patterning and specification in the developing cerebellum, *Annu. Rev. Neurosci.* **18:**385–408.

Hatten, M. E., and Mason, C. A., 1990, Mechanisms of glial-guided neuronal migration *in vitro* and *in vivo, Experientia* **46:**907–916.

Hatten, M. E., Liem, R. K., and Mason, C. A., 1984, Defects in specific associations between astroglia and neurons occur in microcultures of *weaver* mouse cerebellar cells, *J. Neurosci.* **4:**1163–1172.

Hatten, M. E., Liem, R. K., and Mason, C. A., 1986, *Weaver* mouse cerebellar granule neurons fail to migrate on wild-type astroglial processes *in vitro, J. Neurosci.* **6:**2676–2683.

Hawkes, R., Blyth, S., Chockkan, V., Tano, D., Ji, Z., and Mascher, C., 1993, Structural and molecular compartmentation in the cerebellum, *Can. J. Neurol. Sci.* **20**(Supp. 3):S29–S35.

Hedgecock, E. M., and Norris, C. R., 1997, Netrins evoke mixed reactions in motile cells, *Trends Genet.* **13:**251–253.

Heintz, N., Norman, D. J., Gao, W.-Q., and Hatten, M. E., 1993, Neurogenetic approaches to mammalian brain development, *in* "Genome Analysis, Vol. 6: Genome Maps and Neurological Disorders," pp. 19–44, Cold Spring Harbor Laboratory Press, Cold Spring Harbor, NY.

Heller, A. H., Dichter, M. A., and Sidman, R. L., 1983, Anticonvulsant sensitivity of absence seizures in the *tottering* mutant mouse, *Epilepsia* **24:**25–34.

Herrup, K., 1983, Role of *staggerer* gene in determining cell number in cerebellar cortex. I. Granule cell death is an indirect consequence of staggerer gene action, *Brain Res.* **313:**267–274.

Herrup, K., and Busser, J. C., 1995, The induction of multiple cell cycle events precedes target-related neuronal death, *Development* **121:**2385–2395.

Herrup, K., and Mullen, R. J., 1979, *Staggerer* chimeras: Intrinsic nature of Purkinje cell defects and implications for normal cerebellar development, *Brain Res.* **178:**443–457.

Herrup, K., and Wilczynski, S. L., 1982, Cerebellar cell degeneration in the leaner mutant mouse, *Neuroscience* **7:**2185–2196.

Hess, E. J., 1996, Identification of the *weaver* mouse mutation: The end of the beginning, *Neuron* **16:**1073–1076.

Hess, E. J., and Wilson, M. C., 1991, *Tottering* and *leaner* mutations perturb transient developmental expression of tyrosine hydroxylase in embryologically distinct Purkinje cells, *Neuron* **6:**123–132.

Hirano, A., and Dembitzer, H. M., 1975, The fine structure of staggerer cerebellum, *J. Neuropathol. Exp. Neurol.* **34:**1–11.

Hirotsune, S., Takahara, T., Sasaki, N., Hirose, K., Yoshiki, A., Ohashi, T., Kusakabe, M., Murakami, Y., Muramatsu, M., Watanabe, S., Nakao, K., Katsuki, M., and Hayashiyaki, Y., 1995, The *reeler* gene encodes a protein with an EGF-like motif expressed by pioneer neurons, *Nat. Genet.* **10:**77–83.

Hoffarth, R. M., Johnston, J. G., Krushel, L. A., and van der Kooy, D., 1995, The mouse mutation *reeler* causes increased adhesion within a subpopulation of early postmitotic cortical neurons, *J. Neurosci.* **15:**4838–4850.

Hollander, W. F., and Waggie, K. S., 1977, *Meander tail:* A recessive mutant located in chromosome 4 of the mouse, *J. Hered.* **68:**403–406.

Howell, B. W., Hawkes, R., Soriano, P., and Cooper, J. A., 1997, Neuronal position in the developing brain is regulated by *mouse disabled-1, Nature* **389:**733–737.

Joyner, A. L., and Martin, G. R., 1987, *En-1* and *En-2,* two mouse genes with sequence homology to the *Drosophila engrailed* gene: Expression during embryogenesis, *Genes Dev.* **1:**29–38.

Joyner, A. L., Herrup, K., Auerbach, B. A., Davis, C. A., and Rossant, J., 1991, Subtle cerebellar phenotype in mice homozygours for a targeted deletion of the *En-2* homeobox, *Science* **251:**1239–1243.

Kano, M., Hashimoto, K., Chen, C., Abeliovich, A., Aiba, A., Kurihara, H., Watanabe, M., Inoue, Y., and Tonegawa, S., 1995, Impaired synapse elimination during cerebellar development in PKCγ mutant mice, *Cell* **83:**1223–1231.

Kashiwabuchi, N., Ikeda, K., Araki, K., Hirano, T., Shibuki, K., Takayama, C., Inoue, Y., Kutsuwada, T., Yagi, T., Kang, Y., Aizawa, S., and Mishina, M., 1995, Impairment of motor coordination, Purkinje cell synapse formation, and cerebellar long-term depression in GluR/δ2 mutant mice, *Cell* **81:**245–252.

Klein, R., Smeyne, R. J., Wurst, W., Long, L. K., Auerbach, B. A., Joyner, A. L., and Barbacid, M., 1993, Targeted disruption of the trkB neurotrophin receptor gene results in nervous system lesions and neonatal death, *Cell* **75:**113–122.

Kofuji, P., Hofer, M., Millen, K. J., Millonig, J. H., Davidson, N., Lester, H. A., and Hatten, M. E., 1996, Functional analysis of the *weaver* mutant GIRK2 K^+ channel and rescue of *weaver* granule cells, *Neuron* **16:**941–952.

Kohrman, D. C., Plummer, N. W., Schuster, T., Jones, J. M., Jang, W., Burgess, D. L., Galt, J., Spear, B. T., and Meisler, M. H., 1995, Insertional mutation of the motor endplate disease (*med*) locus in mouse chromosome 15, *Genomics* **26:**171–177.

Kohrman, D. C., Harris, J. B., and Meisler, M. M., 1996a, Mutation detection in the *med* and *med^J* alleles of the sodium channel *Scn8a, J. Biol. Chem.* **271:**17576–17581.

Kohrman, D. C., Smith, M. R., Goldin, A. L., Harris, J., and Meisler, M. H., 1996b, A missense mutation in the sodium channel Scn8a is responsible for cerebellar ataxia in the mouse mutant *jolting*, *J. Neurosci.* **16:**5993–5999.

Kose, A., Saito, N., Ito, H., Kikkawa, U., Nishizuka, Y., and Tanaka, C., 1988, Electron microscopic localization of type I protein kinase C in rat Purkinje cells, *J. Neurosci.* **8:**4262–4268.

Krulewski, T. F., Neumann, P. E., and Gordon, J. W., 1989, Insertional mutation in a transgenic mouse allelic with *Purkinje cell degeneration*, *Proc. Natl. Acad. Sci. USA* **86:**3709–3712.

Kuhar, S. G., Feng, L., Vidan, S., Ross, M. E., Hatten, M. E., and Heinzt, N., 1993, Changing patterns of gene expression define four stages of cerebellar granule neuron differentiation, *Development* **117:**97–104.

Landis, S. C., and Mullen, R. J., 1978, The development and degeneration of Purkinje cells in *pcd* mutant mice, *J. Comp. Neurol.* **177:**125–143.

Landis, D. M. D., and Sidman, R. L., 1978, Eletron microscopic analysis of postnatal histogenesis in the cerebellar cortex of *staggerer* mutant mice, *J. Comp. Neurol.* **179:**831–863.

Lane, P. W., 1964, *Mouse News Lett.* **30:**32.

Lane, P. W., 1967, *Mouse News Lett.* **36:**40.

Lane, P. W., Bronson, R. T., and Spencer, C. A., 1992, *Rostral cerebellar malformation (rcm):* A new recessive mutation on chromosome 3 of the mouse, *J. Hered.* **83:**315–318.

Lauritzen, I., De Weille, J., Adelbrecht, C., Lesage, F., Murer, G., Raisman-Vozari, R., and Lazdunski, M., 1997, Comparative expression of the inward rectifier K^+ channel GIRK2 in the cerebellum of normal and *weaver* mutant mice, *Brain Res.* **753:**8–17.

Llinas, R., and Hillman, D. E., 1969, Physiological and morphological organization of the cerebellar circuits in various vertebrates, *in* "Neurobiology of Cerebellar Evolution and Development" (R. Llinas, ed.), pp. 43–73, AMA-ERF Institute of Biomedical Research, Chicago.

Lohof, A., Delhaye-Bouchaud, N., and Mariani, J., 1996, Synapse elimination in the central nervous system: Functional significance and cellular mechanisms, *Rev. Neurosci.* **7:**85–101.

Lomeli, H., Sprengel, R., Laurie, D. J., Kohr, G., Herb, A., Seeburg, P. H., and Wisden, W., 1993, The rat delta-1 and delta-2 subunits extend the excitatory amino acid receptor family, *FEBS Lett.* **315:**318–322.

Maricich, S. M., Soha, J., Trenkner, E., and Herrup, K., 1997, Failed cell migration and death of Purkinje cells and deep nuclear neurons in the *weaver* cerebellum, *J. Neurosci.* **17:**3675–3683.

Martinez, S., and Alvarado-Mallart R.-M., 1989, Rostral cerebellum originates from the caudal portion of the so-called mesencephalic vesicle: A study using chick/quail chimeras, *Eur. J. Neurosci.* **1:**549–560.

Martinez, S., Wassef, M., and Alvarado-Mallart, R.-M., 1991, Induction of a mesencephalic phenotype in the 2-day-old chick prosencephalon is preceded by the early expression of the homeobox gene *en*, *Neuron* **6:**971–981.

Matsui, T., Sashihara, S., Oh, Y., and Waxman, S. G., 1995, An orphan nuclear receptor, mRORα, and its spatial expression in adult mouse brain, *Brain Res. Mol. Brain Res.* **33:**217–226.

Matsuyama, Z., Kawakami, H., Maruyama, H., Izumi, Y., Komure, O., Udaka, F., Kameyama, M., Nishio, T., Kuroda, Y., Nishimura, M., and Nakamura, S., 1997, Molecular features of the CAG repeats of spinocerebellar ataxia 6 (SCA6), *Hum. Mol. Genet.* **6:**1283–1287.

Matysiak-Scholze, U., and Nehls, M., 1997, The structural integrity of RORα isoforms is mutated in *staggerer* mice: Cerebellar coexpression of RORα1 and RORα4, *Genomics* **43:**78–84.

McMahon, A. P., and Bradley, A., 1990, The *Wnt-1 (int-1)* proto-oncogene is required for development of a large region of the mouse brain, *Cell* **62:**1073–1085.

McMahon, A. P., Joyner, A. L., Bradley, A., and McMahon, J. A., 1992, The midbrain–hindbrain phenotype of *Wnt-1-/Wnt-1*-mice results from stepwise deletion of *engrailed*-expressing cells by 9.5 days postcoitum, *Cell* **69:**581–595.

Meier, H., and MacPike, A. D., 1971, Three syndromes produced by two mutant genes in the mouse.

Clinical, pathological, and ultrastructural bases of *tottering, leaner* and heterozygous mice, *J. Hered.* **62:** 297–302.

Messer, A., Plummer-Siegard, J., and Eisenberg, B., 1990, *Staggerer* mutant mouse Purkinje cells do not contain detectable calmodulin mRNA, *J. Neurochem.* **55:**293–302.

Messer, A., Eisenberg, B., and Plummer, J., 1991, The *Lurcher* cerebellar mutant phenotype is not expressed on a *staggerer* mutant background, *J. Neurosci.* **11:**2295–2302.

Miale, I. L., and Sidman, R. L., 1961, An autoradiographic analysis of histogenesis in the mouse cerebellum, *Exp. Neurol.* **4:**277–296.

Millen, K. J., Wurst, W., Herrup, K., and Joyner, A. L., 1994, Abnormal embryonic cerebellar development and patterning of postnatal foliation in two mouse *Engrailed-2* mutants, *Development* **120:**695–706.

Miyata, T., Nakajima, K., Aruga, J., Takahashi, S., Ikenaka, K., Mikoshiba, K., and Ogawa, M., 1996, Distribution of a *reeler* gene-related antigen in the developing cerebellum: An immunohistochemical study with an allogeneic antibody CR-50 on normal and *reeler* mice, *J. Comp. Neurol.* **372:**215–228.

Miyata, T., Nakajima, K., Mikoshiba, K., and Ogawa, M., 1997, Regulation of Purkinje cell alignment by Reelin as revealed with CR-50 antibody, *J. Neurosci.* **17:**3599–3609.

Mjaatvedt, A. E., Citron, M. P., and Reeves, R. H., 1993, High-resolution mapping of *D16led-1, Gart, Gas-4, Cbr, Pcp-4,* and *Erg* on distal mouse chromosome 16, *Genomics* **17:**382–386.

Mullen, R. J., 1982, Analysis of CNS development with mutant mice and chimeras, *in* "Genetic Approaches to Developmental Neurobiology" (Y. Tsukada, ed.), pp. 183–193, Springer-Verlag, Berlin.

Mullen, R. J., Eicher, E. M., and Sidman, R. L., 1976, Purkinje cell degeneration, a new neurological mutation in the mouse, *Proc. Natl. Acad. Sci. USA* **73:**208–212.

Nakagawa, S., Watanabe, M., and Inoue, Y., 1997, Prominent expression of nuclear hormone receptor RORα in Purkinje cells from early development, *Neurosci. Res.* **28:**177–184.

Neumann, P. E., Mueller, G. G., and Sidman, R. L., 1990, Identification and mapping of a mouse gene influencing cerebellar folial pattern, *Brain Res.* **524:**85–89.

Noebels, J. L., and Sidman, R. L., 1979, Inherited epilepsy: Spike-wave and focal motor seizures in the mutant mouse tottering, *Science* **204:**1334–1336.

Norman, D. J., Feng, L., Cheng, S. S., Gubbay, J., Chan, E., and Heintz, N., 1995, The *Lurcher* gene induces apoptotic death in cerebellar Purkinje cells, *Development* **121:**1183–1193.

Nusse, R., and Varmus, H. E., 1982, Many tumors induced by the mouse mammary tumor virus contain a provirus integrated in the same region of the host genome, *Cell* **31:**99–109.

Ogawa, M., Miyata, T., Nakajima, K., Yagyu, K., Seike, M., Ikenaka, K., Yamamoto, H., and Mikoshiba, K., 1995, The *reeler* gene-associated antigen on Cajal-Retzius neurons is a crucial molecule for laminar organization of cortical neurons, *Neuron* **14:**899–912.

Palay, S. F., and Chan-Palay, V., 1974, *Cerebellar Cortex: Cytology and Organization,* Springer-Verlag, Berlin.

Palay, S. F., and Chan-Palay, V., 1982, *The Cerebellum–New Vistas,* Springer-Verlag, Berlin.

Patil, N., Cox, D. R., Bhat, D., Faham, M., Myers, R. M., and Peterson, A. S., 1995, A potassium channel mutation in *weaver* mice implicates membrane excitability in granule cell differentiation, *Nat. Genet.* **11:**126–129.

Phillips, R. J. S., 1960, "*Lurcher.*" A new gene in linkage group XI of the house mouse, *J. Genet.* **57:**35–42.

Pierce, E. T., 1975, Histogenesis of the deep cerebellar nuclei in the mouse: An autoradiographic study, *Brain Res.* **95:**503–518.

Rabacchi, S. A., Bailly, Y., Delhaye-Bouchaud, N., and Mariani, J., 1992, Role of the target in synapse elimination: Studies in cerebellum of developing *Lurcher* mutants and adult chimeric mice, *J. Neurosci.* **12:**4712–4720.

Rakic, P., 1971, Neuron–glia relationships during granule cell migration in developing cerebellar cortex. A Golgi and electron microscopic study in *Macacus rhesus, J. Comp. Neurol.* **141:**283–312.

Rakic, P., and Caviness, V. S., Jr., 1995, Cortical development: View from neurological mutants two decades later, *Neuron* **14:**1101–1104.

Rakic, P., and Sidman, R. L., 1973a, *Weaver* mutant mouse cerebellum: Defective neuronal migration secondary to abnormality of Bergmann glia, *Proc. Natl. Acad. Sci. USA* **70:**240–244.

Rakic, P., and Sidman, R. L., 1973b, Sequence of developmental abnormalities leading to granule cell deficit in cerebellar cortex of *weaver* mutant mice, *J. Comp. Neurol.* **152:**103–132.

Raman, I. M., Sprunger, L. K., Meisler, M. H., and Bean, B. P., 1997, Altered subthreshold sodium currents and disrupted firing patterns in Purkinje neurons of *Scn8a* mutant mice, *Neuron* **19:**881–891.

Ramon y Cajal, S., 1889, Sobre las fibras nerviosas de la capa granulosa del cerebelo, *Rev. Trim. Histol. Norm. Pathol.* **3,4.**

Ramon y Cajal, S., 1911, "Histologie du Susteme Nerveux de l' Homme et des Vertebres, vol. 2" (L. Azoulay, transl.), Maloine, Paris (Reprinted by Consejo Superior de Investigaciones Cientificas, Inst. Ramon y Cajal, Madrid, 1955).

Ramon y Cajal, S., 1960, "Studies on Vertebrate Neurogenesis" (L. Guth, transl.), Thomas, Springfield, IL.

Reeves, R. H., Crowley, M. R., Lorenzon, N., Pavan, W. J., Smeyne, R. J., and Goldowitz, D., 1989, The mouse neurological mutant *weaver* maps within the region of chromosome 16 that is homologous to human chromosome 21, *Genomics* **5:**522–526.

Resibois, A., Cuvelier, L., and Goffinet, A. M., 1997, Abnormalities in the cerebellum and brainstem in homozygous *Lurcher* mice, *Neuroscience* **80:**175–190.

Rezai, Z., and Yoon, C. H., 1972, Abnormal rate of granule cell migration in the cerebellum of "*weaver*" mutant mice, *Dev. Biol.* **29:**17–26.

Riess, O., Schols, L., Bottger, H., Nolte, D., Vieira-Saecker, A. M., Schimming, C., Kreuz, F., Macek, M., Jr., Krebsova, A., Masek, M., Sen, Klockgether, T., Zuhlke, C., and Laccone, F. A., 1997, *SCA6* is caused by moderate CAG expansion in the α1A-voltage-dependent calcium channel gene, *Hum. Mol. Genet.* **6:**1289–1293.

Rocamora, N., Garcia-Ladona, F. J., Palacios, J. M., and Mengod, G., 1993, Differential expression of brain-derived neurotrophic factor, neurotrophin-3 and low-affinity nerve growth factor receptor during the postnatal development of the rat cerebellar system, *Brain Res. Mol. Brain Res.* **17:**1–8.

Rosario, C. M., Yandava, B. D., Kosaras, B., Zurakowski, D., Sidman, R. L., and Synder, E. Y., 1997, Differentiation of engrafted multipotent neural progenitors towards replacement of missing granule neurons in *meander tail* cerebellum may help determine the locus of mutant gene action, *Development* **124:**4213–4224.

Ross, M. E., Fletcher, C., Mason, C. A., Hatten, M. E., and Heintz, N., 1990, *Meander tail* reveals a discrete developmental unit in the mouse cerebellum. *Proc. Natl. Acad. Sci. USA* **87:**4189–4192.

Ryder, E. F., and Cepko, C. L., 1994, Migration patterns of clonally related granule cells and their progenitors in the developing chick cerebellum, *Neuron* **12:**1011–1029.

Saito, N., Kikkawa, U., Nishizuka, Y., and Tanaka, C., 1988, Distribution of protein kinase C-like immunoreactive neurons in rat brain, *J. Neurosci.* **8:**369–382.

Sanguinetti, M. C., and Spector, P. S., 1997, Potassium channelopathies. *Neuropharmacology* **36:**755–762.

Schiffmann, S. N., Bernier, B., and Goffinet, A. M., 1997, Reelin mRNA expression during mouse brain development, *Eur. J. Neurosci.* **9:**1055–1071.

Schwartz, P. M., Borghesani, P. R., Levy, R. L., Pomeroy, S. L., and Segal, R. A., 1997, Abnormal cerebellar development and foliation in BDNF-/- mice reveals a role for neurotrophins in CNS patterning, *Neuron* **19:**269–281.

Segal, R., Romeroy, S., and Stiles, C., 1995, Axonal growth and fasciculation linked to differential expression of BDNF and NT3 receptors in developing cerebellar granule cells, *J. Neurosci.* **15**:4970–4981.

Sheldon, M., Rice, D. S., D'Arcangelo, G., Yoneshima, H., Nakajima, K., Mikoshiba, K., Howell, B. W., Cooper, J. A., Goldowitz, D., and Curran, T., 1997, *Scrambler* and *yotari* disrupt the *disabled* gene and produce a *reeler*-like phenotype in mice, *Nature* **389**:730–733.

Sheppard, A. M., and Pearlman, A. L., 1997, Abnormal reorganization of preplate neurons and their associated extracellular matrix: An early manifestation of altered neocortical development in the *reeler* mutant mouse, *J. Comp. Neurol.* **378**:173–179.

Sidman, R. L., 1968, Development of interneuronal connections in brains of mutant mice, *in* "Physiological and Biochemical Aspects of Nervous Integration" (F. D. Carlson, ed.), pp. 163–193, Prentice-Hall, Englewood Cliffs, NJ.

Sidman, R. L., 1972, Cell interactions in developing mammalian central nervous system, *in* "Third Lepetit Colloquium on Cell Interaction, London" (L. G. Silvestri, ed.), pp. 1–13. North-Holland, Amsterdam.

Sidman, R. L., 1974, Contact interaction among developing mammalian brain cells, *in* "The Cell Surface in Development" (A. A. Moscona, ed.), pp. 221–253, John Wiley & Sons, New York.

Sidman, R. L., 1983, Experimental neurogenetics, *in* "Genetics of Neurological and Psychiatric Disorders" (S. S. Kety, L. P. Rowland, R. L. Sidman, and S. W. Matthysse, eds.), pp. 19–46, Raven Press, New York.

Sidman, R. L., and Rakic, P., 1973, Neuronal migration with special reference to developing human brain: A review, *Brain Res.* **62**:1–35.

Sidman, R. L., Lane, P. W., and Dickie, M. M., 1962, *Staggerer,* a new mutation in the mouse affecting the cerebellum, *Science* **137**:610–612.

Sidman, R. L., Green, M. C., and Appel, S. H., 1965, "Catalog of the Neurological Mutants of the Mouse," Harvard University Press, Cambridge, MA.

Signorini, S., Liao, Y. J., Duncan, S. A., Jan, L. Y., and Stoffel, M., 1997, Normal cerebellar development but susceptibility to seizures in mice lacking G protein-coupled, inwardly rectifying K^+ channel GIRK2, *Proc. Natl. Acad. Sci. USA* **94**:923–927.

Slesinger, P. A., Patil, N., Liao, Y. J., Jan, Y. N., Jan, L. Y., and Cox, D. R., 1996, Functional effects of the mouse *weaver* mutation on G protein-gated inwardly rectifying K^+ channels, *Neuron* **16**:321–331.

Smeyne, R. J., and Goldowitz, D., 1990, Purkinje cell loss is due to a direct action of the *weaver* gene in Purkinje cells: Evidence from chimeric mice, *Brain Res. Dev. Brain Res.* **52**:211–218.

Soha, J. M., and Herrup, K., 1995, Stunted morphologies of cerebellar Purkinje cells in *Lurcher* and *staggerer* mice are cell-intrinsic effects of the mutant genes, *J. Comp. Neurol.* **357**:65–75.

Soriano, E., Alvarado-Mallart, R. M., Dumesnil, N., Del Rio, J. A., and Sotelo, C., 1997, Cajal–Retzius cells regulate the radial glia phenotype in the adult and developing cerebellum and alter granule cell migration, *Cell* **18**:563–577.

Sotelo, C., 1980, Mutant mice and the formation of the cerebellar circuitry, *Trends Neurosci.* **20**:33–35.

Sotelo, C., and Changeux, J.-P., 1974, Bergmann fibers and granule cell migration in the cerebellum of homozygous *weaver* mutant mouse, *Brain Res.* **77**:484–491.

Sotelo, C., and Wassef, M., 1991, Cerebellar development: Afferent organization and Purkinje cell heterogeneity, *Philos. Trans. R. Soc. (Lond.) Ser. B* **331**:307–313.

Surmeier, D. J., Mermelstein, P. G., and Goldowitz, D., 1996, The *weaver* mutation of GIRK2 results in a loss of inwardly rectifying K^+ current in cerebellar granule cells, *Proc. Natl. Acad. Sci. USA* **93**:11191–11195.

Sweet, H. O., Bronson, R. T., Johnson, K. R., Cook, S. A., and Davisson, M. T., 1996, *Scrambler,* a new neurological mutation of the mouse with abnormalities of neuronal migration, *Mamm. Genome* **7**:798–802.

Tessier-Lavigne, M., and Goodman, C. S., 1996, The molecular biology of axon guidance, *Science* **274:**1123–1133.

Thomas, K. R., and Capecchi, M. R., 1990, Targeted disruption of the murine *int-1* proto-oncogene resulting in severe abnormalities in midbrain and cerebellar development, *Nature* **346:**847–850.

Thomas, K. R., Musci, T. S., Neumann, P. E., and Capecchi, M. R., 1991, *Swaying* is a mutant allele of the proto-oncogene *Wnt-1, Cell* **67:**969–976.

Wahlsten, D., and Andison, M., 1991, Patterns of cerebellar foliation in recombinant inbred mice, *Brain Res.* **557:**184–189.

Ware, M. L., Fox, J. W., Gonzalez, J. L., Davis, N. M., Lambert de Rouvroit, C., Russo, C. J., Chua, S. C., Jr., Goffinet, A. M., and Walsh, C. A., 1997, Aberrant splicing of a mouse *disabled* homolog, *mdab1,* in the *scrambler* mouse, *Neuron* **19:**239–249.

Wassef, M., Cholley, B., Heizmann, C. W., and Sotelo, C., 1992, Development of the olivocerebellar projection in the rat: II. Matching of the developmental compartmentations of the cerebellum and inferior olive through the projection map, *J. Comp. Neurol.* **323:**537–550.

Wetmore, C., Ernfors, P., Persson, H., and Olson, L., 1990, Localization of brain-derived neurotrophic factor mRNA to neurons in the brain by in situ hybridization, *Exp. Neurol.* **109:**141–152.

Wetts, R., and Herrup, K., 1982a, Interaction of granule, Purkinje and inferior olivary neurons in Lurcher chimeric mice. I. Qualitative studies, *J. Embryol. Exp. Morphol.* **68:**87–98.

Wetts, R., and Herrup, K., 1982b, Interaction of granule, Purkinje and inferior olivary neurons in Lurcher chimeric mice. II. Granule cell death, *Brain Res.* **250:**358–362.

Wullner, U., Loschmann, P. A., Weller, M., and Klockgether, T., 1995, Apoptotic cell death in the cerebellum of mutant *weaver* and *Lurcher* mice, *Neurosci. Lett.* **200:**109–112.

Wurst, W., Auerbach, A. B., and Joyner, A. L., 1994, Multiple developmental defects in *Engrailed-1* mutant mice: An early mid-hindbrain deletion and patterning defects in forelimbs and sternum, *Development* **120:**2065–2075.

Yamazaki, M., Araki, K., Shibata, A., and Mishina, M., 1992, Molecular cloning of a cDNA encoding a novel member of the mouse glutamate receptor channel family, *Biochem. Biophys. Res. Commun.* **183:**886–892.

Yoon, C. H., 1972, Development mechanism for changes in cerebellum of "*staggerer*" mouse, a neurological mutant of genetic origin, *Neurology* **22:**743–754.

Yoon, C. H., 1976, Pleiotropic effect of the *staggerer* gene, *Brain Res.* **109:**206–215.

Zanjani, H. S., Mariani, J., and Herrup, K., 1990, Cell loss in the inferior olive of the *staggerer* mutant mouse is an indirect effect of the gene, *J. Neurogenetics* **6:**229–241.

Zanjani, H. S., Herrup, K., Guastavino, J. M., Delhaye-Bouchaud, N., and Mariani, J., 1994, Developmental studies of the inferior olivary nucleus in *staggerer* mutant mice, *Brain Res. Dev. Brain Res.* **82:**18–28.

Zheng, C., Heintz, N., and Hatten, M. E., 1996, CNS gene encoding astrotactin, which supports neuronal migration along glial fibers, *Science* **272:**417–419.

Zuo, J., De Jager, P. L., Norman, D. J., and Heintz, N., 1995, Generation of a high-resolution genetic map and a YAC contig of the *Lurcher* locus on mouse chromosome 6, *Genome Res.* **5:**381–392.

Zuo, J., De Jager, P. L., Takahashi, K. A., Jiang, W., Linden, D. J., and Heintz, N., 1997, Neurodegeneration in *Lurcher* mice caused by mutation in δ2 glutamate receptor gene, *Nature* **388:**769–773.

CHAPTER 5

THE ROLE OF NEUROTROPHIC FACTORS IN DEVELOPMENT AND NEURODEGENERATIVE DISORDERS

WILLIAM D. SNIDER and INMACULADA SILOS-SANTIAGO

1. INTRODUCTION

Gene targeting as applied to neuronal growth factors and their receptors has profoundly advanced our understanding of the survival requirements of neurons in both the peripheral and central nervous systems. There is enthusiasm for the idea that this new knowledge will ultimately lead to important therapeutic tools

---

WILLIAM D. SNIDER • Center for the Study of Nervous System Injury, Department of Neurology, Washington University School of Medicine, St. Louis, Missouri 63110. INMACULADA SILOS-SANTIAGO • Laboratory of Developmental Neurogenetics, Department of Molecular Oncology, Bristol-Myers Squibb, PRI, Princeton, New Jersey 08543.

Mouse Models in the Study of Genetic Neurological Disorders, Volume 9 of *Advances in Neurochemistry,* edited by Popko, Kluwer Academic/Plenum Publishers, New York, 1999.

for treating human diseases in which specific populations of neurons degenerate. For the most part, this dream has not yet been realized. Nevertheless, the enhanced understanding of neuronal survival requirements provided as a result of the mouse mutants will undoubtedly impact on our understanding of these diseases in the future. In the current chapter, we focus on the neurotrophin family of molecules and their receptors and the glial cell line-derived neurotrophic family of molecules and their receptors. In published or unpublished work, all of these molecules have been targeted in mutation experiments, and very specific populations of neurons have been shown to require these factors during development. These insights have already led to the identification of gene mutations in two hereditary diseases as well as to clinical trials for treatment of peripheral neuropathies.

2. HISTORICAL REVIEW

The discovery of nerve growth factor (NGF) was a byproduct of the attempt to understand the phenomenon of naturally occurring cell death that had been documented by Victor Hamburger and other investigators in the motor neuron population of the chick spinal cord (Hamburger, 1934, 1958, 1977). Interestingly, the first molecule identified, NGF, had no effect on motor neuron survival but rather was found to influence the survival of sympathetic ganglion cells and neural crest derived sensory neurons (Levi-Montalcini and Booker, 1960; Levi-Montalcini and Angeletti, 1966).

Two terms are in common use to describe NGF and related molecules. The first is "growth factor," which arose from the powerful effects of NGF on neurite outgrowth (which were used to purify the molecule). Subsequently, however, this term has been used for a variety of factors whose predominant effect is to promote proliferation rather than growth per se. This has led to some confusion in the field of neural development, where at later stages the molecules regulate differentiated functions of neurons and not proliferation precursors. Another term in common use is the term "trophic factor," which refers directly to the trophic or nourishing effects of the molecules. Long before NGF was purified, the effects of interruption of neuronal connections between neurons and their targets were recognized and referred to as "trophic." The naming of additional members of the NGF family of neurotrophins reflects this emphasis within the community of neurobiological investigators. In the current work, we refer to these molecules as neurotrophic factors.

3. THE NGF AND GDNF FAMILIES OF NEUROTROPHIC FACTORS

Around the time the first gene targeting experiments were reported, Barde and collaborators succeeded in isolating brain-derived neurotrophic factor

(BDNF) that had been purified from pig brain on the basis of survival effects on sensory neurons (Barde *et al.,* 1982). Unexpectedly, BDNF was found to have significant sequence homology to nerve growth factor (Leibrock *et al.,* 1989), a finding that in short order led to the cloning of additional family members, neurotrophin 3 (*NT-3*) and neurotrophin 4/5 (*NT-4/5*) (Hohn *et al.,* 1990; Berkemeier *et al.,* 1991; Hallböök *et al.,* 1991).

One receptor for NGF was cloned on the basis of binding affinity and was reported by Chao and colleagues in 1986 (Chao *et al.,* 1986). However, it was realized even at that time that there was another component to the NGF receptor. Barbacid and collaborators cloned a proto-oncogene that was a receptor tyrosine kinase fused with tropomyosin control elements termed *trkA* in 1986 (Martín-Zanca *et al.,* 1986). The discovery that the NGF receptor was phosphorylated then led several groups in the early 1990s to the realization that trkA was a component of the NGF receptor (reviewed in Meakin and Shooter, 1992; Barbacid, 1994) (Fig. 1A). Straightforward polymerase chain reaction (PCR) strategies then identified at least two homologous molecules, trkB and trkC (reviewed in Meakin and Shooter, 1992; Barbacid, 1994).

Expression studies showed that these receptors were expressed in rather specific patterns in the peripheral nervous system (PNS), suggesting regulation of specific population by individual family members (Martín-Zanca *et al.,* 1990; Ernfors *et al.,* 1992, 1993; Mu *et al.,* 1993; McMahon *et al.,* 1994; Averill *et al.,* 1995; Molliver *et al.,* 1995; Wright and Snider, 1995). In the central nervous system (CNS), however, quite surprisingly there was widespread expression of both the trkB and trkC (Merlio *et al.,* 1992; Tessarollo *et al.,* 1993; Lamballe *et al.,* 1994; Armanini *et al.,* 1995).

All of the neurotrophins, the receptor tyrosine kinases, and the p75 receptor component have been targeted in null mutation experiments (Lee *et al.,* 1992; reviewed in Snider, 1994; Fariñas and Reichardt, 1996; Snider and Silos-Santiago, 1996). These studies have allowed the functions of each of these molecules with regard to specific neuronal populations to be worked out as well as some of the ligand–receptor interactions that had been uncertain on the basis of *in vitro* studies.

Glial cell line derived neurotrophic factor (GDNF) was identified on the basis of its ability to induce dopamine uptake and survival of mesencephalic dopaminergic neurons (Lin *et al.,* 1993). These neurons were the subject of survival assays because of their importance in the degenerative neurological condition Parkinson's disease. After the cloning of *GDNF,* targeted null mutations (see later) revealed that the phenotype was very similar to the phenotype of null mutations in an orphan receptor tyrosine kinase ret. This observation, as well as other independent work, recently led to the identification of ret as a receptor component for GDNF (Durbec *et al.,* 1996; Jing *et al.,* 1996; Treanor *et al.,* 1996; Trupp *et al.,* 1996).

Initially, there was controversy as to whether ret could bind GDNF. Several

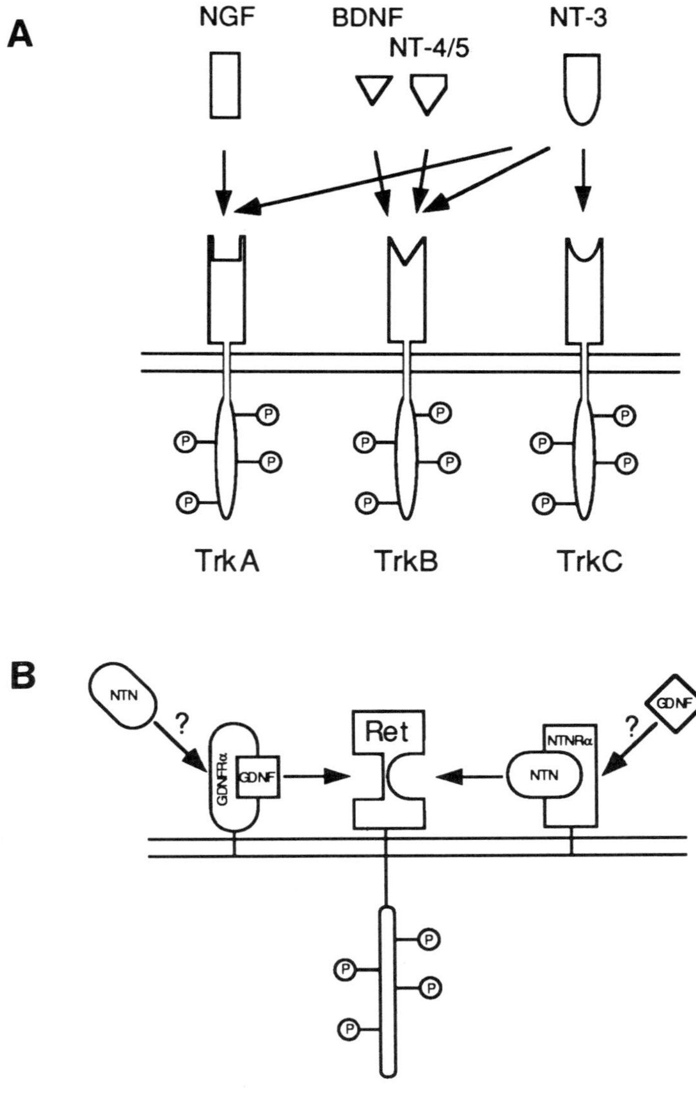

FIGURE 1. Interactions of members of the neurotrophin family (A) and the GDNF family (B) with their receptors.

groups using expression cloning and other strategies identified another component of the GDNF receptor that is a GPI-linked molecule termed GDNF receptor α that appears to confer binding specificity (Jing *et al.*, 1996; Treanor *et al.*, 1996). The discovery of GDNF receptor α sparked excitement as the molecule appeared to be similar in structure to GPI-linked cytokine receptor binding subunits. It provided the first example of a receptor tyrosine kinase that had as an accessory molecule a GPI-linked binding component (Fig. 1B).

Very recently, a second GDNF family member was purified on the basis of survival promoting effects on sympathetic ganglion neurons (Kotzbauer *et al.*, 1996; Creedon *et al.*, 1997). This molecule is present in low abundance and is roughly 50% homologous to GDNF across significant stretches of amino acids. It had not been detected in routine PCR-based strategies to identify new GDNF family members (Fig. 1B). A second binding component that may have specificity for this new factor has also been recently identified (Baloh *et al.*, 1997; Buj-Bello *et al.*, 1997; Klein *et al.*, 1997) (Fig. 1B). *GDNF, nurturin,* and *ret* have all been subjected to gene targeting and the results are quite informative in terms of populations of neurons that require these molecules for survival during development.

4. REGULATION OF SPECIFIC POPULATIONS OF PERIPHERAL NEURONS BY NEUROTROPHIN AND GDNF FAMILY MEMBERS

The original reports of NGF null mutant mice and trkA kinase deficient mice provided striking confirmation of many years of biological work with NGF and with specific antisera that had been developed and utilized in biological experiments (Table 1). Essentially, the entire sympathetic nervous system was eliminated in these mice as well as the entire population of nociceptive sensory neurons which represents roughly 80% of neurons in mammalian dorsal root ganglia (DRG) and in the trigeminal ganglion (Crowley *et al.*, 1994; Smeyne *et al.*, 1994, Silos-Santiago *et al.*, 1995). In addition, some deficits in the central nervous system related to known expression patterns of NGF and the NGF receptor trkA have also been observed (see later). For the most part, within the trkA population, the targeting of NGF gene and the targeting of the *trkA* gene produced results that are in striking accord. The mice do not survive. Most die within 3 weeks of birth due in part to infections related to the inability of the mice to feel pain in any part of their bodies (Crowley *et al.*, 1994; Smeyne *et al.*, 1994).

Targeted mutations of BDNF and trkB (Klein *et al.*, 1993) also resulted in the elimination of specific subsets of peripheral neurons (Table 1). In BDNF null mutant mice and trkB kinase deficient mice 85% of the neurons in the vestibular ganglion are eliminated, a subpopulation of neurons in the cochlear ganglion are eliminated, and significant numbers of neurons in the nodose ganglion which may represent a specific subpopulation are also lost (Ernfors *et al.*, 1994a; Jones *et al.*, 1994; Conover *et al.*, 1995; Liu *et al.*, 1995; Minichiello *et al.*, 1995; Schimmang *et al.*, 1995; Erickson *et al.*, 1996; Silos-Santiago *et al.*, 1997). Again, for the most part, losses observed in the BDNF null mutants and the trkB

TABLE 1. Dependence of Subsets of Peripheral Neurons on Specific Neuronal Growth Factors for Survival

Ganglia	trkA$^{-/-}$	trkB$^{-/-}$	trkC$^{-/-}$	trkB/trkC$^{-/-}$	NGF$^{-/-}$	BDNF$^{-/-}$*	NT4$^{-/-}$	BDNF/NT4$^{-/-}$	NT-3$^{-/-}$*
VII	ND	95[d,e]	20[f,d]	98[d]	ND	48[l,m]	50[m]	94[m]	25[q]
VIII	ND	56–85[d,f,g]	16–30[d,f,g]	58–100[d,g]	ND	76–96[l,m,n,o]	21[m]	82[m]	23–34[n,q]
IX–X	ND	94[d,h]	18[d,]	95[d]	28[k]	43–66[l,m,n,p]	59–61[m,p]	83–90[m,p]	30–47[q,r]
V	70–90[a,i]	33[i]	22[i]	42[i]	ND	27–44[l,n,p]	NS[m,p]	30[m,p]	61–64[q,r]
Cochl	ND	0–20[d,f,g]	51–66[d,f,g]	61–100[d,g]	ND	7[o]	ND	ND	85[o,q]
DRG	70–90[a,b]	NS–20[d,g]	17[d,j]	41–51[d,g]	70[k]	30–34[l,n]	NS[m]	ND	55–78[q,r]
SCG	90–95[a,c]	NS[c]	NS[c]	ND	82[k]	NS[l,n]	NS[m]	NS[m]	45–53[q,r]
V Mes.	ND	ND	ND	ND	ND	39–48[l,n]	ND	ND	52–56[q,r]

Values represent the percentage of neuronal loss in peripheral ganglia of neurotrophins and neurotrophin receptors deficient mice.
*In the *BDNF/NT-3*$^{-/-}$ mice, the cochlear ganglion is absent and 99% of the vestibular neurons are lost (Ernfors *et al.*, 1995).
ND = Not determined.
NS = No significant differences with wild-type mice.
[a]Smeyne *et al.*, 1994.
[b]Silos-Santiago *et al.*, 1995.
[c]Fagan *et al.*, 1996.
[d]Silos-Santiago *et al.*, 1997.
[e]Fritzsch *et al.*, 1997.
[f]Schimmang *et al.*, 1995.
[g]Minichiello *et al.*, 1995.
[h]Erickson *et al.*, 1996.
[i]Piñón *et al.*, 1996.
[j]Klein *et al.*, 1994.
[k]Crowley *et al.*, 1994.
[l]Jones *et al.*, 1994.
[m]Liu *et al.*, 1995.
[n]Ernfors *et al.*, 1994a.
[o]Ernfors *et al.*, 1995.
[p]Conover *et al.*, 1995.
[q]Fariñas *et al.*, 1994.
[r]Ernfors *et al.*, 1994b.

kinase deficient mice appear to correspond with one another. It can be expected that the *trkB* mutation will be more severe because the actions of NT-4 would also be eliminated. There is some evidence of this effect in the nodose ganglion where there are differences in populations of neurons lost in *trkB* null mutations compared to *BDNF* null mutations (Erickson *et al.*, 1996).

NT-4 has also been subjected to gene targeting and the effects of this null mutation are very mild with no indiscernible behavioral deficits and effects on survival (Conover *et al.*, 1995; Liu *et al.*, 1995). However, careful analysis of the nodose ganglion in particular has revealed some subpopulations of neurons that appear to require NT-4 signaling and some that appear to require both NT-4 and BDNF (Erickson *et al.*, 1996) (see also Table 1).

Elimination of NT-3 and trkC have also resulted in clear losses of identifiable populations of neurons (Table 1). Thus, the entire population of neurons in DRG related to sensation of kinesthetic motion in the limbs (proprioceptors) are lost in these mice (Ernfors *et al.*, 1994b; Fariñas *et al.*, 1994; Klein *et al.*, 1994; Tessarollo *et al.*, 1994). Furthermore, most neurons in the spiral ganglion are eliminated (Fariñas *et al.*, 1994; Minichiello *et al.*, 1995; Schimmang *et al.*, 1995; Silos-Santiago *et al.*, 1997). It is likely that other classes of neurons in the DRG, trigeminal ganglia, and nodose ganglia are also lost but the exact types have not been specified.

Finally, *GDNF* and *ret* null mutations also result in specific elimination of subsets of peripheral neurons. The entire sympathetic ganglion is loss in the *ret* null mutations and significant numbers are also lost in GDNF null mutant mice (Durbec *et al.*, 1996; Moore *et al.*,1996). Furthermore, the entire enteric nervous system is lost in both mutants (Schuchardt *et al.*, 1994; Durbec *et al.*, 1996; Moore *et al.*, 1996; Pichel *et al.*, 1996; Sánchez *et al.*, 1996). Whether other populations of peripheral neurons may be affected is currently under study.

A surprise in all of these studies has been the relatively mild impact on the somatic motor neurons that project to skeletal muscles. All of these factors except for NGF have been shown to have survival promoting effects on motor neurons in vitro and in vivo. However, the effects of the null mutations of these molecules and their receptors are modest (Table 2). There appears to be no effect whatever during embryonic development on loss of NT-4, BDNF, or trkB (Ernfors *et al.*, 1994a; Conover *et al.*, 1995; Liu *et al.*, 1995; Silos-Santiago *et al.*, 1997). In older animals the elimination of trkC appears to result in loss of approximately 25% of spinal motor neurons, of which appear to have small soma areas that may represent fusomotor neurons that innervate muscle spindles (Klein *et al.*, 1994). GDNF similarly results in loss of approximately 20% of lumbar motor neurons (Moore *et al.*, 1996; Sánchez *et al.*, 1996) (Table 2). Whether these also represent the fusomotor population is currently under active study. However, it should be emphasized that most motor neurons survive quite nicely in all of these null mutations at least until postnatal day 0.

TABLE 2. Dependence of Motor Neurons on Neuronal Growth Factors for Survival

Nuclei	$trkB^{-/-}$	$trkC^{-/-}$	$trkB/trkC^{-/-}$	$BDNF^{-/-}$	$NT4^{-/-}$	$BDNF/NT4^{-/-}$	$NT\text{-}3^{-/-}$	$GDNF^{-/-}$
Spinal	NS[a]	NS[a]	NS[a]	NS[b,c]	NS[c]	NS[c]	ND	22–31[h,i]
Facial	NS[a]	NS[a]	NS[a]	NS[b,c,d,e]	NS[c,e]	NS[c,e]	NS[f,g]	NS[h,i]
Trigeminal	ND	ND	ND	NS[b]	ND	ND	NS[g]	19–21[h,i]

Values represent the percentage of neuronal loss in different motor nuclei.
ND = Not determined.
NS = No significant differences with wild-type mice.
[a]Silos-Santiago *et al.*, 1997.
[b]Jones *et al.*, 1994.
[c]Conover *et al.*, 1995.
[d]Ernfors *et al.*, 1994a.
[e]Liu *et al.*, 1995.
[f]Ernfors *et al.*, 1994b.
[g]Fariñas *et al.*, 1994.
[h]Moore *et al.*, 1996.
[i]Sánchez *et al.*, 1996.

5. HEREDITARY CONDITIONS ASSOCIATED WITH GROWTH FACTOR RECEPTOR MUTATIONS

Perhaps surprisingly, in view of the striking requirements for neuronal survival of neurotrophins and GDNF family members, only two hereditary conditions in humans are now know to be associated with genetic mutations. Several clinical syndromes are associated with mutations in ret (Edery *et al.,* 1994; Hofstra *et al.,* 1994; Romeo *et al.,* 1994; Carlomagno *et al.,* 1996; Pasini *et al.,* 1996; Eng and Mulligan, 1997). Among these are Hirschsprung's disease, which is known to result from loss of enteric ganglia. It is thought that roughly one third of cases of Hirschsprung's disease are due to mutations in the gene encoding the GDNF receptor ret (Angrist *et al.,* 1996; Ivanchuk *et al.,* 1996; Salomon *et al.,* 1996). A rare syndrome, congenital insensitivity to pain, at least in some families, is now known to be due to mutations in the NGF receptor trkA (Indo *et al.,* 1996). However, perhaps surprisingly, other hereditary neuronal degenerations such as familial dysautonomia, Charcot–Marie–Tooth disease type 2, and other hereditary sensory neuropathies have not yet been linked to mutations in any growth factor or growth factor receptor gene.

Although hereditary conditions associated with deficiencies in these molecules or their receptors are rare, the implications of these survival studies for clinical neurology have been profound. A well-known observation related to NGF is that in addition to regulating survival during development, NGF affects functions of neurons in maturity. Thus, NGF can prevent death after axonal injury and after a variety of toxic insults (see Snider and Johnson, 1989 for a review). Most importantly in the clinical arena, NGF regulates a variety of functions of nociceptive neurons in maturity, particularly those related to inflammation. Thus, NGF is increased in the setting of inflammatory stimuli, nociceptive thresholds are diminished, nociceptive peptides in the DRG are enhanced by NGF, and responsiveness to known inflammatory irritants such as capsaicin is regulated (Lewin *et al.,* 1992; Woolf *et al.,* 1994, 1996, 1997; Safieh-Garabedian *et al.,* 1996). Indeed, the myalgias and other painful perceptions were among the first responses of human volunteers who received injections of NGF as a prelude to clinical trials in using this molecule to treat peripheral neuropathies.

Like NGF, the other members of the neurotrophin family have similar effects in preventing responses to axotomy and in reversing toxic effects on peripheral nerves (McMahon and Priestley, 1995; Anand, 1996; Apfel and Kessler, 1996; Lindsay, 1996; Tomlinson *et al.,* 1996 for reviews). Furthermore, many of these molecules can prevent the axotomy-induced degeneration of motor neurons that occurs in neonates and adults (Oppenheim *et al.,* 1992; Sendtner *et al.,* 1992; Yan *et al.,* 1992, 1993; Henderson *et al.,* 1993; Koliatsos *et al.,* 1993). Thus, NGF is in clinical trials for diabetic neuropathies and neuropathies due to chemotherapeutic agents: NT-3 is in clinical trials for a variety of peripheral neuropathies; and GDNF and BDNF are in clinical trials for motor neuron disease.

6. MULTIPLE TROPHIC FACTORS REGULATE THE DEVELOPMENT OF SOME PERIPHERAL NEURONS

An unexpected finding from the results with the null mutant mice was that many peripheral neurons appeared to require one or more factors for survival during development. It had previously been thought that neurons acquired a survival dependence on a particular factor synthesized in their target fields around the time innervation of the target fields occurred. It was assumed that neurons were trophic factor independent prior to that time and after innervating targets would be regulated by a particular factor throughout development and maturity. Work with the neurotrophins and/or their receptors mutant mice has revealed that this is a far too simplistic view. Perhaps the first clue came from disparate results between work with NT-3 null mutants and trkC kinase deficient mice. Thus, in NT-3 null mutants, about 70% of DRG neurons are lost and 50% of sympathetic ganglion neurons are lost (Ernfors *et al.*, 1994b; Fariñas *et al.*, 1994) whereas in trkC kinase deficient only 20% of DRG neurons are lost and no sympathetic ganglion neurons are lost (Klein *et al.*, 1994; Fagan *et al.*, 1996) (Table 1). In association with these widespread effects of NT-3, the distribution of the NT-3 receptor, trkC, is more diffuse in early development than it is later.

The manner in which NT-3 exerts its effect on neuronal number, however, remains controversial. In association with these widespread effects of NT-3, the distribution of the NT-3 receptor, trkC, is more widespread in early development than it is later. At least three theories are currently in vogue. The first theory is that NT-3 acts primarily at the level of the precursors to regulate survival or proliferation (DiCicco-Bloom *et al.*, 1993; ElShamy and Ernfors, 1996a,b; ElShamy *et al.*, 1996). The second is that NT-3 supports the survival of recently born neurons as they extend axons toward their targets (Fariñas *et al.*, 1996; Wilkinson *et al.*, 1996). The final scenario is that effects of NT-3 may be indirect and that NT-3 may be acting on trkC neurons to synthesize another molecule that then regulates neuronal number.

The precise reasons for the discrepancy in the findings in the NT-3 null mutations and trkC mutations remain incompletely understood. One possibility is that trkA can compensate for trkC in mice that have the trkC mutation, as it is well known that NT-3 at high concentrations can activate trkA *in vitro* (Fig. 1A).

Another emerging idea is that different neurotrophins may regulate different properties of neurons. For example, with sympathetic ganglion neurons, *in vitro,* although NT-3 does not appear to exert effects on survival, it does have potent effects on neurite outgrowth. Thus, it has been proposed that NT-3 (probably acting via trkA) may regulate axonal growth and branching in target fields of trkA positive sympathetic ganglion neurons (Belliveau *et al.,* 1997). Another example of the dual actions of neurotrophins has come from findings that innervation of skin from trigeminal neurons that is eliminated in NGF null mutants and trkA kinase deficient mice is also reduced in NT-3 null mutants. Very surprisingly, this innervation appears to be enhanced in trkC kinase deficient mice (Fundin *et al.,* 1997). The interpretation of these findings is that a cooperative interaction between NT-3 and NGF potentially mediated all through trkA is required for the proper innervation to occur (Fig. 1A).

Finally, there must be cooperative survival between GDNF and NGF in regulating the survival of sympathetic ganglion neurons. Sympathetic ganglia are essentially eliminated in ret, trkA, and NGF null mutant mice (Crowley *et al.,* 1994; Smeyne *et al.,* 1994; Durbec *et al.,* 1996; Fagan et al., 1996). In GDNF null mutant mice, sympathetic ganglion neurons are partially reduced (Moore *et al.,* 1996). Furthermore, sympathetic ganglion neurons strongly express the GDNF receptor ret (Durbec *et al.,* 1996). Thus, it seems almost certain that members of the GDNF family and NGF cooperate in regulating sympathetic ganglion neuron survival during development. The exact time course of survival dependence on the various molecules remains to be worked out. Similar types of interactions, particularly between BDNF and NT-4 but also involving NT-3, have been reported for subpopulations of nodose ganglion neurons (Erickson *et al.,* 1996).

7. REGULATION OF SURVIVAL OF CNS NEURONS

In striking contrast to the PNS, the CNS initially appeared to be normal and all of the neurotrophin null mutant and kinase deficient mice as well as GDNF and ret null mutant mice (Schuchardt *et al.,* 1994; for reviews see Snider, 1994; Fariñas and Reichardt, 1996; Lindsay and Yancopoulos, 1996). However, subsequent work has revealed that substantial abnormalities are present. There are several factors to consider in why the CNS initially appeared normal and in furthering interpreting the neurotrophin dependence of CNS versus PNS neurons. First of all, it would appear that many neurons in the CNS express multiple neurotrophin receptors (see earlier). Although this is also true for some neurons in the PNS, for the most part, the PNS receptor expression patterns appear to be complementary as opposed to the rather uniform pattern observed in the CNS. This expression of multiple receptors by CNS neurons presumably is a function of their need for more complex regulation by extracellular molecules (Snider, 1994). Interestingly, it is now known that CNS neurons have much different survival requirements PNS neurons and some molecular mechanisms have been identified that are responsible for this. For example, CNS neurons appear to be supported by neurotrophins only when their cAMP levels are also regulated by neuronal activity (Meyer-Franke *et al.,* 1995; for a review see Thoenen, 1995). This does not appear to be a requirement for neurotrophin survival of peripheral neurons.

A possible explanation for the paucity of effects of trophic factor gene mutations on CNS neurons is that CNS neurons at every stage of development require multiple neurotrophic factors for regulation of differentiated functions and that elimination of any single factor does not result in noticeable loss.

In this regard, work *in vitro* has recently shown that CNS neurons indeed respond to multiple neurotrophic factors, and in fact different aspects of neuronal function appear to be regulated. The most graphic demonstration of this is the effects of multiple neurotrophins on dendritic arborization of cortical neurons in slice cultures (McAllister *et al.,* 1995, 1996; Katz and Shatz, 1996). BDNF appears to exert a powerful growth promoting effect on dendritic arbors whereas NT-3 remarkably has antagonistic effects on dendrites for some of these neuronal populations. Another possibility that cannot be excluded at the present is that many populations of CNS neurons require factors that have not yet been identified and thus do not require neurotrophins or GDNF family members for survival the way the PNS neurons do.

Nevertheless, survival requirements for CNS neurons are increasingly being identified. The original reports of trkA kinase deficient mice suggested that NGF/trkA signaling is required for the maintenance of some phenotypic aspects of the basal forebrain neurons (Crowley *et al.,* 1994; Smeyne *et al.,* 1994). A more recent and detailed analysis of the trkA mutant mice shows that NGF/trkA signaling is required for the maturation of striatal and basal forebrain cholinergic neurons (Fagan *et al.,* 1997). Furthermore, this study reveals that some basal forebrain cholinergic neurons are eliminated in these mice. These neurons begin to atrophy and/or die around the time the innervate their target fields.

Surprisingly in spite of abundant expression of BDNF and trkB in the brain, no defects were detected initially in the CNS of neonate mice with targeted mutations of BDNF and trkB genes (Snider, 1994; Fariñas and Reichardt, 1996). However, more detailed analyses have shown recently that the absence of the kinase domain in the trkB receptor results in an increase of pycknotic nuclei in different regions of the brain during postnatal life (Minichiello and Klein, 1996; Alcántara *et al.,* 1997). Cell death has been observed in the striatum, cortex, thalamus, hippocampus, and cerebellum (Table 3). The dentate gyrus of the hippocampal region is the most affected area of the CNS (Minichiello and Klein, 1996; Alcántara *et al.,* 1997). Whether any particular subpopulation of neurons is eliminated in these mice remains to be determined. In the dentate gyrus, on the basis of their location, it appears that the majority of the dying cells are those that were generated later in development (Alcántara *et al.,* 1997). Although most of the neurons that die in the CNS of these mice are positive for different calcium-

TABLE 3. Cell Death in the CNS of trkB Kinase Deficient Mice

Hippocampus	
CA3	++
DG	++++
Cortex	
II–III	+++
V–VI	++
Thalamus	
VPM	+
NR	++
Cerebellum	
EGL	+++
IGL	None
PL	None
Motor nuclei	
Facial	None
Spinal	None
Striatum	+++

CA3 = Pyramidal layer CA3; DG = dentate gyrus; II–III and V–VI = cortical layers II–III and V–VI; VPM = nucleus ventroposteromedialis; NR = nucleus reticularis; EGL = external granule cell layer; IGL = internal granule cell layer; PL = Purkinje cell layer.

binding proteins, none of these subpopulations of neurons are completely eliminated. Similarly, in the BDNF null mutant mice both parvalbumin and calbindin positive neurons were decreased in the hippocampus and cortex (Jones *et al.*, 1994). Furthermore, an increase in pyknotic nuclei has been observed in the dentate gyrus of 2-week-old mice (Fariñas, I., personal communication). In the cerebellum, the increase in cell death appears to be restricted to the external granular cell layer in both trkB and trkC kinase deficient mice: this cell death is more dramatic in double mutant mice. Interestingly, the internal granule cell layer does not display a significant number of pyknotic nuclei (Minichiello and Klein, 1996). The absence of active trkB or trkC receptors does not affect the numbers of Purkinje cells: however, their dendritic trees are very atrophic in trkB/trkC double mutant mice. Although the NT-4 null mutants survive to adulthood there are no detailed data about the postnatal development of different regions of the CNS.

As mentioned previously, neurotrophins have been shown to protect motor neurons from axotomy-induced cell death. No somatic motor neurons deficits are detected in newborn mice lacking neurotrophins or their receptors. However, when axotomy is performed in 5-day-old mice there is a decrease in the number of motor neurons that survive in the trkB kinase deficient mice when compared to wild-type or trkC kinase deficient mice. Similarly, in hippocampal slice preparations where some hippocampal neurons are axotomized, an increase in neuronal death has been observed in the trkB kinase deficient mice. These data suggest that trkB signaling protects hippocampal and motor neurons from axotomy-induced cell death (Alcántara *et al.,* 1997).

Finally, GDNF null mutant mice have a normal complement of mesencephalic dopaminergic neurons and locus coeruleus noradrenergic neurons and facial motor neurons at birth. These results are surprising in that neurons in these regions robustly express the GDNF receptor component ret. However, these mice display a small, although significant, deficit in other motor neuron nuclei as the trigeminal, oculomotor, trochlearis, and spinal motor neurons (Moore *et al.,* 1996; Sánchez *et al.,* 1996; Fritzsch and Silos-Santiago, unpublished observations). Whether other populations of neurons may be affected is currently under study. Similar deficits in the CNS have not been yet described in the ret null mutant mice.

A confounding feature of all of these studies is the poor condition of the mice in the postnatal period. Thus, it is difficult to perform the desired studies because of the high postnatal mortality and interpretation is compromised by the fact that the mice are in very poor condition at the time some of these studies are generated. Altogether this problem points up the need for conditional mutants. It can be expected that using *Cre/lox* technology neurotrophins can be eliminated in different parts of the CNS while expression is maintained in the periphery. It is likely that using this technology, viable animals can be produced whose nervous systems are completely devoid of neurotrophins. With this technology it would be possible to generate mice that develop normally until they reach adulthood, and at that point eliminate the gene of interest encoding neurotrophins or their receptors. In view of the many features of neural organization shown to be regulated by neurotrophins *in vitro,* such as axonal branching, synaptic function, posttetanic potentiation, and dendritic arborization, the generation of such animals represents a very important goal in neurotrophin research. Furthermore, other technical advances in gene targeting may allow generation of mice that develop normally during embryonic and postnatal life, and then eliminate neurotrophin signaling during adulthood. These mice may represent a closer model for human neuropathies and neurodegenerative disease states.

8. CONCLUSIONS

The generation of mice with null mutations in loci for trophic factors and their receptors have defined the survival requirements for different populations of neurons in the PNS. This identification of peripheral neurons regulated by different growth factors has led to trials of these molecules as treatments for various types of peripheral neuropathic conditions. So far, results are encouraging for NGF in diabetic neuropathy. In addition, certain hereditary syndromes are known to be due to mutations in growth factor receptors. The role of these molecules in the CNS with regard to degenerative conditions that affect CNS neurons has not been defined by the first round of gene-targeted mice because of the early lethality of the animals. However, within the year, it can be anticipated that the first conditionally mutated neurotrophin genes will be available and these animals will lead to a series of revelations related to the CNS that is as spectacular as what has been observed in the periphery.

REFERENCES

Alcántara, S., Frisén, J., del Río, J. A., Soriano, E., Barbacid, M., and Silos-Santiago, I., 1997, trkB signaling is required for postnatal survival of CNS neurons and protects hippocampal and motor neurons from axotomy-induced cell death, *J. Neurosci.* **17**:3623–3633.

Anand, P., 1996, Neurotrophins and peripheral neuropathy, *Philos. Trans. R. Soc. Lond. Ser. B* **351**:449–454.

Angrist, M., Bolk, S., Halushka, M., Lapchak, P. A., and Chakravarti, A., 1996, Germline mutations in glial cell line-derived neurotrophic factor (*GDNF*) and *RET* in a Hirschsprung disease patient, *Nat. Genet.* **14**:341–344.

Apfel, S. C., and Kessler, J. A., 1996, Neurotrophic factors in the treatment of peripheral neuropathy, *in* "Growth Factors as Drugs for Neurological and Sensory Disorders" (Ciba Foundation Symposium 196), pp. 98–112, Wiley, Chichester.

Armanini, M. P., McMahon, S. B., Sutherland, J., Shelton, D. L., and Phillips, H. S., 1995, Truncated and catalytic isoforms of trkB are co-expresses in neurons of rat and mouse CNS, *Eur. J. Neurosci.* **7**:1403–1409.

Averill, S., McMahon, S. B., Clary, D. O., Reichardt, L. F., and Priestley, J. V., 1995, Immunocytochemical localization of trkA receptors in chemically identified subgroups of adult rat sensory neurons, *Eur. J. Neurosci.* **7**:1484–1494.

Baloh, R. H., Tansey, M. G., Golden, J. P., Creedon, D. J., Heutckeroth, R. O., Keck, C. L., Zimonjic, D. B., Popescu, N. C., Johnson, E. M. Jr., and Milbrandt, J., 1997, TrnR2, a novel receptor that mediates neurturin and GDNF signaling through Ret, *Neuron* **18**:793–802.

Barbacid, M., 1994, The Trk family of neurotrophin receptors, *J. Neurobiol.* **25**:1386–1403.

Barde, Y.-A., Edgar, D., and Thoenen, H., 1982, Purification of a new neurotrophic factor from mammalian brain, *EMBO J.* **1**:549–553.

Belliveau, D. J., Krivko, I., Kohn, J., Lachance, C., Pozniak, C., Rusakov, D., Kaplan, D., and Miller, F. D., 1997, NGF and neurotrophin-3 both activate TrkA on sympathetic neurons but differentially regulate survival and neuritogenesis, *J. Cell Biol.* **136**:375–388.

Berkemeier, L. R., Winslow, J. W., Kaplan, D. R., Nikolics, K., Goeddel, D. V., and Rosenthal, A., 1991, Neurotrophin-5: A novel neurotrophic factor that activates *trk* and *trkB, Neuron* **7**:857–866.

Buj-Bello, A., Adu, J., Piñón, L. G., Thompson, J., Rosenthal, A., Chinchetru, M., Buchman, V. L., and Davies, A. M., 1997, Neurturin responsiveness requires a GPI-linked receptor and the Ret receptor tyrosine kinase, *Nature* **387**:721–724.

Carlomagno, F., De Vita, G., Berlingieri, M. T., de Franciscis, V., Melillo, R. M., Colantuoni, V., Kraus, M. H., Di Fiore, P. P., Fusco, A., and Santoro, M., 1996, Molecular heterogeneity of RET loss of function in Hirschsprung's disease, *EMBO J.* **15**:2717–2725.

Chao, M. V., Bothwell, M. A., Ross, A. H., Koprowski, H., Lanahan, A. A., Buck, C. R., and Sehgal, A., 1986, Gene transfer and molecular cloning of the human NGF receptor, *Science* **232:**518–521.

Conover, J. C., Erickson, J. T., Katz, D. M., Bianchi, L. M., Poueymirou, W. T., McClain, J., Pan, L., Helgren, M., Ip, N. Y., Boland, P., Friedman, B., Wiegand, S., Vejsada, R., Kato, A. C., DeChiara, T. M., and Yancopoulos, G. D., 1995, Neuronal deficits, not involving motor neurons, in mice lacking BDNF and/or NT-4, *Nature* **375:**235–238.

Creedon, D. J., Tansey, M. G., Baloh, R. H., Osborne, P. A., Lampe, P. A., Fahrner, T. J., Heuckeroth, R. O., Milbrandt, J., and Johnson, E. M. Jr., 1997, Neurturin shares receptors and signal transduction pathways with glial cell-derived neurotrophic factor in sympathetic neurons, *Proc. Natl. Acad. Sci. USA* **94:**7018–7023.

Crowley, C., Spencer, S. D., Nishimura, M. C., Chen, K. S., Pitts-Meek, S., Armanini, M. P., Ling, L. H., McMahon, S. B., Shelton, D. L., Levinson, A. D., and Phillips, H. S., 1994, Mice lacking nerve growth factor display perinatal loss of sensory and sympathetic neurons yet develop basal forebrain cholinergic neurons, *Cell* **76:**1–20.

DiCicco-Bloom, E., Friedman, W. J., and Black, I. B., 1993, NT-3 stimulates sympathetic neuroblast proliferation by promoting precursor survival, *Neuron* **11:**1101–1111.

Durbec, P., Marcos-Gutierrez, C. V., Kilkenny, C., Grigoriou, M., Wartiowaara, K., Suvanto, P., Smith, D., Ponder, B., Costantini, F., Saarma, M., Sariola, H., and Pachnis, V., 1996, GDNF signalling through the Ret receptor tyrosine kinase, *Nature* **381:**789–793.

Edery, P., Lyonnet, S., Mulligan, L. M., Pelet, A., Dow, E., Abel, L., Holder, S., Nihoul-Fékété, C., Ponder, B. A. J., and Munnich, A., 1994, Mutations of the *RET* proto-oncogene in Hirschsprung's disease, *Nature* **367:**378–380.

ElShamy, W. M., and Ernfors, P., 1996a, A local action of neurotrophin-3 prevents the death of proliferating sensory neuron precursor cells, *Neuron* **16:**963–972.

ElShamy, W. M., and Ernfors, P., 1996b, Requirements of neurotrophin-3 for the survival of proliferating trigeminal ganglion progenitor cells, *Development* **122:**2405–2414.

ElShamy, W. M., Linnarsson, S., Lee, K.-F., Jaenisch, R., and Ernfors, P., 1996, Prenatal and postnatal requirements of NT-3 for sympathetic neuroblast survival and innervation of specific targets, *Development* **122:**491–500.

Eng, C., and Mulligan, L. M., 1997, Mutations of the *RET* proto-oncogene in the multiple endocrine neoplasia type 2 syndromes, related sporadic tumors, and Hirschsprung disease, *Hum. Mutat.* **9:**97–109.

Erickson, J. T., Conover, J. C., Borday, V., Champagnat, J., Barbacid, M., Yancopoulos, G., and Katz, D. M., 1996, Mice lacking brain-derived neurotrophic factor exhibit visceral sensory neuron losses distinct from mice lacking NT-4 and display a severe developmental deficit in control of breathing, *J. Neurosci.* **16:**5361–5371.

Ernfors, P., Merlio, J. P., and Persson, H., 1992, Cells expressing mRNA for neurotrophins and their receptors during embryonic rat development, *Eur. J. Neurosci.* **4:**1140–1158.

Ernfors, P., Rosario, C. M., Merlio, J.-P., Grant, G., Aldskogius, H., and Perrson, H., 1993, Expression of mRNAs for neurotrophin receptors in the dorsal root ganglion and spinal cord during development and following peripheral or central axotomy, *Mol. Brain Res.* **17:**217–223.

Ernfors, P., Lee, K. F., and Jaenisch, R., 1994a, Mice lacking brain-derived neurotrophic factor develop with sensory deficits, *Nature* **368:**147–150.

Ernfors, P., Lee, K.-F., Kucera, J., and Jaenisch, R., 1994b, Lack of neurotrophin-3 leads to deficiencies in the peripheral nervous system and loss of limb proprioceptive afferents, *Cell* **77:**503–512.

Ernfors, P., Van De Water, T., Loring, J., and Jaenisch, R., 1995, Complementary roles of BDNF and NT-3 in vestibular and auditory development, *Neuron* **14:**1153–1164.

Fagan, A. M., Zhang, H., Landis, S., Smeyne, R. J., Silos-Santiago, I., and Barbacid, M., 1996, TrkA, but not TrkC receptors are essential for survival of sympathetic neurons *in vivo, J. Neurosci.* **16:**6208–6218.

Fagan, A. M., Garber, M., Barbacid, M., Silos-Santiago, I., and Holtzman, D. M., 1997, TrkA is required for the normal development of cholinergic neurons of the striatum and basal forebrain, *J. Neurosci.* **17:**764–7654.

Fariñas, I., and Reichardt, L. F., 1996, Neurotrophic factors and their receptors: Implications of genetic studies, *Semin. Neurosci.* **8:**133–143.

Fariñas, I., Jones, K. R., Backus, C., Wang, X. Y., and Reichardt, L. F., 1994, Severe sensory and sympathetic deficits in mice lacking neurotrophin-3, *Nature* **36:**658–661.

Fariñas, I., Yoshida, C. K., Backus, C., and Reichardt, L. F., 1996, Lack of neurotrophin-3 results in death of spinal sensory neurons and premature differentiation of their precursors, *Neuron* **17:**1065–1078.

Fritzsch, B., Sarai, P. A., Barbacid, M., and Silos-Santiago, I., 1997, Mice with a targeted disruption of the neurotrophin receptor *trk*B lose their gustatory ganglion cells early but do develop taste buds, *Int. J. Dev. Neurosci.* **15:**563–576.

Fundin, B. T., Silos-Santiago, I., Ernfors, P., Fagan, A. M., Aldskogius, H., deChiara, T. M., Philips, H. S., Barbacid, M., Yancopoulos, G. D., and Rice, F. L., 1997, Differential dependency of mechanoreceptors on neurotrophins, trk receptors and p75 LNGFR, *Dev. Biol.* **190:**94–116.

Hallböök, F., Ibañez, C. F., and Persson, H., 1991, Evolutionary studies of the nerve growth factor family reveal a novel member abundantly expressed in *Xenopus* ovary, *Neuron* **6:**845–858.

Hamburger, V., 1934, The effects of wing bud extirpation on the development of the central nervous system of chick embryos, *J. Exp. Zool.* **68:**449–494.

Hamburger, V., 1958, Regression versus peripheral control of differentiation on motor hypoplasia, *Am. J. Anat.* **102:**365–409.

Hamburger, V., 1977, The developmental history of the motorneuron. The F. O. Schmidt lecture in neuroscience, *Neurosci. Res. Prog. Bull.* **15:**1–37.

Henderson, C. E., Camu, W., Mettling, C., Gouin, A., Poulsen, K., Karihaloo, M., Rullamas, J., Evans, T., McMahon, S. B., Armanini, M. T., Berkemeier, L., Phillips, H. S., and Rosenthal, A., 1993, Neurotrophins promote motor neuron survival and are present in embryonic limb bud, *Nature* **363:**266–270.

Hofstra, R. M. W., Landsvater, R. M., Ceccherini, I., Stulp, R. P., Stelwagen, T., Luo, Y., Pasini, B., Höppener, J. W. M., van Amstel, H. K. P., Romeo, G., Lips, J. M., and Buys, C. H. C. M., 1994, a mutation in the *RET* proto-oncogene associated with multiple endocrine neoplasia type 2B and sporadic medullary thyroid carcinoma, *Nature* **367:**375–376.

Hohn, A., Leibrock, J., Bailey, K., and Barde, Y.-A., 1990, Identification and characterization of a novel member of the nerve growth factor/brain-derived neurotrophic factor family, *Nature* **344:**339–341.

Indo, Y., Tsuruta, M., Hayashida, Y., Karim, M. A., Otha, K., Kawano, H., Awaya, Y., and Matsuda, I., 1996, Mutations in the *TRKA*/NGF receptor gene in patients with congenital insensitivity to pain with anhidrosis, *Nat. Genet.* **13:**485–488.

Ivanchuk, S. M., Myers, S. M., Eng, C., and Mulligan, L. M., 1996, *De nova* mutation of GDNF, ligand for the RET/GDNFR-α receptor complex, in Hirschsprung disease, *Hum. Mol. Genet.* **5:**2023–2026.

Jing, S., Wen, D., Yu, Y., Holst, P. L., Luo, Y., Tamir, R., Antonio, L., Hu, Z., Cupples, R., Louis, J.C., Hu, S., Altrock, B. W., and Fox, G. M., 1996, GDNF-induced activation of the ret protein tyrosine kinase is mediated by GDNFR-α, a novel receptor for GDNF, *Cell* **85:**1113–1124.

Jones, K. R., Fariñas, I., Backus, C., and Reichardt, L. F., 1994, Targeted disruption of the BDNF gene perturbs brain and sensory neuron development but not motor neuron development, *Cell* **76:**989–999.

Katz, L. C., and Shatz, C. J., 1996, Synaptic activity and the construction of cortical circuits, *Science* **274:**1133–1138.

Klein, R., Smeyne, R. J., Wrurst, W., Long, L. K., Auerbach, B. A., Joyner, A. L., and Barbacid, M., 1993, Targeted disruption of the *trk*B neurotrophin receptor gene results in nervous system lesion and neonatal death, *Cell* **75:**1–20.

Klein, R., Silos-Santiago, I., Smeyne, R. J., Lira, S. A., Brambilla, R., Bryant, S., Zhang, L., Snider, W. D., and Barbacid, M., 1994, Disruption of the neurotrophin-3 receptor gene *trk*C eliminates Ia muscle afferents and results in abnormal movements, *Nature* **368:**249–251.

Klein, R. D., Sherman, D., Ho, W.-H., Stone, D., Bennet, G. L., Moffat, B., Vandlen, R., Simmons, L., Gu, Q., Hongo, J.-A., Devaux, B., Poulsen, K., Armanini, M., Nozaki, C., Asai, N., Goddard, A., Phillips, H., Henderson, C. E., Takahashi, M., and Rosenthal, A., 1997, A GPI-linked protein that interacts with Ret to form a candidate neurturin receptor, *Nature* **387:**717–721.

Koliatsos, V. E., Clatterbuck, R. E., Winslow, J. W., Cayouette, M. H., and Price, D., 1993, Evidence that brain-derived neurotrophic factor is a trophic factor for motor neurons *in vivo, Neuron* **10:**359–367.

Kotzbauer, P. T., Lampe, P. A., Heuckeroth, R. O., Golden, J. P., Creedon, D. J., Johnson, E. M., and Milbrandt, J., 1996, Neurturin, a relative of glial-cell-line-derived neurotrophic factor, *Nature* **384:**467–470.

Lamballe, F., Smeyne, R. J., and Barbacid, M., 1994, Developmental expression of *trk*C, the neurotrophin-3 receptor, in the mammalian nervous system, *J. Neurosci.* **14:**14–28.

Lee, K.-L., Li, E., Huber, J., Landis, S. C., Sharpe, A. H., Chao, M. V., and Jaenisch, R., 1992, Targeted mutation of the gene encoding the low affinity NGF receptor p75 leads to deficits in the peripheral sensory nervous system, *Cell* **69:**737–749.

Leibrock, J., Lottspeich, F., Hohn, A., Hofer, M., Hengerer, B., Masiakowski, P., Thoenen, H., and Barde, Y.-A., 1989, Molecular cloning and expression of brain-derived neurotrophic factor, *Nature* **341:**149–152.

Levi-Montalcini, R., and Angeletti, P. U., 1966, Immunosympathectomy, *Pharmacol. Rev.* **18:**619–628.

Levi-Montalcini, R., and Booker, B., 1960, Destruction of the sympathetic ganglia in mammals by an antiserum to a nerve-growth protein, *Proc. Natl. Acad. Sci. USA* **46:**384–391.

Lewin, G. R., Ritter, A. M., and Mendell, L. M., 1992, On the role of nerve growth factor in the development of myelinated nociceptors. J. Neurosci. **12:**1896–1905.

Lin, L. F., Doherty, D. H., Lile, J. D., Bektesh, S., and Collins, F., 1993, GDNF: A glial cell line-derived neurotrophic factor for midbrain dopaminergic neurons, *Science* **260:**1130–1132.

Lindsay, R. M., 1996, Therapeutic potential of the neurotrophins and neurotrophin-CNTF combinations in peripheral neuropathies and motor neuron diseases, *in* "Growth Factors as Drugs for Neurological and Sensory Disorders" (Ciba Foundation Symposium 196), pp. 39–53, Wiley, Chichester.

Lindsay, R. M., and Yancopoulos, G. D., 1996, GDNF in a bind with known orphan: Accessory implicated in new twist, *Neuron* **17:**571–574.

Liu, X., Ernfors, P., Wu, H., and Jaenisch, R., 1995, Sensory but not motor neuron deficits in mice lacking NT-4 and BDNF, *Nature* **375:**238–241.

Martin-Zanca, D., Hughes, S. H., and Barbacid, M., 1986, A human oncogene formed by the fusion of truncated tropomyosin and protein tyrosine kinase sequences, *Nature* **319:**743–748.

Martin-Zanca, D., Barbacid, M., and Parada, L. F., 1990, Expression of the *trk* proto-oncogene is restricted to the sensory cranial and spinal ganglia of neural crest origin in mouse development, *Genes Dev.* **4:**683–694.

McAllister, A. K., Lo, D. C., and Katz, L. C., 1995, Neurotrophins regulate dendritic growth in developing visual cortex, *Neuron* **15:**791–803.

McAllister, A. K., Katz, L. C., and Lo, D. C., 1996, Neurotrophin regulation of cortical dendritic growth requires activity, *Neuron* **17:**1057–1064.

McMahon, S. B., and Priestley, J. V., 1995, Peripheral neuropathies and neurotrophic factors, Animal models and clinical perspectives, *Curr. Opin. Neurobiol.* **5:**616–624.

McMahon, S. B., Armanini, M. P., Ling, L. H., and Phillips, H. S., 1994, Expression and coexpression of trk receptors in subpopulations of adult primary sensory neurons projecting to identified peripheral targets, *Neuron* **12:**1161–1171.

Meakin, S. O., and Shooter, E. M., 1992, The nerve growth factor family of receptors, *TINS* **15:**323–331.

Merlio, J. P., Ernfors, P., Jaber, M., and Persson, H., 1992, Molecular cloning of rat *trk*C and distribution of cells expressing messenger RNAs for members of the *trk* family in the rat central nervous system, *Neurosci.* **51:**513–532.

Meyer-Franke, A., Kaplan, M. R., Pfrieger, F. W., and Barres, B. A., 1995, Characterization of the signaling interactions that promote the survival and growth of developing retinal ganglion cells in culture, *Neuron* **15:**805–819.

Minichiello, L., and Klein, R., 1996, TrkB and trkC neurotrophin receptors cooperate in promoting survival of hippocampal and cerebellar granule neurons, *Genes Dev.* **10:**2849–2858.

Minichiello, L., Piehl, F., Vazquez, E., Schimmang, T., Hokfelt, T., Represa, J., and Klein, R., 1995, Differential effects of combined trk receptor mutations on dorsal root ganglion and inner ear sensory neurons, *Development* **121:**4068–4075.

Molliver, D. C., Radeke, M. J., Feinstein, S. C., and Snider, W. D., 1995, Presence or absence of TrkA protein distinguishes subsets of small sensory neurons with unique cytochemical characteristics and dorsal horn projections, *J. Comp. Neurol.* **361:**404–416.

Moore, M. W., Klein, R. D., Fariñas, I., Sauer, H., Armanini, M., Phillips, H., Reichardt, L. F., Ryan, A. M., Carver-Moore, K., and Rosenthal, A., 1996, Renal and neuronal abnormalities in mice lacking GDNF, *Nature* **382:**76–79.

Mu, X., Silos-Santiago, I., Carroll, S. L., and Snider, W. D., 1993, Neurotrophin receptor genes are expressed in distinct patterns in developing dorsal root ganglia, *J. Neurosci.* **13:**4029–4041.

Oppenheim, R. W., Prevette, D., Quin-Wei, Y., and Yan, Q., 1992, Brain-derived neurotrophic factor (BDNF) rescues developing avian motorneurons from cell death, *Nature* **360:**755–757.

Pasini, B., Ceccherini, I., and Romeo, G., 1996, *RET* mutations in human diseases, *TIG* **12:** 138–144.

Pichel, J. G., Shen, L., Sheng, H. Z., Granholm, A.-C., Drago, J., Grinberg, A., Lee, E. J., Huang, S. P., Saarma, M., Hoffer, B. J., Sariola, H., and Westphal, H., 1996, Defects in enteric innervation and kidney development in mice lacking GDNF, *Nature* **382:**73–76.

Piñón, L. G. P., Minichiello, L., Klein, R., and Davies, A. M., 1996, Timing of neuronal death in *trk*A, *trk*B and *trk*C mutant embryos reveals development changes in sensory neuron dependence on Trk signalling, *Development* **122:**3255–3261.

Romeo, G., Ronchetto, P., Luo, Y., Barone, V., Seri, M., Ceccherini, I., Pasini, B., Bocciardi, R., Lerone, M., Kääriäinen, H., and Martucciello, G., 1994, Point mutations affecting the tyrosine kinase domain of the *RET* proto-oncogene in Hirschsprung's disease, *Nature* **367:**377–378.

Safieh-Garabedian, B., Poole, S., Allchorne, A., Kanaan, S., Saade, N., and Woolf, C. J., 1996, Zinc reduces the hyperalgesia and upregulation of NGF and IL-1 beta produced by peripheral inflammation in the rat, *Neuropharmacology* **35:**599–603.

Salomon, R., Attié, T., Pelet, A., Bidaud, C., Eng, C., Amiel, J., Sarnacki, S., Goulet, O., Ricour, C., Nihoul-Fékété, C., Munnich, A., and Lyonnet, S., 1996, Germline mutations of the RET ligand GDNF are not sufficient to cause Hirschsprung disease, *Nat. Genet.* **14:**345–347.

Sánchez, M. P., Silos-Santiago, I., Frisén, J., He, B., Lira, S. A., and Barbacid, M., 1996, Renal agencies and the absence of enteric neurons in mice lacking GDNF, *Nature* **382:**70–73.

Schimmang, T., Minichiello, L., Vázquez, E., San José, I., Giraldez, F., Klein, R., and Represa, J., 1995, Developing inner ear sensory neurons require TrkB and TrkC receptors for innervation of their peripheral targets, *Development* **121:**3381–3391.

Schuchardt, A., D'Agati, V., Larsson-Blomberg, L., Costantini, F., and Pachnis, V., 1994, Defects in the kidney and enteric nervous system of mice lacking the tyrosine kinase receptor ret, *Nature* **367:**380–383.

Sendtner, M., Holtmann, B., Kolbeck, R., Thoenen, H., and Barde, Y.-A., 1992, Brain-derived neurotrophic factor prevents the death of motorneurons in newborn rats after nerve section, *Nature* **360**:757–759.

Silos-Santiago, I., Molliver, D. C., Ozaki, S., Smeyne, R. J., Fagan, A. M., Barbacid, M., and Snider, W. D., 1995, Non-trkA-expressing small DRG neurons are lost in trkA deficient mice. *J. Neurosci.* **15**:5929–5942.

Silos-Santiago, I., Fagan, A. M., Garber, M., Fritzsch, B., and Barbacid, M., 1997, Severe sensory deficits but normal CNS development in newborn mice lacking trkB and trkC tyrosine protein kinase receptors, *Eur. J. Neurosci.* **9**:2045–2056.

Smeyne, R. J., Klein, R., Schnapp, A., Long, L. K., Bryant, S., Lewin, A., Lira, S., and Barbacid, M., 1994, Severe sensory and sympathetic neuropathies in mice carrying a disrupted Trk/NGF receptor gene, *Nature* **368**:246–249.

Snider, W. D., 1994, Functions of the neurotrophins during nervous system development: What the knockouts are teaching us, *Cell* **77**:627–638.

Snider, W. D., and Johnson, E. M. Jr., 1989, Neurotrophic molecules, *Ann. Neurol.* **26**:489–506.

Snider, W. D., and Silos-Santiago, I. S., 1996, Dorsal root ganglion neurons require functional neurotrophin receptors for survival during development, Philos. Trans. R. Soc. Lond. Ser. B **351**:395–403.

Tessarollo, L., Tsoulfas, P., Martín-Zanca, D., Gilbert, D. J., Jenkins, N. A., Copeland, N. G., and Parada, L. F., 1993, trkC, a receptor for neurotrophin-3, is widely expressed in the developing nervous system and in non-neuronal tissues, *Development* **118**:463–475.

Tessarollo, L., Vogel, K. S., Palko, M. E., Reid, S. W., and Parada, L. F., 1994, Targeted mutation in the neurotrophin-3 gene results in loss of muscle sensory neurons, *Proc. Natl. Acad. Sci. USA* **91**:11844–11848.

Thoenen, H., 1995, Neurotrophins and neuronal plasticity, *Science* **270**:593–598.

Tomlinson, D. R., Fernyhough, P., and Diemel, L. T., 1996, Neurotrophins and peripheral neuropathy, *Philos. Trans. R. Soc. Lond. Ser. B* **351**:455–462.

Treanor, J. J. S., Goodman, L., de Sauvage, F., Stone, D. M., Poulsen, K. T., Beck, C. D., Gray, C., Armanini, M., Pollock, R. A., Hefti, F., Phillips, H. S., Goddard, A., Moore, M. W., Buj-Bello, A., Davies, A. M., Asai, N., Takahashi, M., Vandlen, R., Henderson, C. E., and Rosenthal, A., 1996, Characterization of a multicomponent receptor for GDNF, *Nature* **382**:80–83.

Trupp, M., Arenas, E., Fainzilber, M., Nilsson, A.-S., Sieber, A.-A., Grigoriou, M., Kilkenny, C., Salazar-Grueso, E., Pachnis, V., Arumäe, U., Sariola, H., Saarma, M., and Ibañez, C. F., 1996, Functional receptor for GDNF encoded by the *c-ret* proto-oncogene, *Nature* **381**:785–789.

Wilkinson, G. A., Fariñas, I., Backus, C., Yoshida, C. K., and Reichardt, L. F., 1996, Neurotrophin-3 is a survival factor *in vivo* for early mouse trigeminal neurons, *J. Neurosci.* **16**:7661–7669.

Woolf, C. J., Safieh-Garabedian, B., Ma, Q. P., Crilly, P. , and Winter, J., 1994, Nerve growth factor contributes to the generation of inflammatory sensory hypersensitivity, *Neuroscience* **62**:327–331.

Woolf, C. J., Ma, Q. P., Allchorne, A., and Poole, S., 1996, Peripheral cell types contributing to the hyperalgesic action of nerve growth factor in inflammation, *J. Neurosci.* **15**:2716–2723.

Woolf, C. J., Allchorne, A., Safieh-Garbedian, B., and Poole, S., 1997, Cytokines, nerve growth factor and inflammatory hyperalgesia: The contribution of tumor necrosis factor alpha, *Br. J. Pharmacol.* **121**:417–424.

Wright, D. E., and Snider, W. D., 1995, Neurotrophin receptor mRNA expression defines distinct populations of neurons in rat dorsal root ganglion, *J. Comp. Neurol.* **361**:321–333.

Yan, Q., Elliott, J., and Snider, W. D., 1992, Brain-derived neurotrophic factor rescues spinal motor neurons from axotomy-induced cell death, *Nature* **360**:753–755.

Yan, Q., Elliott, J. L., Matheson, C., Sun, J., Zhang, L., Mu, X., Rex, K. L., and Snider, W. D., 1993, Influences of neurotrophins on mammalian motoneurons *in vivo, J. Neurobiol.* **24**:1555–1577.

CHAPTER 6

TRANSGENIC MICE WITH NEUROFILAMENT ABNORMALITIES

CARMEN LAMPRON, SÉBASTIEN COUILLARD-DESPRÉS, and JEAN-PIERRE JULIEN

1. INTRODUCTION

Abnormal accumulation of neurofilaments in the cell body and proximal axon of motor neurons (often called spheroids or Lewy bodies) is a pathological hallmark of amyotrophic lateral sclerosis (ALS), an adult-onset motor neuron disease (Carpenter, 1968; Hirano *et al.*, 1984; Schmidt *et al.*, 1987). There are two forms of ALS, the familial (FALS; 10% of cases) and sporadic forms (90% of cases). Mutations in the copper/zinc superoxide dismutase (*SOD1*) gene were found to be responsible for ~20% of familial cases (Brown, 1995). The mechanism in-

---

CARMEN LAMPRON, SÉBASTIEN COUILLARD-DESPRÉS, and JEAN-PIERRE JULIEN • *Centre for Research in Neuroscience, McGill University, The Montréal General Hospital Research Institute, Montréal H3G 1A4, Canada.*

Mouse Models in the Study of Genetic Neurological Disorders, Volume 9 of *Advances in Neurochemistry,* edited by Popko, Kluwer Academic/Plenum Publishers, New York, 1999.

volved in selective death of motor neurons by *SOD1* mutations is not yet fully understood, and for more than 95% of ALS cases the etiologies remain unknown. The striking similarities in the clinical and pathological features of sporadic ALS and FALS suggest that similar mechanisms of disease may occur. Although neurofilament accumulations have been widely viewed as a consequence of neurodegeneration, recent transgenic mice and genetic studies provided evidence for neurofilament involvement in motor neuron disease. Further support for a key role of neurofilaments in ALS came from the discovery of codon deletions in the phosphorylation domain of the neurofilament heavy protein (NF-H) associated with some sporadic cases. Here we review the phenotypes occurring in various transgenic mouse models with neurofilament depositions and discuss the possible mechanisms underlying neurofilament-induced pathology.

2. NEUROFILAMENT STRUCTURE AND FUNCTION

Neurofilaments are formed by the copolymerization of three intermediate filament proteins, NF-L (61 kDa), NF-M (90 kDa), and NF-H (110 kDa) (Hoffman and Lasek, 1975). The three neurofilament subunits share with other members of the intermediate filament family a central domain of approximately 310 amino acids, which is involved in the formation of coiled-coil structures (Fig. 1). The current model of a 10-nm filament structure is that two coiled-coil dimers of protein subunits line up in a staggered fashion to form an antiparallel tetramer (Steinert and Roop, 1988). Eight tetramers packed together in a helical array are required to make a ropelike 10-nm filament. Neurofilaments are obligate heteropolymers requiring NF-L with either NF-M or NF-H for polymer formation (Ching and Liem, 1993; Lee *et al.,* 1993). There is evidence for the existence of two types of heterotetramers in neurofilaments, one containing NF-L and NF-M and the other containing NF-L and NF-H (Cohlberg *et al.,* 1995).

Although neurofilament proteins are present in the vast majority of neurons in the nervous system, they are particularly abundant in large myelinated axons of the peripheral nervous system that originate from motor and sensory neurons. In addition to conferring a mechanical support to the cell, neurofilaments play a role in mediating the caliber of large myelinated axons, and they are thereby a determination of the conduction velocity. The unequivocal proof of neurofilament involvement in the control of axonal caliber was recently provided from the analysis of animals lacking axonal neurofilaments including a quivering quail mutant deficient in NF-L protein (Yamasaki *et al.,* 1991), transgenic mice expressing a *NF-H/lacZ* fusion construct (Eyer and Peterson, 1994), and NF-L deficient mice that were recently generated in our laboratory (Zhu *et al.,* 1997). All these

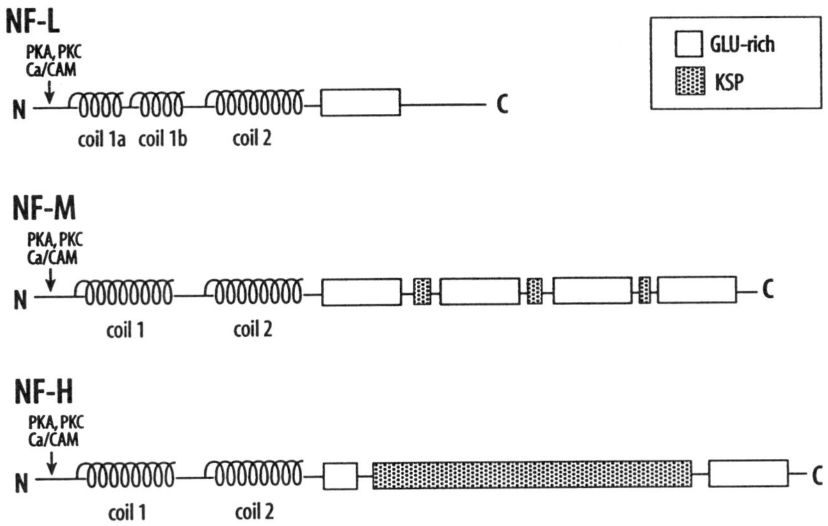

FIGURE 1. Structure of neurofilament proteins. The three neurofilament subunits have a central domain of approximately 310 amino acids that is involved in the formation of coiled-coil structures. The current model of an intermediate filament is that two coiled-coil dimers of protein subunit line up in a staggered fashion to form an antiparallel tetramer. Eight tetramers packed together are required to make a ropelike 10-nm filament. Neurofilaments are obligate heteropolymers. There is evidence for the existence of two types of heterotetramers in neurofilaments, one containing NF-L and NF-M and the other containing NF-L and NF-H.

animals showed a dramatic hypotrophy of axons due to the scarcity of neurofilaments. Our analysis of ventral root axons in mice lacking NF-L protein revealed a reduction of two- to threefold in the caliber of myelinated axons (Fig. 2).

There is evidence that the axonal caliber can be modulated not only by the number of neurofilaments but also by local changes in neurofilament phosphorylation. The majority of phosphorylation sites are located in the long carobxy terminal tail domain of the NF-M and NF-H proteins that form side-arm projections at the periphery of the filament (Hisanaga and Hirokawa, 1990). Moreover, the tail domain of NF-H is quite remarkable, with a high content of charged amino acids and multiple repeats of Lys–Ser–Pro (KSP) that account for the unusually high ratio of phosphoserine residues in this protein (Julien and Mushynski, 1982; Julien *et al.*, 1988; Lees *et al.*, 1988). Support for this view comes from *Trembler* mice and hypomyelinating transgenic mice in which reduced levels of NF-H phosphorylation resulted in a decrease in the axon caliber (de Waegh *et al.*, 1992; Cole *et al.*, 1994).

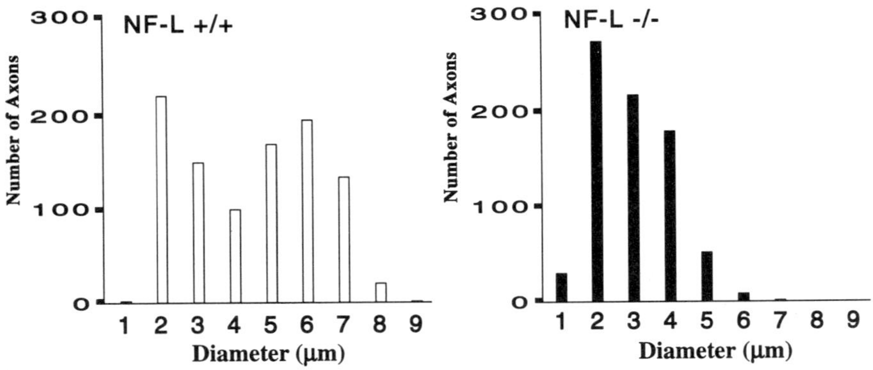

FIGURE 2. Decreased axonal caliber of myelinated axons from the ventral root of mice with targeted disruption of the *NF-L* gene.

3. TRANSGENIC MOUSE MODELS WITH NEUROFILAMENT ACCUMULATIONS

In the past few years, several transgenic mouse models bearing neurofilament accumulations have been generated by different laboratories. It is noteworthy that abnormal neurofilament accumulations can be formed by overexpressing any single neurofilament protein, as illustrated in Table 1. Our laboratory generated several transgenic mouse lines expressing human NF-H proteins by microinjecting large genomic clones bearing the normal human *NF-H* gene. Two normal NF-H alleles exist in the human population; one allele bears 43 KSP phosphorylation repeats and the other allele has 44 KSP repeats (Figlewicz *et al.,* 1994). Both alleles were found capable to induce motor neuron disease when overexpressed in transgenic mice. A modest two- to threefold overexpression of the human NF-H protein with 43 KSP repeats in mice provoked a late-onset motor neuron disease, characterized by the presence of aberrant neurofilament accumulations in spinal motor neurons (Côté *et al.,* 1993). Figure 3 shows similar neurofilamentous swellings in spinal motor neurons from transgenic mice overexpressing the *NF-H* gene coding for 44 KSP repeats. Motor neuron dysfunction in NF-H transgenic mice progresses with age and is accompanied by the atrophy and slow degeneration of distal axons, resulting in secondary muscle degeneration (Collard *et al.,* 1995). Our interpretation is that an excess of human NF-H protein changes neurofilament properties, resulting in a retardation of axonal transport and in the aberrant aggregation of neurofilaments in cell body and proximal axon.

Surprisingly, transgenic mice overexpressing a four- to fivefold increase in the wild-type mouse NF-H protein did not develop a motor neuron disease

TABLE 1. Transgenic Mice with Abnormal Neurofilament Accumulations

Transgene	Neuronal populations affected	Axonal degeneration	References
Overexpression of neurofilament protein			
Human NF-H	• Spinal motor neurons and DRG neurons	Yes, motor axons	Côté *et al.*, 1993; Collard *et al.*, 1995
Mouse NF-H	• Spinal motor neurons and DRG neurons	No	Marszalek *et al.*, 1996
Human NF-M	• Cortical neurons and forebrain neurons	?	Vickers *et al.*, 1994
MSV/mouse NF-M	• Spinal motor neurons and DRG neurons	No	Wong *et al.*, 1995a
MSV/mouse NF-L	• Spinal motor neurons and DRG neurons	Yes, motor axons	Xu *et al.*, 1993
Human NF-L	• Thalamic neurons and cortical neurons	?	Ma *et al.*, 1995
Low level expression of mutant neurofilament protein			
MSV/mutant NF-L	• Spinal motor neurons and DRG neurons	Yes, motor axons	Lee *et al., 1994*
Mouse NF-H/lacZ	Perikarya of CNS and PNS neurons	No	Eyer and Peterson, 1994
Other mouse models			
SOD1 mutant hG93A	• Prominent vacuoles in motor neurons and spheroids in proximal motor axons	Yes motor axons	Gurney *et al.*, 1994; Tu *et al.*, 1996
BAPG1 knockout or *(dt/dr)*	• Sensory neurons	Yes, DRG axons	Yang *et al.*, 1996

(Marszalek *et al.*, 1996). Nonetheless, such overexpression of the mouse NF-H proteins resulted in a retarded axonal transport of neurofilaments, accumulation of neurofilaments in motor neurons, and atrophy of myelinated axons (Marszalek *et al.*, 1996). There are considerable sequence differences between the phosphorylation tail domain of the mouse and human NF-H proteins that may account for the different effects of these two proteins. Mouse NF-H protein has more KSP repeats but its number of KSPXK, a consensus sequence for the cdk5 kinase, occurs only 9 times whereas in human NF-H it occurs 34 times (Pant and Veeranna, 1995). Furthermore, the human NF-H exhibits a high degree of homology with the NF-H from rabbit, an animal species highly susceptible to neurofilamentous pathology induced by aluminum (Strong, 1994). Thus, the enhanced deleterious effect of the human NF-H in mice is likely due to the properties of its

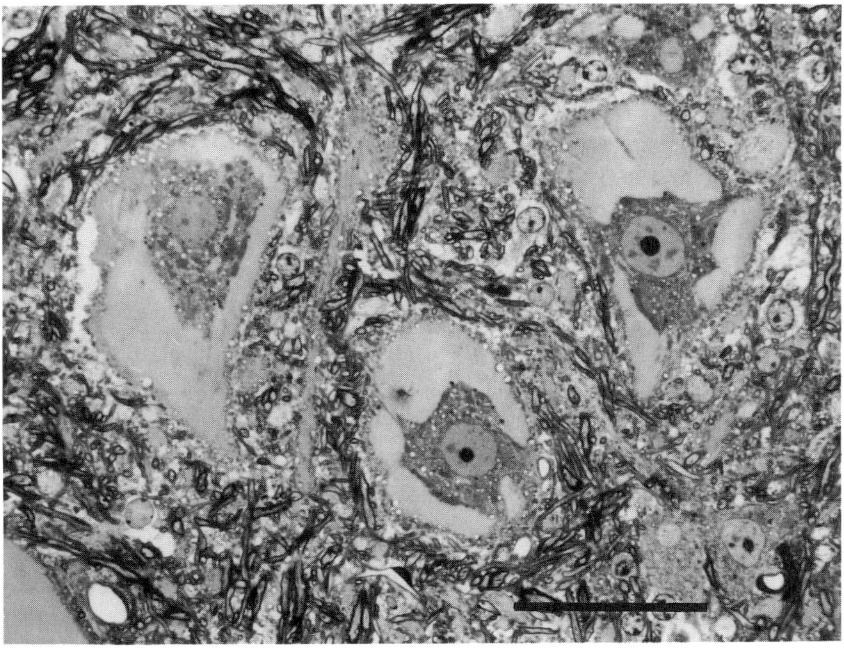

FIGURE 3. Light microscopy showing the accumulation of neurofilaments in spinal motor neurons of transgenic mice overexpressing the human gene coding for the NF-H protein having 44 KSP repeats. Bar = 50 μm.

tail phosphorylation domain, a region that can affect filament packing density and axonal transport (de Waegh *et al.*, 1992).

High-level expression of the wild-type mouse NF-L protein in transgenic mice achieved by the use of the strong viral promotor from murine sarcoma virus (MSV) provoked an early-onset motor neuron disease accompanied by massive accumulation of neurofilaments in spinal motor neurons and by muscle atrophy (Xu *et al.*, 1993). A more severe phenotype was produced by expressing an assembly-disrupting NF-L mutant having a leucine to proline substitution residue near the end of the conserved rod domain (Lee *et al.*, 1994). Although no such NF-L mutations have been reported in human ALS, similar mutations in keratins are the cause of a severe form of genetic skin diseases (Fuchs, 1994). Expression of this mutant NF-L protein at only 50% the endogenous NF-L levels was sufficient to induce within 4 weeks massive neurofilament accumulations in spinal motor neurons and was accompanied by death of motor neurons, neuronophagia, and severe denervation atrophy of the skeletal muscle.

These transgenic studies with the mouse *NF-L* and human *NF-H* genes provided the first demonstration that neurofilament abnormalities can provoke neuronal degeneration and death. But the cellular selectivity is remarkable. Degeneration is restricted to spinal motor neurons even though the NF-L or NF-H transgenes are expressed throughout the nervous system. However, it is noteworthy that these transgenic mouse models are not absolute replicas of human ALS in that motor neurons from the brain cortex did not develop neurofilamentous pathology. It is likely that this is due to the smaller size and thereby lower neurofilament content of mouse cortex neurons as compared to the corresponding neurons in human. Clearly, additional unknown cellular factors may also contribute to the deleterious effects of neurofilament deposits because large neurofilament accumulations were well tolerated within dorsal root ganglia (DRG) neurons (Côté *et al.,* 1993; Xu *et al.,* 1993; Lee *et al.,* 1994).

As shown in Table 1, some types of neurofilament depositions appear to be less toxic than others. Not all transgenic mouse lines with neurofilament abnormalities developed motor neuron disease. For instance, no overt phenotypes occurred in transgenic mice expressing a mouse *NF-H/lacZ* fusion gene even though massive neurofilament aggregates were detected in neuronal perikarya throughout the nervous system (Eyer and Peterson, 1994). Also, large neurofilamentous swellings in DRG and motor neurons of transgenic mice overexpressing a NF-M transgene were found, but no axonal degeneration was observed (Wong *et al.,* 1995a). Moreover, transgenic mice with moderate overexpression of the human *NF-L* (Ma *et al.,* 1995) or *NF-M* genes (Vickers *et al.,* 1994) appeared normal, although a detailed analysis of the mice revealed the presence of abnormal perikaryal immunoreactivity in some populations of brain neurons. It has not yet been documented whether these accumulations of neurofilaments cause neurodegeneration with age.

4. DEFECTIVE AXONAL TRANSPORT IN NF-H TRANSGENIC MICE

Axonal transport studies with transgenic mice overexpressing human or mouse NF-H proteins provided clues on how disorganized neurofilaments can contribute to disease. Pulse-labeling with [^{35}S]methionine of spinal motor neurons in NF-H transgenics demonstrated that an excess of NF-H protein (either human or mouse NF-H) slowed down the axonal transport rates of neurofilament proteins (Collard *et al.,* 1995; Marszalek *et al.,* 1996). In addition, impairment of axonal transport in motor neurons from mice overexpresssing the human NF-H protein was also evident for tubulin and actin, two cytoskeletal proteins of the slow moving components (Collard *et al.,* 1995). Further microscopic analysis of these transgenic mice confirmed a paucity of cytoskeletal elements, mitochondria and smooth endoplasmic reticulum in degenerating motor axons (Collard *et al.,* 1995). From these data, it has been proposed that the abnormal accumulation of neurofilaments can interfere with the normal delivery of components required for axonal maintenance and that a similar mechanism may contribute to disease in human ALS (Collard *et al.,* 1995; Cleveland, 1996). This mechanism of disease, illustrated in Fig. 4, is in agreement with the recent report of a marked increase of neurofilaments, mitochondria, or lysosomes in the axon hillock of motor neurons in ALS patients whereas a decreased content of cytoplasmic organelles was detected in the initial axon segment from chromatolytic neurons (Sasaki and Iwata, 1996).

The exact mechanism by which disorganized neurofilaments can interfere with axonal transport is not fully understood. Moreover, it is unclear why some types of neurofilamentous accumulations are more detrimental than others. As mentioned previously, in some transgenic mouse models the presence of large neurofilament aggregates in neuronal perikarya was well tolerated. A simplistic view is that some types of neurofilament accumulations can physically impede axonal transport. However, no simple rule can be set as intoxication by β,β′-iminodipropionitrile (IDPN) also induces neurofilamentous accumulations in proximal motor axons, yet without any effect on fast axonal transport (Griffin *et al.,* 1983). An alternative explanation is that disorganized neurofilaments could

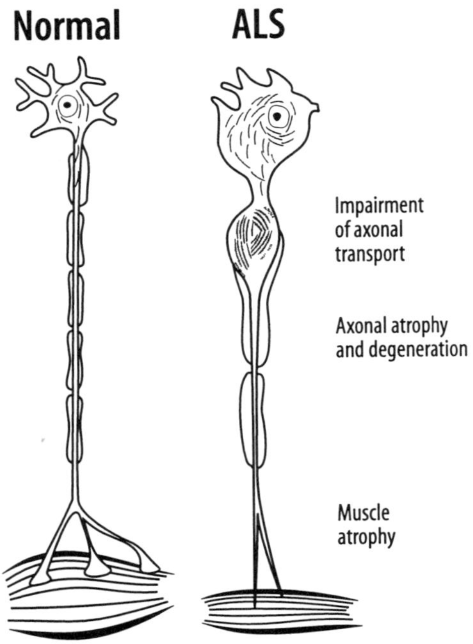

FIGURE 4. Model of motor neuron degeneration by neurofilament-induced disruption of axonal transport. The abnormal accumulation of neurofilaments can interfere with the delivery of components required for axonal maintenance including cytoskeletal elements and mitochondria.

alter microtubule functions in axonal transport. NF-H can bind *in vitro* the carboxy-terminal of tubulin but the *in vivo* significance of these interactions remains to be investigated (Miyasaka *et al.,* 1993).

5. FACTORS THAT CAN POTENTIALLY INDUCE THE ACCUMULATION OF NEUROFILAMENTS

Impairment of axonal transport by disorganized neurofilaments is a pathological mechanism consistent with several aspects of ALS. It can explain in part the cellular selectivity of the disease. Large motor neurons represent a class of neurons more susceptible to develop neurofilament abnormalities because of their high content of neurofilament proteins. Moreover, there is a retardation in the slow axonal transport of cytoskeletal elements during aging (McQuarrie *et al.,* 1989), a factor that can predispose to the disease. Although ALS is a disease of many etiological factors, neurofilament depositions may be key intermediates contributing to neurodegeneration. Many factors can potentially lead to the accumulation of neurofilaments including deregulation of neurofilament protein synthesis, defective transport, abnormal phosphorylation and proteolysis, and other protein modifications. A list of potential factors is provided in Table 2.

Further evidence for neurofilament involvement in ALS came from the discovery of codon deletions in the KSP phosphorylation domain of the human *NF-H.* Figlewicz and her colleagues (1994) have identified two deletion mutations in the *NF-H* gene from 5 of 356 sporadic cases analyzed. The results are summarized in Fig. 5. Recently, another laboratory (A. Al-Chalabi, J. F. Powel, C. Russ, and P. N. Leigh, London, personal communication) also detected two novel deletions in this region of the NF-H tail in two out of 196 ALS patients. Again, no NF-H variant was detected by this group in 188 controls. The combined data indicate that *NF-H* mutations may account for a small percentage of ALS cases (~1.3%). So far, the search for mutations in the KSP repeat domain of NFH (Rooke *et al.,* 1996) or in all coding regions of NF-L, NF-M, or NF-H (Vechio *et al.,* 1996) have failed to reveal mutations linked to disease in > 100 FALS cases. Therefore, one can conclude that mutations in neurofilament genes are responsible for only a very small fraction of ALS cases.

Although the mechanism underlying disease induced by NF-H mutations remains to be elucidated, our working hypothesis is that codon deletions in the NF-H tail domain could alter the properties of neurofilaments. It is possible that changes in the KSP repeats of NF-H could alter the cross-bridging between neurofilaments, resulting in a tendency to form aggregates. Alternatively, the mutations could modify the normal interactions of NF-H tail domain with microtubules (Miyasaka *et al.,* 1993), resulting in an impairment of axonal transport. To address this problem, work is now in progress in our laboratory to generate

TABLE 2. Factors That Can Potentially Induce Neurofilament Accumulations

Factor	Evidence	References
Deregulation of neurofilament gene expression	• Transgenic mice overexpressing NF-L, NF-M, or NF-H	Côté *et al.*, 1993; Xu *et al.*, 1993; Lee *et al.*, 1994; Marszalek *et al.*, 1996
	• Decreased ratio of *NF-L* to *NF-H* mRNAs in a dog model HCSMA	Muma and Cork, 1993
Neurofilament gene mutations	• Codon deletions in the *NF-H* gene of ALS patients	Figlewicz *et al.*, 1994
	• Severe motor neuron disease in transgenic mice expressing a NF-L assembly-disrupting mutant	Lee *et al.*, 1994
Alterations in neurofilament-associated proteins	• Degeneration of sensory neurons in mice lacking a BPAG1 isoform	Yang *et al.*, 1996
Altered neurofilament phosphorylation	• Aggregation of neurofilaments modulated by PKC inhibitor	Carter *et al.*, 1996
	• Hyperphosphorylation of NF-H protein in neuronal perikarya by stress-activated protein kinase	Giasson and Mushynski, 1996
SOD1 mutations	• Neurofilament inclusions in FALS patients and in transgenic mice expressing *SOD1* mutants	Rouleau *et al.*, 1996; Tu *et al.*, 1995
Toxic agents such as IDPN, hexanedione, and acrylamide	• Neurofilament accumulations in animal models and cultured cells	Griffin *et al.*, 1983
Aluminum intoxication	• Neurofibrillary pathology in rabbits	Strong, 1994

transgenic mice expressing human NF-H mutants into a null mouse NF-H background.

Mutations in the KSP-repeat domain emphasize the importance of the NF-H phosphorylation domain. In four ALS cases, loss of a lysine residue in the sequence SPVKEE converted this consensus phosphorylation sequence for the cdk5 kinase (Pant and Veeranna, 1995) into a phosphorylation site pertinent to a stress-activated kinase (SAPK) (Giasson and Mushynski, 1996). Agents that activate SAPKγ are capable of inducing abnormal hyperphosphorylation of the NF-H protein in perikarya of cultured rat DRG neurons (Giasson and Mushynski, 1996). This finding raises the possibility that phosphorylation due to stress may lead to neurofilament-induced pathology. However, it remains to be established

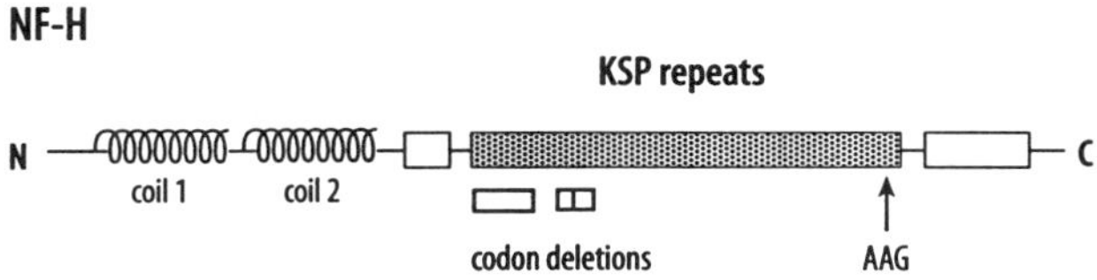

Normal:	43 KSP
Normal:	44 KSP
ALS	(1case): 102 bp deletion including 5 KSP
ALS	(4 cases): 3 bp deletion coding for K in SPVKEE
ALS	(1 case): 24 bp deletion
ALS	(1 case): 18 bp deletion

FIGURE 5. Codon deletions in the *NF-H* gene of some ALS patients. There are two normal forms of the human *NF-H* gene, one with 43 KSP and the other with 44 KSP repeats (Figlewicz *et al.,* 1994). To date, the combined results from the screening of 562 sporadic ALS cases by two groups (our group and Leigh's group in London) revealed codon deletions in the *NF-H* gene of seven unrelated ALS patients diagnosed as sporadic cases. In three ALS patients, the deletions consisted of 102 bp, 24 bp, and 18 bp resulting in the loss of one or more KSP phosphorylation motifs. Four other ALS cases shared the same 3 bp deletion for the codon AAG (K) in the phosphorylation sequence SPVKEE. No such mutations were found in ~500 control samples.

that activation of SAPKγ can provoke *in vivo* abnormal phosphorylation and accumulation neurofilaments.

Disorganization of neurofilaments can also result from defects in proteins interacting with neurofilaments as revealed recently by the analysis of mice lacking the neuronal isoform of BPAG1, a protein that crosslinks actin and neurofilaments (Yang *et al.,* 1996). The BPAG1 tail was found to selectively associate with the NF-H protein in transfection studies in COS cells. The BPAG1 belongs to a group of large coiled-coil proteins that includes desmoplakin and plectin. A neuronal splice form, BPAG1n, is expressed specifically in sensory neurons. The targeted disruption of the BPAG1 gene in mice produced a selective degeneration and death of sensory neurons identical to *dystonia musculorum* (*dt/dt*), a well-known autosomal recessive mouse mutant. The BPAG1 deficient or *dt/dt* mice are characterized by the presence of abnormal neurofilament aggregates in sensory axons (Yang *et al.,* 1996). A defect in axonal transport has been suggested as a possible mechanism of disease, but it remains unclear whether neurofilament disorganization directly contributes to the sensory neuron degenerative process in the BPAG1 mutant mouse. The *wobbler* mouse is another mutant of progressive neurodegeneration that also develops neurofilament accumulations in spinal motor neurons (Table 1). The *wobbler* gene has not yet been identified but it does not map to any of the neurofilament gene loci (des Portes *et al.,* 1994). Of particular interest is the recent finding of a three- to fourfold increase of NF-M mRNA levels in the *wobbler* mouse affected neurons (Pernas-Alonso, 1996). This mouse mutant will provide another useful model to investigate neurofilaments as possible intermediates of disease.

6. A LINK BETWEEN *SOD1* AND NEUROFILAMENTS?

Mutations in the copper/zinc superoxide dismutase (*SOD1*) gene coding for a ubiquitous metalloprotein are responsible for ~20% of FALS cases (Brown, 1995). Copper is the essential cofactor of *SOD1* for the conversion of the superoxide anion ($\cdot O_2^-$) to hydrogen peroxide (H_2O_2), which is then detoxified to water and oxygen by catalase and glutathione peroxidase. So far, more than 40 different *SOD1* mutations have been identified in FALS, with the majority inherited in an autosomal dominant fashion (Brown, 1995). Most *SOD1* mutations are missense point mutations and the current view is that they cause ALS through a mechanism involving not a loss of superoxide dismutase activity but rather a gain of detrimental function (Brown, 1995). Evidence in favor of this view was provided from transgenic mouse studies. Transgenic mice overexpressing the G93A (Gurney *et al.*, 1994), G37R (Wong *et al.*, 1995b), and G85R (Ripps *et al.*, 1995) *SOD1* mutants developed motor neuron disease, whereas SOD activities in these mice were not reduced. In addition, mice homozygous for the targeted disruption of the *SOD1* gene did not develop motor neuron disease (Reaume *et al.*, 1996).

To date, three potential mechanisms have been proposed to explain the toxicity of *SOD1* mutants. Beckman and his colleagues (1993) suggested a mechanism in which the mutations would render the copper in the active site of *SOD1* more accessible to peroxynitrite to form a nitroniumlike intermediate that can nitrate proteins on tyrosine residues. In another mechanism, *SOD1* mutations would increase the enzyme peroxidase activity and produce more hydroxyl radicals from hydrogen peroxide as substrate than the wild-type *SOD1* (Wideau-Pazos *et al.*, 1996). A third mechanism is inferred from the discovery that *SOD1* mutations can alter the activity of calcineurin, a phosphatase under the control of Ca^{2+} and calmodulin. Wang and his colleagues (1996) have recently reported that normal *SOD1* protects calcineurin from inactivation, probably by preventing oxidation of the iron at its catalytic site. Their results suggest that *SOD1* could play a role in signal transduction phosphorylation cascades. Therefore, changes in phosphatase activity or increases in protein nitration or in peroxidase activity resulting from *SOD1* mutations are three nonmutually exclusive mechanisms that can contribute to damage of the various cellular components. But why do *SOD1* mutations selectively provoke degeneration of motor neurons in *SOD1*-induced ALS?

There is growing evidence for a link between *SOD1* mutations and the formation of neurofilament depositions. The presence of neurofilament accumulations in motor neurons has recently been described in FALS cases with different *SOD1* mutations (Rouleau *et al.*, 1996) and neurofilament inclusions resembling those found in ALS have been observed in transgenic mice expressing *SOD1* mutants (Table 1; Tu *et al.*, 1996). Moreover, it is logical that abundant proteins with long half-lives in motor neurons, such as neurofilament proteins, would be favored targets of damage by *SOD1* mutants. Support for the nitration hypothesis was provided by immunohistochemical detection of nitrotyrosine residues in NF-L associated with neurofilamentous spheroids within motor neurons of FALS cases (Chou *et al.*, 1996). It was hypothesized that tyrosine nitration of NF-L may impede its assembly and lead to the formation of disorganized filaments (Chou *et al.*, 1996). Reactive nitrogen species such as peroxynitrite might also create crosslinks by the formation of dityrosine and thereby induce neurofilament aggregation. Other types of oxidative damage by altered *SOD1* activity are also possible. For instance, protein crosslinks may occur through a copper-mediated oxidation of sulfhydryl groups or a production of carbonyls on lysine residues. Indeed, carbonyl-related modifications of NF-H protein has been reported in the neurofibrillary pathology of Alzheimer's disease (Smith *et al.*, 1995) but it remains to be demonstrated in ALS. Finally, the recent finding that calcineurin activity is modulated in part by *SOD1* raises the interesting possibility that *SOD1* mutations could lead to anomalous phosphorylation of neurofilament proteins.

7. PROSPECTS

The recent lines of evidence for a key role of neurofilaments in motor neuron disease should stimulate research on the potential factors and signaling pathways that could lead to their abnormal accumulations in neurons. Factors that can potentially induce neurofilament disorganization are described in Table 2. Also of particular importance are the current investigations aiming to clarify the possible oxidative modifications or other posttranslational modifications of neurofilament proteins by *SOD1* mutants. Another aspect that merits further attention is the apparent paradox that *NF-L* mRNA levels in ALS are reduced by 60% in motor neurons with neurofilament accumulations (Bergeron *et al.*, 1994). A decreased ratio of *NF-L* to *NF-H* mRNA levels has also been reported in a dog model of motor neuron disease (Muma and Cork, 1993). These findings raise the possibility that alterations in the stoichiometry of the NF-L, NF-M, and NF-H subunits might lead to disorganization of neurofilaments and perhaps to disease. Transgenic mouse studies recently published by Couillard-Després *et al.* (1998) and by Williamson *et al.* (1998) demonstrate that the levels of neurofilament proteins can affect the progression of disease caused by SOD1 mutations. In this regard, the use of knockout mice for neurofilament genes should provide a powerful approach to study how reduced levels of subunits can affect neurofilament metabolism and neuronal function.

Although current evidence suggests that toxicity of neurofilament depositions involves impairment of axonal transport, it remains to be explored why certain types of neurofilament accumulations are more detrimental than others in specific cell types. The species differences in the *NF-H* properties and the location of mutations found in the *NF-H* gene of some ALS patients emphasizes the importance of the KSP phosphorylation domain. In this regard, the transgenic mouse approach should be useful to address the effects of these mutations on the organization and metabolism of neurofilaments. In addition, mating of *NF-L* deficient mice that lack neurofilaments with the various mouse models of neurodegeneration, such as transgenic mice expressing *SOD1* mutants, BPAG1 knockout mice, and *wobbler* mice, should provide a unique approach to determine to what extent neurofilaments contribute to diseases of different primary causes. It is hoped that further insights into the mechanisms of neurofilament-induced pathologies will lead to the development of novel therapeutic strategies.

ACKNOWLEDGMENTS

We thank Dr. N. Leigh (London) for communicating his unpublished results of ALS screening and Dr. Jürgen Meier for reading the manuscript. This work was supported by the Medical Research Council of Canada (MRC) and the ALS association (USA). J.-P. J. has a MRC senior scholarship.

REFERENCES

Beckman, J. S., Carson, M., Smith, C. D., and Koppenol, W. H., 1993, ALS, SOD and peroxynitrite, *Nature* **364:**584.

Bergeron, C., Beric-Maskarel, K., Muntasser, S., Weyer, L., Somerville, M. J., and Percy, M., 1994, Neurofilament light and polyadenylated mRNA levels are decreased in amyotrophic lateral sclerosis motor neurons, *J. Neuropathol. Exp. Neurol.* **53:**221–230.

Brown, R. H., 1995, Amyotrophic lateral sclerosis: Recent insights from genetics and transgenic mice, *Cell* **80:**687–692.

Carpenter, S., 1968, Proximal axonal enlargement in motor neuron disease, *Neurology* **18:**841–851.

Carter, J. E., Gallo, J.-M., Anderson, V. E. R., Anderton, B. H., and Robertson, J., 1996, Aggregation of neurofilaments in NF-L transfected neuronal cells. Regeneration of the filamentous network by a protein kinase C inhibitor, *J. Neurochem.* **67:**1997–2004.

Ching, G. Y., and Liem, R., 1993, Assembly of type IV neuronal intermediate filaments in nonneuronal cells in the absence of preexisting cytoplasmic intermediate filaments, *J. Cell Biol.* **122:**1323–1335.

Chou, S. M., Wang, H. S., and Komai, K., 1996, Colocalization of NOS and SOD1 in neurofilament accumulation within motor neurons of amyotrophic lateral sclerosis: An immunohistochemical study, *J. Chem. Neuroanat.* **10**:249–258.

Cleveland, D. W., 1996, Neuronal growth and death: Order and disorder in the axoplasm, *Cell* **84**:663–666.

Cohlberg, J. A., Hajarian, H., Tran, T., Alipourjeddi, P., and Noveen, A., 1995, Neurofilament protein heterotetramers as assembly intermediates, *J. Biol. Chem.* **270**:9334–9339.

Cole, J. S., Messing, A., Trojanowski, J. Q., and Lee, V. M.-Y., 1994, Modulation of axon diameter and neurofilaments by hypomyelinating Schwann cells in transgenic mice, *J. Neurosci.* **14**:6956–6966.

Collard, J. F., Côté, F., and Julien, J.-P., 1995, Defective axonal transport in a transgenic mouse model of amyotrophic lateral sclerosis, *Nature* **375**:61–64.

Côté, F., Collard, J.-F., and Julien, J.-P., 1993, Progressive neuronopathy in transgenic mice expressing the human neurofilament heavy gene: A mouse model of amyotrophic lateral sclerosis, *Cell* **73**:35–46.

Couillard-Després, S., Zhu, Q., Wong, P., Price, D., Cleveland, D. W., and Julien, J.-P., 1998, Protective effect of neurofilament heavy gene overexpression in motor neuron disease induced by mutant superoxide dismutase, *Proc. Natl. Acad. Sci.* **95**:9626–9630.

des Portes, V., Coulpier, M., Melki, J., and Dreyfus, P. A., 1994, Early detection of mouse *wobbler* mutation: A model of pathological motoneurone death, *Neuroreport* **5**:1861–1864.

de Waegh, S. M., Lee, V. M.-Y., and Brady, S. T., 1992, Local modulation of neurofilament phosphorylation, axonal caliber, and slow axonal transport by myelinating Schwann cells, *Cell* **68**:451–463.

Eyer, J., and Peterson, A., 1994, Neurofilament-deficient axons and perikaryal aggregates in viable transgenic mice expressing a neurofilament-β-galactosidase fusion protein, *Neuron* **12**:389–405.

Figlewicz, D. A., Krizus, A., Martinoli, M. G., Meininger, V., Dib, M., Rouleau, G. A., and Julien, J. P., 1994, Variants of the heavy neurofilament subunit are associated with the development of amyotrophic lateral sclerosis, *Hum. Mol. Genet.* **3**:1757–1761.

Fuchs, E., 1994, Intermediate filaments and disease: Mutations that cripple cell strength, *J. Cell Biol.* **125**:511–516.

Giasson, B. I., and Mushynski, W. E., 1996, Aberrant stress-induced phosphorylation of perikaryal neurofilaments, *J. Biol. Chem.* **271**:30404–30409.

Griffin, J. W., Fahnestock, K. E., Price, D. L., and Cork, L. C., 1983, Cytoskeletal disorganization induced by local application of β,β′-iminodipropionitrile and 2,5-hexandione, *Ann. Neurol.* **14**:55–61.

Gurney, M. E., Pu, H., Chiu, A. Y., Dal Canto, M. C., Polchow, C. Y., Alexander, D. D., Caliendo, J., Hentati, A., Kwon, Y. W., Deng, H.-X., Chen, W., Zhai, P., Sufit, R. L., and Siddique, T., 1994, Motor neuron degeneration in mice that express a human Cu, Zn superoxide dismutase mutation, *Science* **264**:1772–1775.

Hirano, A., Donnenfeld, H., Sasaki, S., and Nakano, I., 1984, Fine structural observations of neurofilamentous changes in amyotrophic lateral sclerosis, *J. Neuropathol. Exp. Neurol.* **43**:461–470.

Hisanaga, S., and Hirokawa, N., 1990, Molecular architecture of neurofilament. II. Reassembly process of neurofilament-L protein in vitro, *J. Mol. Biol.* **211**:871–882.

Hoffman, P. N., and Lasek, R. J., 1975, The slow component of axonal transport. Identification of major structural polypeptides of the axon and their generality among mammalian neurons, *J. Cell Biol.* **66**:351–366.

Julien, J.-P., and Mushynski, W. E., 1982, Multiple phosphorylation sites in mammalian neurofilament polypeptides, *J. Biol. Chem.* **257**:10467–10470.

Julien, J.-P., Côté, F., Beaudet, L., Sidky, M., Flavell, D., Grosveld, F., and Mushynski, W., 1988,

Sequence and structure of the mouse gene coding for the largest neurofilament subunit, *Gene* **68:**307–314.

Lee, M. K., Xu, Z., Wong, P. C., and Cleveland, D. W., 1993, Neurofilaments are obligate heteropolymers in vivo, *J. Cell Biol.* **122:**1337–1350.

Lee, M. K., Marszalek, J. R., and Cleveland, D. W., 1994, A mutant neurofilament subunit causes massive, selective motor neuron death: Implications for the pathogenesis of human motor neuron disease, *Neuron* **13:**975–988.

Lees, J. F., Shneidman, P. S., Skuntz, S. F., Carden, M. J., and Lazzarini, R. A., 1988, The structure and organization of the human heavy neurofilament subunit (NF-H) and the gene encoding it, *EMBO J.* **7:**1947–1955.

Ma, D., Descarries, L., Julien, J. P., and Doucet, G., 1995, Abnormal perikaryal accumulation of neurofilament light protein in the brain of mice transgenic for the human protein: Sequence of postnatal development, *Neuroscience* **68:**135–149.

Marszalek, J. R., Williamson, T. L., Lee, M. K., Xu, Z., Hoffman, P. N., Crawford, T. O., and Cleveland, D. W., 1996, Neurofilament subunit NF-H modulates axonal diameter by selectively showing neurofilament transport, *J. Cell Biol.* **135:**711–724.

McQuarrie, I. G., Brady, S. T., and Lasek, R. J., 1989, Retardation in the slow axonal transport of cytoskeletal elements during maturation and aging, *Neurobiol. Aging* **10:**359–365.

Miyasaka, H., Okabe, S., Ishiguro, K., Uchida, T., and Hirokawa, N., 1993, Interaction of the tail domain of high molecular weight subunits of neurofilaments with the COOH-terminal region of tubulin and its regulation by tau protein kinase II, *J. Biol. Chem.* **268:**22695–22702.

Muma, N. A., and Cork, L. C., 1993, Alternations in neurofilament mRNA in hereditary canine spinal muscular atrophy, *Lab. Invest.* **69:**436–442.

Pant, H., and Veeranna, 1995, Neurofilament phosphorylation, *Biochem. Cell Biol.* **73:**575–592.

Pernas-Alonso, R., Schaffner, A. E., Perrone-Capano, C., Orlando, A., Morelli, F., Hansen, C. T., Barker, J. L., Esposito, B., Cacucci, F., and di Porzio, U., 1996, Early upregulation of medium neurofilament gene expression in developing spinal cord of the *wobbler* mouse mutant, *Mol. Brain. Res.* **38:**267–275.

Reaume, A. G., Elliott, J. L., Hoffman, E. K., Kowall, N. W., Ferrante, R. J., Siwek, D. F., Wilcox, H. M., Flood, D. G., Beal, M. F., Brwon, R. H., Scott, R. W., and Snider, W. D., 1996, Motor neurons in Cu/Zn superoxide dismutase-deficient mice develop normally but exhibit enhanced cell death after axonal injury, *Nat. Genet.* **13:**43–47.

Ripps, M. E., Huntley, G. W., Hof, P. R., Morrison, J. H., and Gordon, J. W., 1995, Transgenic mice expressing an altered murine superoxide dismutase gene provide an animal model of amyotrophic lateral sclerosis, *Proc. Natl. Acad. Sci. USA* **92:**689–693.

Rooke, K., Figlewicz, D. A., Han, F. Y., and Rouleau, G. A., 1996, Analysis of the KSP repeat of the neurofilament heavy subunit in familial amyotrophic lateral sclerosis, *Neurology* **46:**789–790.

Rouleau, G. A., Clark, A. W., Rooke, K., Pramatarova, A., Krizus, A., Suchowersky, O., Julien, J.-P., and Figlewicz, D., 1996, *SOD1* mutation is associated with accumulation of neurofilaments in amyotrophic lateral sclerosis, *Ann. Neurol.* **39:**128–131.

Sasaki, S., and Iwata, M., 1996, Impairment of fast axonal transport in the proximal axons of anterior horn neurons in amyotrophic lateral sclerosis, *Neurology* **47:**535–540.

Schmidt, M. L., Carden, M. J., Lee, V. M.-Y., and Trojanowski, J. Q., 1987, Phosphate dependent and independent neurofilament epitopes in the axonal swellings of patients with motor neuron disease and controls, *Lab. Invest.* **56:**282–294.

Smith, M. A., Rudnicka-Nawrot, M., Richey, P. L., Praprotnik, D., Mulvihill, P., Miller, C. A., Sayre, L. M., and Perry, G., 1995, Carbonyl-related posttranslational modification of neurofilament protein in the neurofibrillay pathology of Alzheimer's disease, *J. Neurochem.* **64:**2660–2666.

Steinert, P. M., and Roop, D. R., 1988, Molecular and cellular biology of intermediate filaments, *Annu. Rev. Biochem.* **57:**593–625.

Strong, M. J., 1994, Aluminum neurotoxicity: An experimental approach to the induction of neurofilamentous inclusions, *J. Neurol. Sci.* **124:**20–26.

Tu, P.-H., Raju, P., Robinson, K. A., Gurney, M. E., Trojanowski, J. Q., and Lee, V. M.-Y., 1996, Transgenic mice carrying a human mutant superoxide dismutase transgene develop neuron cytoskeletal pathology resembling human amyotrophic lateral sclerosis lesions, *Proc. Natl. Acad. Sci. USA* **93:**3155–3160.

Vechio, J. D., Bruijn, L. I., Xu, Z., Brown, R. H., and Cleveland, D. W., 1996, Sequence variants in human neurofilament proteins: Absence of linkage to familial amyotrophic lateral sclerosis, *Ann. Neurol.* **40:**603–610.

Vickers, J. C., Morrison, J. H., Friedrich, V. L., Elder, G. A., Perl, D. P., Katz, R. N., and Lazzarini, R. A., 1994, Age-associated and cell-type-specific neurofibrillary pathology in transgenic mice expressing the human midsized neurofilament subunit, *J. Neurosci.* **14:**5603–5612.

Wang, X., Culotta, V. C., and Klee, C. B., 1996, Superoxide dismutase protects calcineurin from inactivation, *Nature* **383:**434–437.

Wideau-Pazos, M., Goto, J. J., Rabizadeh, S., Gralla, E. B., Roe, J. A., Lee, M. K., Valentine, J. S., and Bredesen, D. E., 1996, Altered reactivity of superoxide dismutase in familial amyotrophic lateral sclerosis, *Science* **271:**515–518.

Williamson, T. L., Bruijn, L. I., Zhu, Q., Anderson, K. L., Anderson, S. D., Julien, J.-P., and Cleveland, D. W., 1998, Absence of neurofilaments reduces the selective vulnerability of motor neurons and slows disease caused by a familial ALS-linked SOD1 Mutant, *Proc. Natl. Acad. Sci.* **95:**9631–9636.

Wong, P. C., Marszalek, J., Crawford, T. O., Xu, Z., Hsieh, S.-T., Griffin, J. W., and Cleveland, D. W., 1995a, Increasing neurofilament subunit NF-M expression reduces axonal NF-H, inhibits radial growth, and results in neurofilamentous accumulation in motor neurons, *J. Cell Biol.* **130:**1413–1422.

Wong, P. C., Pardo, P. C., Borchelt, D. R., Lee, M. K., Copeland, N. G., Jenkins, N. A., Sisodia, S. S., Cleveland, D. W., and Price, D. L., 1995b, An adverse property of a familial ALS-linked *SOD1* mutation causes motor neuron disease characterized by vacuolar degeneration of mitochondria, *Neuron* **14:**1105–1116.

Xu, Z., Cork, L. C., Griffin, J. W., and Cleveland, D. W., 1993, Increased expression of neurofilament subunit NF-L produces morphological alterations that resemble the pathology of human motor neuron disease, *Cell* **73:**23–33.

Yamasaki, H., Itakura, C., and Mizutani, M., 1991, Hereditary hypotrophic axonopathy with neurofilament deficiency in a mutant strain of the Japanese quail, *Acta Neuropathol.* **82:**427–434.

Yang, Y., Dowling, J., Yu, Q.-C., Kouklis, P., Cleveland, D. W., and Fuchs, E., 1996, An essential cytoskeletal linker protein connecting actin filaments to intermediate filaments, *Cell* **86:**655–665.

Zhu, Q., Couillard-Després, S., and Julien, J.-P., 1997, Delayed regeneration of myelinated axons in mice lacking neurofilaments, *Exp. Neurol.* **148:**299–316.

CHAPTER 7

MOUSE MODELS OF AMYOTROPHIC LATERAL SCLEROSIS

TONI L. WILLIAMSON
and DON W. CLEVELAND

Amyotrophic lateral sclerosis (ALS) is a fatal disease in which degeneration of upper and lower motor neurons leads to progressive muscle weakness. The most common form of adult motor neuron disease in humans, ALS typically initiates in middle to late life, leading to paralysis and death within 3–5 years. Approximately 10% of ALS cases are inherited in an autosomal dominant fashion. These familial ALS (FALS) cases are generally thought to be clinically indistinguishable from sporadic ALS (Emery and Holloway, 1983; Li *et al.*, 1988), although there are exceptions (see later). The disease usually begins asymmetrically in one limb, most commonly the leg, and then appears to spread to involve contiguous groups of motor neurons. Many affected neurons in ALS patients show cytoskeletal pathology in the form of neurofilament accumula-

TONI L. WILLIAMSON • Ludwig Institute for Cancer Research, University of California, San Diego, La Jolla, California 92093. DON W. CLEVELAND • Ludwig Institute for Cancer Research and Departments of Medicine and Neuroscience, University of California, San Diego, La Jolla, California 92093.

Mouse Models in the Study of Genetic Neurological Disorders, Volume 9 of *Advances in Neurochemistry,* edited by Popko, Kluwer Academic/Plenum Publishers, New York, 1999.

tions, both within the cell bodies and in proximal axons (Carpenter, 1968; Chou and Fakadej, 1971; Hirano *et al.,* 1984a,b). Motor neuron loss is also accompanied by reactive gliosis (Leigh *et al.,* 1991), ubiquitin-positive inclusions (Leigh *et al.,* 1991), and apparent fragmentation of the Golgi (Gonatas *et al.,* 1992).

1. SOD1 ACTIVITY AND ALS

It is estimated that 15–20% of familial ALS patients have missense mutations in the gene encoding Cu/Zn superoxide dismutase 1 (SOD1) (Rosen *et al.,* 1993; Brown, 1995; Siddique and Deng, 1996). Currently, more than 45 missense mutations in SOD1 have been reported (Siddique and Deng, 1996). SOD1 is a metalloenzyme of 153 amino acids that catalyzes the formation of hydrogen peroxide through dismutation of superoxide anions (Fridovich, 1995). SOD1 is thought to be expressed in all eukaryotic cells and its activity is believed to play a crucial role in protection against oxygen radical-induced cellular damage (Halliwell, 1994; Yu, 1994).

With the discovery of mutations in SOD1 as a primary cause of disease, the key question is how do mutations in SOD1 (a highly abundant, ubiquitous enzyme) lead to late-onset motor neuron degeneration? An initial hypothesis was that SOD1-linked FALS resulted from free radical damage due to decreased *SOD1* activity (Bowling *et al.,* 1993; Deng *et al.,* 1993; Orrell *et al.,* 1995). Consistent with this hypothesis, diminished SOD1 activity was found to lead to increased neuronal cell death in a variety of cell culture models (Rothstein *et al.,* 1994a; Greenlund *et al.,* 1995; Troy *et al.,* 1996). Further, SOD1 activity was reported to be reduced by 50–60% in red blood cells, lymphocytes, and brain from individuals with SOD1 mutations (Bowling *et al.,* 1993; Deng *et al.,* 1993; Robberecht *et al.,* 1994). However, *in vitro* studies quickly showed that various FALS-linked SOD1 mutants have different specific activities and at least some mutant polypeptides (e.g., glycine substituted to arginine at position 37—G37R) retain full specific activity (Borchelt *et al.,* 1994). In addition, although it was argued that some mutations dominantly reduce the activity of the wild-type subunits (Orrell *et al.,* 1995), DNA transfection revealed that this is certainly not a general property, with several mutants failing dominantly to affect the activity or stability of wild-type SOD1 (Borchelt *et al.,* 1995). Perhaps the most relevant observation was the finding that neither age at onset nor rapidity of disease progression correlate with SOD1 activity levels (Cleveland *et al.,* 1995). These collective observations called into question the idea that reduction in SOD1 activity plays a critical role in FALS.

1.1. Transgenic Mouse Models Reveal SOD1 Mutations Cause Disease from a Novel Toxic Property, Not Loss of Activity

1.1.1. Disease Arising from Mutation G93A

To examine experimentally how SOD1 mutations do cause selective motor neuron death, several groups constructed transgenic mice that express specific SOD1 mutations. In the first of these models, Gurney *et al.* (1994) produced one transgenic mouse line expressing FALS-linked SOD1 mutation G93A that developed motor neuron degeneration that both clinically and pathologically resembled ALS. The affected G93A line became paralyzed in one or more limbs as a result of loss of spinal cord motor neurons and died at ~5–6 months of age. Overexpression of wild-type human SOD1 generated no phenotype. The results with the one affected line were suggestive that this SOD1 mutant may confer a dominant property (sometimes called a "gain of function") that plays a key role in the pathogenesis of ALS, although, as the authors correctly noted, the possibility could not be eliminated that gene disruption at the site of transgene insertion in the one affected line could have been responsible for the disease phenotype.

The pathological changes in the G93A line have been documented extensively, although some confusion remains as to exactly how the disease progresses. Abnormally phosphorylated neurofilaments were initially reported in the cell bodies of most of the remaining motor neurons at end stage (Gurney *et al.*, 1994), but these are much less striking than reported in human disease (e.g., Fig. 3E). Rather than cell bodies bloated with masses of tangled neurofilaments, the cell bodies in the G93A mice were reduced in size and there was no evidence of excessive neurofilament accumulation. A subsequent report found that the earliest and most prominent change of large motor neurons in the spinal cord in the G93A mice was microvesiculation (Dal Canto and Gurney, 1994, 1995). These vacuoles were present both in motor neuron cell bodies and proximal axons, and apparently originated from dilation of rough endoplasmic reticulum and degenerating mitochondria. Others subsequently concluded from studying the same mice that neurofilamentous accumulation and vacuolization occur concomitantly and that neurofilament accumulation is the most prominent pathological feature in these mice (Tu *et al.,* 1996). In mice expressing lower levels of the G93A mutant, vacuolar pathology is reported to be minimal, while motor neuron atrophy, astrocytosis, and the presence of numerous ubiquitinated Lewy bodies are the main pathological features, along with axonal swellings containing neurofilaments (Dal Canto and Gurney, 1997).

1.1.2. Disease Arising from Mutation G37R

Rapidly following the report of disease in an initial G93A mutant mouse line was an analysis of a set of transgenic mice accumulating another of the familial ALS-associated mutant proteins (G37R). As this mutant retains full specific activity and was shown to be present with only a modestly reduced level in lymphocytes from patients (Borchelt *et al.,* 1994), it seemed at the outset highly unlikely that for this mutant disease could arise from loss of activity. Indeed, four lines of mice expressing the mutant between 4 and 12 times the level of the endogenous mouse SOD1 polypeptide in spinal cord each developed severe, progressive motor neuron disease (Wong *et al.,* 1995). At lower levels of mutant accumulation, pathology was restricted to lower motor neurons, whereas higher levels caused more severe abnormalities and affected a variety of other neuronal populations. Like the case for mice expressing G93A, the most obvious cellular abnormality was the presence in axons and dendrites of membrane-bounded vacuoles (Fig. 1C), which appeared at the electron microscopic level to be derived from degenerating mitochondria. However, in the G37R mice the vacuoles were strongly enriched in the proximal axons and dendrites with the neuronal cell bodies largely spared, in contrast with G93A mice in which the motor neuron cell bodies were primarily affected (Dal Canto and Gurney, 1994). Since multiple lines of mice expressing wild-type human SOD1 at up to 14 times normal mouse SOD1 polypeptide levels (with a corresponding increase in SOD1 activity) do not develop an overt phenotype, the disease in mice expressing the G37R mutant SOD1 must arise from the acquisition of an adverse property by the mutant enzyme, rather than elevation or loss of SOD1 activity.

1.1.3. Disease Arising from Mutation G85R

In both the G93A and G37R mouse models of familial ALS-linked SOD1 mutations, very high levels of mutant protein were required to provoke disease.

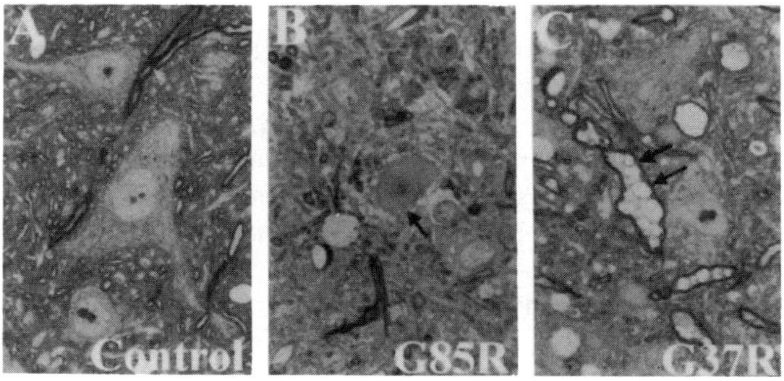

FIGURE 1. Different SOD1 mutations show different pathological changes. (A) Ventral horn of wild-type mouse, showing normal motor neurons. (B) Ventral horn of a mouse expressing the G85R SOD1 mutant, showing a Lewy body-like inclusion (*arrow*). (C) Ventral horn of a mouse expressing the G37R SOD1 mutant, showing highly vacuolar pathology. Proximal axons in the ventral spinal cord have swellings that are full of membrane-bound organelles/vesicles (*arrows*). For color representation of this image, see insert at right.

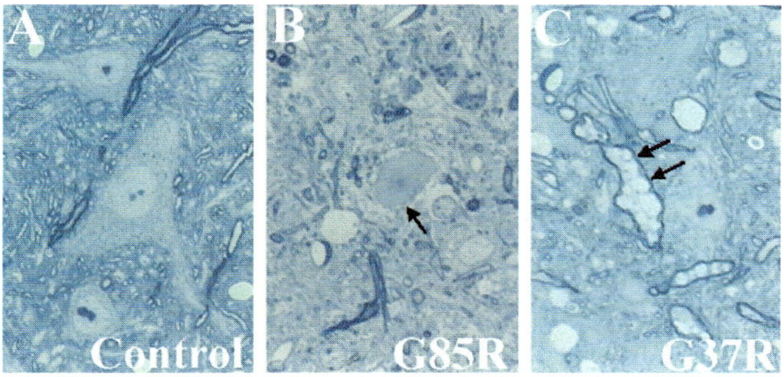

FIGURE 1.

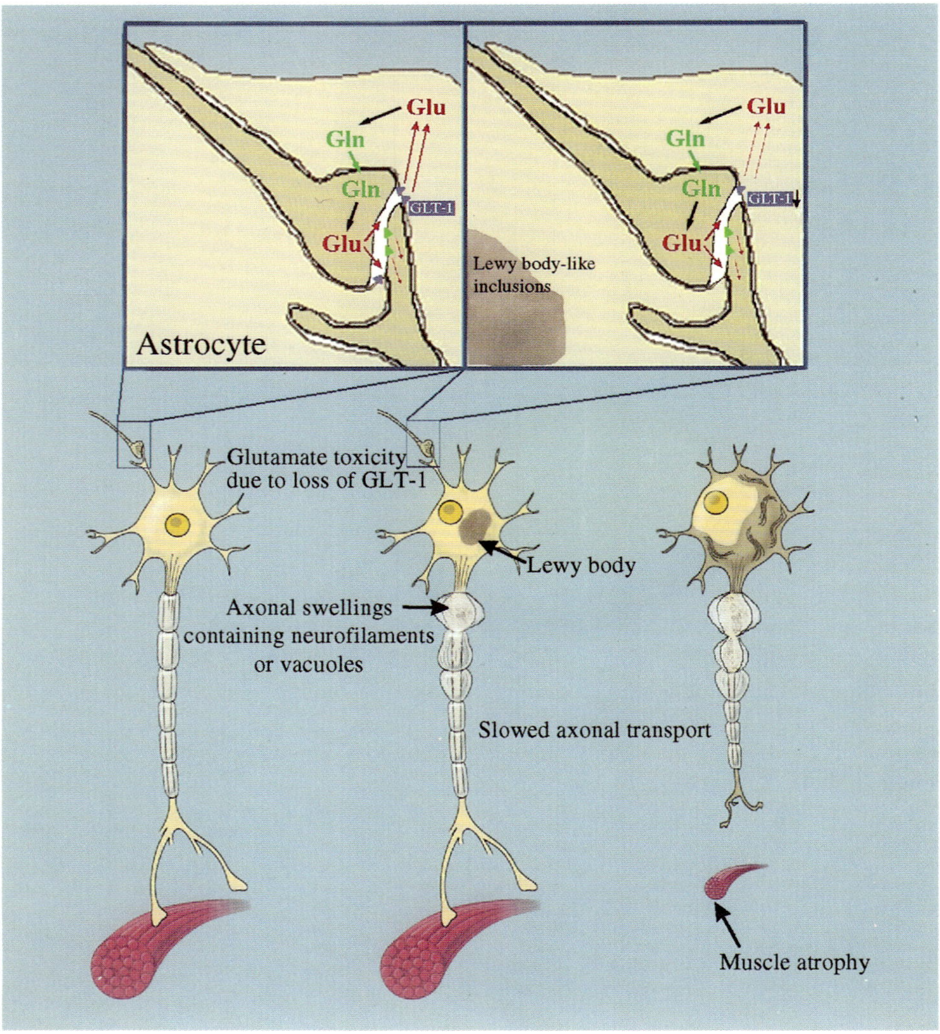
Glu
Gln
Gln
Glu
GLT-1
Astrocyte
Glu
Gln
Gln
Glu
GLT-1
Lewy body-like inclusions
Glutamate toxicity due to loss of GLT-1
Lewy body
Axonal swellings containing neurofilaments or vacuoles
Slowed axonal transport
Muscle atrophy

FIGURE 5.

This raised an obvious question of the degree to which disease requiring these excessive levels of mutant subunits was faithfully mimicking the human disease pathway. An initial reassurance arose from the report (Ripps *et al.*, 1995) of one mouse line that developed a very rapidly progressing motor neuron disease from expressing a mutation (G86R) in mouse SOD1 that is homologous in position to the familial ALS-linked human G85R mutation. Disease arose in this mouse without apparent elevation or loss of SOD1 activity. This finding was significantly extended by analysis of seven lines of mice that accumulate SOD1 mutant G85R to levels between 20% and 100% of the endogenous wild-type mouse SOD1 (e.g., levels appropriate for dominantly inherited disease) and invariably developed motor neuron disease characterized by an extremely rapid clinical progression after onset. In all cases, there was no change in the endogenous SOD1 protein or activity levels. The level of human G85R FALS-linked mutant SOD1 necessary to cause disease is ~20-fold less than the amount required to mediate disease by G93A (Gurney *et al.*, 1994) or G37R (Wong *et al.*, 1995).

The most striking pathologic finding in transgenic mice expressing human G85R SOD1 is the presence of numerous inclusions (Fig. 1B), which immunostaining with antibodies to glial fibrillary acidic protein (GFAP) revealed to be within astrocytes. These inclusions appear prior to, or contemporaneous with, similar aggregates in motor neurons. By end-stage, when there is a twofold elevation in SOD1 protein in whole spinal cord extracts and a 60% loss of large axons, inclusions are approximately 10 times more abundant in astrocytes than in neurons. This indicates that for this mutation, astroglia are a major target of SOD1-mediated damage. Strikingly, both the dense core and the periphery of both astrocytic and neuronal inclusions are intensely reactive with antibodies to SOD1. While no autopsy material is available for familial cases with the G85R mutation (N. Siddique and T. Siddique, personal communication), similar neuronal SOD1-positive inclusions have been described in some familial ALS patients with mutation A4V (Shibata *et al.*, 1996b) and inclusions with similar SOD1 immunoreactivity in astrocytes have also been reported (Kato *et al.*, 1997) in a familial ALS patient with a 2 bp deletion in the 126th codon of SOD1, leading to a frameshift and truncation of the final 27 amino acids (Kato *et al.*, 1996). The demonstration that SOD1-containing inclusions in astrocytes are early abnormalities and increase markedly in abundance after disease onset provides strong support for the view that at least some of the familial ALS-linked SOD1 mutants mediate disease by direct affects on astrocytes, in both human and murine disease.

1.2. Complete Deletion of the *SOD1* Gene Does Not Cause Motor Neuron Disease in Mice

Further evidence showing that ALS mediated by SOD1 mutation arises from the gain of a toxic property(ies) of the mutants was provided through the use of homologous recombination to produce a complete deletion of the SOD1 gene in mice (Reaume *et al.*, 1996). Whereas SOD1 is essential for efficient aerobic growth of yeast (Liu *et al.*, 1992), mice completely deficient in SOD1 develop normally without overt phenotype. These mice show no changes in stride length or running activity and showed no tremor or paralysis typical of the SOD1 mutant mice. No vacuolization, cell loss, or other pathological changes were observed in the brain or spinal cord. Brain tissues from SOD1-deficient mice do not show any increases in either lipid peroxidation or protein carbonyl content, two common end products of oxidative damage. However, SOD1-deficient mice did exhibit marked vulnerability to motor neuron loss after axonal injury, suggesting that SOD1 is required under physiologically stressful conditions.

The absence of overt disease in mice completely lacking SOD1, combined with the progressive disease arising in mice expressing FALS-linked mutations G93A and G37R (which have significantly elevated SOD1 activity levels) or G85R which has normal SOD1 activity levels), demonstrates that disease in mice arises from a toxic property of the mutant subunits rather than a loss of SOD1 activity.

1.3. Identifying the Toxic Property of the FALS-Linked SOD1 Mutants: An Aberrant Property of One *FALS*-Linked *SOD1* Mutant Is a Catalyzed Increase in Nitrotyrosine

The obvious question emerging from these initial transgenic mouse efforts is to identify the toxic property(ies) of the FALS-linked SOD1 mutants. One initial proposed mechanism is that the mutant subunits unfold slightly, allowing increased access of peroxynitrite (the spontaneous reaction product of superoxide and nitric oxide) to the enzyme-bound copper, which in turn uses peroxynitrite to nitrate tyrosine residues on proteins (Beckman *et al.*, 1993). A second hypothesis is that mutants catalyze the formation of hydroxyl radicals from hydrogen peroxide (via the Fenton Reaction). *In vitro,* at least two mutants have been reported to act as peroxidases, using the usual end product hydrogen peroxide as a substrate (Wiedau-Pazos *et al.*, 1996), albeit this evidence has been directly challenged (Singh *et al.*, 1998).

To test whether tyrosine nitration or use of hydrogen peroxide as a substrate was a feature of disease generated in mice, levels of nitrotyrosine and markers for hydroxyl radical formation have been measured in two lines of transgenic mice that develop progressive motor neuron disease from expressing human familial ALS-linked SOD1 mutation G37R (Bruijn *et al.*, 1997a). Relative to normal mice or mice expressing high levels of wild-type human SOD1, free 3-nitrotyrosine levels were found to be elevated by up to two- to threefold in spinal cord and brainstem, coincident with the earliest pathological abnormalities, and remained elevated throughout progression of disease. However, immunoblotting of spinal cord extracts using an antibody that recognizes nitrotyrosine (Beckman *et al.,* 1994) failed to reveal an increase in protein-bound nitrotyrosine during any stage of *SOD1*-mutant-mediated disease in mice or at end stage in spinal cord extracts from patients with sporadic ALS or SOD1-mediated (mutation A4V) FALS. Neither could an elevation of nitrotyrosine immunoreactivity be seen by immunocytochemistry at any stage of disease in mice. The inability to detect nitration even of highly abundant proteins such as neurofilaments and GFAP, despite a sensitivity capable of detecting signals from as little as 2 ng of *in vitro* nitrated standards, indicates that the stoichiometry of protein-bound nitration must be very low, if present at all.

On the other hand, immunostaining with an antibody to nitrotyrosine has suggested increased staining in motor neurons from sporadic ALS spinal cords (Abe *et al.,* 1995, 1997; Chou *et al.,* 1996). Given the notorious difficulties in using immunocytochemistry as a quantitative method, these are initial intriguing findings, but demonstration of the specificity of the staining (e.g., by blocking antibody binding with nitrotyrosine) remains for the future, as does identification of any protein that may be affected by nitration.

To address the question of whether mutant SOD1 acts as a peroxidase, salicylate trapping has been used to measure the levels of hydroxyl radicals in mice expressing the G37R mutation (Bruijn *et al.,* 1997a). This revealed no difference in hydroxyl radical levels in comparing G37R mice with control animals. In addition, a search for lipid peroxidation by measuring levels of the peroxidation product malondialdehyde (MDA) (Bruijn *et al.,* 1997a) revealed no increase in MDA levels in G37R mice.

To test the proposal that the toxicity of SOD1 mutants may arise from oxidative stress, mice expressing the G93A SOD1 mutant were supplemented with high dietary levels of vitamin E (Gurney *et al.,* 1996). Vitamin E can protect membranes

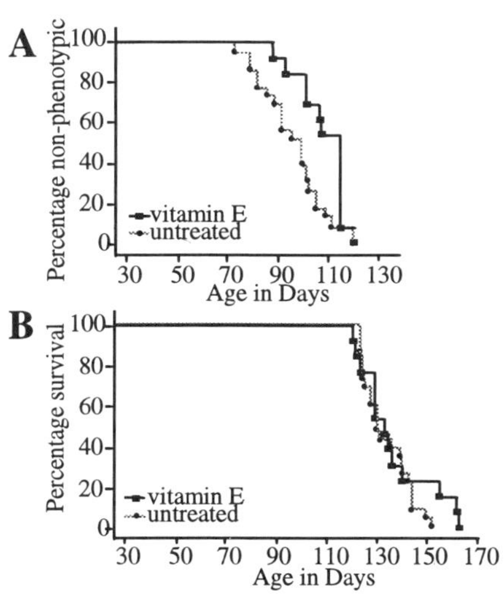

FIGURE 2. Disease onset was significantly delayed in SOD1 mutant mice (G93A) fed a diet supplemented with vitamin E (A), but survival was not affected (B). [Data redrawn from Gurney *et al.* (1996).]

against oxidative injury by donating a hydrogen atom to lipid peroxyl radicals (McCray, 1985). Although it was unclear whether serum levels of vitamin E were elevated, supplementation of vitamin E produced a small delay in the onset of disease (Fig. 2A). A disappointing feature, however, was that despite delayed onset, vitamin E did not affect overall survival (Fig. 2B).

1.4. Toxicity from a Familial ALS-Linked SOD1 Mutant Is Independent of SOD1 Activity and Wild Type SOD1 Protein and Is Not Alleviated by Raising SOD1 Activity

As outlined above, transgenic mice expressing familial amyotrophic lateral sclerosis (ALS)-linked mutations in superoxide dismutase (SOD1) have shown that motor neuron death arises from a mutant-mediated toxic property(ies). To test the mechanism of toxicity, the influence of wild type SOD1 activity and protein was examined (Bruijn *et al.,* 1998) by mating one line of $SOD1^{G85R}$ expressing mice (Bruijn *et al.,* 1997) to mice either deleted in the endogenous SOD1 genes (Reaume *et al.,* 1996) or carrying a transgene that expresses five times the normal wild type SOD1 protein level (Wong *et al.,* 1996). An important aspect of experimental design was selection of the $SOD1^{G85R}$ mouse model: the $SOD1^{G85R}$ polypeptide retains only about 10% of the activity of the wild type and causes disease when accumulated to levels between 20% and 100% of the endogenous SOD1 polypeptide. Hence, elimination of the endogenous SOD1 leads to an approximately tenfold reduction in overall SOD1 activity, while inclusion of the transgene encoding the wild type human SOD1 elevates overall activity to five times the normal level.

The results from this mating experiment were unambiguous: both elimination and elevation of wild type SOD1 were found to have no effect on mutant-mediated disease. This outcome has direct therapeutic implications. That disease was not accelerated in the complete absence of wild type SOD1 argues directly against the hypothesis that toxicity arises even in part from a reduction in normal SOD1 activity. More importantly, simce chronic, intracellular elevation of wild type SOD1 activity does not affect disease onset or progression, the use of SOD mimetics is an unlikely therapy. Indeed, the absence of influence of SOD1 activity on mutant toxicity challenges the idea that toxicity derives from (or is even linked to) oxidative stress arising from superoxide.

These mating experiences also showed that motor neuron degeneration almost certainly did not arise from more efficient use by this SOD1 mutant of two previously proposed aberrant substrates, peroxynitrite (Beckman *et al.,* 1993) or H_2O_2 (Wiedau-Pazos *et al.,* 1997). Peroxynitrite production (apparently formed through spontaneous reaction of nitric oxide with superoxide) should be significantly elevated by the absence of wild type SOD1 and probably lowered by a more rapid removal of superoxide in mice with a sixfold elevation of wild type SOD1, thereby accelerating or slowing disease, respectively. This was clearly not the case for mutant $SOD1^{G85R}$. Similarly, H_2O_2 produced in the active site by dismutation of superoxide would probably be the most important substrate for aberrant catalysis were the mutant to use it as a substrate. A sixfold molar excess of wild type SOD1 would obviously reduce the access of the mutant to superoxide (through competition and/or by accelerated catalytic destruction of it), while elimination of wild type SOD1 would strongly enhance it (since much more superoxide would be cycled through the mutant enzyme). Both would predict changes in the amount of H_2O_2 produced by the mutants and hence the rate of damage, but this was not found. When combined with *in vitro* evidence questioning whether the mutants use H_2O_2 as a substrate (Singh *et al.*, 1998), the proposal that toxicity arises from use of H_2O_2 as an aberrant substrate seems unlikely.

2. LINKING MOTOR NEURON GROWTH AND DEATH: NEUROFILAMENTS, AXONAL DISORGANIZATION, AND MOTOR NEURON DISEASE

Neurofilaments, assembled from NF-L (65 kDa), NF-M (95 kDa), and NF-H (115 kDa), are the major intermediate filaments in many types of mature neurons and are particularly abundant in large myelinated axons. All three neurofilament subunits can coassemble into filaments, although assembly *in vivo* requires expression of NF-L and a substoichiometric amount of either NF-M or NF-H (Ching and Liem, 1993; Lee *et al.*, 1993). Neurofilaments become abundant axonal components only after axons have reached their targets, concomitant with radial growth. Since the initial visualization of neurofilaments by the great neuroanatomists of the 19th century (e.g., Cajal, 1903), mounting evidence has pointed to their importance as the primary determinants of the large increase in axonal diameter that takes place following stable synapse formation. For human motor neurons that elongate with diameters under 1 μm, but achieve final calibers of up to 14 μm, this radial growth phase represents a more than 100-fold increase in cell volume.

The first direct evidence that neurofilaments are a primary determinant of axonal caliber came from the identification of a mutant Japanese quail (*quiver* of *Quv*) that accumulates no neurofilaments in axons or perikarya due to a premature translation termination in NF-L (Ohara *et al.*, 1993). This results in no detectable NF-L protein, total absence of neurofilaments, and axons that fail to grow radially and that have reduced conduction velocity (Ohara *et al.*, 1993; Sakaguchi *et al.*, 1993). Additional evidence comes from a transgenic mouse line expressing an NF-H subunit with almost the entire 120 kDa β-galactosidase polypeptide attached to its carboxy-terminal tail. Expression of this mutant NF-H blocks transport of all neurofilaments into axons and severely inhibits radial growth (Eyer and Peterson, 1994).

In addition to playing a structural role during normal radial growth, abnormal accumulations of neurofilaments are a common early pathological hallmark in patients with sporadic ALS [(Carpenter, 1968; Chou and Fakadej, 1971; Hirano *et al.*, 1984a); reviewed in (Hirano, 1991)] and familial ALS (Hirano *et al.*, 1984b). Two types of morphological changes are associated with this accumulation of filaments early in the degenerative process: first, focal accumulation with this accumulation of filaments early in the degenerative process: first, focal accumulation of filaments in cell bodies, often visualized by light microscopy as hyaline inclusions (Schochet *et al.*, 1969; Chou and Fakadej, 1971) and neurofilamentous swellings in axons (Carpenter, 1968; Cork *et al.*, 1982; Hirano *et al.*, 1984a; Cork *et al.*, 1988) and second, a diffuse accumulation of neurofilaments throughout cell bodies, axons, and dendrites. This latter type of accumulation is commonly seen in neurons with dispersed endoplasmic reticulum (Cork *et al.*, 1982; Hirano *et al.*, 1984a,b; Wiley *et al.*, 1987), a phenomenon known as chromatolysis. These observations led to the hypothesis that abnormalities in neurofilament organization may be involved in the pathogenesis of ALS. Indeed, recent efforts have demonstrated that disease in the now classic report of FALS (Hirano *et al.*, 1984b), in which neurofilament misaccumulation is the prominent pathologic hallmark, arises from SOD1 mutation A4V (Shibata *et al.*, 1996a), by far the most abundant *SOD1* mutation in North America. Neurofilamentous accumulations have also been found arising from a second SOD1 mutation, I113T (Rouleau *et al.*, 1996).

Direct evidence for the ability of aberrant neurofilaments to cause motor neuron disease came initially from transgenic mice expressing three to four times the normal levels of wild-type mouse NF-L (Xu *et al.*, 1993) or wild type human NF-H (Cote *et al.*, 1993). Both have massive accumulations of neurofilaments leading to swollen perikarya with eccentrically located nuclei. Neurofilament accumulation is accompanied by proximal axon swellings, motor neuron dysfunction, and neurogenic atrophy of skeletal muscle. However, these mice do not exhibit extensive motor neuron death (Xu *et al.*, 1993; Collard *et al.*, 1995).

2.1. Mutations in Neurofilaments Can Cause ALS-like Disease in Mice

To examine directly whether aberrant neurofilament subunits could be a direct cause of motor neuron disease that mimicked the selective killing of motor neurons found in human ALS, several lines of transgenic mice were constructed to express a point mutation at the end of the rod domain of NF-L (Lee *et al.*, 1994). Transgenic mice showed abnormal gait, reduced activity, and weakness in both upper and lower limbs as early as 18 days after birth. These abnormalities progressively increased in severity and, in most lines, progressed to death.

Examination of spinal cord from animals expressing this transgene revealed loss of motor neurons of the lumbar (Fig. 3B) and cervical spinal cord compared to wild-type littermates (Fig. 3A). As in many ALS cases (see Fig.

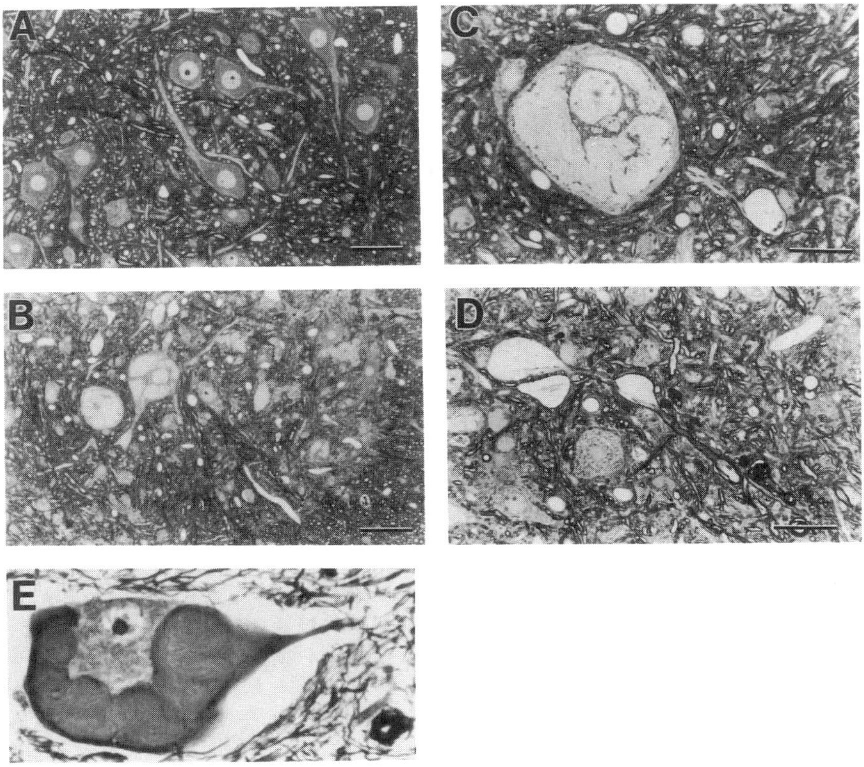

FIGURE 3. Expression of NF-L containing a point mutation causes motor neuron cell death. (A) Ventral horn from wild-type mouse, showing normal motor neurons. (B) Ventral horn of transgenic mouse, showing degenerating motor neurons and axonal swellings. High-power micrographs show (C) a swollen motor neuron cell body with a proximal axon swelling, and (D) a single axon containing multiple axonal swellings. (E) High-power micrograph showing neurofilament accumulations in a motor neuron from an ALS patient. [Reprinted with permission from Hirano (1988).]

3E), the surviving ventral horn motor neurons were chromatolytic and were grossly distended with material that stained lightly with toluidine blue (Fig. 3C), but stained intensely with silver (Lee *et al.,* 1994). Moreover, as reported frequently for ALS (see e.g., Carpenter, 1968; Hirano *et al.,* 1984a,b), neurofilament-rich swellings of proximal axons were frequently found in the initial unmyelinated segment and in the root exit zones (e.g., Fig. 3D). Examination of ventral roots showed a dramatic loss of motor axons reminiscent of ALS (Lee *et al.,* 1994). A particularly interesting finding from these studies is that only the largest, neurofilament-rich, motor axons are at risk in sporadic ALS [(Kawamura *et al.,* 1981); Fig. 4A], in SOD1-mediated

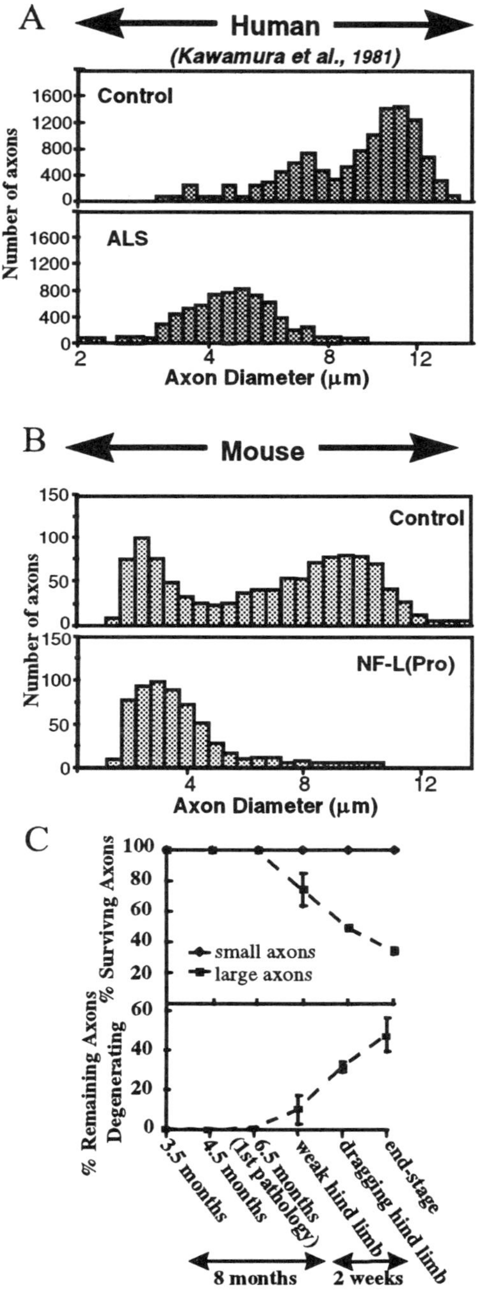

FIGURE 4. Graphs showing the selective loss of large-caliber motor neurons in human ALS (A), in mice expressing the NF-L (Pro) transgene (B) and in mice expressing mutant SOD1 (C). The human data are replotted from Kawamura *et al.*, 1981. The mouse data are replotted from Lee *et al.*, 1994 and Bruijn *et al.*, 1997b.

disease in mice [(Bruijn *et al.,* 1997b); Fig. 4C] and in mice expressing this neurofilament mutant [(Lee *et al.,* 1994); Fig. 4B].

These mouse models show that neurofilament mutations can cause motor neuron degeneration in mice and suggested that neurofilament mutations might be responsible for some ALS cases. While no mutations in neurofilament genes have been found in over 200 familial ALS patients (Vechio *et al.,* 1996; Rooke *et al.,* 1996), deletions of 1 or 34 codons in the KSP repeat of NF-H in 5 of 356 sporadic ALS patients (Figlewicz *et al.,* 1994), but not in an even larger number of DNAs. The sum of these findings suggests that these NF-H alterations represent risk factors, but are not sufficient to be sole causes of human disease. SSCP (single-strand conformational polymorphism) analysis coupled with DNA sequencing was used to search for mutations in the entirety of human NF-L, NF-M, and NF-H genes from 100 FALS patients and 75 sporadic ALS patients (Vechio *et al.,* 1996). This revealed six polypeptide sequence variants in the rod and tail domains of NF-L, NF-M, and NF-H. However, all were found at a comparable frequency in DNA from normal individuals and no variant cosegregated with disease. In another analysis of the NF-H KSP domain, no variants were found in any of 117 FALS cases (Rooke *et al.,* 1996).

2.2. Neurofilaments and SOD1 Mutant-Mediated Disease: Absence of Neurofilaments or Increased Perikaryal Levels of Neurofilaments Slows SOD1-Mediated Disease in Mice

The frequent neurofilamentous pathology in sporadic and familial ALS, the proof that primary defects can provoke motor neuron disease in mice, and the discovery of mutations in the NF-H gene in several cases of sporadic ALS disease together suggest a central role for neurofilaments in ALS. By mating animals with the NF-L gene disrupted with mice that develop disease from expressing $SOD1^{G85R}$, how or whether neurofilaments contribute to disease caused by this familial ALS-linked SOD1 mutant was examined. This revealed (Williamson *et al.,* 1998) that onset and progression of disease were significantly slowed (by five weeks), while selectivity of mutant-mediated toxicity for motor neurons was reduced. In NF-L deleted animals, levels of the two remaining neurofilament subunits NF-M and NF-H are markedly reduced in axons, but are elevated in motor neuron cell bodies. Thus, while this experiment proved that neither perikaryal nor axonal neurofilaments are essential for SOD1-mediated disease, it also demonstrated that the assembled axonal neurofilaments represent one factor conferring selective vulnerability to motor neurons, and that their absence slows $SOD1^{G85R}$ mutant-mediated toxicity.

To further investigate the role of neurofilaments in motor neuron disease caused by SOD1 mutations, transgenic mice expressing $SOD1^{G37R}$ were mated with two different transgenic lines expressing high levels of wild type human NF-H (Couillard-Depres *et al.*, 1998). The influence of increased levels of NF-H is to trap a large proprotion of neurofilaments in the neuronal perikarya, producing axons with fewer neurofilaments and cell bodies with increased levels. This resulted in a striking slowing of SOD1-mutant mediated disease, with overall life span extended by up to 65%. To date, this slowing of SOD1-mediated disease represents by far the most effective alleviation of disease and suggests that productive therapies might be produced by agents that increase perikaryal levels of neurofilament proteins.

3. AXONAL TRANSPORT

One possible mechanism for the pathogenesis of ALS is a defect that either directly or indirectly impairs slow axonal transport, leading to aberrant accumulation of neurofilaments in the proximal axon and inhibiting further transport of neurofilaments and potentially other substrates. In view of this, degeneration may arise from the simple mechanism of axonal strangulation. Transport is comprised of two components: movement of vesicles and mitochondria by kinesin and related proteins (fast transport) and movement of the major structural components of the neuron, many enzymes, and other cytoplasmic proteins (slow transport). Slow transport can be divided into two components based on rate of movement: SCa (~0.5 mm/day) containing neurofilament proteins, tubulin, and actin, and SCb (~1–2 mm/day), containing tubulin, actin, and other cytoplasmic proteins. The mechanism(s) of slow transport [and motor(s) that power it] is not known.

The demonstration that mice with disruptions in the NF-L gene or transgenic mice expressing increased levels of NF-H (see 2.2. above) can strongly affect SOD1-mediated disease raised the key question as to the underlying mechanism. The known neurofilament-dependent slowing of axonal transport during postnatal development and the aberrant perikaryal and axonal aggregates of neurofilaments common to both sporadic and SOD1-mediated ALS initially suggested that an important aspect of toxicity may arise from SOD1-mediated damage to transport. This has now been tested for two familial ALS-linked SOD1 mutations, $SOD1^{G37R}$ and $SOD1^{G85R}$ (Williamson and Cleveland, 1998). In both cases, selective retardation of tubulin transport was found before the onset of axonal degeneration and for 6–8 months prior to the appearance of clinical symptoms. For mutant $SOD1^{G85R}$, this is the earliest detectable abnormality, arising at least 4–6 months berfore any detectable pathological changes. Thus, SOD1 mutant-mediated damage to the cargoes or machinery of slow transport is an early feature of toxicity to motor neurons.

4. NEUROTROPHIC FACTORS, Bcl-2, AND APOPTOSIS

Apoptotic cell death occurs during the normal development of the nervous system and is regulated by neurotrophic factors. Treatment with neurotrophic factors has been shown to prevent the motor neuron cell death that can be induced by axotomy [ciliary neurotrophic factor (CNTF)—Sendtner *et al.*, 1990; brain-derived neurotrophic factor (BDNF)—Sendtner *et al.*, 1992a; Yan *et al.*, 1994; glial cell line derived neurotrophic factor (GDNF)—Oppenheim *et al.*, 1995; Yan *et al.*, 1995]. In addition, overexpression of the proto-oncogene Bcl-2 has been shown to protect motor neurons from axotomy-induced death (Dubois-Dauphin *et al.*, 1994) and from permanent ischemia (Martinou *et al.*, 1994). Although high levels of Bcl-2 expression in mice result in increased numbers of motor neurons, the protection afforded against axotomy-induced cell death is also seen in mice that express Bcl-2 only after birth and therefore have normal numbers of motor neurons (Martinou *et al.*, 1994). Owing to these protective effects, neurotrophic factors and anti-apoptotic proteins such as Bcl-2 have been suggested as potential therapeutic agents for ALS.

4.1. *Pmn* Mice, *wobbler* Mice, and CNTF/GDNF

Several mouse models have been used to test the efficacy of neurotrophins in preventing motor neuron degeneration. These include *pmn* mice (progressive motor neuropathy), which develop weakness in the hindlimbs during the third week of life and die at ~6 weeks of age (Schmalbruch *et al.*, 1991). Motor neurons show demyelination (Schmalbruch *et al.*, 1991) and degeneration of 30% of facial nucleus motor neurons (Sendtner *et al.*, 1992b). Another mutant mouse, *wobbler,* carries an inherited autosomal recessive mutation that leads to the progressive degeneration of spinal cord and brainstem motor neurons (Duchen and Strich, 1968; LaVail *et al.*, 1987). Early experiments to treat these mutant mice with neurotrophic factors were very promising. For example, implantation of CNTF expressing cells resulted in improvement of motor function and increased life span in *pmn* mice (Sendtner *et al.*, 1992b), although tumor formation in these mice prevented examination of long-term effects. CNTF injections also showed progression of disease in *wobbler* mice (Mitsumoto *et al.*, 1994a). Even more promising was the finding that cotreatment of *wobbler* mice with injections of CNTF and BDNF arrested disease progression and prevented loss of any motor axons (Mitsumoto *et al.*, 1994b). However, treatment of *pmn* mice with CNTF or GDNF resulted in increased survival of motor neurons, but little (for GDNF) or no effect (for CNTF) on axon numbers (Sagot *et al.*, 1995b, 1996). Neither growth factor increased the life span of the *pmn* mice. In these studies, cells expressing the neurotrophic factor were enclosed in a polymer envelope to prevent uncontrolled proliferation of the transplanted cells, which may have delivered less neurotrophin than could be administered by injection. One particularly promising strategy for delivering neurotrophic factors to motor neurons is that of adenovirus-mediated gene transfer. An NT-3 expressing adenovirus vector injected intramuscularly produced a 50% increase in life span in *pmn* mice (Hasse *et al.,* 1997). In addition, treated animals showed a reduced loss of motor axons and improved muscle function.

4.2. *Pmn* and *wobbler* Mice and Bcl-2

Transgenic mice overexpressing *Bcl-2* have also been mated to *pmn* and *wobbler* mice to examine whether increased *Bcl-2* can affect disease progression. *Pmn* mice showed increased motor neuron survival (i.e., cell body number) in the presence of the *Bcl-2* transgene, but showed no increase in axon number or life span (Sagot *et al.,* 1995a). This result is very reminiscent of the results obtained with CNTF or GDNF described previously. In *wobbler* mice, expression of the *Bcl-2* transgene had no effect on disease progression (Coulpier *et al.,* 1996).

4.3. FALS-Linked SOD1 Mice and Apoptosis Inhibitors

An obvious weakness to the early reports using neurotrophins or apoptosis inhibitors to study disease progression in *pmn* or *wobbler* mice was the absence of evidence that the disease pathway in these mice is relevant to human disease. Mice expressing human FALS-linked SOD1 mutations provide a much better model of human ALS. Overexpression of Bcl-2 in SOD1 mutant mice slows disease onset, prolongs the life of these mice, and attenuates the spinal cord motor neuron degeneration (Kostic *et al.,* 1997) (Table 1). The duration of disease is unaffected. SOD1 mutant mice have also been mated to mice expressing a dominant negative inhibitor of interleukin-1β-converting enzyme (ICE), a

TABLE 1. Expression of Antiapoptotic Proteins Prolonging the Life of Mice Expressing Mutant SOD1

Experiment	Genotype	Onset/days	End stage/days	Duration/days
Bcl-2 overexpression	G93A/-	170 ± 3	240 ± 2	70 ± 5
	G93A/bcl-2	203 ± 6	275 ± 4	72 ± 8
Dominant negative inhibitor	G93A/-	238 ± 4	249 ± 4	12 ± 2
of ICE	G93A/M17Z	243 ± 6	270 ± 4	27 ± 4

Mice expressing the G93A SOD1 mutant were crossed with mice overexpressing *Bcl-2* or mice expressing a dominant inhibitor of interleukin-1β converting enzyme (ICE). Expression of *Bcl-2* delayed onset of disease, but did not alter duration of disease. Expression of an ICE inhibitor slows progression of disease, but does not affect onset of disease.

protease (caspase) involved in apoptosis. This ICE mutant protein has been shown to inhibit trophic factor-induced apoptosis in dorsal root ganglion neurons and to reduce brain injury in response to ischemia (Friedlander *et al.*, 1997b). Mice that express both mutant SOD1 and mutant ICE show no change in the onset of disease compared to SOD1 alone, but the duration of disease is increased (Table 1) (Friedlander *et al.*, 1997a). That Bcl-2 extends the age of disease onset, but not progression, while the mutant ICE slows progression without affecting onset, suggests that Bcl-2 and the mutant ICE mediate their protective effects through different pathways. While an initial view might have been that Bcl-2 may slow motor neuron death due to its antiapoptotic effect, a second possibility is that it affects disease onset through its known antioxidant properties (Kane *et al.*, 1993; Zhong *et al.*, 1993). Interestingly, removing Bcl-2 from mice by homologous recombination has no effect on the survival of motor, sensory, or sympathetic neurons during the period of naturally occurring cell death, possibly due to redundancy with other Bcl-2-like components. In addition, Bcl-2 is not required for the protective effects of BDNF and CNTF on axotomized motor neurons. However, Bcl-2 appears to be crucial for survival of neurons in the postnatal period after programmed cell death (Michaelidis *et al.*, 1996). Bcl-2 null mice show an extensive loss of neurons in postnatal life.

5. GLUTAMATE EXCITOTOXICITY

L-Glutamate is the major excitatory neurotransmitter of the mammalian central nervous system (Nichols and Atwell, 1990). However, glutamate and related excitatory amino acids are also known to be powerful neurotoxins. When the presynaptic terminal is depolarized, glutamate is released into the synaptic cleft in a Ca^{2+}-dependent manner. Glutamate activates ionotropic and metabotropic glutamate receptors. Glutamate is then cleared from the synaptic cleft by glutamate transporters in the surrounding astrocytes and neurons. Four high-affinity sodium-dependent glutamate transporters have been identified and cloned: EAAT1 (GLAST), EAAT2 (GLT-1), EAAT3 (EAAC1), and EAAT4 (Kanai and Hediger, 1992; Pines *et al.*, 1992; Arriza *et al.*, 1994; Fairman *et al.*, 1995). Immunohistochemistry, *in situ* hybridization, and western blot analysis (Danbolt *et al.*, 1992; Levy *et al.*, 1993; Rothstein *et al.*, 1994b) have shown that EAAC1 is expressed in neurons, with particularly high expression in hippocampus, cerebellum, and basal ganglia; GLAST is expressed in neurons and glia, with strong staining in the molecular layer of the cerebellum, while GLT-1 is expressed in astrocytes in spinal cord and all brain regions.

The first evidence to implicate glutamate toxicity in ALS was the finding that the levels of glutamate are elevated in cerebrospinal fluid from ALS patients (Plaitakis *et al.*, 1988; Rothstein *et al.*, 1990). High-affinity glutamate transport is also decreased in synaptosomal fractions from motor cortex and spinal cord in ALS patients (Rothstein *et al.*, 1992; Shaw *et al.*, 1994), owing to a selective decrease in the GLT-1 transporter (Rothstein *et al.*, 1992; Shaw *et al.*, 1994), owing to a selective decrease in the GLT-1 transporter (Rothstein *et al.*, 1995). Organotypic cultures from postnatal rat spinal cord have been used to show that glutamate transporter inhibitors lead to slow neurotoxicity and can be protected from this toxicity by non-*N*-methyl-D-aspartate receptor antagonists or by glutamate release or synthesis inhibitors (riluzole and gabapentin) (Rothstein *et al.*, 1993; Rothstein and Kuncl, 1995). To test directly whether loss of glutamate transporters can cause motor neuron deficits, antisense oligonucleotides to astroglial glutamate transporters were administered to rats using an intraventricular cannular, resulting in progressive paresis beginning in the hindlimbs (Rothstein *et al.*, 1996). Disruption of the mouse GLT-1 gene by homologous recombination, produced mice with no detectable GLT-1 protein (Tanaka *et al.*, 1997). Glutamate uptake in these mice was reduced to less than 6% of normal levels in cortical synaptosomes, suggesting the GLT-1 is the major cortical glutamate transporter. In the absence of GLT-1, mice appeared normal at birth, but gained weight slowly and died prematurely (50% survival after 6 weeks) as a result of spontaneous seizures. Although motor neurons were not examined in this initial study, the early death of mice lacking GLT-1 precludes any conclusions regarding its role in ALS, a late-onset disease. However, in mice expressing an FALS-linked mutant SOD1, the level of GLT-1 in spinal cord was decreased by 50% in end-stage mice, despite an increased number of astrocytes (Bruijn *et al.*,1997b). This is consistent with depletion of GLT-1 as a contributor.

6. SUMMARY/CONCLUSIONS

Mice expressing the human FALS-linked SOD1 mutations have proven that mutations in SOD1 cause ALS due to the gain of a toxic property. The nature of this toxic property remains unresolved. An early aspect of this toxicity is the slowing of selected cargoes of slow axonal transport. Toxicity does not arise even in part from reduction in SOD1 activity and chronic elevation of wild type SOD1 activity does not ameliorate SOD1 mutant-mediated disease, making therapeutic strategies to increase SOD1 levels using SOD1 mimetics highly unlikely to succeed. Loss of the astroglial glutamate transporter GLT-1 has been observed in both human ALS and in mice expressing SOD1 mutants, despite increased numbers of astrocytes. Treatment of other mouse models of motor neuron degeneration with various neurotrophins has been successful in slowing or even arresting disease progression. These experiments have yet to be tried on mice expressing human SOD1 mutations, which are arguably much better models of human ALS. Expression of the antiapoptotic proteins Bcl-2 and mutant ICE do slow onset or

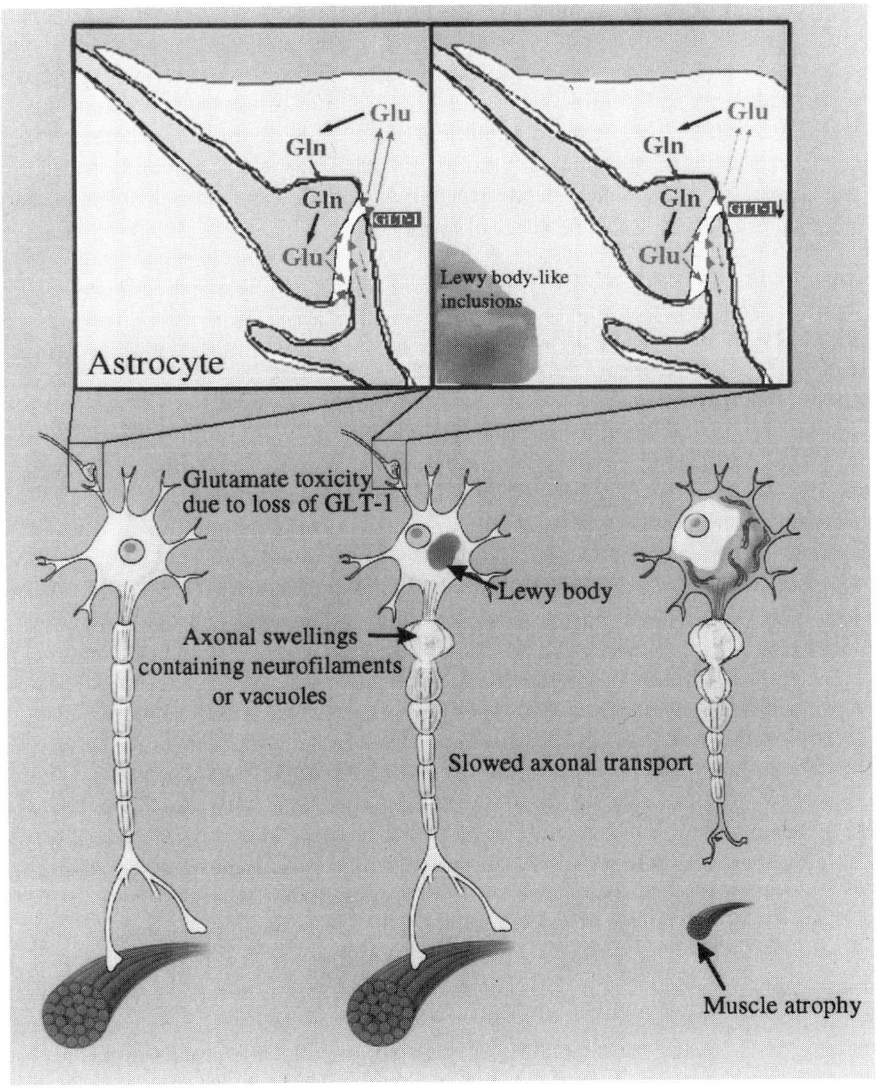

FIGURE 5. Diagrammatic representation of disease progression in mouse models of ALS. High-power diagram shows transport of glutamate from synaptic cleft into surrounding astrocytes (not shown at low power). Lewy body-like inclusions are found in astrocytes and neurons in *G85R SOD1* mice, while axonal vacuole accumulation is found in G37R mice. G93A mice show both types of pathology, although vacuoles are more prevalent in cell bodies than in axons. For color representation of this image, see insert facing page 141.

progression of disease in mice expressing the G93A SOD1 mutant, suggesting that a search for small molecule inhibitors of the caspases essential for the apoptotic pathway may represent a new approach for therapeutic intervention. Reduction in axonal neurofilaments and elevation in perikaryal filaments slows SOD1-mediated disease extending life span by up to six months. Such changes in neurofilament content represent the most effective known alleviation of disease and support the idea that a productive therapy could be the use of agents that incresse perikaryal levels of neurofilaments.

REFERENCES

Abe, K., Pan, L. H., Watanabe, M., Kato, T., and Itoyama, Y., 1995, Induction of nitrotyrosine-like immunoreactivity in the lower motor neurons of amyotrophic lateral sclerosis, *Neurosci. Lett.* **199:**152–154.

Abe, K., Pan, L.-H., Watanabe, M., Kato, T., and Itoyama, Y., 1997, Upregulation of protein-tyrosine nitration in the anterior horn cells of amyotrophic lateral sclerosis, *Neurol. Res.* **19:**124–128.

Arriza, J. L., Fairman, W. A., Wadiche, J. I., Murdoch, G. H., Kavanaugh, M. P., and Amara, S. G., 1994, Functional comparisons of three glutamate transporter subtypes cloned from human motor cortex, *J. Neurosci.* **14:**5559–5569.

Avraham, K. B., Schickler, M., Sapoznikov, D., Yarom, R., and Groner, Y., 1988, Down's syndrome: Abnormal neuromuscular junction in tongue of transgenic mice with elevated levels of human Cu/Zn superoxide dismutase, *Cell* **54:**823–829.

Beckman, J. S., Carson, M., Smith, C. D., and Koppenol, W. H., 1993, ALS, SOD and peroxynitrite, *Nature* **364:**584.

Beckman, J. S., Ye, Y. Z., Anderson, P. G., Chen, J., Accavotti, M. A., Tarpey, M. M., and White, C. R., 1994, Extensive nitration of protein tyrosines in human atherosclerosis detected by immunohistochemistry *Biol. Chem. Hoppe. Seyler* **375:**81–88.

Borchelt, D. R., Lee, M. K., Slunt, H. S., Guarnieri, M., Xu, Z. S., Wong, P. C., Brown, R. H., Jr., Price, D. L., Sisodia, S. S., and Cleveland, D. W., 1994, Superoxide dismutase 1 with mutations linked to familial amyotrophic lateral sclerosis possesses significant activity, *Proc. Natl. Acad. Sci. USA* **91:**8292–8296.

Borchelt, D. R., Guarnieri, M., Wong, P. C., Lee, M. K., Slunt, H. S., Xu, Z. S., Sisodia, S. S., Price, D. L., and Cleveland, D. W., 1995, Superoxide dismutase 1 subunits with mutations linked to familial amyotrophic lateral sclerosis do not affect wild-type subunit function. *J. Biol. Chem.* **270:**3234–3238.

Bowling, A. C., Schulz, J. B., Brown, R. H., Jr., and Beal, M. F., 1993, Superoxide dismutase activity, oxidative damage, and mitochondrial energy metabolism in familial and sporadic amyotrophic lateral sclerosis, *J. Neurochem.* **61:**2322–2325.

Brown, R. H., 1995, Amyotrophic lateral sclerosis: Recent insights from genetics and transgenic mice, *Cell* **80:**687–692.

Bruijn, L. I., Beal, M. F., Becher, M. W., Schulz, J. B., Wong, P. C., Price, D. L., and Cleveland, D. W., 1997a, Elevated free nitrotyrosine levels, but not protein-bound nitrotyrosine or hydroxyl radicals, throughout amyotrophic lateral sclerosis (ALS)-like disease implicate tyrosine nitration as an aberrant *in vivo* property of one familial ALS-linked superoxide dismutase 1 mutant, *Proc. Natl. Acad. Sci. USA* **94:**7606–7611.

Bruijn, L. I., Houseweart, M. K., Kato, S., Amderson, K. L., Anderson, S. D., Ohama, E., Reaume, A. G., Scott, R. W., and Cleveland, D. W., 1998, Aggregation and motor neuron toxicity of an ALS-linked SOD1 mutant independent from wild-type SOD1, *Science* **281**:1851–1854.

Bruijn, L. I., Becher, M. W., Lee, M. K., Anderson, K. L., Jenkins, N. A., Copeland, N. G., Sisodia, S. S., Rothstein, J. D., Borchelt, D. R., Price, D. L., and Cleveland, D. W., 1997b, ALS-linked SOD1 mutant G85R mediates damage to astrocytes and promotes rapidly progressive disease with SOD1-containing inclusions, *Neuron* **18**:327–338.

Cajal, S. R., 1903, Embryogenesis of neurofibrils, *Trab. Lab Invest. Biol. Universidad Madrid* **2**:219–225.

Carpenter, S., 1968, Proximal axonal enlargement in motor neuron disease, *Neurology* **18**:841–851.

Ching, G. Y., and Liem, R. K., 1993, Assembly of type IV neuronal intermediate filaments in nonneuronal cells in the absence of preexisting cytoplasmic intermediate filaments, *J. Cell Biol.* **122**:1323–1335.

Chou, S. M., and Fakadej, A. V., 1971, Ultrastructure of chromatolytic motor neurons and anterior spinal roots in a case of Werdnig–Hoffman disease, *J. Neuropathol. Exp. Neurol.* **30**:42–55.

Chou, S. M., Wang, H. S., and Taniguchi, A., 1996, Role of SOD-1 and nitric oxide/cyclic GMP cascade on neurofilament aggregation in ALS/MND, *J. Neurol. Sci.* **139**(Suppl):16–26.

Cleveland, D. W., Laing, N., Hurse, P. V., and Brown, R. H., 1995, Toxic mutants in Charcot's sclerosis, *Nature* **378**:342–343.

Collard, J. F., Cote, F., and Julien, J. P., 1995, Defective axonal transport in a transgenic mouse model of amyotrophic lateral sclerosis, *Nature* **375**:61–64.

Cork, L. C., Griffin, J. W., Choy, C., Padula, C. A., and Price, D. L., 1982, Pathology of motor neurons in accelerated hereditary canine spinal muscular atrophy, *Lab. Invest.* **46**:89–99.

Cork, L. C., Troncoso, J. C., Klavano, G. G., Johnson, E. S., Sternberger, L. A., Sternberger, N. H., and Price, D. L., 1988, Neurofilamentous abnormalities in motor neurons in spontaneously occurring animal disorders, *J. Neuropathol. Exp. Neurol.* **47**:420–431.

Cote, F., Collard, J. F., and Julien, J. P., 1993, Progressive neuronopathy in transgenic mice expressing the human neurofilament heavy gene: A mouse model of amyotrophic lateral sclerosis, *Cell* **73**:35–46.

Couillard-Depres, S., Zhu, Q., Wong, P. C., Price, D. L., Cleveland, D. W., and Julien, J. P., 1998, Protective effect of neurofilament heavy gene overexpression in motor neuron disease induced by mutant superoxide dismutase, *Proc. Natl. Acad. Sci. U.S.A.* **95**:9626–9630.

Coulpier, M., Junier, M.-P., Peschanski, M., and Dreyfus, P. A., 1996, Bcl-2 sensitivity differentiates two pathways for motoneuronal death in the wobbler mutant mouse, *J. Neurosci.* **16**:5897–5904.

Dal Canto, M. C., and Gurney, M. E., 1994, Development of central nervous system pathology in a murine transgenic model of human amyotrophic lateral sclerosis, *Am. J. Pathol.* **145**:1271–1279.

Dal Canto, M. C., and Gurney, M. E., 1995, Neuropathological changes in two lines of mice carrying a transgene for mutant human Cu,Zn SOD, and in mice overexpressing wild type SOD: A model of familial amyotrophic lateral sclerosis (FALS), *Brain Res.* **676**:25–40.

Dal Canto, M. C., and Gurney, M. E., 1997, A low expressor line of transgenic mice carrying a mutant human Cu,Zn superoxide dismutase (SOD1) gene develops pathological changes that most closely resemble those in human amyotrophic lateral sclerosis, *Acta Neuropathol.* **93**:537–550.

Danbolt, N. C., Storm-Mathisen, J., and Kanner, B. I., 1992, An Na/K coupled L-glutamate transporter purified from rat brain is located in glial cell processes, *Neuroscience* **51**:295–310.

Deng, H.-X., Hentati, A., Tainer, J. A., Iqbal, Z., Cayabyab, A., Hung, W.-Y., Getzoff, E. D., Hu, P., Herzfeldt, B., Roos, R. P., Warner, C., and Siddique, T., 1993, Amyotrophic lateral sclerosis and structural defects in Cu,Zn superoxide dismutase, *Science* **261**:1047–1051.

Dubois-Dauphin, M., Frankowski, H., Tsujimoto, Y., Huarte, J., and Martinou, J. C., 1994, Neonatal motoneurons overexpressing the bcl-2 protooncogene in transgenic mice are protected from axotomy-induced cell death, *Proc. Natl. Acad. Sci. USA* **91:**3309–3313.

Duchen, L. W., and Strich, S. J., 1968, An hereditary motor neuron disease with progressive denervation of muscle in the mouse: The mutant "wobbler," *J. Neurol. Neurosurg. Psychiatry* **31:**535–542.

Emery, A. E. H., and Holloway, S., 1983, Familial motor neuron diseases, *Adv. Neurol.* **36:**139–147.

Eyer, J., and Peterson, A., 1994, Neurofilament-deficient axons and perikaryal aggregates in viable transgenic mice expressing a neurofilament-beta-galactosidase fusion protein, *Neuron* **12:**389–405.

Fairman, W. A., Vandenberg, R. J., Arriza, J. L., Kavanaugh, M. P., and Amara, S. G., 1995, An excitatory amino-acid transporter with properties of a ligand-gated chloride channel, *Nature* **375:**599–603.

Figlewicz, D. A., Krizus, A., Martinoli, M. G., Meininger, V., Dib, M., Rouleau, G. A., and Julien, J. P., 1994, Variants of the heavy neurofilament subunit are associated with the development of amyotrophic lateral sclerosis, *Hum. Mol. Genet.* **3:**1757–1761.

Fridovich, I., 1995, Superoxide radical and superoxide dismutases, *Annu. Rev. Biochem.* **64:**97–112.

Friedlander, R. M., Brown, R. H., Gagliardini, V., Wang, J., and Yuan, J., 1997a, Inhibition of KE slows ALS in mice, *Nature* **388:**31.

Friedlander, R. M., Gagliardini, V., Hara, H., Fink, K. B., Li, W., MacDonald, G., Fishman, M. C., Greenberg, A. H., Moskowitz, M. A., and Yuan, J., 1997b, Expression of a dominant negative nutant of interleukin-IB-converting enzyme in transgenic mice provents neuronal cell death induced by trophic factor withdrawal and ischemic brain injury, *J. Exp. Med.* **185:**933–940.

Gonatas, N. K., Stieber, A., Mourelatos, Z., Chen, Y., Gonatas, J. O., Appel, S. H., Hays, A. P., Hickey, W. F., and Hauw, J. J., 1992, Fragmentation of the Golgi apparatus of motor neurons in amyotrophic lateral sclerosis, *Am. J. Pathol.* **140:**731–737.

Greenlund, L. J., Deckwerth, T. L., and Johnson, E. M., Jr., 1995, Superoxide dismutase delays neuronal apoptosis: A role for reactive oxygen species in programmed neuronal death, *Neuron* **14:**303–315.

Gurney, M. E., Pu, H., Chiu, A. Y., Dal Canto, M. C., Polchow, C. Y., Alexander, D. D., Caliendo, J., Hentati, A., Kwon, Y. W., and Deng, H. X., 1994, Motor neuron degeneration in mice that express a human Cu,Zn superoxide dismutase mutation, *Science* **264:** 1772–1775.

Gurney, M. E., Cutting, F. B., Zhai, P., Doble, A., Taylor, C. P., Andrus, P. K., and Hall, E. D., 1996, Benefit of vitamin E, riluzole, and gabapentin in a transgenic model of familial amyotrophic lateral sclerosis, *Ann Neurol.* **39:**147–157.

Halliwell, B., 1994, Free radicals, antioxidants, and human disease: Curiosity, cause, or consequence? *Lancet* **344:**721–724.

Hasse, G., Kennel, P., Pettmann, B., Vigne, E., Akli, S., Revah, F., Schmalbruch, H., and Kahn, A., 1997, Gene therapy of murine motor neuron disease using adenoviral vectors for neurotrophic factors, *Nat. Med.* **3:**429–436.

Hirano, A., 1988, "Color Atlas of Pathology of the Nervous System," Igaku-shoin, New York.

Hirano, A., 1991, Cytopathology of amyotrophic lateral sclerosis, in "Advances in Neurology, Vol. 56. Amyotrophic Lateral Sclerosis and Other Motor Neuron Diseases" (L. P. Rowland, ed.), pp. 91–101. Raven Press, New York.

Hirano, A., Donnenfeld, H., Sasaki, S., and Nakano, I., 1984a, Fine structural observations of neurofilamentous changes in amyotrophic lateral sclerosis, *J. Neuropathol. Exp. Neurol.* **43:**461–470.

Hirano, A., Nakano, I., Kurland, L. T., Mulder, D. W., Holley, P. W., and Saccomanno, G., 1984b, Fine structural study of neurofibrillary changes in a family with amyotrophic lateral sclerosis, *J. Neuropathol. Exp. Neurol.* **43:**471–480.

Kanai, Y., and Hediger, M. A., 1992, Primary structure and function characterization of a high-affinity glutamate transporter, *Nature* **360:**467–471.

Kane, D. J., Sarafian, T. A., Anton, R., Hahn, H., Gralla, E. B., Valentine, J. S., Ord, T., and Bredesen, D. E., 1993, Bcl-2 inhibition of neural death: Decreased generation of reactive oxygen species, *Science* **262:**1274–1277.

Kato, S., Shimono, M., Watanabe, Y., Nakashima, K., Takahashi, K., and Ohama, E., 1996, Familial amyotrophic lateral sclerosis with a two base pair deletion in superoxide dismutase 1 gene: Multisystem degeneration with intracytoplasmic hyaline inclusions in astrocytes, *J. Neuropathol. Exp. Neurol. 55:*1089–1101.

Kato, S., Hayashi, H., Nakashima, K., Nanba, E., Kato, M., Hirano, A., Nakano, I., Asayama, K., and Ohama, E., 1997, Pathological characterization of astrocytic hyaline inclusions in familial amyotrophic lateral sclerosis, *Am. J. Pathol.* **151:**611–620.

Kawamura, Y., Dyck, P. J., Shimono, M., Okazaki, H., Tateishi, J., and Doi, H., 1981, Morphometric comparison of the vulnerability of peripheral motor and sensory neurons in amyotrophic lateral sclerosis, *J. Neuropathol. Exp. Neurol.* **40:**667–675.

Kostic, V., Jackson-Lewis, V., deBilbao, F., Dubois-Dauphin, M., and Przenborski, S., 1997, Bcl-2 prolonging life in a transgenic mouse model of a familial amyotrophic lateral sclerosis, *Science* **277:**559–562.

LaVail, J. H., Koo, E. H., and Dekker, N. P., 1987, Motoneuron loss in abducens nucleus of wobbler mice, *Brain Res.* **404:**127–132.

Lee, M. K., Xu, Z., Wong, P. C., and Cleveland, D. W., 1993, Neurofilaments are obligate heteropolymers *in vivo, J. Cell Biol.* **122:**1337–1350.

Lee, M. K., Marszalek, J. R., and Cleveland, D. W., 1994, A mutant neurofilament subunit causes massive, selective motor neuron death: Implications for the pathogenesis of human motor neuron disease, *Neuron* **13:**975–988.

Leigh, P. N., Whitwell, H., Garofalo, O., Buller, J., Swash, M., Martin, J. E., Gallo, J. M., Weller, R. O., and Anderton, B. H., 1991, Ubiquitin-immunoreactive intraneuronal inclusions in amyotrophic lateral sclerosis. Morphology, distribution, and specificity, *Brain* **114:**775–788.

Levy, L. M., Lehre, K. P., Rolstad, B., and Danbolt, N. C., 1993, A monoclonal antibody raised against an Na/K coupled L-glutamate transporter purified from rat brain confirms glial cell localization, *FEBS Lett.* **317:**79–84.

Li, T.-M., Alberman, E., and Swash, M., 1988, Comparison of sporadic and familial disease amongst 580 cases of motor neuron disease, *J. Neurol. Neurosurg. Psychiatry* **51:**778–784.

Liu, X. F., Elashvili, I., Gralla, E. B., Valentine, J. S., Lapinskas, P., and Culotta, V. C., 1992, Yeast lacking superoxide dismutase. Isolation of genetic suppressors, *J. Biol. Chem.* **267:**18298–18302.

Marszalek, J. R., Williamson, T. L., Lee, M. K., Xu, Z.-S., Crawford, T. O., Hoffman, P. N., and Cleveland, D. W., 1996, Neurofilament subunit NF-H modulates axonal diameter by affecting the rate of neurofilament transport, *J. Cell Biol.* **135:**711–724.

Martinou, J. C., Dubois-Dauphin, M., Staple, J. K., Rodriguez, I., Frankowski, H., Missotten, M., Albertini, P., Talabot, D., and Pietra, C., 1994, Overexpression of BCL-2 in transgenic mice protects neurons from naturally occurring cell death and experimental ischemia, *Neuron* **13:**1017–1030.

McCray, P. B., 1985, Vitamin E: Interactions with free radicals and ascorbate, *Annu. Rev. Nutr.* **5:**323–340.

Michaelidis, T. M., Sendtner, M., Cooper, J. D., Airaksinen, M. S., Holtman, B., Meyer, M., and Theonen, H., 1996, Inactivation of bcl-2 results in progressive degeneration of motor neurons, sympathetic and sensory neurons during early postnatal development, *Neuron* **17:**75–89.

Mitsumoto, H., Ikeda, K., Holmlund, T., Greene, T., Cedarbaum, J. M., Wong, V., and Lindsay, R.,

1994a, The effects of ciliary neurotrophic factor on motor dysfunction in wobbler mouse motor neuron disease, *Ann. Neurol.* **36:**142–148.

Mitsumoto, H., Ikeda, K., Klinkosz, B., Cedarbaum, J. M., Wong, V., and Lindsay, R. M., 1994b, Arrest of motor neuron disease in wobbler mice cotreated with CNTF and BDNF, *Science* **265:**1107–1110.

Nicholls, D., and Atwell, D., 1990, The release and uptake of excitatory amino acids, *Trends Pharmacol. Sci.* **11:**462–468.

Ohara, O., Gahara, Y., Miyake, T., Teraoka, H., and Kitamura, T., 1993, Neurofilament deficiency in quail caused by nonsense mutation in neurofilament-L gene, *J. Cell Biol.* **121:**387–395.

Oppenheim, R. W., Houenou, L. J., Johnson, J. E., Lin, L. F., Li, L., Lo, A. C., Newsome, A. L., Prevette, D. M., and Wang, S., 1995, Developing motor neurons rescued from programmed and axontomy-induced cell death by GDNF, *Nature* **373:**344–346.

Orrell, R., Belleroche, J., Marklund, S., Bowe, F., and Hallewell, R., 1995, A novel SOD mutant and ALS, *Nature* **374:**504–505.

Peled-Kamar, M., Lotem, J., Wirguin, I., Weiner, L., Hermalin, A., and Groner, Y., 1997, Oxidative stress mediates impairment of muscle function in transgenic mice with elevated level of wild-type Cu/Zn superoxide dismutase, *Proc. Natl. Acad. Sci. USA* **94:**3883–3887.

Pines, G., Danbolt, N. C., Bjoras, M., Zhang, Y., Bendahan, A., Eide, L., Koepsell, H., Storm-Mathisen, J., Seeberg, E., and Kanner, B. I., 1992, Cloning and expression of a rat brain L-glutamate transporter, *Nature* **360:**464–467.

Plaitakis, A., Constantakakis, E., and Smith, J., 1988, The neuroexcitotoxic amino acids glutamate and aspartate are altered in the spinal cord and brain in amyotrophic lateral sclerosis, *Ann. Neurol.* **24:**446–449.

Reaume, A. B., Elliott, J. L., Hoffman, E. K., Kowall, N. W., Ferrante, R. J., Siwek, D. F., Wilcox, H. M., Flood, D. G., Beal, M. F., Brown, R. H., Scott, R. W., and Snider, W. D., 1996, Motor neurons in Cu/Zn superoxide dismutase-deficient mice develop normally but exhibit enhanced cell death after axonal injury, *Nat. Genet.* **13:**43–47.

Ripps, M. E., Huntley, G. W., Hof, P. R., Morrison, J. H., and Gordon, J. W., 1995, Transgenic mice expressing an altered murine superoxide dismutase gene provide an animal model of amyotrophic lateral sclerosis, *Proc. Natl. Acad. Sci. USA* **92:**689–693.

Robberecht, W., Sapp, P., Viaene, M. K., Rosen, D., McKenna-Yasek, D., Haines, J., Horvitz, R., Theys, P., and Brown, R., 1994, Cu/Zn superoxide dismutase activity in familial and sporadic amyotrophic lateral sclerosis, *J. Neurochem.* **62:**384–387.

Rooke, K., Figlewicz, D. A., Han, F., and Rouleau, G. A., 1996, Analysis of the KSP repeat of the neurofilament heavy subunit in familial amyotrophic lateral sclerosis, *Ann. Neurol.* **46:**789–790.

Rosen, D. R., Siddique, T., Patterson, D., Figlewicz, D. A., Sapp, P., Hentati, A., Donaldson, D., Goto, J., O'Regan, J. P., Deng, H. X., and Brown, R., 1993, Mutations in Cu/Zn superoxide dismutase gene are associated with familial amyotropic lateral sclerosis, *Nature* **362:**59–62.

Rothstein, J. D., and Kuncl, R. W., 1995, Neuroprotective strategies in a model of chronic glutamate-mediated motor neuron toxicity, *J. Neurochem.* **65:**643–651.

Rothstein, J. D., Tsai, G., Kuncl, R. W., Clawson, L., Cornblath, D. R., Drachman, D. B., Pestronk, A., Stauch, B. L., and Coyle, J. T., 1990, Abnormal excitatory amino acid metabolism in amyotrophic lateral sclerosis, *Ann. Neurol.* **28:**18–25.

Rothstein, J. D., Martin, L. J., and Kuncl, R. W., 1992, Decreased glutamate transport by the brain and spinal cord in amyotrophic lateral sclerosis, *N. Engl. J. Med.* **326:**1464–1468.

Rothstein, J. D., Jin, L., Dykes-Hoberg, M., and Kuncl, R. W., 1993, Chronic inhibition of glutamate uptake produces a model of slow neurotoxicity, *Proc. Natl. Acad. Sci. USA* **90:**6591–6595.

Rothstein, J. D., Bristol, L. A., Hosler, B., Brown, R. H., Jr., and Kuncl, R. W., 1994a, Chronic inhibition of superoxide dismutase produces apoptotic death of spinal neurons, *Proc. Natl. Acad. Sci. USA* **91:**4155–4159.

Rothstein, J. D., Martin, L., Levey, A. I., Dykes-Hoberg, M., Jin, L., Wu, D., Nash, N., and Kuncl, R. W., 1994b, Localization of neuronal and glial glutamate transporters, *Neuron* **13:**713–725.

Rothstein, J. D., Van Kammen, M., Levey, A. I., Martin, L. J., and Kuncl, R. W., 1995, Selective loss of glial glutamate transporter GLT-1 in amyotrophic lateral sclerosis, *Ann. Neurol.* **38:**73–84.

Rothstein, J. D., Dykes-Hoberg, M., Pardo, C. A., Bristol, L. A., Jin, L., Kuncl, R. W., Kanai, Y., Hediger, M. A., Wang, Y., Schielke, J., and Welty, D., 1996, Knockout of glutamate transporters reveals a major role for astroglial transport in excitotoxicity and clearance of glutamate, *Neuron* **16:**674–686.

Rouleau, G. A., Clark, A. W., Rooke, K., Pramatarova, A., Krizus, A., Suchowersky, O., Julien, J.-P., and Figlewicz, D., 1996, SOD1 mutation is associated with accumulation of neurofilaments in amyotrophic lateral sclerosis, *Ann. Neurol.* **39:**128–131.

Sagot, Y., Dubois-Dauphin, M., Tan, S. A., de Bilbao, F., Aebischer, P., Martinou, J. C., and Kato, A. C., 1995a, Bcl-2 overexpression prevents motoneuron cell body loss but not axonal degeneration in a mouse model of a neurodegenerative disease, *J. Neurosci.* **15:**7727–7733.

Sagot, Y., Tan, S. A., Baetge, E., Schmalbruch, H., Kato, A. C., and Aebischer, P., 1995b, Polymer encapsulated cell lines genetically engineered to release ciliary neurotrophic factor can slow down progressive motor neuronopathy in the mouse, *Eur. J. Neurosci.* **7:**1313–1322.

Sagot, Y., Tan, S. A., Hammang, J. P., Aebischer, P., and Kato, A. C., 1996, GDNF slows loss of motoneurons but not axonal degeneration or premature death of pmn/pmn mice, *J. Neurosci.* **16:**2335–2341.

Sakaguchi, T., Okada, M., Kitamura, T., and Kawasaki, K., 1993, Reduced diameter and conduction velocity of myelinated fibers in the sciatic nerve of a neurofilament-deficient mutant quail, *Neurosci. Lett.* **153:**65–68.

Schmalbruch, H., Jensen, H.-J. S., Bjrg, M., Kamieniecka, Z., and Kurland, L., 1991, A new mouse mutant with progressive motor neuronopathy, *J. Neuropathol. Exp. Neurol.* **50:**192–204.

Schochet, S. S., Hardman, J. M., Ladewig, P. P., and Earle, K. M., 1969, *Intraneuronal conglomerates in sporadic motor neuron disease. A light and electron microscopic study, Arch. Neurol.* **20:**548–553.

Sendtner, M., Kreutzberg, G. W., and Theonen, H., 1990, Ciliary neurotrophic factor prevents the degeneration of motor neurons after axotomy, *Nature* **345:**440–441.

Sendtner, M., Holtman, B., Kolbeck, R., Theonen, H., and Borde, Y.-A., 1992a, Brain-derived neurotrophic factor prevents the death of motoneurons in newborn rats after nerve section, *Nature* **360:**757–759.

Sendtner, M., Schmalbruch, H., Stocku, K. A., Carroll, P., Kreutzberg, G. W., and Theonen, H., 1992b, Ciliary neurotrophic factor prevents degeneration of motor neurons in mouse mutant progressive motor neuronopathy, *Nature* **358:**502–504.

Shaw, P. J., Chinnery, R. M., and Ince, P. G., 1994, D-Aspartate binding sites in the normal human spinal cord and changes in motor neuron disease: A quantitative autoradiographic study, *Brain Res.* **655:**195–201.

Shibata, N., Hirano, A., Kobayashi, M., Siddique, T., Deng, H.-X., Hung, W.-Y., Kato, T., and Asayama, K., 1996a, Intense superoxide dismutase-1 immunoreactivity in intracytoplasmic hyaline inclusions of familial amyotrohpic lateral sclerosis with posterior column involvement, *J. Neuropathol. Exp. Neurol.* **55:**481–490.

Shibata, N., Hirano, A., Kobayashi, M., Siddique, T., Deng, H. X., Hung, W. Y., Kato, T., and Asayama, K. 1996b, Intense superoxide dismutase-1 immunoreactivity in intracytoplasmic hyaline inclusions of familial amyotrophic lateral sclerosis with posterior column involvement, *J. Neuropathol. Exp. Neurol.* **55:**481–490.

Siddique, T., and Deng, H. X., 1996, Genetics of amyotrophic lateral sclerosis, *Hum. Mol. Genet,* **5** (Spec No.):1465–1470.

Singh, R. J., Karoui, H., Gunther, M. R., Beckman, J. S., Mason, R. P., and Kalyanaraman, B., 1998, Reexamination of the mechanism of hydroxyl radical adducts formed from the reaction between familial amyotrophic lateral sclerosis-associated Cu,Zn superoxide dismutase mutants and H_2O_2. *Proc. Natl. Acad. Sci. U.S.A.* **95:**6675–6680.

Tanaka, K., Watase, K., Manabe, T., Yamada, K., Watanabe, M., Takahashi, K., Iwama, H., Nishikawa, T., Ichihara, N., Kikuchi, T., Okuyama, S., Kawashima, N., Hori, S., Takimoto, M., and Wada, K., 1997, Epilepsy and exacerbation of brain injury in mice lacking the glutamate transporter GLT-1, *Science* **276:**1699–1702.

Troy, C. M., Derossi, S., Prochiantz, A., Greene, L. A., and Shelanski, M. L., 1996, Down-regulatoin of SOD1 leads to cell death by the NO-peroxynitrite pathway, *J. Neurosci.* **16:**253–261.

Tu, P. H., Raju, P., Robinson, K. A., Gurney, M. E., Trojanowski, J. Q., and Lee, V. M.-Y., 1996, Transgenic mice carrying a human mutant superoxide dismutase transgene develop neuronal cytoskeletal pathology resembling human amyotrophic lateral sclerosis lesions, *Proc. Natl. Acad. Sci. USA* **93:**3155–3160.

Vechio, J. D., Bruijn, L. I., Xu, Z., Brown, R. H., Jr., and Cleveland, D. W., 1996, Sequence variants in human neurofilament proteins: Absence of linkage to familial amyotrophic lateral sclerosis, *Ann. Neurol.* **40:**603–610.

Wiedau-Pazos, M., Goto, J. J., Rabizadeh, S., Gralla, E. B., Roe, J. A., Lee, M. K., Valentine, J. S., and Bredesen, D. E., 1996, Altered reactivity of superoxide dismutase in familial amyotrophic lateral sclerosis [see comments], *Science* **271:**515–518.

Wiley, C. A., Love, S., Skoglund, R. R., and Lampert, P. W., 1987, Infantile neurodegenerative disease with neuronal accumulation of phosphorylated neurofilaments, *Acta Neuropathol. (Berl.)* **72:**369–376.

Williamson, T. L., Bruijn, L. I., Zhu, Q., Anderson, K. L., Anderson, S. D., Julien, J.-P., and Cleveland, D. W., 1998, Absence of neurofilaments reduces the selective vulnerability of motor neurons and slows disease caused by a familial ALS-linked SOD1 mutant, *Proc. Natl. Acad. Sci. U.S.A.* **95:**9631–9636.

Williamson, T. L., and Cleveland, D. W., Slowing of axonal transport is a very early event in the selective toxicity of ALS-linked SOD1 mutants to motor neurons, *Nature Neurosci.,* in press.

Wong, P. C., Pardo, C. A., Borchelt, D. R., Lee, M. K., Copeland, N. G., Jenkins, N. A., Sisodia, S. S., Cleveland, D. W., and Price, D. L., 1995, An adverse property of a familial ALS-linked SOD1 mutation causes motor neuron disease characterized by vacuolar degeneration of mitochondria, *Neuron* **14:**1105–1116.

Xu, Z., Cork, L. C., Griffin, J. W., and Cleveland, D. W., 1993, Increased expression of neurofilament subunit NF-L produces morphological alterations that resemble the pathology of human motor neuron disease, *Cell* **73:**23–33.

Yan, Q., Matheson, C., Lopez, O. T., and Miller, J. A., 1994, The biological responses of axotomized adult motoneurons to brain derived neurotrophic factor, *J. Neurosci.* **14:**5281–5291.

Yan, Q., Matheson, C., and Lopez, O. T., 1995, *In vivo* neurotrophic effect of GDNF on neonatal and adult facial motor neurons, *Nature* **373:**341–344.

Yu, B. P., 1994, Cellular defenses against damage from reaactive oxygen species, *Physiol. Rev.* **74:**138–162.

Zhong, L. T., Sarafian, T., Kane, D. J., Charles, A. C., Mah, S. P., Edwards, R. H., and Bredesen, D. E., 1993, bcl-2 inhibits death of central neural cells induced by multiple agents, *Proc. Natl. Acad. Sci. USA* **90:**4533–4537.

CHAPTER 8

TRANSGENIC MOUSE MODELS OF CAG TRINUCLEOTIDE REPEAT NEUROLOGIC DISEASES

HARRY T. ORR and HUDA Y. ZOGHBI

1. INTRODUCTION

During the past few years a new class of mutation has burst upon the human genetics scene, resulting in a rethinking of some basic principles of Mendelian genetics. This form of mutation, termed unstable trinucleotide repeats, is a dynamic mutation that is produced by heritable unstable DNA. For reasons not understood, these dynamic mutations seem to effect predominately genes whose function is critical for neuromuscular tissue. To date 12 diseases have been characterized as caused by the expansion of an unstable trinucleotide repeat.

HARRY T. ORR • Departments of Laboratory Medicine and Pathology, and Biochemistry, and Institute of Human Genetics, University of Minnesota, Minneapolis, Minnesota 55455. *HUDA Y. ZOGHBI • Departments of Pediatrics, and Molecular and Human Genetics, and Howard Hughes Medical Institute, Baylor College of Medicine, Houston, Texas 77030.*

Mouse Models in the Study of Genetic Neurological Disorders, Volume 9 of *Advances in Neurochemistry,* edited by Popko, Kluwer Academic/Plenum Publishers, New York, 1999.

Among these are fragile XA (Kremer *et al.*, 1991; Verkerk *et al.*, 1991; Yu *et al.*, 1991) and fragile XE (Knight *et al.*, 1993) syndromes, myotonic dystrophy (Brook *et al.*, 1992; Mahadevan *et al.*, 1992), spinal and bulbar muscular atrophy (SBMA) (LaSpada *et al.*, 1991), Huntington disease (HD) (Huntington's Disease Research Collaborative Group, 1993), Friedreich's ataxia (Campuzano *et al.*, 1996), dentatrubral–pallidoluysian atrophy (DRPLA) (Koide *et al.*, 1994; Nagafuchi *et al.*, 1994)/Haw River syndrome (Burke *et al.*, 1994), and the spinocerebellar ataxias SCA1 (Orr *et al.*, 1993), SCA2 (Imbert *et al.*, 1996; Pulst *et al.*, 1996; Sanpei *et al.*, 1996), SCA3/Machado-Joseph disease (MJD) (Kawaguchi *et al.*, 1994), SCA6 (Zhuchenko *et al.*, 1997), and SCA7 (Stevanin *et al.*, 1996).

The type of triplet repeat and its location within the affected gene varies between the different disorders. Yet, eight loci responsible for neurological disease contain unstable CAG repeats that are located within the coding region of the involved gene. These CAG-based disorders are *SBMA, DRPLA, HD, SCA1, SCA2, SCA3/MJD, SCA6* and *SCA7* (Table 1). Variation in the number of repeats at these loci results in a change in the length of a polyglutamine tract in their respective protein. Outside of the polyglutamine tract, these proteins are unrelated. Furthermore, subcellular localization studies suggest differing functions for these proteins. However, there are a number of similarities among these CAG repeats that have led to some generalizations. Most notable among these are the concepts that instability and expansions are via a common mechanism and that expansion of the CAG repeat into the affected size range produces a gain-of-function alteration in the protein that in turn is the basis of pathogenesis. Thus, in each disorder, an increase in length of a polyglutamine tract confers a toxic property to the protein, causing dysfunction and death of a select population of vulnerable nerve cells. When considering models of pathogenesis, this latter point is critical. In each of the CAG triplet based disorders described to date, the affected gene is expressed widely throughout the nervous system and peripheral tissues. Yet, in each disease only select populations of nerve cells appear to be susceptible to the deleterious effects of the expanded polyglutamine tract. Furthermore, the cells affected vary between each disorder. For a model of pathogenesis to be viable, this cell specificity of pathogenesis must be explained.

2. PATHOLOGY AND NEUROLOGICAL ALTERATION IN TRANSGENIC MICE WITH EXPANDED CAG TRACTS

As part of efforts to understand the mechanism of pathogenesis of the CAG trinucleotide diseases, investigators have begun to establish transgenic mouse models for several of these disorders (Table 2). These model systems provide an excellent opportunity to examine at the molecular and cellular levels the course

TABLE 1. CAG Trinucleotide Disorders

Disease	Gene product function/ subcellular location	CAG repeat size range		Cellular pathology
		Unaffected	Affected	
Spinal & Bulbar Muscular Atrophy (SBMA)	Androgen receptor/cytoplasmic & nuclear	12–34	40–62	Severe loss of anterior horn cells & neurons in cranial motor nuclei V, VI, VII & XII.
Huntington Disease (HD)	unknown/cytoplasmic (nuclear?)	6–37	35–121	Severe loss of neurons in the caudate/putamen. Loss of globus pallidus, sustania nigra, subthalamic nuclei, ventrolateral thalamic nucleus & cerebral cortex.
Spinocerebellar Ataxia Type 1 (SCA1)	unknown/ predominantly nuclear, some cytoplasmic	6–44	40–81	Severe Purkinje cell loss. Loss of neurons in olivary, pontine, dentate nuclei, & motor cranial nuclei IX, X, & XII.
Machado–Joseph Disease (MJD/ SCA3)	unknown	12–40	61–84	Severe loss of neurons in globus pallidus, red, subthalamic, pontine, & dentate nuclei. Loss of Purkinje cells.
Spinocerebellar Ataxia Type 2 (SCA2)	unknown	15–29	35–59	Severe Purkinje cell loss. Loss of neurons in olivary, & pontine nuclei, & substantia nigra
Spinocerebellar Ataxia Type 6 (SCA6)	voltage-dependent Ca channel	4–16	21–27	Severe Purkinje cell loss. Moderate loss of granule cells, dentate nucleus & inferior olivary neurons.
Spinocerebellar Ataxia Type 7 (SCA7)	unknown	unknown	unknown	
Dentatrubal-Pallidoluysian Atrophy (DRPLA)	unknown	7–34	54–70	Severe loss of neurons in the globus pallidus, subthalamic nucleus, red nucleus, & dentate nucleus. Loss of neurons in the caudate/putamen.

TABLE 2. Published Reports of CAG Transgenic Mice

Transgene (disease)	Promotor	Repeat size	RNA expression	Protein expression	Pathology/ Phenotype	Repeat instability		Reference
						Germline	Somatic	
Mx-AR, 24 (SBMA)	Mx1 (intracellular anti-influenza protein)	24	muscle, brain, kidney (RT-PCR)	kidney & lung (western)	normal/normal	stable	stable	Bingham *et al.*, 1995
Mx-AR^{exp}-1	Mx1	45	n.d.[a] (RT-PCR)	kidney & lung (western)	normal/normal	stable	stable	Ibid.
Mx-AR^{exp}-2	Mx1	45	brain, muscle, spinal cord, kidney (RT-PCR)	muscle, spinal cord, kidney, lung (western)	normal/normal	stable	stable	Ibid.
NSE-AR-1	neuron-specific enolase	24	n.d.[a] (RT-PCR)	neg (western)	normal/normal	stable	stable	Ibid.
NSE-AR-2	′	24	brain, muscle (RT-PCR)	neg. (western)	normal/normal	stable	stable	Ibid.
NSE-AR^{exp}	′	45	brain (RT-PCR)	neg. (western)	normal/normal	stable	stable	Ibid.
HD-44 (HD)	CMV	44	brain (northern)	neg. (western)	normal/normal	stable	stable	Goldberg *et al.*, 1996
HA-Q79C (MJD)	L7/Pcp-2 Purkinje cell Specific)	79 (protein fragment)	n.d.[a]	n.d.[a]	severe cerebellar atrophy/ataxia	stable	n.a.[b]	Ibid.

HA-Q79	L7/Pcp-2	79 (protein fragment)	n.d.[a]	n.d.[a]	n.d./ataxia	stable	n.a.[b]	Ibid.
HA-MJD79	L7/Pcp-2	79 (intact fragment)	n.d.[a]	n.d.[a]	normal/normal	stable	n.a.[b]	Ibid.
HA-Q35C	L7/Pcp-2	35 (protein fragment)	n.d.[a]	n.d.[a]	normal/normal	stable	n.a.[b]	Ibid.
PS-30 (SCA1)	L7/Pcp-2	30	cerebellum (northern)	cerebellum (western)	normal/normal	stable	n.a.[b]	Burright *et al.*, 1995
PS-82	L7/Pcp-2	82	cerebellum (northern)	neg. (western)	Purkinje cell loss/ ataxia	stable	n.a.[b]	Ibid.
HD-R6/0 (HD)	unknown	130–145	neg. (RT-PCR)	n.d.[a]	normal/normal	uncertain	stable	Mangiarini *et al.*, 1996; 1997
HD-R6/1	unknown	110–120	wide spread (RT-PCR)	neg. (western)	normal/movement disorders, death	unstable	unstable	Ibid.
HD-R6/2	unknown	130–149	wide spread (RT-PCR)	wide spread (western)	normal/movement disorders, death	unstable	unstable	Ibid.
HD-R6/5	unknown	125–159	wide spread (RT-PCR)	wide spread (western)	normal/normal	unstable	unstable	Ibid.

[a]Not detected.
[b]Not analyzed.

of disease *in vivo*. The results of these studies are beginning to provide a basis for understanding the factors important in repeat instability, the selective nerve cell vulnerability, the toxic properties of proteins with expanded polyglutamine tracts, and the temporal and spatial aspect of each disease.

2.1. Spinal and Bulbar Muscular Atrophy

Spinal and bulbar muscular atrophy (SBMA or Kennedy's disease) is an X-linked motor neuron disease caused by the expansion of an unstable CAG trinucleotide within the androgen receptor gene (LaSpada *et al.,* 1991). On unaffected chromosomes the *AR* gene contains from 11 to 33 CAG repeats while on SBMA affected chromosomes there are 38 to 66 repeats in the *AR* gene. Males affected with SBMA have a progressive limb weakness and swallowing problems as adults. Death usually results from complications caused by the loss of neurons within cranial nerve nuclei in the brainstem. SBMA pathology is characterized by the loss of lower motor neurons and primary sensory ganglion cells. Secondary changes in striated muscle and peripheral nerve are found. Females heterozygote for an expanded *AR* CAG repeat are asymptomatic. Importantly, males with a deletion of the *AR* gene do not develop SBMA. Rather they develop testicular feminization. Thus, the motor neuron disease associated with CAG repeat expansion in SBMA is unlikely to be due to a loss of AR function. However, the fact that SBMA affected males are often sterile and present with some gynecomastia suggests that expansion of the AR CAG repeat does result in some loss of AR function.

AR CAG repeats within the unaffected size range do not change in length upon transmission from parent to offspring. In contrast, AR CAG repeats in the affected size range often change in length, either contracting or expanding, upon transmission from generation to generation. An expanded AR CAG repeat changes in size more frequently when transmitted from an affected male than when inherited from a female carrier. A relationship between the size of the AR CAG repeat and clinical severity of SBMA exists. The severity of the clinical presentation of SBMA increases with increasing length of the expanded CAG repeat.

Bingham and colleagues (1995) have established several lines of transgenic mice expressing human cDNA *AR* transgenes with either 24 CAGs (unaffected allele) or 45 CAGs (affected allele). Expression of the *AR* transgene, directed from either the mouse Mx or neural specific enolase promoters, was detected at the mRNA and protein levels, although at a lower level than the expression of the endogenous receptor. Interestingly, overexpression of wild-type *AR* yielded mice having a greater grip strength, suggesting that the *AR* is important for motor neuron function. However, transgenic mice expressing an *AR* transgene with 45 CAG repeats failed to develop any pathological signs. Furthermore, the expanded allele was transmitted between generations without any change in repeat length. As suggested by these investigators, the fact that only a low level expression of the *AR* cDNA transgene was achieved and that an *AR*-affected allele having a relative few number of repeats, 45 CAGs, was used most likely explain the lack of an altered phenotype and repeat instability in these transgenic mice.

2.2. Machado–Joseph Disease/Spinocerebellar Ataxia Type 3

Machado–Joseph disease (MJD) is an autosomal dominant spinocerebellar degeneration, originally described in Portuguese families. MJD presents clinically with cerebellar ataxia, dystonia or rigidity, ophthalmoplegia, and sometimes with bulging eyes. The areas of the brain that demonstrate pathology in MJD include the globus pallidus, subthalamic nucleus, red nucleus, pontine nucleus, and the dentate nucleus (Maruyama *et al.,* 1995). Purkinje cells of the cerebellar cortex are affected to a lesser degree. MJD is also caused by the expansion of a CAG repeat located within the coding region of the *MJD1* gene on chromosome 14q32.1 (Kawaguchi *et al.,* 1994). Unaffected chromosomes have from 13 to 36 CAG repeats in the *MJD1* gene, whereas MJD chromosomes have from 61 to 84 CAG repeats. The *MJD1* repeat length is inversely correlated with disease severity and age at onset. There is also a greater degree of CAG repeat expansion in paternal transmission of affected alleles than in maternal transmissions.

To investigate the molecular events responsible for MJD/SCA3, Ikeda and colleagues (1996) expressed MJD1 protein in a transfected cell line and in Purkinje cells of transgenic mice. These experiments utilized two basic forms of the MJD protein, the intact MJD protein and a segment of the protein that consisted primarily of the polyglutamine tract. When an intact MJD protein, with either a normal number of glutamine residues (35) or an expanded number (79), was expressed in transfected COS cells or in Purkinje cells of transgenic mice, no alterations in cell viability or phenotype were seen. In contrast, experiments with a MJD fragment containing the polyglutamine stretch yield different results. Again a fragment containing 35 glutamine residues had no effect. However, in cell lines and in Purkinje cells, expression of a fragment from MJD1 with 79 glutamine residues caused cell death. Interestingly, analyses of the expressed polyglutamine peptide by western blotting and immunocytochemistry in COS cells indicated that the fragment containing 79 glutamines was covalently modified such that it formed into large complexes.

Transgenic mice in which the polyglutamine peptide with 79 residues was expressed in Purkinje cells developed abnormal neurological behaviors that consisted of a wide-based hindlimb stance, frequent falling when moving, and inability to rear on their hindlegs. These signs developed as early as the fourth week after birth. Pathologically, cerebella from mice expressing the 79 residues fragment were severely atrophic. The size of the cerebellum was one eighth that of a littermate control. All three layers of the cerebellar cortex exhibited substantial cellular loss. The molecular layer was shrunk, Purkinje cells were for the most part absent, and the granule layer was also substantially reduced and demonstrated significant cell loss.

Clearly, this study with the MJD1 protein demonstrates that a peptide containing a number of glutamine residues from the affected range of *MJD* alleles can cause cell death. The authors also presented data consistent with the idea that such a peptide forms covalent aggregates. The question that needs to be considered is whether these observations are relevant to the pathogenesis of MJD or any of the other CAG-based disorders. One point of concern is the fact that the cerebellar cortex is relatively spared in comparison to its involvement in the CAG-based inherited ataxias, i.e., SCA1 and SCA2. Ikeda *et al.* (1996) suggest that their finding of substantial cerebellar pathology in transgenic mice expressing only the 79 residues fragment and not in mice expressing an intact MJD1 protein with 79 glutamines could be explained by the involvement of proteases that function to release polyglutamine fragments in a cell- and protein-specific fashion. However, there is no direct evidence in support of this idea. As discussed later, there are data from *HD* transgenic mice that are consistent with involvement of polyglutamine fragments in pathogenesis and data from *SCA1* transgenic mice that as of now do indicate the presence of a pathogenic fragment.

2.3. Huntington Disease

Huntington disease (HD) is the most widely recognized of the CAG-based disorders, having a relatively high frequency of 1 in 10,000 among individuals of Western European ancestry. HD is a progressive motor, psychiatric, and cognitive disorder that is caused by the expansion of CAG trinucleotide repeat within the coding region of a novel gene located on chromosome 4p16.3 (Huntington's Disease Research Collaborative Group, 1993). Individuals affected with HD carry one allele of the *HD* gene with an expanded CAG repeat between 36 and 120 units. By contrast, unaffected chromosomes have between 6 and 34 CAG triplets within the *HD* gene. Like the other CAG-based disorders unaffected *HD* alleles have CAG repeats that are inherited in a stable Mendelian fashion whereas expanded, affected *HD* alleles have CAG repeats that are unstable during parent-to-offspring transmission. The length of the expanded repeat is inversely correlated to age at onset and disease severity (Duyao *et al.* 1993). Length of an expanded *HD* repeat seems also to be directly correlated with intergenerational instability. Increasing repeat length results in alleles with increasing intergenerational instability.

How does the expansion of the CAG in the *HD* gene cause disease? Like the other CAG-based disorders it is clear that the protein product, designated huntingtin, of an expanded *HD* allele is expressed. The huntingtin protein consists of about 3140 amino acids, depending on the number of glutamine residues because the polymorphic CAG repeat is translated into an uninterrupted stretch of glutamine residues. An elongated stretch of glutamine residues appears to alter the electrophoretic properties of huntingtin on sodium dodecyl sulfate (SDS) polyacrylamide gels (Trottier *et al.*, 1995; Persichetti *et al.*, 1995). Pulse-chase experiments indicate that an expanded allele of huntingtin does not undergo increased turnover and is not cleaved to smaller fragments (Persichetti *et al.*, 1996). Most studies indicate that the huntingtin protein is localized to the cytoplasm and is sometimes found in association with the cytoskeleton, microtubules, or synaptic vesicles (Aronin *et al.*, 1995; DiFiglia *et al.*, 1995; Sharp *et al.*, 1995). However, nuclear localization for huntingtin in transfected cells has been reported (De Rooij *et al.*, 1996). Consistent with this observation, a functional nuclear localization peptide motif has been isolated from huntingtin (Bessert *et al.*, 1995).

Several approaches have been used to establish transgenic mice suitable for gaining insight into the physiological role of the huntingtin protein and a mouse model of HD. The first approaches used homologous recombination to disrupt the murine *Hdh* gene. Three groups have created null mutant mice using this strategy. In the first study, the murine *Hdh* gene was disrupted by the insertion of the neomycin (*neo*) gene into exon 5 (Nasir *et al.*, 1995). This insertion yielded a mutant *Hdh* gene in which the 5′ half of exon 5 plus the 3′ portion of intron 4 were deleted and replaced by the *neo* gene. This disruption resulted in a premature stop codon in exon 6 of *Hdh*. Using antibodies directed against the amino terminus of huntingtin, a protein of appropriate size, 20 kDa, was detectable on western blots of extracts prepared from brains of mice heterozygous for this targeted disruption.

Mice homozygous for this disruption of the *Hdh* gene died before embryonic day 8.5 and failed to form somites and undergo organogenesis. Curiously, mice heterozygous for the *Hdh* disruption displayed an increase in motor activity and cognitive deficits. In addition, these mice appeared to have neuropathological changes in the subthalamic nucleus. However, there were no histological alterations detectable in the caudate/putamen, a region of prominent pathology in HD, or in the globus pallidus, substantia nigra, hippocampus, and motor cortex.

Two other groups have also generated *Hdh* mutant mice by homologous recombination. In one study, exons 4 and 5 of the *Hdh* gene were replaced by the *neo* gene (Duyao *et al.*, 1995). Western blot analysis of homozygous null ES cells revealed a complete lack of huntingtin expression and no evidence for the expression of a smaller fragment of the protein. In the other study, a targeted deletion of the *Hdh* promoter region and exon 1 was generated (Zetlin *et al.*, 1995). Although not examined directly, this mutation will most likely be a null mutation. Like the previous study, both of these studies found that mice homozygous for *Hdh* disruption die around embryonic day 8.5. However, neither study was able to identify behavioral or pathological alterations in mice heterozygous for the *Hdh* targeted mutation. Thus, it remains unclear as to the significance of the behavioral alterations found by Nasir *et al.* (1995). In particular whether some aspect of the symptoms associated with HD are due to a loss of function. It is intriguing to note that the study by Nasir *et al.* (1995) did generate heterozygous mice that synthesize a fragment of huntingtin while the other two studies generated mutations that are truly null. The phenotypic and histological changes found in the heterozygous mice in the first study might be explained by the production of truncated form of murine huntingtin.

It is clear from these three studies that utilized a targeted disruption of *Hdh* approach that the huntingtin protein is essential for development of mice beyond embryonic day 8.5. Furthermore, it is also highly unlikely that the dominant expansion of CAG repeats observed in HD patients, in which embryonic lethality is not found, induces a simple loss of function in huntingtin. However, a very interesting relationship was found by Zetlin *et al.* (1996) between the cell death in *Hdh* homozygous null embryos and cell death that occurs normally in developing mouse embryos. During normal early embryogenesis a substantial number of cells die within the distal region of the developing epiblast, which forms an epithelium or epithelia surrounding the proamniotic cavity. At embryonic day 6.5 the *Hdh* gene is expressed only in the epiblast. Of considerable interest is the observation that in *Hdh* null embryos cell death is not random, but occurs in the same subpopulation of epiblast cells that die normally during embryogenesis. This observation raises the intriguing suggestion that huntingtin may function to oppose the extent of programmed cell death. If so, there may be an important mechanistic relationship between the increased programmed cell death seen in embryonic cells of *Hdh* null mice and the loss of striatal cells seen in HD patients.

Recently, *HD* transgenic mice have also been generated using yeast artificial chromosomes (YACs) with either 600 kb or 350 kb of genomic DNA that contain the entire human *HD* gene with 18 CAG repeats (Hodgson *et al.,* 1996). Transgene temporal and spatial patterns of expression seem to be identical to that of endogenous mouse *Hdh.* This point is dramatically demonstrated by the fact that expression of HD from a YAC rescues the embryonic lethal effects of homozygous null *Hdh* mutations. Thus, the HD YAC transgene is very likely expressed during development in a manner very similar to endogenous Hdh. These studies also demonstrate that the function of the human HD protein during development is similar to that of the murine protein. The fact that the YAC HD transgene had an expression pattern similar to that of the endogenous gene strongly suggests that a YAC transgenic strategy will be very effective for establishing mouse models of HD that are more likely to mimic the human disease.

Initial attempts to generate transgenic mice with a gain-in-function mutation in huntingtin and an altered neurological phenotype were met with little success. In one study, a human HD cDNA transgene containing the CMV promoter and 44 CAG repeats was used to produce transgenic mice by pronuclear injection of single cell embryos (Goldberg *et al.*, 1996). After several months, transgenic animals failed to develop any signs of neurological of histological alterations. Although high levels of transgene mRNA were found, no evidence of transgene protein expression could be found. Subsequent analysis of the transgene revealed the presence of a frameshift mutation within the coding region that resulted in a termination codon at residue 727 prior to the glutamine tract. Analysis of the transmission of this transgene from parents to offspring found that the 44 CAG repeat was stable. Although this study reveals little regarding the pathogenesis of HD, the data do support the concept that protein expression from a mutant allele is required for pathogenesis and that the neurodegeneration is not mediated by DNA protein interactions involving an expanded allele or through the presence of RNA transcribed from an expanded allele. Moreover, the absence of intergenerational instability of the CAG repeat when included within a cDNA transgene indicates that additional genomic sequences are critical for repeat instability. However, some caution is warranted in assessing these experiments because the transgenic mice carried an expanded repeat of only 44 CAGs. Mice could be less sensitive than humans to the genetic and phenotypic effects of CAG repeat expansion. Thus, to generate an effect seen in humans with a given CAG repeat length, longer repeats may need to be introduced into the mouse genome.

In support of this idea are transgenic mice that develop a progressive neurological phenotype and that were established by the introduction of a 1.9-kb genomic fragment from the human *HD* gene containing the 5′ upstream region and exon 1 (Mangiarini *et al.*, 1996). A single founder animal was obtained that apparently was a germline chimera with one set of germ cells containing two integration events and a second set containing three integration events. After backcrossing this founder, segregation of the integrations produced four transgenic lines that carried CAG repeat expansions ranging from about 115 to 150 repeats. Thus, these animals carry *HD* CAG repeat expansions that are considerably longer than those that produce juvenile forms of HD in human. In three of these lines, the transgene was widely expressed at both RNA and protein levels.

These *HD* transgenic mice display a progressive neurological phenotype consisting of dyskinesia of limbs when held by the tail, tremor, repetitive stereotypic movements, and seizures. Age at onset varies between lines 6 six to 15 weeks with death usually a few weeks after onset. Transgenic mice also developed a progressive decrease in body weight. Histopathological studies revealed mild changes in the liver as the only consistent alteration. Brains from transgenic mice were found to be about 20% smaller, through all central nervous system (CNS) areas, than those of nontransgenic littermates. Although localized basal ganglian pathology characteristic of HD was not detected in these transgenic mice, aspects of their movement disorder and general reduction of brain size are suggestive of HD.

The *HD* transgenic studies of Mangiarini *et al.* (1996) are consistent with the work of Ikeda *et al.* (1996), suggesting that the a fragment of an expanded allele carrying the polyglutamine tract is sufficient to generate disease. However, the question remains as to the relevance of these models to the corresponding human disease. In particular, previous protein studies do not support the concept that an expanded HD protein is cleaved in tissues from affected individuals. In the case of HD it will be critical to determine if the expression of a transgene carrying an intact HD gene with a comparable number of repeats results in a phenotype similar to that seen with the exon 1 construct reported by Mangiarini *et al.* (1996). From the data presented by Mangiarini *et al.* it is unclear whether expression was due to a promoter within flanking genomic sequences at the site of integration or from elements contained with the 1 kb of 5′ upstream sequence from the *HD* gene included in the transgene. Thus, transgenic mice need to be developed in which expression of the transgene is driven from the endogenous *HD* promoter. The results of these additional transgenic experiments should help to determine the extent to which the polyglutamine tract is sufficient for the development of HD and possibly help to identify additional elements that function to establish the cell specificity of pathology characteristic of HD.

2.4. Spinocerebellar Ataxia Type 1

Spinocerebellar ataxia type 1 (SCA1) is the first of a growing series of autosomal dominant cerebellar ataxias found to be caused by the expansion of an unstable CAG trinucleotide located within the coding region of a novel gene (Orr *et al.*, 1993; Banfi *et al.*, 1994). Patients affected with SCA1 develop gait ataxia, dysarthria, and nystagmus. As the disease progress other signs of cerebellar dysfunction such as dysmetria and hypotonia become apparent. In the late stages of the disease, degeneration of several cranial nerves in the brain stem causes facial weakness and bulbar dysfunction. Death usually results from problems with choking and poor cough reflex due to loss of bulbar function. Neuropathologic alterations in SCA1 include severe loss of cerebellar Purkinje cells and dentate nucleus neurons. Neuronal degeneration also occurs in the inferior olive, pontine nucleus, and cranial nerve nuclei III, IV, IX, X, and XII.

The gene affected in SCA1 is located on human chromosome 6p23 and the mutation responsible for disease is the expansion of a CAG trinucleotide repeat (Orr *et al.*, 1993). The *SCA1* CAG repeat is highly polymorphic in the population and typically contains 6–44 repeat units on unaffected chromosomes and 40–82 repeat units on affected chromosomes. Normal alleles with 21 or more repeats are interrupted by one to three CAT units. On the hand, all disease alleles have a continuous CAG repeat tract (Chung *et al.*, 1993). An expanded *SCA1* CAG repeat is unstable and is susceptible to contractions and expansions when transmitted form parent to offspring. As with the other diseases associated CAG repeats, there is a paternal bias toward expansion upon transmission. Somewhat unique to the *SCA1* CAG repeat is the observation that repeat instability in SCA1 is more complex than a simple variation in repeat number. The loss of a CAT trinucleotide interruption seems to predispose the *SCA1* CAG repeat tract to intergenerational instability.

The protein encoded by the *SCA1* gene, ataxin-1, contains between 792 and 869 amino acids depending on the number of CAG repeats. The *SCA1* transcript is 11 kb in length and widely expressed. The murine *Sca1* gene is highly homologous to the human *SCA1* gene (Banfi *et al.*, 1996). However, like murine homologs of the other CAG-containing genes mutated in human diseases, the murine *Sca1* gene encodes a reduced number of glutamine residues, two, at the site of the polyglutamine tract in ataxin-1. *In situ* hybridization studies indicate that during murine cerebellar development there is a transient burst of *SCA1* expression around postnatal day 14 within Purkinje cells, a prominent site of pathology in SCA1. This is the developmental stage in mouse when the cerebellar cortex becomes physiologically functional. Immunohistochemical analyses show that ataxin-1 is expressed prominently within the nucleus of Purkinje cells and to a lesser extent within the cytoplasm.

Initial attempts to establish a transgenic model of SCA1 utilized the strong Purkinje cell-specific regulatory elements from the *Pcp-2* gene to direct expression of the human *SCA1* cDNA transgenes (Burright *et al.*, 1995). Two transgene constructs, one with an interrupted CAG tract of 30 repeats and a second with an uninterrupted CAG tract of 82 repeats, were injected into mouse embryos. Six transgenic lines were established with each construct. None of the transgenic lines carrying the 30 CAG repeat construct developed any signs of an altered neurobehavior or histopathology past 18 months of age even though transgene expression within Purkinje cells was very high as evident from RNA and protein analyses. In contrast, animals from five of the six transgenic lines varying the 82 CAG repeat *SCA1* transgene developed severe ataxia and pathological changes within their cerebellar Purkinje cells. The single transgenic line that did not develop these alterations failed to express the transgene. These studies demonstrate that pathological changes are induced specifically by the expression of an intact expanded allele of *SCA1*.

To gain further insight into the progression disease in he *SCA1* transgenic mice, animals have been examined for neurological alterations using several behavioral tests and for morphological changes using routine histology and immunohistochemistry (Clark *et al.*, 1997). Transgenic animals of the B05/ + *SCA1* transgenic line develop a progressive loss of cerebellar function. Results of behavioral tests demonstrate that at 5 weeks of age cerebellar impairment appears to be limited to a decreased ability to improve motor performance as assessed by the rotating rod apparatus. At this age, the B05/ + animals perform as well as the wild-type littermate controls on the first day of trials on the rotating rod; impairment was noted in the B05/ + mice only on successive days of trials. The latter observation indicates that a training or learning phase is required for the deficiency in B05/ + to manifest itself on the rotating rod. The absence of gait abnormalities, and normal motor activity, balance, and coordination as assessed by the bar cross and open field tests, support the conclusion that the impairment on the rotating rod is due to a decreased ability of the B05/ + mice to learn the task as opposed to impairment in motor activity, fine motor control, or coordination.

With increasing age, as the cerebellar impairments of B05/ + mice worsen to reach a stage of severe ataxia, deficiencies in motor activity and gait become apparent. By 1 year of age when there is substantial loss of cerebellar function, B05/ + mice were never able to match the performance of wild-type animals on the rotating rod even on the first day of trials, and do not demonstrate any ability to improve their performance with training. Furthermore, 1-year-old B05/ + mice were unable to perform any normal behavior on the bar-cross apparatus, typically falling off immediately upon being placed on the wide platform. These results suggest that in the B05/ + mice cerebellar dysfunction can be divided into two phases. In the first phase, dysfunction is limited to an impairment of motor learning. At a later stage, impairment advances to a point where motor activity and coordination become abnormal and severe ataxia ensues.

The morphological alterations that were noted in the B05/ + mice, after they were ataxic by home cage behavior at 12 weeks, developed after the cerebellar cortex has matured and, therefore, were not a result of maldevelopment induced by expression of the mutant transgene. There were no histological alterations during the first 3 weeks of postnatal life. At the time of decreased motor learning on the rotating rod apparatus, a loss of proximal dendritic branches and a decrease in the number of dendritic spines was observed in Purkinje cells of B05/ + mice, suggesting that the expression of mutant ataxin-1 affects the maintenance of dendrites and spines. At the time that ataxia, as assessed by home cage behavior, was established in the B05/ + mice (12–15 weeks) there was little evidence of loss of Purkinje cells, but there were significant alterations in dendritic and perikaryal morphology as well as evidence of perikaryal heterotopia that was not seen in younger animals. These changes then became more widespread and severe as the animals aged.

Another striking feature of B05/ + transgenic mice that followed the onset of ataxia was the presence of numerous Purkinje cells with their perikarya heterotopically located in the intermediate levels of the molecular layer. Histologic examination of young animals revealed no heterotopia during development or in the early stages of the disease: however, heterotopic cells were frequent by the time of overt ataxia. Therefore, the heterotopia cannot be explained by abnormal migration of Purkinje cells during development. A more likely explanation for the occurrence of heterotopic Purkinje cells is based on the appearance of the dendritic changes earlier in the course of the disease. Many Purkinje cells have evidence of simplification of the proximal dendrites with loss of branches and dendritic spines as early as 6 weeks of age. Loss of synaptic input in proximal parts of the dendritic tree, close to the perikaryon, would make it difficult for the cell to generate an action potential. A compensatory mechanism to maintain electrical activity in the Purkinje cell might be to retract the nonfunctional denuded dendritic trunk. Because the morphologically preserved distal dendrites maintain contact with parallel fibers in the superficial molecular layer, the only way the shortening of the proximal dendritic trunk can be effected is by movement of the perikaryon into the molecular layer with compensatory elongation of the axon. This scenario also provides an explanation for the increased frequency of Purkinje cells with two primary dendrites, and invokes a mechanism that includes the resorption of a denuded primary dendrite to a major branch point. Purkinje cells with this morphology were frequently seen heterotopically located within the molecular layer of older B05/ + animals.

Several postmortem studies of the cerebellum in patients with SCA1 have shown structural abnormalities of Purkinje cells as well as cell loss. Using Golgi techniques and immunohistochemical methods Ferrer *et al.* (1994) and Koeppen, 1991 have described dendritic simplification with loss of spines, similar to what we describe here in the B05/ + transgenic mice. A pathologic hallmark of SCA1 is the occurrence of frequent proximal axonal dilatations (torpedoes), a feature not found in the B05/ + *SCA1* mice. It is clear from autopsy studies of SCA1 patients that morphological alterations antedate cell death in at least some of the Purkinje cells of these patients. Unfortunately, the results of these investigations have been difficult to interpret for several reasons. Because of the complexity of the disease process and the involvement of multiple neuronal populations, it is difficult to distinguish between transsynaptic effects and events in the cell that are the direct consequence of expression of a mutant *SCA1* gene. In addition, autopsy studies are typically performed late in the course of the disease, when cell degeneration is more extensive. Targeting of the *SCA1* mutation to a single cell type known to be involved in the native disease, the Purkinje cell, has allowed an evaluation of the relative effects of structural alterations and cellular loss on the development of ataxia in the *SCA1* transgenic mice.

In the B05/ + line onset of ataxia occurred at a time when loss of Purkinje cells was barely detectable. This observation, coupled with the numerous morphological abnormalities found in Purkinje cells from B05/ + transgenic animals, indicates that expression of the mutant form of ataxin-1, in a cell that is vulnerable in the human disease, can lead to cellular dysfunction sufficient to induce ataxia without causing death of the affected neuronal population. Therefore, it is likely that in the *SCA1* transgenic mice the disease is not caused by cell loss. Rather, loss of Purkinje cells seen at later stages of disease is most likely the result of the dysfunction induced at an earlier stage.

3. CAG REPEAT INSTABILITY IN TRANSGENIC MICE

An intriguing aspect of unstable trinucleotide repeat mutations is the fact that they have to date been observed only in humans. Thus, an important question was whether expanded human CAG repeats would manifest instability once introduced into mouse DNA and would this instability be similar to that seen for the unstable trinucleotide repeats in humans. At this point, it would be useful to make a few general comments regarding repeat instability in humans. Instability of CAG repeats consists of both contractions and expansions, and these alterations always occur as changes in the number of unit triplets. Only those repeats that have expanded into the affected size range exhibit detectable instability. Repeat instability can be characterized as either somatic, variation in repeat number between somatic tissues or with age of an individual, or as germline, variation of repeat number between generations or within germ cells of an individual. In the case of the CAG trinucleotide repeats in humans, an expanded unstable allele is far more unstable in germ cells or between generations than in somatic cells (MacDonald *et al.*, 1993; Zühlke *et al.*, 1993; Telenius *et al.*, 1994; Leeflang *et al.*, 1995). Furthermore, compared to the non-CAG trinucleotides, unstable CAG trinucleotides have a fairly modest degree of instability. Changes when they occur tend to be fewer than 10 repeats and on rare occasions may be an increase of 20–30 repeats. Finally, there is a strong parental bias in CAG repeat instability. There is a significant bias toward expansion upon paternal transmission of an unstable CAG allele. This bias is nicely illustrated by the fact that juvenile onset cases of Huntington disease and spinocerebellar ataxia type 1 develop when a mutant allele is inherited from the father.

Initial studies that utilized transgenic mice to study expanded alleles of human CAG disease associated genes found that the trinucleotide repeats were absolutely stable and did not undergo any change in length during intergenerational transmission. As described previously, these early studies included transgenic mice with an androgen receptor cDNA with 45 CAGs (Bingham *et al.*, 1995), HD transgenics with a cDNA carrying 44 CAGs (Goldberg*et al.*, 1996), SCA1 mice with 82 CAGs in a cDNA transgene (Burright *et al.*, 1995), and MJD transgenic mice carrying cDNAs with 79 repeats (Ikeda *et al.*, 1996). In all of these studies, the expanded CAG tract was located within a cDNA construct. Thus, one explanation for the lack of repeat instability in these studies might be that repeat length alone is not sufficient to induce instability. Perhaps some other *cis*-linked genomic factors are required. The involvement of *cis* elements in repeat instability would explain the identification of founder haplotypes for unstable alleles of HD, MJD (Takiyama *et al.,* 1995), and fragile X syndrome (Richards *et al.,* 1992). One *cis*-element that does impact on trinucleotide repeat instability is an origin of replication (Maurer *et al.,* 1996).

In a very recent study, Mangiarini *et al.* (1997) produced transgenic mice carrying 130–150 CAGs within a very small portion of the *HD* gene, exon 1. In three of four transgenic lines characterized, the CAG repeat was unstable both in somatic tissue with increasing age and in germline tissue upon transmission to offspring. The one line in which the repeat was stable failed to express the transgene, suggesting that repeat instability in this model system requires transcription. In addition, these HD transgenic animals carry a repeat greater than 130 units, a longer repeat than used by others. Thus, in the mouse the threshold of repeat length for instability may be substantially longer than seen in humans. However, this point remains unclear because a CTG repeat length of 55 in the *MD* gene was shown to be unstable in transgenic mice, albeit at a low level (Gourdon *et al.,* 1997).

Does the CAG repeat instability observed in the HD transgenic mice correlate with that found in humans? The data from this transgenic study are limited at this time so some aspects of repeat instability found in humans cannot be assessed. The HD transgenic mice, at least in one line, did demonstrate a tendency to increase on male transmissions compared to female transmissions. A paternal bias toward expansion is an important characteristic of an expanded human *HD* allele and all other disease associated CAG repeats. One point of concern relative to the appropriateness of these transgenic mice as a model of instability seen in humans is the considerable degree of CAG repeat instability seen in brain and somatic tissues of the transgenic lines. In some cases, somatic instability appears to be greater than that found in the germline. This is in contrast to the instability found for *HD* alleles in human. Repeat instability or mosaicism is much more dramatic in human sperm than in human somatic tissues (MacDonald *et al.,* 1993; Zühlke *et al.,* 1993; Telenius *et al.,* 1994; Leeflang *et al.,* 1995).

4. CONCLUSIONS AND CLOSING COMMENTS

Transgenic mouse models in which an altered neurological phenotype is apparent have been obtained for three of the CAG-based disorders, SCA1, MJD, and HD. It is hoped these mouse models will provide insight into the pathogenesis of this interesting class of human diseases. There is little question based on the genetic evidence from both humans and transgenic mice that CAG repeat expansion includes a gain-of-function alteration in the corresponding gene product. For the disease phenotype to be elicited, the protein carrying an expanded polyglutamine tract must be synthesized. Other models of pathogenesis based on changes at the DNA or RNA levels are not supported by the evidence accumulating. Although it is tempting, based on the many similarities found among the CAG based disorders, to suggest a common mechanism of pathogenesis among these diseases there is no direct evidence to support this concept. Unfortunately, the mechanism(s) of pathogenesis remain unclear. Yet, much speculation has taken place regarding possible mechanisms of CAG-based disease pathogenesis.

Probably the mechanism of pathogenesis supported to the greatest degree by data from transgenic mice is the concept of toxic polyglutamine peptide fragment. This idea is consistent with the MJD and certain HD transgenic mice studies. However, there is no evidence from human studies that support a toxic peptide scenario. To the contrary, protein studies using human tissue from HD patients found no difference in the stability of the huntingtin protein. A second mechanism of pathogenesis that has surfaced is one that involves the disruption of energy production. A comprise in energy production could result from an interaction between the glycolytic enzyme glyceraldehyde-3-phosphate dehydrogenase (GAPDH) and polyglutamine proteins (Burke *et al.*, 1996; Koshy *et al.*, 1996). GAPDH is an interesting candidate for involvement in neurodegenerative disease because it is essential in glycolysis and impairment of energy production has often been considered as an attractive mechanism for several neurodegenerative disorders including HD. However, analysis of ataxin-1 and GAPDH interactions failed to demonstrate that this interaction is dependent on length of the polyglutamine tract. In addition, the cellular levels of the polyglutamine proteins that are associated with disease are far less than the level of GAPDH in neurons. These two points detract substantially from the viability of the disruption of GAPDH as a mechanism for polyglutamine pathogenesis. A final mechanism of pathogenesis that should be considered is one that involves the disruption of nuclear function. From the outset, it has been appreciated that many proteins rich in glutamine residues are localized to the nucleus. Polyglutamine tracts are found in several nuclear proteins important in neural development, e.g., prospero and couch potato. Clearly, the androgen receptor function is nuclear and immunohistochemical studies indicate that the ataxin-1 protein is predominantly nuclear in localization. Although most studies indicate that the huntingtin protein is cytoplasmic, there are reports of it being localized to the nucleus and containing a nuclear localization peptide motif. Thus, it would seem prudent to keep open the possibility that some or may be all of the CAG based disorders develop due to the disruption of a nuclear function.

While the usefulness of mouse models for studying the factors important for CAG repeat instability remains an open question, it is encouraging that studies are beginning to find some evidence of repeat instability in mouse models. Clearly, further analyses of CAG repeat instability in mice are required before a definitive conclusion can be drawn regarding whether mice will be able to serve as a model for CAG repeat instabilities. Perhaps the generation of mouse models using YACs or homologous recombination to insert expanded CAG tracts within the murine gene will prove to be a useful approach for instability studies.

In closing, it is important to emphasize that by its very nature the use of mouse models to examine human CAG trinucleotide based disorders is very likely to consist of some inaccuracies. Similar disorders have yet to be found as natural diseases of the mouse. In addition, the life span of a mouse is considerably shorter than that of humans, even those affected with juvenile forms of CAG diseases. Therefore, each mouse model of a human CAG based disease will most likely include some features that may accurately represent those found in the human disease and other aspects that are unique to each individual mouse model and do not apply to human disease or other mouse models. Furthermore, given the difficulties presented by undertaking both pathogenesis and repeat instability studies in human, it is very likely that the use of mouse models will remain a major focus for research on the CAG trinucleotide repeat diseases.

REFERENCES

Aronin, N., Chase, K., Young, C., Sapp, E., Schwarz, C., Matta, N., Kornreich, R., Landwehrmeyer, B., Bird, E., Beal, M. F., Vonsattel, J.-P., Smith, T., Carraway, R., Boyce, F. M., Young, A. B., Penney, J. B., and DiFiglia, M., 1995, CAG expansion affects the expression of mutant Huntingtin in the Huntington's disease brain, *Neuron* **15:**1193–1201.

Banfi, S., Servadio, A., Chung, M.-Y., Kwiatkowski, T. J., Jr., McCall, A. E., Duvick, L. A., Shen, Y., Roth, E. J., Orr, H. T., and Zoghbi, H. Y., 1994, Identification and characterization of the gene causing type 1 spinocerebellar ataxia, *Nat. Genet.* **7:**513–519.

Banfi, S., Servadio, A., Chung, M.-Y., Capozzoli, F., Duvick, L. A., Elde, R., Zoghbi, H. Y., and Orr, H. T., 1996, cloning and developmental expression analysis of the murine homolog of the spinocerebellar ataxia type 1 gene (*Sca1*), *Hum. Mol. Genet.* **5:**33–40.

Bessert, D. A., Gutridge, K. L., Dunbar, J. C., and Carlock, L. R., 1995, The identification of a functional nuclear localization signal in the Huntington disease protein, *Mol. Brain Res.* **33:**165–173.

Bingham, P. M., Scott, M. O., Wang, S., McPhaul, M. J., Wilson, E. M., Garbern, J. Y., Merry, D. E., and Fischbeck, K. H., 1995, Stability of an expanded trinucleotide repeat in the androgen receptor gene in transgenic mice, *Nat. Genet.* **9:**191–196.

Brook, J. D., McCurrach, M. E., Harley, H. G., Buckler, A. J., Church, D., Aburatani, H., Hunter, K., Stanton, V. P., Thirion, J.-P., Hudson, T., Sohn, R., Zemelman, B., Snell, R. G., Rundle, S. A., Crow, S., Davies, J., Shelbourne, P., Buxton, J., Jones, C., Juvonen, V., Johnson, K., Harper, P. S., Shaw, D. J., and Housman, D. E., 1992, Molecular basis of myotonic dystrophy: Expansion of a trinucleotide (CTG) repeat at the 3′ end of a transcript encoding a protein kinase family member, *Cell* **86:**799–808.

Burke, J. R., Wingfield, M. S., Lewis, K. E., Roses, A. D., Lee, J. E., Hulette, C., Perical-Vance, M. A., and Vance, J. M., 1994, The Haw River syndrome: Dentatorubropallidoluysian atrophy (DRPLA) in an African-American family, *Nat. Genet.* **7:**521–524.

Burke, J. R., Enghild, J. J., Martin, M. E., Jou, Y.-S., Myers, R. M., Roses, A. D., Vance, J. M., and Strittmatter, W. J., 1996, Huntingtin and DRPLA proteins selectively interact with the enzyme GAPDH, *Nat. Med.* **2:**347–350.

Burright, E. N., Clark, H. B., Servadio, A., Matilla, T., Feddersen, R. M., Yunis, W. S., Duvick, L. A., Zoghbi, H. Y., and Orr, H. T., 1995, *SCA1* transgenic mice: A model for neurodegeneration caused by an expanded CAG trinucleotide repeat, *Cell* **82**:937–948.

Campuzano, V., Montermini, L., Moltò, M. D., Pianese, L., Cossée, M., Cavalcanti, F., Monos, E., Rodius, F., Duclos, F., Monticelli, A., Zara, F., Cañizares, J., Koutnikova, H., Bidichandani, S. I., Gellera, C., Brice, A., Trouillas, P., De Michele, G., Filla, A., De Frutos, R., Palau, F., Patel, P. I., Di Donato, S., Mandel, J.-L., Cocozza, S., Koenig, M., and Pandolfo, M., 1996, Friedreich's ataxia: Autosomal recessive disease caused by an intronic GAA triplet repeat expansion *Science* **271**;1423–1427.

Chung, M.-y., Ranum, L. P. W., Duvick, L. A., Servadio, A., Zoghbi, H. Y., and Orr, H. T., 1993, Evidence for a mechanism predisposing to intergenerational CAG repeat instability in spinocerebellar ataxia type 1, *Nat. Genet.* **5**:254–258.

Clark, H. B., Burright, E. N., Yunis, W. S., Larson, S., Wilcox, C., Hartman, B., Zoghbi, H. Y., and Orr, H. T., 1997, Purkinje cell expression of a mutant allele of SCA1 in transgenic mice leads to disparate effects on motor behaviors, followed by a progressive cerebellar dysfunction and histological alteration, *J. Neurosci.* **17**:7385–7395.

De Rooij, K. E., Dorsman, J. C., Smoor, M. A., Den Dunnen, J. T., and Van Ommen, G.-J. B., 1996, Subcellular localization of the Huntington's disease gene product in cell lines by immunofluorescence and biochemical subcelluar fractionation, *Hum. Mol. Genet.* **5**:1093–1099.

DiFiglia, M., Sapp, E., Chase, K., Schwartz, C., Meloni, A., Young, C., Martin, E., Vonsattel, J.-P., Carraway, R., Reeves, S. A., Boyce, F. M., and Aronin, N., 1995, Huntingtin is a cytoplasmic protein associated with vesicles in human and rat brain brain neurons, *Neuron* **14**:1075–1081.

Duyao, M., Ambrose, C., Myers, R., Novelletto, A., Persichetti, F., Frontali, M., Folstein, S., Ross, C., Franz, M., Abbott, M., Gray, J., Conneally, P., Young, A., Penny, J., Hollingsworth, Z., Shoulson, I., Lazzarini, A., Falek, A., Koroshetz, M., Sax, D., Bird, E., Vonsattel, J., Bonilla, E., Alvir, J., Conde, J. B., Cha, J.-H., Dure, L., Gomez, F., Ramos, M., Sanchez-Ramos, J., Snodgrass, S., de Young, M., Wexler, N., Moscowitz, C., Penchaszadeh, G., MacFarlane, H., Anderson, M., Jenkins, B., Srinidhi, J., Barnes, G., Guesella, J., and MacDonald, M., 1993, Trinucleotide repeat length instability and age of onset on Huntington's disease, *Nat. Genet.* **4**:387–392.

Duyao, M. P., Auerbach, A. B., Ryan, A., Persichetti, F., Barnes, G. T., McNeil, S. M., Ge, P., Vonsattel, J.-P., Guesella, J. F., Joyner, A. L., and MacDonald, M. E., 1995, Inactivation of the mouse Huntington's disease gene homolog *Hdh, Science* **269**:407–410.

Ferrer, I., Genis, D., Dávalus, A., Bernadó, L., Sant, F., and Serano, T., 1994, The Purkinje cell in olivopontocerebellar atrophy. A golgi and immunohistochemical study, *Neuropathol. Appl. Neurobiol.* **20**:38–46.

Goldberg, Y. P., Kalchman, M. A., Metzler, Nasir, J., Zeisler, J., Graham, R., Koide, H. B., O'Kusky, J., Sharp, A. H., Ross, C. A., Jirik, F., and Hayden, M. R., 1996, Absence of disease phenotype and intergenerational stability of the CAG repeat in transgenic mice expressing the human Huntington disease transcript *Hum. Mol. Genet.* **5**:177–185.

Gourdon, G., Radvanyi, R., Lia, A.-S., Duros, C., Blanche, M., Abitbol, M., Junien, C., and Hofmann-Radvanyi, H., 1997, Moderate intergenerational and somatic instability of a 55-CTG repeat in transgenic mice, *Nat. Genet.* **15**:190–192.

Hodgson, J. G., Smith, D. J., McCutcheion, K., Koide, H. B., Nishiyama, K., Dinulos, M. B., Stevens, M. E., Bissada, N., Nasir, J., Kanazawa, I., Disteche, C. M., Rubin, E. M., and Hayden, M. R., 1996, Human huntingtin derived from YAC trangenes compensates for loss of murine huntingtin by rescue of the embryonic lethal phenotype, *Hum. Mol. Genet.* **5**:1875–1885.

Huntington's Disease Collaborative Research Group, 1993, A novel gene containing a trinucleotide repeat that is expanded and unstable on Huntington's disease chromosomes, *Cell* **72**:971–983.

Ikeda, H., Tamaguchi, M., Sugai, S., Aze, Y., Narumiya, S., and Kakizuka, A., 1996, Expanded polyglutamine in the Machado–Joseph disease protein induces cell death *in viro* and *in vivo*, *Nat. Genet.* **13:** 196–202.
Imbert, G., Saudou, F., Yvert, G., Devys, D., Trottier, Y., Garnier, J.-M., Weber, C., Mandel, J.-L., Cancel, G., Abbas, N., Dürr, A., Didierjean, O., Stevanin, G., Agid, Y., and Brice, A., 1996, Cloning of the gene for spinocerebellar ataxia 2 reveals a locus with high sensitivity to expanded CAG/glutamine repeats, *Nat. Genet.* **14:**285–291.
Kawaguchi, Y., Okamoto, T., Taniwaki, M., Aizawa, M., Inoue, M., Katayama, S., Kawakami, H., Nakamura, S., Nishimura, M., Akiguchi, I., Kimura, J., Narumiya, S., and Kakizuka, A., 1994, CAG expansions in a novel gene for Machado–Joseph disease at chromosome 14q32.1 *Nat. Genet.* **8:**221–227.
Knight, S. J. L., 1993, Trinucleotide repeat amplification and hypermethylation of a CpG island in FRAXE mental retardation, *Cell* **74:**127–134.
Koeppen, A. H., 1991, The Purkinje cell and its afferents in human hereditary ataxia, *J. Neuropath. Expt. Neurol.* **50:**505–514.
Koide, R., Ikeuchi, T., Onodera, O., Tanaka, H., Igarashi, S., Endo, K., Takahashi, H., Kondo, R., Ishikawa, A., Hayashi, T., Saito, M., Tomoda, A., Miike, T., Naito, H., Ikuta, F., and Tsuji, S., 1994, Unstable expansion of CAG repeat in hereditary dentatorubral–pallidoluysian atrophy (DRPLA), *Nat. Genet.* **6:**9–13.
Koshy, B., Matilla, T., Burright, E. N., Merry, D. E., Fischbeck, K. H., Orr, H. T., and Zoghbi, H. Y., 1996, Spinocerebellar ataxia type-1 and spinobulbar muscular atrophy gene products interact with glyceraldehyde-3-phosphate dehydrogenase, *Hum. Mol. Genet.* **9;**1311–1318.
Kremer, E. J., Pritchard, M., Lynch, M., Yu, S., Holman, K., Baker, E., Warren, S. T., Schlessinger, D., Sutherland, G. R., and Richards, R. I., 1991, Mapping of DNA instability of the fragile X to a trinucleotide repeat sequence p(CCG)n, *Science* **252;**1711–1714.
LaSpada, A. R., Wilson, E. M., Lubahn, D. B., Harding, A. E., and Fischbeck, K. H., 1991, Androgen receptor gene mutations in X-linked spinal and bulbar muscular atrophy, *Nature* **352:**77–79.
Leeflang, E. P., Zhang, L., Tavaré, S., Hubert, R., Srinidhi, J., MacDonald, M. E., Myers, R. H., de Young, M., Wexler, N. S., Gusella, J. F., and Arnheim, N., 1995, Single sperm analysis of the trinucleotide repeats in the Huntington's disease gene: Qualification of the mutation frequency spectrum, *Hum. Mol. Genet.* **4:**1519–1526.
MacDonald, M. E., Barnes, G., Srinidhi, J., Duyao, M. P., Ambrose, C. M., Myers, R. H., Gray, J., Conneally, P. M., Young, A., Penny, J., Shoulson, I., Hollingsworth, Z., Koroshetz, W., Bird, E., Vonssattel, J. P., Bonilla, E., Moscowitz, C., Penchaszadh, G., Brzustowicz, L., Alvir, J., Conde, J. B., Cha, J.-H., Dure, L., Gomez, F., Ramos-Arroyo, M., Sanchez-Ramos, J., Snodgrass, S. R., de Young, M., Wexler, N. S., MacFarlane, H., Anderson, M. A., Jenkins, B., and Gusella, J. F., 1993, Gametic but not somatic instability of CAG repeat length in Huntington's disease, *J. Med. Genet.* **30:**982–986.
Mahadevan, M., Tsilfidis, C., Sabourin, L., Shutler, G., Amemiya, C., Jansen, G., Neville, C., Narang, M., Barceló, J., O'Hay, K., Leblond, S., Earle-MacDonald, J., de Jong, P. J., Wieringa, B., and Korneluk, R. G., 1992, Myotinc dystrophy mutation. An unstable CTG repeat in the 3′ translated region of the gene, *Science* **255:**1253–1255.
Mangiarini, L., Sathasivam, K., Seller, M., Cozens, B., Harper, A., Hetherington, C., Lawton, M., Trottier, Y., Lehrach, H., Davies, S. W., and Bates, G. P., 1996, Exon 1 of the *HD* gene with an expanded CAG repeat is sufficient to cause a progressive neurological phenotype in transgenic mice, *Cell* **87:**493–506.
Mangiarini, L., Sathasivam, K., Mahal, A., Mott, R., Seller, M., and Bates G. P., 1997, Instability of highly expanded CAG repeats in mice transgenic for the Huntington's disease mutation, *Nat. Genet.* **15:**197–200.
Maruyama, H., Nakamura, S., Matsuyama, Z., Sakai, T., Doyu, M., Sobue, G., Seto, M., Tsujihata,

M., Oh-i, T., Nisho, T., Sunohara, N., Talahashi, R., Hayashi, M., Nishino, I., Ohtake, T., Oda, T., Nishimura, M., Saida, T., Matsumoto, H., Baba, M., Kawaguchi, Y., Kakizuka, A., and Kawakami, H., 1995, Molecular features of the CAG repeats and clinical manifestations of Machado–Joseph disease, *Hum. Mol. Genet.* **4:**807–812.

Maurer, D. J., O'Callaghan, B. L., and Livingston, D. M., 1996, Orientation dependence of trinucleotide CAG repeat instability in yeast, *Mol. Cell. Biol.* **16:**6617–6622.

Nagafuchi, S., Yanagisawa, H., Sato, K., Shirayama, T., Ohsaki, E., Bundo, M., Takeda, T., Tadokoro, K., Konda, I., Murayama, N., Tanaka, Y., Kikushima, H., Umino, K., Kurosawa, H., Furukawa, T., Nihei, K., Inoue, Y., Sano, A., Komure, O., Takahashi, M., Yoshizawa, T., Kanazawa, I., and Yamada, M., 1994, Dentatorubral and pallidoluysian atrophy expansion of an unstable CAG trinucleotide on chromosome 12p, *Nat. Genet.* **6;**14–18.

Nasir, J., Floresoc, S. B., O'Kusky, J. R., Diewert, V. M., Richman, J. M., Zeisler, J., Borowski, A., Marth, J. D., Phillips, A. G., and Hayden, M. R., 1995, Targeted disruption of the Huntington's disease gene results in embryonic lethality and behavioral and morphological changes in heterozygotes, *Cell* **81:**811–823.

Orr, H. T., Chung, M.-y., Banfi, S., Kwiatkowski, T. J., Jr., Servadio, A., Beaudet, A. L., McCall, A. E., Duvick, L. A., Ranum, L. P. W., and Zoghbi, H. Y., 1993, Expansion of an unstable trinucleotide CAG repeat in spinocerebellar ataxia type 1, *Nat. Genet.* **4:**221–226.

Persichetti, F., Ambrose, C. M., Ge, P., McNeil, S. M., Srinidhi, J., Anderson, M. A., Jenkins, B., Barnes, G. T., Duyao, M. P., Kanaley, L., Wexler, N. S., Myers, R. H., Bird, E. D., Vonsattel, J.-P., MacDonald, M. E., and Gusella, J. F., 1995, Normal and expanded Huntington's disease gene alleles produce distinguishable proteins due to translation across the CAG repeat, *Mol. Med.* **1:**374–383.

Persichetti, F., Carlee, L., Faber, P. W., McNeil, S. M., Ambrose, C. M., Srinidhi, J., Anderson, M. A., Barnes, G. T., Gusella, J. F., and MacDonald, M. E., 1996, Differential expression of normal and mutant Huntington's disease gene alleles, *Neurobiol. Dis.* **3:**183–190.

Pulst, S.-M., Nechiporuk, A., Nechiporuk, T., Gispert, S., Chen, X.-N., Lopes-Cendes, I., Pearlman, S., Starkman, S., Orozco-Diaz, G., Lunkes, A., Dejong, P., Rouleau, G. A., Auburger, G., Korenberg, J. R., Figueroa, C., and Sahba, S., 1996, Moderate expansion of a normally biallelic trinucleotide repeat in spinocerebellar ataxia type 2, *Nat. Genet.* **14:**269–276.

Richards, R. I., Holamn, K., Firend, K., Kremer, E., Hillen, D., Staples, A., Brown, W. T., Goonewardena, P., Tarleton, J., Schwartz, C., and Sutherland, G. R., 1992, Evidence of founder chromosomes in fragile X syndrome. *Nat. Genet.* **1:**257–260.

Sanpei, K., Takano, H., Igarashi, S., Sato, T., Oyake, M., Sasaki, H., Wakisaka, A., Tashiro, K., Ishida, Y., Ikeuchi, T., Koide, R., Saito, M., Sato, A., Tanaka, T., Hanyu, S., Takiyama, Y., Nishizawa, M., Simizu, N., Nomura, Y., Segawa, M., Iwabuchi, K., Eguchi, I., Tanaka, H., Takahashi, H., and Tsuji, S., 1996, Identification of the spinocerebellar ataxia type 2 gene using a direct identification of repeat expansion and cloning technique, DIRECT, *Nat. Gene.* **14:**277–284.

Sharp, A. H., Loev, S. J., Schilling, G., Li, S.-H., Li, X.-J., Bao, J., Wagster, M. V., Kotzuk, J. A., Steiner, J. P., Lo, A., Hedreen, J., Sisodia, S., Snyder, S. H., Dawson, T. M., Ryugo, D. K., and Ross, C. A., 1995, Widespread expression of Huntington's disease gene (IT15) protein product, *Neuron* **14:**1065–1074.

Stevanin, G., Trottier, Y., Cancel, G., Dürr, A., David, G., Didierjean, O., Bürk, K., Imbert, G., Saudou, F., Abada-Bendib, M., Gourfinkel-An, I., Benomar, A., Abbas, N., Klockgether, T., Grid, D., Agid, Y., Mandel, J.-L., and Brice, A., 1996, Screening for proteins with polyglutamine expansions in autosomal dominant cerebellar ataxias, *Hum. Mol. Genet.* **5:**1887–1892.

Takiyama, Y., Igarashi, S., Rogaeva, E. A., Endo, K., Rogaev, E. I., Tanaka, H., Sherrington, R., Sanpei, K., Liang, Y., Saito, M., Tsuda, T., Takano, H., Ikeda, M., Lin, C., Chi, H., Kennedy, J.

L., Lang, A. E., Wherrett, J. R., Segawa, M., Nomura, Y., Yuasa, T., Weissenbach, J., Yoshida, M. Nishizawa, M., Kidd, K. K., Tsuji, S., and St George-Hyslop, P. H., 1995, Evidence for intergenerational instability in the CAG repeat in the MJD1 gene and for conserved haplotypes at flanking markers amongst Japanese and Caucasian subjects with Machado-Joseph disease, *Hum. Mol. Genet.* **4:**1137–1146.

Telenius, H., Kremer, B., Goldberg, Y. P., Theilmann, J., Andrew, S. E., Zeisler, J., Adam, S., Greenberg, C., Ives, E. J., Clarke, L. A., and Hayden, M. R., 1994, Somatic and gonadal mosaicism of the Huntington disease gene CAG repeat in brain and sperm, *Nat. Genet.* **6:**409–414.

Trottier, Y., Devys, D., Imbert, G., Saudou, F., An, I., Lutz, Y., Weber, C., Agid, Y., Hirsch, E. C., and Mandel, J.-L., 1995, Cellular localization of the Huntington's disease protein and discrimination of the normal and mutated form, *Nat. Genet.* **10:**104–110.

Verkerk, A. J. M. H., Pieretti, M., Sutcliffe, J. S., Fu, Y., Kuhl, D. P. A., Pizzuti, A., Reiner, O., Richards, S., Victoria, M. F., Zhang, F., Eussen, B. E., van Ommen, G.-J. B., Blonden, L. A. J., Riggins, G. J., Chastain, J. L., Concept, C. B., Galjaard, H., Caskey, C. T., Nelson, D. L., Oostra, B. A., and Warren, S. T., 1991, Identification of a gene (*FMR-1*) containing a CGG repeat coincident with a fragile X breakpoint cluster region exhibiting length variation in fragile X syndrome, *Cell* **65:**905–914.

Yu, S., Pritchard, M., Kremer, E., Lunch, M., Nancarrow, J., Baker, E., Holman, K., Mulley, J. C., Warren, S. T., Schlessinger, D., Sutherland, G. R., and Richards, R. I., 1991, Fragile X genotype characterized by an unstable region of DNA, *Science* **252;**1179–1181.

Zetlin, S., Liu, J.-P., Chapman, D. L., Papaioannou, V. E., and Efstratiadis, A., 1995, Increased apoptosis and early embryonic lethality in mice nullizygous for the Huntington's disease gene homologue *Nat. Genet.* **11:**155–163.

Zhuchenko, O., Bailey, J., Bonnen, P., Ashizawa, T., Stockton, D. W., Amos, C., Dobyns, W., Subramony, S. H., Zoghbi, H. Y., and Lee, C. C., 1997, Autosomal dominant cerebellar ataxia (SCA6) associated with small polyglutamine expansions in the α_{1A}-voltage-dependent calcium channel, *Nat. Gene.* **15:**62–69.

Zühlke, C., Riess, O., Bockel, B., Lange, H., and Thies, U., 1993, Mitotic stability and meiotic variability of the $(CAG)_n$ repeat in the Huntington disease gene, *Hum. Mol. Genet.* **2:**2063–2067.

CHAPTER 9

ALZHEIMER'S DISEASE AND GENETICALLY ENGINEERED ANIMAL MODELS

DAVID R. BORCHELT, PHILIP C. WONG, SANGRAM S. SISODIA, and DONALD L. PRICE

1. INTRODUCTION

With the growth of the aging population (Olshansky *et al.*, 1993), there has been a significant increase in the prevalence of disorders that affect the elderly; the most common major disabilities in this age group are impairments in cognitive

DAVID R. BORCHELT and PHILIP C. WONG • Department of Pathology and Division of Neuropathology, The Johns Hopkins University School of Medicine, Baltimore, Maryland 21205-2196. SANGRAM S. SISODIA • Department of Pharmacological and Physiological Sciences, University of Chicago, Chicago, Illinois 60637. DONALD L. PRICE • Departments of Pathology, Neurology, and Neuroscience and Division of Neuropathology, The Johns Hopkins University School of Medicine, Baltimore, Maryland 21205-2196.

Mouse Models in the Study of Genetic Neurological Disorders, Volume 9 of *Advances in Neurochemistry,* edited by Popko, Kluwer Academic/Plenum Publishers, New York, 1999.

and memory processes. The most frequent cause of these problems is Alzheimer's disease (AD) (McKhann *et al.,* 1984; Khachaturian, 1985; Katzman, 1997). The prevalence of AD, ~10% in persons >65 years of age (e.g., 4 million individuals in the United States) (Pfeffer *et al.,* 1987; Bachman *et al.,* 1993), is increasing because of significant shifts in life expectancy and demographic parameters. In 1900, 1% of the world's population was >65 years of age; in 1992, that figure was 6.2%; and it is estimated that in 2050 ~25% of the population will be >65 years of age (Olshansky *et al.,* 1993). Because of the prevalence and magnitude of the morbidity and mortality associated with AD, it is imperative to understand the etiologies and pathogeneses of this disease and to develop effective treatments to ameliorate its course and prevent its occurrence.

In the past, investigators have used spontaneously occurring models (e.g., aged nonhuman primates) or experimentally induced models of disease (e.g., retrograde degeneration of basal forebrain neurons induced by fimbria–fornix axotomy) to study features that resemble those identified in individuals with AD. However, as a result of advances in transgenic (Tg) technologies (Bedell *et al.,* 1997a,b), investigators have begun to create, for the first time, genetically engineered models of AD and other neurodegenerative disorders (Bedell *et al.,* 1997a, Bedell *et al.,* 1997b). These model systems provide excellent opportunities to examine the genes and proteins implicated in pathogenic pathways; toxic properties of mutant proteins; the character and spatial/temporal evolution of the disease; the reasons for selective vulnerability of specific subsets of neurons; mechanisms of nerve cell degeneration; and the efficacies of new experimental therapies. Examples of the use of these genetic modeling strategies are illustrated by Tg mouse models of: FALS [*G93A, G37R,* and *G85R* superoxide dismutase 1 (SOD1) mutations] (Gurney *et al.,* 1994; Games *et al.,* 1995; Wong *et al.,* 1995; Hsiao *et al.,* 1996; Bruijn *et al.,* 1997; Borchelt *et al.,* 1997); Gerstmann-Sträussler-Scheinker syndrome, a prion disorder (*PrP102* mutation) (Hsiao *et al.,* 1990); Spinocerebellar ataxia-1 (SCA1) (ataxin-1, GAG expanded repeat) (Burright et al., 1995); and Huntington's disease (huntingtin, CAG expanded repeats) (Mangiarini *et al.,* 1996; Davies *et al.,* 1997; Scherzinger *et al.,* 1997). In each of these cases, Tg mice that overexpress mutant transgenes develop some features that resemble the human disease (Price *et al.,* 1998).

In this chapter, we review the clinical syndrome of AD; the neuropathology (including vulnerable neuronal systems, cytoskeletal alterations, amyloidogenesis, and mechanisms that lead to dysfunction/death of nerve cells); the risk factors, particularly genetic influences, that occur in AD; and some of the recent advances that have been made in studies of genetically engineered animal models relevant to autosomal dominant familial AD (FAD) (Price *et al.,* 1996; Price and Sisodia, 1997, Price *et al.,* 1998; Price *et al.,* in press).

2. ALZHEIMER'S DISEASE

2.1. Clinical Syndrome

The clinical syndrome of AD, the major cause of senile dementia, involves a decline in cognitive abilities, particularly memory, speed of performance, and problem solving (McKhann *et al.,* 1984; Khachaturian, 1985). The clinical course of AD may vary considerably, but most patients show declines in mental status, usually at a rate of 3–4 points per year on the Blessed Information–Memory–Concentration Test or the Blessed Dementia Index (Blessed *et al.,* 1968) and 2–3 points per year on the Mini Mental State Examination (Folstein *et al.,* 1975). Memory abnormalities are usually associated with impairments in language, calculation, visuospatial perceptions, judgment, and behavior (McKhann *et al.,* 1984; Khachaturian, 1985; Katzman *et al.,* 1988; Katzman *et al.,* 1989), and, in the late stage of AD, patients are profoundly demented and usually die of intercurrent illnesses.

The diagnosis of AD is based on clinical history, physical examination, neuropsychological testing, and a variety of diagnostic assessments used to exclude other causes of dementia and to determine the possibility or probability of a diagnosis of AD (McKhann *et al.,* 1984; Khachaturian, 1985; Atack, 1989). For diagnostic research purposes, investigators have begun to assess the value of determining levels of β-amyloid (Aβ) and tau protein in body fluids and establishing apolipoprotein E (apoE) genotypes. However, at present, these assessments are not appropriate for routine diagnostic evaluations (Roses, 1997). Neuroimaging studies, including computed tomography (CT) and magnetic resonance imaging (MRI), can identify potentially treatable diseases and disclose abnormalities (particularly in the medial temporal lobe) that may have high predictive value for establishing an accurate diagnosis of AD. Positron emission tomography (PET) and single-photon emission computed tomography (SPECT) show decreased regional glucose metabolism and blood flow in the parietal and temporal lobes, with involvement of other cortical areas at later stages. However, with the exception of brain biopsy, there are currently no tests that definitively establish the diagnosis of AD in living individuals.

2.2. Neuropathology

Correlative neuropathological–neurochemical studies have shown that symptoms and signs of AD are associated with abnormalities in populations of neurons in the amygdala, hippocampus, neocortex, anterior thalamus, basal forebrain cholinergic system, and monoaminergic brain stem systems (i.e., locus coeruleus, raphe complex). Cellular alterations include amyloid deposits, dystrophic neurites, senile plaques (i.e., neurites displayed around a core of amyloid, usually associated with astrocytes and microglia), neurofibrillary tangles (NFTs), and neuropil threads.

In 1985, the National Institute on Aging (NIA) and other agencies sponsored a conference with the goal of establishing autopsy criteria for the diagnosis of AD (Khachaturian, 1985), based primarily on the number of senile plaques in neocortex, adjusted by age, factoring in the clinical history and the number of NFTs. Several years later, the Consortium to Establish a Registry for AD (CERAD) was established as a joint effort of many medical and research institutions to standardize the clinical and pathological diagnoses of AD. CERAD has formulated a useful algorithm for the diagnosis of AD based on a semiquantitative assessment of senile (neuritic) plaques in neocortex, the age of the patient, and the clinical history of dementia (Mirra *et al.,* 1991, 1993, 1997). At present, the CERAD criteria do not include NFTs in the diagnostic algorithm nor do they include lesions in the hippocampus, entorhinal cortex, or neocortex. More recently, a panel of neuropathologists outlined a new consensus recommendation for the postmortem diagnosis of AD that deals with some of these issues (Hyman and Trojanowski, 1997).

2.2.1. Vulnerable Neuronal Systems

Correlative neuropathological–neurochemical studies have shown that clinical symptoms/signs in cases of AD are associated with abnormalities that involve neuronal populations in the amygdala, hippocampus, neocortex, anterior thalamus, basal forebrain cholinergic system, and monoaminergic brain stem systems (i.e., locus coeruleus, raphe complex) (Whitehouse *et al.,* 1982; Hyman *et al.,* 1984, 1986, 1990; Kemper, 1984; Vogels *et al.,* 1990; Braak and Braak, 1991, 1994; Braak *et al.,* 1996a,b).

Braak and Braak (1991, 1994) have presented evidence for six stages of AD. Analyses of the distributions of patterns of NFTs, neuropil threads, and neurites in plaques suggest that specific architectonic units are selectively damaged in a predictable sequence. Initial lesions occur in the transentorhinal and entorhinal cortices, followed by involvement of the hippocampus and, subsequently, the neocortex. In the transentorhinal stages (I, II), NFT and neuropil threads are present in the transentorhinal area (stage I) and the CA field of hippocampus (stage II). In the limbic stages (III, IV), hippocampus and basolateral amygdala become involved as severe disease occurs in transentorhinal and entorhinal regions as well as the limbic nuclei of the thalamus. In the isocortical stages of disease (V, VI), there is initial involvement of the basal portion of the isocortex with subsequent cellular pathology involving the association areas.

The cellular abnormalities involving neural circuits have profound clinical consequences presumed to be the result of compromised synaptic interactions in affected regions of the brain (Perdahl *et al.*, 1984; Davies *et al.*, 1987; Hamos *et al., 1989; DeKosky and Scheff, 1990; Masliah et al.*, 1990, 1991, 1992; Terry *et al.*, 1991; Scheff and Price, 1993; Sze *et al.*, 1997). For example, synaptic abnormalities in the hippocampus correlate with the severity of the memory deficits in individuals with AD (Sze *et al.*, 1997). Alterations in the basal forebrain cholinergic system also contribute to these memory difficulties and may be a factor in the attention deficits that occur in these patients. Higher cognitive deficits (e.g., disturbances in language, movement, and judgment) are related to pathology in the neocortex, whereas the psychiatric disturbances that occur in some patients may reflect the involvement of the limbic cortex, amygdala, thalamus, and monoaminergic systems.

2.2.2. Cytoskeletal Pathology in Neurons

Many affected neurons exhibit NFTs (neurofibrillary lesions in cell bodies and proximal axons), neuropil threads (filamentous accumulations in dendrites), and dystrophic neurites (filament-containing neuronal processes, particularly in distal axons) (Feany and Dickson, 1996). The principal ultrastructural components of NFTs are intracellular accumulations of poorly soluble straight and paired helical filaments comprised principally of hyperphosphorylated isoforms of tau, a low molecular weight, microtubule-associated protein (Goedert *et al.*, 1991; Lee *et al.*, 1991; Lee and Trojanowski, 1992; Dickson, 1997). The tau species in NFTs are in a state in which they are unable to bind microtubules. Thus, abnormal tau can alter the stability of microtubules, with subsequent effects on intracellular transport, cellular geometry, and neuronal viability. It is uncertain whether phosphorylation is necessary or sufficient for the formation of these filaments because recent *in vitro* studies have shown that exposure of nonphosphorylated three- and four-repeat recombinant tau to sulfated glycosaminoglycans can lead to the formation of paired and straight helical filaments, respectively (Gómez-Isla *et al.*, 1997). These findings suggest that interactions of tau and glycosaminoglycans may play an important role in the formation of NFTs *in vivo*. Other NFT-associated epitopes include ubiquitin, microtubule-associated protein (MAP-2), neurofilament proteins (particularly phosphorylated epitopes of the 200-kDa protein), and, possibly, Aβ (Anderton *et al.*, 1982; Masters *et al.*, 1985; Cork *et al.*, 1986; Grundke-Iqbal *et al.*, 1986; Kosik *et al.*, 1986; Perry *et al.*, 1987). Eventually, NFT-bearing cells die, possibly by mechanisms that involve apoptotic pathways (Su *et al.*, 1994; Smale *et al.*, 1995; Troncoso *et al.*, 1996).

Neurites and neuropil threads, enlarged distal axons (neurites) and dendrites (threads), also contain accumulations of paired helical filaments and 15-nm straight filaments (Wisniewski and Terry, 1973; Anderton *et al.,* 1982; Kosik *et al.,* 1986; Perry *et al.,* 1987; Goedert *et al.,* 1991; Lee *et al.,* 1991; Martin *et al.,* 1991; McKee *et al.,* 1991; Perry *et al.,* 1991; Masliah *et al.,* 1992). Because NFTs, neurites, and neuropil threads contain similar cytoskeletal constituents, it is likely that the genesis of these fibrillar inclusions is the result of common mechanisms.

2.2.3. *Amyloidogenesis*

A histological hallmark of AD is the presence of senile plaques in amygdala, hippocampus, and neocortex (McKhann *et al.,* 1984; Khachaturian, 1985; Probst *et al.,* 1987, 1991). Thioflavin S/Congo red-positive amyloid is comprised principally of Aβ, a ~ 4-kDa peptide derived from the amyloid precursor protein (APP). Aβ species are deposited in the neural parenchyma (as diffuse or compact amyloid in plaques) and around blood vessels (congophilic angiopathy) (Glenner and Wong, 1984; Roher *et al.,* 1986; Iwatsubo *et al.,* 1994; Lemere *et al.,* 1996). Aβ-containing peptide species include Aβ40 and Aβ42 (43), with the longer forms of Aβ42 deposited early. Aβ42 is the principal component in diffuse or compact amyloid plaques (Iwatsubo *et al.,* 1994). These Aβ fibrillar aggregates act as a nidus for subsequent deposits of other proteins, including α_1-antichymotrypsin, components of the complement cascade, apoE and J, cholinesterases, etc. (Abraham *et al.,* 1988, 1989; Johnson *et al.,* 1992; Mesulam *et al.,* 1992; Wisniewski and Frangione, 1992; Schmechel *et al.,* 1993; Strittmatter *et al.,* 1993; Mesulam and Geula, 1994). Deposits of Aβ and these other proteins may play roles in recruitment of astrocytes and microglial cells to these sites. These cells can produce a variety of peptides that can participate in pathogenic processes occurring in the parenchyma. Aspects of the amyloidogenesis are reviewed with the biology of APP (see later).

2.2.4. *Mechanisms Leading to Dysfunction/Death of Neurons*

These processes are not well understood but may involve Aβ toxicity (Yankner *et al.,* 1989; Kowall *et al.,* 1991; Pike *et al.,* 1991; Yoshikawa *et al.,* 1992), abnormalities of the endosomal/lysosomal system (Nixon and Cataldo, 1995; Cataldo *et al.,* 1996, 1997), excitotoxicity (Greenamyre and Young, 1989; Koh *et al.,* 1990; Beal, 1992a,b), the activation of proteases (e.g., calpain) (Saito *et al.,* 1993), and apoptotic processes (Su *et al.,* 1994; Troncoso *et al.,* 1996). For example, methods to tag DNA fragments and to visualize the nuclear morphology of cells in the brains of individuals with AD have disclosed histochemical and morphological features consistent with apoptosis of neurons and astrocytes (Su *et al.*, 1994; Troncoso *et al.*, 1996).

3. PRINCIPAL RISK FACTORS

The principal identified risk factors for AD are genes and age.

3.1. Genes

Approximately 10% of cases of AD are familial and show autosomal dominant inheritance; subsets of these cases are linked to mutations of specific genes, including those encoding APP and presenilin 1 and 2 (PS1/PS2) (Bird, 1994; Schellenberg, 1995; Haass, 1997; Hardy, 1997). Late-onset disease is associated with the presence of the *apoE4* allele (Roses, 1995).

3.1.1. APP

APP is a type-I integral membrane glycoprotein encoded by a gene on chromosome 21 containing the Aβ region, comprised of 28 amino acids of the ectodomain and 11–14 amino acids of the adjacent transmembrane domain of APP (Glenner and Wong, 1984; Masters *et al.*, 1985). APP molecules mature through the constitutive secretory pathway and are modified by the addition of both *N*- and *O*-linked carbohydrates, phosphate, and sulfate moieties (Oltersdorf *et al.*, 1989; Hung and Selkoe, 1994). Varying levels of newly synthesized APP molecules appear at the cell surface (Haass *et al.*, 1992; Sisodia, 1992); some of these molecules are cleaved endoproteolytically between positions 16 and 17 of Aβ by APP "α-secretase" (Esch *et al.*, 1990; Sisodia *et al.*, 1990; Anderson *et al.*, 1991) to release the ectodomain of APP, including residues 1–16 of Aβ, into the medium. This α-secretase cleavage within the Aβ domain precludes the formation of Aβ. However, Aβ can be generated by several pathways, not mutually exclusive, that involve the endoproteolytic cleavage of APP by hitherto uncharacterized activities, termed "β-" and "γ-secretase," which generate the N- and C-termini of the peptide, respectively. In one model, β-secretase cleavage occurs within endocytic compartments, and γ-secretase cleavage of the residual ~100 amino acid membrane-bound fragment occurs virtually simultaneously with the formation and release of Aβ. Although the majority of Aβ appears to be generated in the endosomal recycling pathway, a fraction of secreted Aβ species is generated in a secretory pathway [i.e., in endoplasmic reticulum (Cook *et al.*, 1997; Hartmann *et al.*, 1997)]. Moreover, it is now clear that ~90% of secreted Aβ peptides are Aβ40, a relatively benign soluble form of the peptide, whereas the remaining ~10% of secreted Aβ peptides are Aβ42 and Aβ43.

Physicochemical studies indicate that Aβ42 and/or Aβ43, rather than Aβ40, are more fibrillogenic peptides (Burdick *et al.*, 1992) and provide the substrate for amyloid deposits (Jarrett *et al.*, 1993). Immunocytochemical studies performed with antibodies uniquely specific for Aβ40 or Aβ42 in the cortices of individuals with AD or with Down's syndrome have demonstrated that Aβ deposition begins with Aβ42 (43) and not with Aβ40 (Iwatsubo *et al.*, 1994, 1995; Shinkai *et al.*, 1995; Lemere *et al.*, 1996).

In a small fraction of cases of FAD, mutations have been documented in the gene encoding APP (Chartier-Harlin *et al.*, 1991; Goate *et al.*, 1991; Naruse *et al.*, 1991; Hendricks *et al.*, 1992; Mullan *et al.*, 1992). In some families, the valine residue at position 717 (four amino acids downstream of the C-terminus of Aβ42) is substituted with either isoleucine, glycine, or phenylalanine (Chartier-Harlin *et al.*, 1991; Goate *et al.*, 1991; Naruse *et al.*, 1991). Cells that express APP harboring 717 mutations do not secrete higher levels of Aβ but do secrete a higher fraction of longer Aβ peptides (i.e., extending to Aβ residue 42) relative to cells that express wild-type (wt) APP (Suzuki *et al.*, 1994). In two large, related, early-onset AD families from Sweden, Lys–Met codons immediately N-terminal to Asp 1 of Aβ are substituted with Asn–Leu (Mullan *et al.*, 1992). Cells that express *APP* harboring the "Swedish" substitutions (*APPswe*) secrete higher levels of Aβ40 and Aβ42 as compared to cells expressing wt constructs (Citron *et al.*, 1992; Thinakaran *et al.*, 1996). Mutations in codons 692 and 693 within the Aβ peptide lead to presenile dementia and cerebral hemorrhages (Levy *et al.*, 1990; Van Broeckhoven *et al.*, 1990; Hendricks *et al.*, 1992). The substitution of Gly for Ala at amino acid 692 (Hendricks *et al.*, 1992) diminishes the efficiency of α-secretase processing and increases the microheterogeneity/hydrophobicity of secreted Aβ species (Haass *et al.*, 1994). The mutation leading to a Glu-to-Gln substitution at position 693 of APP in hereditary cerebral hemorrhage with amyloid, Dutch types (HCHWA-D) (corresponding to amino acid 22 of Aβ) is associated with an increased propensity of a synthetic Aβ peptide to aggregate into fibrils (Wisniewski *et al.*, 1991).

3.1.2. Presenilins

PS1 and PS2, encoded by genes on chromosomes 14 and 1, respectively, are highly homologous proteins (Sherrington *et al.*, 1995) predicted to contain eight transmembrane helices (Doan *et al.*, 1996). The N-terminus, loop, and C-terminus of PS1 are oriented toward the cytoplasm (Doan *et al.*, 1996; De Strooper *et al.*, 1997). Although PS1 is synthesized as an ~42- to 43-kDa polypeptide, the preponderant PS1-related species that accumulates in cultured mammalian cells and in the brains and systemic tissues of rodents, *PS1* Tg mice, primates, and humans are ~27- to 28-kDa N-terminal and ~16- to 17-kDa C-terminal derivatives (Lee *et al.*, 1996; Thinakaran *et al.*, 1996; Podlisny *et al.*, 1997). Epitope mapping studies suggest that PS1 is cleaved within a region that encompasses amino acids 260–320, a domain in which >50% of identified FAD-linked PS1 mutations occur. Recent studies suggest that cleavage may be heterogeneous, occurring between amino acids 292 and 299 (Podlisny *et al*,. 1997). These results are consistent with the demonstration that the FAD-linked PS1ΔE9 variant, which lacks amino acids 290–319, fails to be cleaved (Thinakaran *et al.*, 1996). The accumulation of ~17 kDa and ~27 kDa human-specific PS1 derivatives in the brains of Tg mice that express human PS1 is highly regulated and saturable (Thinakaran *et al.*, 1996; Lee *et al.*, 1997): levels of PS1 derivatives are disproportionate to levels of Tg-derived mRNA or full-length human PS1. The stoichiometry of accumulated ~17-kDa and ~27-kDa PS1 fragments is ~1:1 in non-Tg and Tg mouse brains; this ratio is independent of the level of Tg-derived human PS1 mRNA expression (Thinakaran *et al.*, 1996). Because of the paucity of accumulated full-length PS and the generality of PS processing across tissues, it has been suggested that PS fragments represent the mature and functional forms of PS1 (Thinakaran *et al.*, 1996; Ratovitski *et al.*, 1997). Mechanism(s) involved in the regulation of the levels of accumulated PS1 derivatives have not been established, but the abundance of fragments appears to be coordinately influenced by competition for limiting cellular factors (Thinakaran *et al.*, 1997).

PS1 and PS2 mRNA are expressed in a variety of peripheral tissues and in the brain (Rogaev *et al.*, 1995; Sherrington *et al.*, 1995; Kovacs *et al.*, 1996; Lee *et al.*, 1996; Levey *et al.*, 1997). Although the structural conservation and relatively ubiquitous expression patterns of PS1 and PS2 mRNA suggest some degree of functional redundancy, differences exist in relative levels of expression, suggesting that PS1 and PS2 may play different roles in tissue- or development-specific processes. In the brains of adult mammals, both PS1 and PS2 transcripts are expressed in many neuronal populations (Lee *et al.*, 1996), and mRNA are also present in glial cells (Rogaev *et al.*, 1995; Sherrington *et al.*, 1995; Lee *et al.*, 1996). Notably, nerve cells known to be at risk in AD (i.e., neurons in hippocampal CA fields, the medial and cortical amygdala, and neocortex) express PS and APP transcripts at high levels, whereas neurons in regions less prone to AD-associated pathology express PS and APP transcripts at more variable levels (Lee *et al.*, 1996). Light microscopic immunocytochemical studies of rodent (Lee *et al.*, 1996), nonhuman primate (Lah *et al.*, 1997), and human brains using antibodies selective for the N- or C-terminal PS1 fragments revealed that PS1 was present in all brain regions, with the strongest labeling in neurons and the neuropil, including axons and dendrites (Lah *et al.*, 1997). Immunoreactivity was most robust in neurons of the neocortex, hippocampus, and basal forebrain cholinergic system (Lah *et al.*, 1997).

Although the biological functions of PS are not yet well understood, insight regarding PS function emerged with the discovery of a homologous gene in *C. elegans*, termed *sel-12;* mutant alleles of *sel-12* were uncovered as suppressors of a hypomorphic multivulval phenotype in *C. elegans* linked to hyperactivity of the *C. elegans* Notch homologues lin12 and glp1 (Levitan, Greenwald, 1995). Notch and lin12/glp1 are transmembrane receptors required for the specification of cell fate and lateral inhibition during development (Artavanis-Tsakonas *et al.,* 1995). The extremely high amino acid homology between PS and *sel-12* led to the prediction that related proteins would be functionally interchangeable. Consistent with this hypothesis, an egg-laying (*egl*) defect associated with loss of *sel-12* function in *C. elegans* is rescued efficiently by the expression of human PS1 and PS2; the rescue efficiency of human PS is essentially indistinguishable in Tg worms that express *sel-12* (Levitan *et al.,* 1996; Baumeister *et al.,* 1997). The *egl* defect was only weakly rescued in Tg worms that express several human FAD-linked PS1 variants (Levitan *et al.,* 1996; Baumeister *et al.,* 1997), suggesting that *PS1* missense variants behaved as loss-of-function alleles.

Mutations in the *PS1* gene appear to be the cause of the majority of cases of early-onset AD (St George-Hyslop *et al.,* 1992; Sherrington *et al.,* 1995). Approximately 50% of mutations occur within or immediately adjacent to the predicted loop domain (Sherrington *et al.,* 1995; Doan *et al.,* 1996). Cases with the Glu → 280 Ala mutation show massive deposits of Aβ42 in many brain regions (Lemere *et al.,* 1996). The *PS2* gene shows substantial homology to *PS1,* and *PS2* mutations have been reported to cause autosomal dominant AD in Volga German kindreds and in an Italian pedigree (Levy-Lahad *et al.,* 1995; Levy-Lahad *et al.,* 1995; Li *et al.,* 1995; Rogaev *et al.,* 1995).

The mechanisms by which mutations in *PS1/PS2* predispose individuals to FAD are not clear, but recent studies indicate that the mutation in *PS1* influences levels of Aβ42: plasma and conditioned media from fibroblasts obtained from carriers of *PS1/PS2* mutations have elevated levels of Aβ42 species as compared to samples from unaffected family members (Scheuner *et al.,* 1996); *in vitro* studies of transfected cell lines expressing *PS1* mutations (*A246E, M146L,* or *ΔE9*) have increased ratios of Aβ42:Aβ40 as compared to ratios in media of cells expressing wt *PS1* (Borchelt *et al.,* 1996). Moreover, as outlined later, the brains of young Tg mice coexpressing a chimeric *APPswe* and a FAD-linked *PS1* variant show elevated Aβ42/Aβ40 ratios as compared to ratios in the brains of Tg mice expressing *APPswe* alone or TG mice coexpressing wt human *PS1* and *APPswe* (Borchelt *et al.,* 1996). At 9–12 months of age, mice with the double mutation show an accelerated deposition of amyloid as compared to mice that express single Tgs (Borchelt *et al.,* 1997). Thus, one mechanism by which these mutant *PS1* cause AD is to influence APP processing to increase the extracellular concentration of Aβ42 peptides prone to deposition/aggregation.

3.1.3. ApoE

This ~34-kDa glycoprotein, encoded by a gene on chromosome 19, carries cholesterol and other lipids in the blood (Weisgraber *et al.,* 1994). There are three alleles at the single *ApoE* locus: *ApoE3* has a cysteine at position 112 and an arginine at position 158; *ApoE4* has arginine at both positions; and *ApoE2* has cysteine at both positions. The *ApoE3* allele is most common in the general population (frequency 0.78); the allelic frequency of *ApoE4* is 0.14. However, in clinic-based studies, patients with late-onset disease (>65 years of age) have an *ApoE4* allelic frequency of 0.50. The risk for AD and the age of onset of disease are influenced by the presence of *ApoE4* (Ordovas *et al.*, 1987; Mahley, 1988; Corder *et al.*, 1993; Poirier *et al.*, 1993; Rebeck *et al.*, 1993; Saunders *et al.*, 1993; Strittmatter *et al.*, 1993; Chartier-Harlin *et al.*, 1994; Galasko *et al.*, 1994; Goedert *et al.*, 1994; Katzman, 1994; Nalbantoglu *et al.*, 1994; Roses, 1994, 1995; Roses *et al.*, 1994; Tsai *et al.*, 1994; Hendrie *et al.*, 1995; van Duijn *et al.*, 1995; Gómez-Isla *et al.*, 1996). The *apoE4* allele appears to be associated with an increased Aβ burden (Rebeck *et al.*, 1993; Schmechel *et al.*, 1993; Nalbantoglu *et al.*, 1994; Gearing *et al.*, 1996; Gómez-Isla, *et al.*, 1996).

Available evidence indicates that the presence of *apoE4* enhances the risk for AD in a dose-dependent fashion and may influence the rate of progression (Hyman, 1996; Farlow, 1997). In a prospective longitudinal study, Gómez-Isla and colleagues (Gómez-Isla *et al.*, 1996) compared the age at onset, duration, and rate of progression of disease in patients with different *apoE* genotypes. The *E4* allele was associated with early onset but no change in the rate of progression of dementia. NFTs were correlated with clinical assessments of both the duration and severity of dementia but not with *apoE* genotype, whereas Aβ deposition was elevated in association with the presence of *apoE4* (Gómez-Isla *et al.*, 1996). The impact of *apoE* genotype seems to be greatest on Aβ deposition, but it has not yet been defined whether *apoE* influences production, aggregation, fibrillogenesis, or clearance. It should be emphasized that many individuals reach old age without developing dementia in spite of the inheritance of either one or two of the *apoE4* alleles (Hyman *et al.*, 1996). Thus, *ApoE4* is neither necessary nor sufficient for disease. Although *apoE* clearly acts as a susceptibility factor for AD, the mechanisms of this increased vulnerability to disease are not well understood.

3.2. Age

Although many older individuals remain intellectually intact and some make outstanding contributions in late life, significant numbers of the elderly develop overt dementia (Bachman *et al.*, 1993; Amaducci and Tesco, 1994), and it is now established that AD is the most common cause of dementia occurring in mid-to-late life.

4. GENETICALLY ENGINEERED ANIMAL MODELS

4.1. Transgenic Mice

4.1.1. APP Transgenic Mice

To generate animal models of Aβ amyloidogenesis and the associated histopathology of AD, many groups have created Tg mice that express wt APP, FAD-linked APP variants, C-terminal fragments of APP, and Aβ (Kammesheidt *et al.*, 1992; Neve *et al.*, 1992; Buxbaum *et al.*, 1993; Lamb, *et al.*, 1993, 1997; Higgins *et al.*, 1994; Games *et al.*, 1995; Hsiao *et al.*, 1995, 1996; LaFerla *et al.*, 1995; Moran *et al.*, 1995; Masliah *et al.*, 1996; Irizarry *et al.*, 1997; Borchelt *et al.*, 1997; Nalbantoglu *et al.*, 1997). Although early efforts were disappointing because mice did not exhibit the cellular abnormalities characteristic of AD, more recent work has been very successful. There now exist multiple lines of Tg mice that show features of AD. To illustrate the approach, we review some of the studies of mice that show the neuropathological features of human AD (Games *et al.*, 1995; Hsiao *et al.*, 1996; Borchelt, *et al.*, 1997).

In the first successful effort, the PDGF β-promoter was used to drive the expression of a human APP minigene that encodes the FAD-linked APP (*717V→F*) mutation in an outbred strain; the construct contained portions of APP introns 6–8 that allow alternative splicing of exons 7 and 8. Levels of human APP mRNA and protein significantly exceeded levels of endogenous APP. The transcripts encoded the three major splicing variants, particularly the KPI-coding form, and levels of the transgene product were ~4–5 times higher than those of endogenous APP (Games *et al.*, 1995). Brains showed diffuse Aβ deposits and plaques with dystrophic neurites displayed around Aβ cores (Masliah *et al.*, 1996). This work was the first to reproduce some of the cellular phenotypes of AD in mice.

Our efforts and those of Dr. Hsiao and co-workers in this area were very instructional. In collaboration with Dr. Hsiao, we produced multiple lines of Tg FVB mice that express a variety of human and mouse APP transgenes (wt and mutant) placed under the transcriptional control of the hamster prion gene promoter (Hsiao *et al.*, 1995). FVB/N mice were chosen for these studies because they are an inbred strain, have an easily injectable and prominent pronucleus in single-cell embryos, are highly fertile, and have been used successfully in producing Tg models of prion diseases (Hsiao *et al.*, 1994). The expression of Tg-encoded APP was largely copy-number dependent (Hsiao *et al.*, 1995). Mice with >30 copies produced 3–5 times the endogenous levels of APP. Tg mice expressing these high levels of APP showed behavioral abnormalities including inactivity, agitation, neophobia, and seizures; and diminished glucose utilization in cortical–limbic areas (there was significant gliosis in these brain regions). These abnormalities appeared earlier with increasing levels of APP in the brain. No extracellular amyloid was detected. A similar neurological disorder develops naturally in older non-Tg FVB mice, and our work indicates that this age- and strain-related phenotype is exacerbated by the presence of the APP Tg, leading to premature death. Thus, despite achieving high levels of the Tg product (up to 5 times endogenous), the FVB strain is not ideal for attempts to produce mutant APP Tg mice with AD-type pathology.

When it became apparent that achieving high levels of expression was key to producing models of autosomal disease, Dr. Hsiao and colleagues and our group used similar strategies to produce lines of Tg mice in which the hamster prion protein (PrP) promoter overexpressed human *APP-695.swe* (Hsiao *et al.*, 1996; Borchelt *et al.*, 1997).

For these studies, we made two important changes in our approaches: we used a newly modified vector; and we introduced the transgene into another strain of mice. The *SHaPrP.neo* vector, which we used initially (Hsiao *et al.*, 1895), is a large 45-kb cosmid; hence, genetic manipulation of this vector is difficult. To overcome this difficulty, we obtained a much smaller fragment of the mouse PrP genome (*phgPrP*) (Dr. Charles Weissmann, University of Zurich) in which the second 10-kb intron and 15 kb of 3′ untranslated sequence were removed; this vector expresses PrP in a virtually identical pattern as a full-length genomic clone (*cos6.11*). The open reading frame of the coding exon in the *phgPrP* plasmid was replaced with a unique *Xho*I restriction endonuclease site, allowing for insertion of a cDNA that encodes a murine 695-amino acid isoform of the APP with a "humanized" Aβ domain and mutations (*K595N, M596L*) linked to Swedish pedigrees of FAD (Mullan *et al.*, 1992). Using this strategy, we have recently produced two lines of mice with Aβ deposits (Borchelt *et al.*, 1997). In the brains of these lines of mice, levels of Aβ40 and Aβ42 were elevated, the ratio of Aβ42:40 was increased, and dystrophic neurites and Aβ deposits were conspicuous in amygdala, hippocampus, and cortex (Hsiao *et al.*, 1996; Borchelt *et al.*, 1997). Although there were abnormalities in the neuropil of the hippocampus, there was no evidence of neuronal loss in CA1 (Irizarry *et al.*, 1997). Dr. Hsiao tested Tg mice on two memory tests (spatial reference and alternation tasks); at 9–10 months of age, animals were impaired on these tasks. The interpretation of these observations has been challenged (Routtenberg, 1997) and defended (Hsiao *et al.*, 1997).

Using a different approach, Drs. Lamb and Gearhart and colleagues produced Tg mice containing yeast artificial chromosomes (YACs) that encompass the entire 400-kDa human *APP* gene with a FAD mutation (Lamb *et al.*, 1997). The levels of Aβ peptides were increased in the brain, and there was a diminution in levels of α-secretase-generated soluble APP derivatives. In addition, levels of the longer Aβ peptides (i.e., Aβ1–42, 43) were elevated in the YAC Tg mice expressing the *APPV717* mutation, whereas *APPswe* Tg mice showed increased levels of all Aβ peptides (Lamb *et al.,* 1997).

4.1.2. PS1 and APP Tg Mice

A number of lines of *PS1* mutant mice have been studied. The brains of young Tg mice that express human *PS1* harboring FAD-linked mutations show an increased ratio of Aβ42:40 (Borchelt *et al.,* 1996, 1997; Duff *et al.,* 1996; Thinakaran *et al.,* 1996; Citron *et al.,* 1997). To date, our *PS1* mutant mice have not developed a behavioral phenotype or amyloid deposits.

In our studies, we coexpressed *A246E* human *PS1* and a chimeric mouse/human *APP695* harboring a human Aβ domain as well as mutations (*K595N, M596L*) linked to *APPswe* FAD pedigrees (Borchelt *et al.,* 1996). At 12-months of age, Tg animals that coexpress *A246E* human *PS1* and *APPswe* contained numerous amyloid deposits (Borchelt *et al.,* 1997), many of which were associated with dystrophic neurites and reactive astrocytes. Parallel analyses of brains from age-matched animals that express *APP695.swe* alone or mice that express *A246E* human *PS1* alone were free of amyloid deposits (Borchelt *et al.,* 1997). These observations indicate that *A246E* human *PS1* acts synergistically with *APPswe* to accelerate the rate of amyloid deposition. Our findings suggest that the principal mechanism by which mutations in *PS1* causes disease is through elevating extracellular concentrations of Aβ1–42 and thereby accelerating the deposition of amyloid.

4.2. Gene-Targeted Mice

4.2.1. APP Knockout Mice

When compared to hemizygous APP or wt littermates, homozygous APP knockout mice were fertile and viable but exhibited subtle decreases in locomotor activity and forelimb grip strength as well as reactive astrogliosis (Zheng *et al.,* 1995). The absence of substantial phenotypes in APP knockout mice may be related to functional redundancy provided by homologous amyloid precursor-like proteins (APLP1 and APLP2), molecules expressed at high levels with developmental and cellular distributions similar to APP (Wasco *et al.,* 1992, 1993; Slunt *et al.,* 1994).

4.2.2. PS1 Knockout Mice

Homozygous mutant mice failed to survive beyond the early postnatal period (Shen *et al.,* 1997; Wong *et al.,* 1997). The most striking phenotype observed in *PS1−/−* embryos was a severe perturbation in the development of the axial skeleton and ribs. The failed development of the axial skeleton in *PS1−/−* animals was traced to defects in somitogenesis; in E8.5 and E9.5 embryos, somites were irregularly shaped and misaligned along the entire length of the neural tube and largely absent at the caudalmost regions. The abnormal somite patterns in *PS1−/−* embryos are highly reminiscent of somite segmentation defects described in mice with functionally inactivated *Notch1* and *Dll1* (encoding a Notch ligand) alleles (Conlon *et al.,* 1995; Hrabe de Angelis *et al.,* 1997). Remarkably, the expression of mRNA that encodes *Notch1* and *Dll1* is reduced considerably in the presomitic mesoderm of PS1−/− mice (Wong *et al.,* 1997). In addition, all PS1−/− embryos exhibited intraparenchymal hemorrhages after day 11 of gestation. It has also been reported that, in the brains of PS1−/− mice, the ventricular zone is thinner by day 14.5 and that massive neuronal loss in specific subregions is apparent after day 16.5. Shen and colleagues (Shen *et al.,* 1997) have interpreted these observations to indicate that PS1 is required for normal neurogenesis and neuronal survival.

5. CONCLUSIONS

During the past decade, significant progress has been made in understanding the biology of AD, particularly in defining genes implicated in FAD, and defining the character, evolution, and mechanisms of the cellular abnormalities in the brain. The discovery that mutations in genes encoding APP, PS1, and PS2 are linked to FAD has ushered in a new and exciting era of research aimed at clarifying the relationships of genetic abnormalities to the pathogenesis of AD. Moreover, as exemplified by studies of the roles of *apoE* allele types in late-onset AD, other genetic factors undoubtedly play significant roles in late-onset disease. Finally, very significant progress has been made in producing Tg models of FAD, in targeting genes implicated in FAD, and in the analyses of these models. We are particularly enthusiastic about the utility of these animal models. Recent work was clearly demonstrated that Tg mice expressing mutant human FAD-linked genes show behavioral impairments, Aβ42 deposits, dystrophic neurites, and senile plaques. It is now possible to coexpress two transgenes so that investigators can examine the effects of introducing human *apoE2, -3,* or *-4* alleles into Tg mice expressing wt or mutant *APP* or *PS1/2* transgenes. Finally, we believe that these Tg models will be of great value for testing novel therapies that, if efficacious in these model systems, can be introduced rapidly into clinical trials for patients with AD.

ACKNOWLEDGMENTS

The authors gratefully acknowledge discussions with Drs. Claudia H. Kawas, Juan C. Troncoso, Karen Hsiao, John D. Gearhart, Bruce T. Lamb, Michael K. Lee, Gopal Thinakaran, Lee J. Martin, Vassilis E. Koliatsos, Neal G. Copeland, and Nancy A. Jenkins.

This work was supported by grants from the U.S. Public Health Service (AG05146, NS20471, AG14248) as well as the Adler Foundation, the Alzheimer's Association, the Develbiss fund, and the American Health Assistance Foundation, as well as gifts from Merck, Sharp & Dohme. Drs. Price, Borchelt, and Wong are the recipients of a Leadership and Excellence in Alzheimer's Disease (LEAD) award (AG 07914); Dr. Price is a recipient of two Javits Neuroscience Investigator Awards (NS 10580).

REFERENCES

Abraham, C. R., Selkoe, D. J., and Potter, H., 1988, Immunocytochemical identification of the serine protease inhibitor α_1-antichymotrypsin, in the brain amyloid deposits of Alzheimer's disease, *Cell* **52:**487–501.

Abraham, C. R., and Selkoe, D. J., Potter, H., Price, D. L., and Cork, L. C., 1989, α_1-Antichymotrypisn is present together with the β-protein in monkey brain amyloid deposits, *Neuroscience* **32:**715–720.

Amaducci, L., and Tesco, G., 1994, Aging as a major risk for degenerative diseases of the central nervous system, *Curr. Opin. Neurol.* **7:**283–286.

Anderson, J. P., Esch, F. S., Keim, P. S., Sambamurti, K., Lieberburg, I., and Robakis, N. K., 1991, Exact cleavage site of Alzheimer amyloid precursor in neuronal PC-12 cells, *Neurosci. Lett.* **128:**126–128.

Anderton, B. H., Breinburg, D., Downes, M. J., Green, P. J., Tomlinson, B. E., Ulrich, J., Wood, J. N., and Kahn, J., 1982, Monoclonal antibodies show that neurofibrillary tangles and neurofilaments share antigenic determinants, *Nature* **298:**84–86.

Artavanis-Tsakonas, S., Matsuno, K., and Fortini, M. E., 1995, Notch signaling, *Science* **268:**225–232.

Atack, J. R., 1989, Cerebrospinal fluid neurochemical markers in Alzheimer's disease, *in* "Biological Markers of Alzheimer's Disease," (F. Boller, R., Katzman, A. Rascol, J.-L. Signoret, and Y. Christen, eds.), pp. 1–160, Springer-Verlag, Berlin.

Bachman, D. L., Wolf, P. A., Linn, R. T., Knoefel, J. E., Cobb, J. L., Belander, A. J., White, L. R., and D'Agostino, R. B., 1993, Incidence of dementia and probably Alzheimer's disease in a general population: The Framingham study, *Neurology* **43:**515–519.

Baumeister, R., Leimer, U., Zweckbronner, I., Jakubek, C., Grünberg, J., and Haass, C., 1997, Human presenilin-1, but not familial Alzheimer's disease (FAD) mutants, facilitate *Caenorhabditis elegans* Notch signalling independently of proteolytic processing, *Genes Function* **1:**149–159.

Beal, M. F., 1992a, Does impairment of energy metabolism result in excitotoxic neuronal death in neurodegenerative illnesses? *Ann. Neurol.* **31:**119–130.

Beal, M. F., 1992b, Role of excitotoxicity in human neurological disease, *Curr. Opin. Neurobiol.* **2**:657–662.

Bedell, M. A., Jenkins, N. A., and Copeland, N. G., 1997a, Mouse models of human disease. Part I: Techniques and resources for genetic analysis in mice, *Genes Dev.* **11**:1–10.

Bedell, M. A., Largaespada, D. A., Jenkins, N. A., and Copeland, N. G., 1997b, mouse models of human disease. Part II: Recent progress and future directions, *Genes Dev.* **11**:11–43.

Bird, T. D., 1994, Clinical genetics of familial Alzheimer disease, *in* "Alzheimer Disease," (R. D. Terry, R. Katzman, and K. L. Bick, eds.), pp. 65–74, Raven Press, New York.

Blessed, G., Tomlinson, B. E., and Roth, M., 1968, The association between quantitative measures of dementia and of senile change in the cerebral grey matter of elderly subjects, *Br. J. Psychiatry* **114**:797–811.

Borchelt, D. R., Thinakaran, G., Eckman, C. B., Lee, M. K., Davenport, F., Ratovitski, T., Prada, C.-M., Kim, G., Seekins, S., Yager, D., Slunt, H. H., Wang, R., Seeger, M., Levey, A. I., Gandy, S. E., Copeland, N. G., Jenkins, N. A., Price, D. L., and Younkin, S. G., Sisodia, S. S., 1996, Familial Alzheimer's disease-linked presenilin 1 variants elevate Aβ1-42/1–40 ratio in vitro and in vivo, *Neuron* **17**:1005–1013.

Borchelt, D. R., Ratovitski, T., Van Lare, J., Lee, M. K., Gonzales, V. B., Jenkins, N. A., Copeland, N. G., Price, D. L., and Sisodia, S. S., 1997, Accelerated amyloid deposition in the brains of transgenic mice co-expressing mutant presenilin 1 and amyloid precursor proteins, *Neuron* **19**:939–945.

Braak, H., and Braak, E., 1991, Alzheimer's disease affects limbic nuclei of the thalamus, *Acta Neuropathol.* **81**:261–268.

Braak, H., and Braak, E., 1994, Pathology of Alzheimer's disease, *in* "Neurodegenerative Diseases" (D. B. Calne, ed.), pp. 585–613, W. B. Saunders, Philadelphia.

Braak, H., Braak, E., Bohl, J., and Reintjes, R., 1996a, Age, neurofibrillary changes, Aβ-amyloid and the onset of Alzheimer's disease, *Neurosci. Lett.* **210**:87–90.

Braak, H., Braak, E., Yilmazer, D., de Vos, R. A. I., Jansen, E. N. H., and Bohl, J., 1996b, Pattern of brain destruction in Parkinson's and Alzheimer's diseases, *J. Neural Transm.* **103**:455–490.

Bruijn, L. I., Becher, M. W., Lee, M. K., Anderson, K. L., Jenkins, N. S., Copeland, N. G., Sisodia, S. S., Rothstein, J. D., Borchelt, D. R., Price, D. L., and Cleveland, D. W., 1997, ALS-linked SOD1 mutant G85R mediates damage to astrocytes and promotes rapidly progressive disease with SOD1-containing inclusions, *Neuron* **18**:327–338.

Burdick, D., Soreghan, B., Kwon, M., Kosmoski, J., Knauer, M., Henschen, A., Yates, J., Cotman, C., and Glabe, C., 1992, Assembly and aggregation properties of synthetic Alzheimer's A4/β amyloid peptide analogs, *J. Biol. Chem.* **267**:546–554.

Burright, E. N., Clark, H. B., Servadio, A., Matilla, T., Feddersen, R. M., Yunis, W. S., Duvick, L. A., Zoghbi, H. Y., and Orr, H. T., 1995, *SCA1* transgenic mice: A model for neurodegeneration caused by an expanded CAG trinucleotide repeat, *Cell* **82**:937–948.

Buxbaum, J. D., Christensen, J. L., Ruefli, A. A., Greengard, P., and Loring, J. F., 1993, Expression of APP in brains of transgenic mice containing the entire human *APP* gene, *Biochem. Biophys. Res. Commun.* **197**:639–645.

Cataldo, A. M., Hamilton, D. J., Barnett, J. L., Paskevich, P. A., and Nixon, R. A. 1996, Properties of the endosomal–lysosomal system in the human central nervous system: Disturbances mark most neurons in populations at risk to degenerate in Alzheimer's disease, *J. Neurosci.* **1**:186–199.

Cataldo, A. M., Barnett, J. L., Pieroni, C., and Nixon, R. A., 1997, Increased neuronal endocytosis and protease delivery to early endosomes in sporadic Alzheimer's disease: Neuropathologic evidence for a mechanism of increased β-amyloidogenesis, *J. Neurosci.* **17**:6142–6151.

Chartier-Harlin, M.-C., Crawford, F., Houlden, H., Warren, A., Hughes, D., Fidani, L., Goate, A., Rossor, M., Roques, P., Hardy, J., and Mullan, M., 1991, Early-onset Alzheimer's disease caused by mutations at codon 717 of the β-amyloid precursor protein gene, *Nature* **353**:844–846.

Chartier-Harlin, M.-C., Parfitt, M., Legrain, S., Perez-Tur, J., Brousseau, T., Evans, A., Berr, C., Vidal, O., Roques, P., Gourlet, V., 1994, Apolipoprotein E, epsilon 4 allele as a major risk factor for sporadic early and late-onset forms of Alzheimer's disease: Analysis of the 19q13.2 chromosomal region, *Hum. Mol. Genet.* **3**:569–574.

Citron, M., Oltersdorf, T., Haass, C., McConlogue, L., Hung, A. Y., Seubert, P., Vigo-Pelfrey, C., Lieberburg, I., and Selkoe, D. J., 1992, Mutation of the β-amyloid precursor protein in familial Alzheimer's disease increases β-protein production, *Nature* **360**:672–674.

Citron, M., Westaway, D., Xia, W., Carlson, G., Diehl, T., Levesque, G., Johnson-Wood, K., Lee, M., Seubert, P., Davis, A., Kholodenko, D., Motter, R., Sherrington, R., Perry, B., Yao, H., Strome, R., Lieberburg, I., Rommens, J., Kim, S., Schenk, D., Fraser, P., St. George Hyslop, P., and Selkoe, D. J., 1997, Mutant presenilins of Alzheimer's disease increase production of 42-residue amyloid β-protein in both transfected cells and transgenic mice, *Nature Med.* **3**:67–72.

Conlon, R. A., Reaume, A. G., and Rossant, J., 1995, *Notch 1* is required for the coordinate segmentation of somites, *Development* **121**:1533–1545.

Cook, D. G., Forman, M. S., Sung, J. C., Leight, S., Kolson, D. L., Iwatsubo, T., Lee, V. M. Y., and Doms, R. W., 1997, Alzheimer's Aβ (1–42) is generated in the endoplasmic reticulum/intermediate compartment of NT2N cells, *Nat. Med.* **3**:1021–1023.

Corder, E. H., Saunders, A. M., Strittmatter, W. J., Schmechel, D. E., Gaskell, P. C., Small, G. W., Roses, A. D., Haines, J. L., and Pericak-Vance, M. A., 1993, Gene dose of apolipoprotein-E type 4 allele and the risk of Alzheimer's disease in late onset families, *Science* **261**:921–923.

Cork, L. C., Sternberger, N. H., Sternberger, L. A., Casanova, M. F., Struble, R. G., and Price, D. L., 1986, Phosphorylated neurofilament antigens in neurofibrillary tangles in Alzheimer's disease. *J. Neuropathol. Exp. Neurol.* **45**:56–64.

Davies, C. A., Mann, D. M. A., Sumpter, P. Q., and Yates, P. O., 1987, A quantitative morphometric analysis of the neuronal and synaptic content of the frontal and temporal cortex in patients with Alzheimer's disease, *J. Neurol. Sci.* **78**:151–164.

Davies, S. W., Turmaine, M., Cozens, B. A., DiFiglia, M., Sharp, A. H., Ross, C. A., Scherzinger, E., Wanker, E. E., Mangiarini, L., and Bates, G. P., 1997, Formation of neuronal intranuclear inclusions underlies the neurological dysfunction in mice transgenic for the HD mutation, *Cell* **90**:537–548.

DeKosky, S. T., and Scheff, S. W., 1990, Synapse loss in frontal cortex biopsies in Alzheimer's disease: Correlation with cognitive severity, *Ann. Neurol.* **27**:457–463.

De Strooper, B., Beullens, M., Contreras, B., Levesque, L., Craessaerts, K., Cordell, B., Moechars, D., Bollen, M., Fraser, P., St. George Hyslop, P., and Vanleuven, F., 1997, Phosphorylation, subcellular localization, and membrane orientation of the Alzheimer's disease-associated presenilins, *J. Biol. Chem.* **272**:3590–3598.

Dickson, D. W., 1997, Neurodegenerative diseases with cytoskeletal pathology: A biochemical classification, *Ann. Neurol.* **42**:541–544.

Doan, A., Thinakaran, G., Borchelt, D. R., Slunt, H. H., Ratovitsky, T., Podlisny, M., Selkoe, D. J., Seeger, M., Gandy, S. E., Price, D. L., and Sisodia, S. S., 1996a, Protein topology of presenilin 1, *Neuron* **17**:1023–1030.

Doan, A., Thinakaran, G., Lanahan, A., Price, D. L., Sisodia, S. A., and Worley, P. E., 1996b, Identification and characterization of a presenilin 1 interacting protein. *Soc. Neurosci. Abstr.* **22**:728

Duff, K., Eckman, C., Zehr, C., Yu, X., Prada, C.-M., Perez-Tur, J., Hutton, M., Buee, L., Harigaya, Y., Yager, D., Morgan, D., Gordon, M. N., Holcomb, L., Refolo, L., Zenk, B., Hardy, J., and Younkin, S., 1996, Increased amyloid-β42 (43) in brains of mice expressing mutant presenilin 1, *Nature* **383**:710–713.

Esch, F. S., Keim, P. S., Beattie, E. C., Blacher, R. W., Culwell, A. R., Oltersdorf, T., McClure, D., and Ward, P. J., 1990, Cleavage of amyloid β peptide during constitutive processing of its precursor *Science* **248**:1122–1124

Farlow, M. R., 1997, Alzheimer's disease: Clinical implications of the apolipoprotein E genotype, *Neurology* **48:**30–S34

Feany, M. B., and Dickson, D. W., 1996, Neurodegenerative disorders with extensive tau pathology: A comparative study and review, *Ann. Neurol.* **40:**139–148.

Folstein, M. F., Folstein, S. E., and McHugh, P. R., 1975, "Minimental state." A practical method for grading the cognitive state of patients for the clinician, *J. Psychiatr. Res.* **12:**189–198.

Galasko, D., Saitoh, T., Xia, Y., Thal, L. J., Katzman, R., Hill, L. R., and Hansen, L., 1994, The apolipoprotein E allele ϵ 4 is overrepresented in patients with the Lewy body variant of Alzheimer's disease, *Neurology* **44:**1950–1951.

Games, D., Adams, D., Alessandrini, R., Barbour, R., Berthelette, P., Blackwell, C., Carr, T., Clemens, J., Donaldson, T., Gillespie, F., Guido, T., Hagopian, S., Johnson-Wood, K., Khan, K., Lee, M., Leibowitz, P., Lieberburg, I., Little, S., Masliah, E., McConlogue, L., Montoya-Zavala, M., Mucke, L., Paganini, L., Penniman, E., Power, M., Schenk, D., Seubert, P., Snyder, B., Soriano, F., Tan, H., Vitale, J., Wadsworth, S., Wolozin, B., and Zhao, J., 1995, Alzheimer-type neuropathology in transgenic mice overexpressing V717F β-amyloid precursor protein, *Nature* **373:**523–527.

Gearing, M., Mori, H., and Mirra, S. S., 1996, Aβ-peptide length and apolipoprotein E genotype in Alzheimer's disease, *Ann. Neurol.* **39:**395–399.

Glenner, G. G., and Wong, C. W., 1984a, Alzheimer's disease: Initial report of the purification and characterization of a novel cerebrovascular amyloid protein, *Biochem. Biophys. Res. Commun.* **120:**885–890.

Glenner, G. G., and Wong, C. W., 1984b, Alzheimer's disease and Down's syndrome: Sharing of a unique cerebrovascular amyloid fibril protein, *Biochem. Biophys. Res. Commun.* **122:**1131–1135.

Goate, A., Chartier-Harlin, M.-C., Mullan, M., Brown, J., Crawford, F., Fidani, L., Giuffra, L., Haynes, A., Irving, N., James, L., Mant, R., Newton, P., Rooke, K., Roques, P., Talbot, C., Pericak-Vance, M., Roses, A., Williamson, R., Rossor, M., Owen, M., and Hardy, J., 1991, Segregation of a missense mutation in the amyloid precursor protein gene with familial Alzheimer's disease, *Nature* **349:**704–706.

Goedert, M., Sisodia, S. S., and Price, D. L., 1991, Neurofibrillary tangles and β-amyloid deposits in Alzheimer's disease, *Curr. Opin. Neurobiol.* **1:**441–447.

Goedert, M., Strittmatter, W. J., and Roses, A. D., 1994, Risky apolipoprotein in brain, *Nature* **372:**45–46.

Gómez-Isla, T., West, H. L., Rebeck, G. W., Harr, S. D., Growdon, J. H., Locascio, J. J., Perls, T. T., Lipsitz, L. A., and Hyman, B. T., 1996, Clinical and pathological correlates of apolipoprotein E ϵ4 in Alzheimer disease, *Ann. Neurol.* **39:**62–70.

Gómez-Isla, T., Hollister, R., West, H., Mui, S., Growdon, J. H., Petersen, R. C., Parisi, J. E., and Hyman, B. T., 1997, Neuronal loss correlates with but exceeds neurofibrillary tangles in Alzheimer's disease, *Ann. Neurol.* **41:**17–24.

Greenamyre, J. T., and Young, A. B., 1989, Excitatory amino acids and Alzheimer's disease, *Neurobiol. Aging* **10:**593–602.

Grundke-Iqbal, I., Iqbal, K., Quinlan, M., Tung, Y.-C., Zaidi, M. S., and Wisniewski, H. M., 1986, Microtubule-associated protein tau. A component of Alzheimer paired helical filaments, *J. Biol. Chem.* **261:**6084–6089.

Gurney, M. E., Pu, H., Chiu, A. Y., Dal Canto, M. C., Polchow, C. Y., Alexander, D. D., Caliendo, J., Hentati, A., Kwon, Y. W., Deng, H.-X., Chen, W., Zhai, P., Sufit, R. L., and Siddique, T., 1994, Motor neuron degeneration in mice that express a human Cu, Zn superoxide dismutase mutation, *Science* **264:**1772–1775.

Haass, C., 1997, Presenilins: Genes for life and death, *Neuron* **18:**687–690.

Haass, C., Koo, E. H., Mellon, A., Hung, A. Y., and Selkoe, D. J., 1992, Targeting of cell-surface β-amyloid precursor protein to lysosomes: Alternative processing into amyloid-bearing fragments, *Nature* **357**:500–503.

Haass, C., Hung, A. Y., Selkoe, D. J., and Teplow, D. B., 1994, Mutations associated with a locus for familial Alzheimer's disease result in alternative processing of amyloid β-protein precursor, *J. Biol. Chem.* **269**:17741–17748.

Hamos, J. E., DeGennaro, L. J., and Drachman, D. A., 1989, Synaptic loss in Alzheimer's disease and other dementias, *Neurology* **39**:355–361.

Hardy, J., 1997, Amyloid, the presenilins and Alzheimer's disease, *Trends Neurosci.* **20**:154–159.

Hartmann, T., Bieger, S. C., Brühl, B., Tienari, P. J., Ida, N., Allsop, D., Roberts, G. W., Masters, C. L., Dotti, C. G., Unsicker, K., and Beyreuther, K., 1997, Distinct sites of intracellular production for Alzheimer's disease Aβ40–42 amyloid peptides, *Nat. Med.* **3**:1016–1020.

Hendricks, L., van Duijn, C. M., Cras, P., Cruts, M., Van Hul, W., van Harskamp, F., Warren, A., McInnis, M. G., Antonarakis, S. E., Martin, J.-J., Hofman, A., and Van Broeckhoven, C., 1992, Presenile dementia and cerebal haemorrhage linked to a mutation at codon 692 of the β-amyloid precursor protein gene, *Nat. Genet.* **1**:218–221.

Hendrie, H. C., Hall, K. S., Hui, S., Unverzagt, F. W., Yu, C. E., Lahiri, D. K., Sahota, A., Farlow, M., Musick, B., Class, C. A., Brashear, A., Burdine, V. E., Osuntokun, B. O., Ogunniyi, A. O., Gureje, O., Baiyewu, O., and Schellenberg, G. D., 1995, Apolipoprotein E genotypes and Alzheimer's disease in a community study of elderly African Americans, *Ann. Neurol.* **37**:118–120.

Higgins, L. S., Holtzman, D. M., Rabin, J., Mobley, W. C., and Cordell, B., 1994, Transgenic mouse brain histopathology resembles early Alzheimer's disease, *Ann. Neurol.* **35**:598–607.

Hrabe, de Angellis, M., McIntyre, J., and Gossler, A., 1997, Maintenance of somite borders in mice requires the *Delta* homologue *Dll1, Nature* **386**:717–721.

Hsiao, K. K., Scott, M., Foster, D., Groth, D., F., DeArmond, S. J., and Prusiner, S. B., 1990, spontaneous neurodegeneration in transgenic mice with mutant prion protein, *Science* **250**:1587–1590.

Hsiao, K. K., Groth, D., Scott, M., Yang, S.-L., Serban, H., Rapp, D., Foster, D., Torchia, M., DeArmond, S. J., and Prusiner, S. B., 1994, Serial transmission in rodents in neurodegeneration from transgenic mice expressing mutant prion proteins, *Proc. Natl. Acad. Sci. USA* **91**:9126–9130.

Hsiao, K. K., Borchelt, D. R., Olson, K., Johannsdottir, R., Kitt, C., Yunis, W., Xu, S., Eckman, C., Younkin, S., Price, D., Iadecola, C., Clark, H. B., and Carlson, G., 1995, Age-related CNS disorder and early death in transgenic FVB/N mice overexpressing Alzheimer amyloid precursor proteins. *Neuron* **15**:1203–1218.

Hsiao, K., Chapman, P., Nilsen, S., Eckman, C., Harigaya, Y., Younkin, S., Yang, F., and Cole, G., 1996, Correlative memory defícts, Aβ elevation and amyloid plaques in transgenic mice, *Science* **274**:99–102.

Hsiao, K., Chapman, P., Nilsen, S., Eckman, C., Harigaya, Y., Younkin, S., Yang, F., and Cole, G., 1997, Measuring memory in a mouse model of Alzheimer's disease, *Science* **277**:840–841.

Hung, A. Y., and Selkoe, D. J., 1994, Selective ectodomain phosphorylation and regulated cleavage of β-amyloid precursor protein, *EMBO J.* **13**:534–542.

Hyman, B. T., 1996, Alzheimer's disease or Alzheimer's diseases? Clues from molecular epidemiology *Ann. Neurol.* **40** 135–136.

Hyman, B. T., and Trojanowski, J. Q., 1997, Editorial on consensus recommendations for the postmortem diagnosis of Alzheimer disease from the National Institute on Aging and the Reagan Institute Working Group on diagnostic criteria for the neuropathological assessment of Alzheimer disease, *J. Neuropathol. Exp. Neurol.* **56**;1095–1097.

Hyman, B. T., Van Hoesen, G. W., Damasio, A. R., and Barnes, C. L., 1984, Alzheimer's disease: Cell-specific pathology isolates the hippocampal formation, *Science* **225**:1168–1170.

Hyman, B. T., Van Hoesen, G. W., Kromer, L. J., and Damasio, A. R., 1986, Perforant pathway changes and the memory impairment of Alzheimer's disease, *Ann. Neurol.* **20:**472–481.

Hyman, B. T., Van Hoesen, G. W., and Damasio, A. R., 1990, Memory-related neural systems in Alzheimer's disease: An anatomic study, *Neurology* **40:**1721–1730.

Hyman, B. T., Gomez-Isla, T., Briggs, M., Chung, H., Nichols, S., Kohout, F., and Wallace, R., 1996, Apolipoprotein E and cognitive change in an elderly population, *Ann. Neurol.* **40:**55–66.

Irizarry, M. C., McNamara, M., Fedorchak, K., Hsiao, K., and Hyman, B. T., 1997, App_{Sw} transgenic mice develop age-related Aβ deposits and neuropil abnormalities, but no neuronal loss in CA1, *J. Neuropathol. Exp. Neurol.* **56:**965–973.

Iwatsubo, T., Odaka, A., Suzuki, N., Mizusawa, H., Nukina, N., and Ihara, Y., 1994, Visualization of Aβ42(43)-positive and Aβ40-positive senile plaques with end-specific Aβ-monoclonal antibodies: Evidence that an initially deposited Aβ species is Aβ1–42 (43), *Neuron* **13:**45–53.

Iwatsubo, T., Mann, D. M. A., Odaka, A., Suzuki, N., and Ihara, Y., 1995, Amyloid β protein (Aβ) deposition: Aβ42 (43) precedes Aβ40 in Down syndrome, *Ann. Neurol.* **37:**294–299.

Jarrett, J. T., Berger, E. P., and Lansbury, P. T., Jr., 1993, The carboxy terminus of the β amyloid protein is critical for the seeding of amyloid formation: Implications for the pathogenesis of Alzheimer's disease, *Biochemistry* **32:**4693–4697.

Johnson, S. A., Lampert-Etchells, M., Pasinetti, G. M., Rozovsky, I., and Finch, C. E., 1992, Complement mRNA in the mammalian brain: Responses to Alzheimer's disease and experimental brain lesioning, *Neurobiol. Aging* **13:**641–648.

Kammesheidt, A., Boyce, F. M., Spanoyannis, A. F., Cummings, B. J., Ortegón, M., Cotman, C., Vaught, J. L., and Neve, R. L., 1992, Deposition of β/A4 immunoreactivity and neuronal pathology in transgenic mice expressing the carboxyterminal fragment of the Alzheimer amyloid precursor in the brain, *Proc. Natl. Acad. Sci. USA* **89:**10857–10861.

Katzman, R., 1994, Apolipoprotein E and Alzheimer's disease, *Curr. Opin. Neurobiol.* **4:**703–707.

Katzman, R., 1997, The aging brain. Limitations in our knowledge and future approaches, *Arch. Neurol.* **54:**1201–1205.

Katzman, R., Terry, R., DeTeresa, R., Brown, T., Davis, P., Fuld, P., Renbing, X., and Peck, A., 1988, Clinical, pathological, and neurochemical changes in dementia: A subgroup with preserved mental status and numerous neocortical plaques, *Ann. Neurol.* **23:**138–144.

Katzman, R., Aronson, M., Fuld, P., Kawas, C., Brown, T., Morganstern, H., Frischman, W., Gidez, L., Elder, H., and Ooi, W. L., 1989, Development of dementing illnesses in an 80-year old volunteer cohort, *Ann. Neurol.* **25:**317–324.

Kemper, T., 1984, Neuroanatomical and neuropathological changes in normal aging and in dementia, *In* "Clinical Neurology of Aging" (M. L. Albert, ed.), pp. 9–52, Oxford University Press, New York.

Khachaturian, Z., 1985, Diagnosis of Alzheimer's disease, *Arch. Neurol.* **42:**1097–1105.

Koh, J.-Y., Yang, L. L., and Cotman, C. W., 1990, β-Amyloid protein increases the vulnerability of cultured cortical neurons to exctitoxic damage, *Brain Res.* **553:**315–320.

Kosik, K. S., Joachim, C. L., and Selkoe, D. J., 1986, Microtubule-associated protein (tau) is a major antigenic component of paired helical filaments in Alzheimer disease, *Proc. Natl. Acad. Sci. USA* **83:**4044–4048.

Kovacs, D. M., Fausett, H. J., Page, K. J., Kim, T.-W., Moir, R. D., Merriam, D. E., Hollister, R. D., Hallmark, O. G., Mancini, R., Felsenstein, K. M., Hyman, B. T., Tanzi, R. E., and Wasco, W., 1996, Alzheimer-associated presenilins 1 and 2; Neuronal expression in brain and localization to intracellular membranes in mammalian cells, *Nat. Med.* **2:**224–229.

Kowall, N. W., Beal, M. F., Busciglio, J., Duffy, L. K., and Yankner, B. A., 1991, An *in vivo* model for the neurodegenerative effects of β amyloid and protection by substance P, *Proc. Natl. Acad. Sci. USA* **88:**7247–7251.

LaFerla, F. M., Tinkle, B. T., Bieberich, C. J., Haudenschild, C. C., and Jay, G., 1995, The Alzheimer's Aβ peptide induces neurodegeneration and apoptotic cell death in transgenic mice, *Nat. Genet.* **9:**21–30.

Lah, J. J., Heilman, C. J., Nash, N. R., Rees, H. D., Yi, H., Counts, S. E., and Levey, A. I., 1997, Light and electron microscopic localization of presenilin-1 in primate brain, *J. Neurosci.* **17:**1971–1980.

Lamb, B. T., Sisodia, S. S., Lawler, A. M., Slunt, H. H., Kitt, C. A., Kearns, W. G., Pearson, P. L., Price, D. L., and Gearhart, J. D., 1993, Introduction and expression of the 400 kilobase *precursor amyloid protein* gene in transgenic mice, *Nat. Genet.* **5:**22–30.

Lamb, B. T., Call, L. M., Slunt, H. H., Bardel, K. A., Lawler, A. M. Eckman, C. B., Younkin, S. G., Holtz, G., Wagner, S. L., Price, D. L., Sisodia, S. S., and Gearhart, J. D., 1997, altered metabolism of familial Alzheimer's disease-linked amyloid precursor protein variants in yeast artificial chromosome transgenic mice, *Hum. Mol. Genet.* **6:**1535–1541.

Lee, V. M. Y., and Trojanowski, J. Q., 1992, The disordered neuronal cytoskeleton in Alzheimer's disease, *Curr. Opin. Neurobiol.* **2:**653–656.

Lee, V. M. Y., Balin, B. J., Otvos, L., Jr., and Trojanowski, J. Q., 1991, A68: A major subunit of paired helical filaments and derivatized forms of normal tau, *Science* **251:**675–678.

Lee, M. K., Borchelt, D. R., Kim, G., Thinakaran, G., Slunt, H. H., Ratovitski, T., Martin, L. J., Kittur, A., Gandy, S., Levey, A. I., Jenkins, N., Copeland, N., Price, D. L., and Sisodia, S. S., 1997, Hyperaccumulation of FAD-linked presenilin 1 variants *in vivo, Nat. Med.* **3:**756–760.

Lee, M. K., Slunt, H. H., Martin, L. J., Thinakaran, G., Kim, G., Gandy, S. E., Seeger, M., Koo, E., Price, D. L., and Sisodia, S. S., 1996, Expression of presenilin 1 and 2 (PS1 and PS2) in human and murine tissues, *J. Neurosci.* **16:**7513–7525.

Lemere, C. A., Lopera, F., Kosik, K. S., Lendon, C. L., Ossa, J., Saido, T. C., Yamaguchi, H., Ruiz, A., Martinez, A., Madrigal, L., Hincapie, L., Arango, J. C. L., Anthony, D. C., Koo, E. H., Goate, A. M., Selkoe, D. J., and Arango, J. C. V., 1996a, The E280A presenilin 1 Alzheimer mutation produces increased Aβ42 deposition and severe cerebellar pathology, *Nat. Med.* **2:**1146–1150.

Lemere, C. A., Blusztajn, J. K., Yamaguchi, H., Wisniewski, T., Saido, T. C., and Selkoe, D. J., 1996b, Sequence of deposition of heterogeneous amyloid β-peptides and APO E in Down syndrome: Implications for initial events in amyloid plaque formation, *Neurobiol. Dis.* **3:**16–32.

Levey, A. I., Heilman, C. J., Lah, J. J., Nash, N. R., Rees, H. D., Wakai, M., Mirra, S. S., Rye, D. B., Nochlin, D., Bird, T. D., and Mufson, E. J., 1997, Presenilin-1 protein expression in familial and sporadic Alzheimer's disease, *Ann. Neurol.* **41;**742–753.

Levitan, D., and Greenwald, I., 1995, Facilitation of *lin-12*-mediated signalling by *sel-12,* a *Caenorhabditis elegans* S182 Alzheimer's disease gene, *Nature* **337:**351–354.

Levitan, D., Doyle, T. G., Brousseau, D., Lee, M. K., Thinakaran, G., Slunt, H. H., Sisodia, S. S., and Greenwald, I., 1996, Assessment of normal and mutant human presenilin function in *Caenorhabditis elegans, Proc. Natl. Acad. Sci. USA* **93:**14940–14944.

Levy, E., Carman, M. D., Fernandez-Madrid, I. J., Power, M. D., Lieberburg, I., van Duinen, S. G., Bots, G. T. A. M., Luyendijk, W., and Frangione, B., 1990, Mutation of the Alzheimer's disease amyloid gene in hereditary cerebral hemorrage, Dutch type, *Science* **248:**1124–1126.

Levy-Lahad, E., Wasco, W., Poorkaj, P., Romano, D. M., Oshima, J., Pettingell, W. H., Yu, C.-E., Jondro, P. D., Schmidt, S. D., Wang, K., Crowley, A. C., Fu, Y.-H., Guenette, S. Y., Galas, D., Nemens, E., Wijsman, E. M., Bird, T. D., Schellenberg, G. D., and Tanzi, R. E., 1995a, Candidate gene for the chromosome 1 familial Alzheimer's disease locus, *Science* **269:**973–977.

Levy-Lahad, E., Wijsman, E. M., Nemens, E., Anderson, L., Goddard, K. A. B., Weber, J. L., Bird, T. D., and Schellenberg, G. D., 1995b, A familial Alzheimer's disease locus on chromosome 1, *Science* **269:**970–973.

Li, J., Ma, J., and Potter, H., 1995, Identification and expression analysis of a potential familial

Alzheimer disease gene on chromosome 1 related to *AD3, Proc. Natl. Acad. Sci. USA* **92:**12180–12184.

Mahley, R. W., 1988, Apolipoprotein E: Cholesterol transport protein with expanding role in cell biology, *Science* **240:**622–630.

Mangiarini, L., Sathasivam, K., Seller, M., Cozens, B., Harper, A., Hetherington, C., Lawton, M., Trottier, Y., Lehrach, H., Davies, S. W., and Bates, G. P., 1996, Exon 1 of the *HD* gene with an expanded CAG repeat is sufficient to cause a progressive neurological phenotype in transgenic mice, *Cell* **87:**493–506.

Martin, L. J., Sisodia, S. S., Koo, E. H., Cork, L. C., Dellovade, T. L., Weidemann, A., Beyreuther, K., Masters, C., and Price, D. L., 1991, Amyloid precursor protein in aged nonhuman primates, *Proc. Natl. Acad. Sci. USA* **88:**1461–1465.

Masliah, E., Terry, R. D., Alford, M., and DeTeresa, R., 1990, Quantitative immunohistochemistry of synaptophysin in human neocortex: An alternative method to estimate density of presynaptic terminals in paraffin sections, *J. Histochem. Cytochem.* **38:**837–844.

Masliah, E., Hansen, L., Albright, T., Mallory, M., and Terry, R. D., 1991, Immunoelectron microscopic study of synaptic pathology in Alzheimer's disease, *Acta Neuropathol.* **81:**428–433.

Masliah, E., Ellisman, M., Carragher, B., Mallory, M., Young, S., Hansen, L., DeTeresa, R., and Terry, R. D., 1992, Three-dimensional analysis of the relationship between synaptic pathology and neuropil threads in Alzheimer disease, *J. Neuropathol. Exp. Neurol.* **51:**404–414.

Masliah, E., Sisk, A., Mallory, M., Mucke, L., Schenk, D., and Games, D., 1996, Comparison of neurodegenerative pathology in transgenic mice overexpressing V717F β-amyloid precursor protein and Alzheimer's disease, *J. Neurosci.* **16;**5795–5811.

Masters, C. L., Multhaup, G., Simms, G., Pottgiesser, J., Martins, R. N., and Beyreuther, K., 1985, Neuronal origin of a cerebral amyloid: Neurofibrillary tangles of Alzheimer's disease contain the same protein as the amyloid of plaque cores and blood vessels, *EMBO J.* **4:**2757–2763.

Masters, C. L., Simms, G., Weinman, N. A., Multhaup, G., McDonald, B. L., and Beyreuther, K., 1985, Amyloid plaque core protein in Alzheimer disease and Down syndrome, *Proc. Natl. Acad. Sci. USA* **82:**4245–4249.

McKee, A. C., Kosik, K. S., and Kowall, N. W., 1991, Neuritic pathology and dementia in Alzheimer's disease, *Ann. Neurol.* **30:**156–165.

McKhann, G., Drachman, D., Folstein, M., Katzman, R., Price, D., and Stadlan, E. M., 1984, Clinical diagnosis of Alzheimer's disease: Report of the NINCDS-ADRDA Work Group under the auspices of the Department of Health and Human Services Task Force on Alzheimer's Disease, *Neurology* **34:**939–944.

Mesulam, M.-M., and Geula, C., 1994, Butyrylcholinesterase reactivity differentiates the amyloid plaques of aging from those of dementia, *Ann. Neurol.* **36:**722–727.

Mesulam, M., Carson, K., Price, B., and Geula, C., 1992, Cholinesterases in the amyloid angiopathy of Alzheimer's disease, *Ann. Neurol.* **31:**565–569.

Mirra, S. S., Heyman, A., McKeel, D., Sumi, S. M., Crain, B. J., Brownlee, L. M., Vogel, F. S., Hughes, J. P., van Belle, G., Berg, L., 1991, The consortium to establish a registry for Alzheimer's disease (CERAD). Part II. Standardization of the neuropathologic assessment of Alzheimer's disease, *Neurology* **41:**479–486.

Mirra, S. S., Hart, M. H., and Terry, R. D., 1993, Making the diagnosis of Alzheimer's disease. A primer for practicing pathologists, *Arch. Pathol. Lab. Med.* **117:**132–144.

Mirra, S. S., Gearing, M., and Nash, F., 1997, Neuropathologic assessment of Alzheimer's disease, *Neurology* **49:**S14–S16.

Moran, P. M., Higgins, L. S., Cordell, B., and Moser, P. C., 1995, Age-related learning deficits in transgenic mice expressing the 751-amino acid isoform of human β-amyloid precursor protein, *Proc. Natl. Acad. Sci. USA* **92:**5341–5345.

Mullan, M., Crawford, F., Axelman, K., Houlden, H., Lillius, L., Winblad, B., and Lannfelt, L., 1992,

A pathogenic mutation for probably Alzheimer's disease in the APP gene at the N-terminus of β-amyloid, *Nat. Genet.* **1**:345–347.

Nalbantoglu, J., Gilfix, B. M., Bertrand, P., Robitaille, Y., Gauthier, S., Rosenblatt, D. S., and Poirier, J., 1994, Predictive value of apolipoprotein E genotyping in Alzheimer's disease: Results of an autopsy series and an analysis of several combined studies, *Ann. Neurol.* **36**:889–895.

Nalbantoglu, J., Tirado-Santiago, G., Lahsaïni, A., Poirier, J., Goncalves, O., Verg, G., Momoli, F., Welner, S. A., Massicotte, G., Julien, J.-P., and Shapiro, M. L., 1997, Impaired learning and LTP in mice expressing the carboxy terminus of the Alzheimer amyloid precursor proteins, *Nature* **387**:500–505.

Naruse, S., Igarashi, S., Kobayashi, H., Aoki, K., Inuzuka, T., Kaneko, K., Shimizu, T., Iihara, K., Kojima, T., Miyatake, T., and Tsuji, S., 1991, Mis-sense mutation Val → Ile in exon 17 of amyloid precursor protein gene in Japanese familial Alzheimer's disease, *Lancet* **337**:978–979.

Neve, R. L., Kammesheidt, A., and Hohmann, C. F., 1992, Brain transplants of cells expressing the carboxyl-terminal fragment of the Alzheimer amyloid protein precursor cause specific neuropathology in vivo, *Proc. Natl. Acad. Sci. USA* **89**:3448–3452.

Nixon, R. A., and Cataldo, A. M., 1995, The endosomal–lysosomal system of neurons: New roles, *Trends Neurosci.* **18**:489–496.

Olshansky, S. J., Carnes, B. A., and Cassel, C. K., 1993, The aging of the human species, *Sci. Am.* **268**:46–52.

Oltersdorf, T., Fritz, L. C., Schenk, D. B., Lieberburg, I., Johnson-Wood, K. L., Beattie, E. C., Ward, P. J., Blacher, R. W., Dovey, H. F., and Sinha, S., 1989, The secreted form of the Alzheimer's amyloid precursor protein with the Kunitz domain is protease nexin-II, *Nature* **341**:144–147.

Ordovas, J. M., Litwack-Klein, L., Wilson, P. W. F., Schaefer, M. M., and Schaefer, E. J., 1987, Apolipoprotein E isoform phenotyping methodology and population frequency with identification of apoE1 and apoE5 isoforms, *J. Lipid Res.* **28**:371–380.

Perdahl, E., Adolfsson, R., Alafuzoff, I., Albert, K. A., Nestler, E. J., Greengard, P., and Winblad, B., 1984, Synapsin I (Protein I) in different brain regions in senile dementia of Alzheimer type and in multiinfarct dementia, *J. Neural Transm.* **60**:133–141.

Perry, G., Friedman, R., Shaw, G., and Chau, V., 1987, Ubiquitin is detected in neurofibrillary tangles and senile plaque neurites of Alzheimer disease brains, *Proc. Natl. Acad. Sci. USA* **84**:3033–3036.

Perry, G., Kawai, M., Tabaton, M., Onorato, M., Mulvihill, P., Richey, P., Morandi, A., Connolly, J. A., and Gambetti, P., 1991, Neuropil threads of Alzheimer's disease show a marked alteration of the normal cytoskeleton, *J. Neurosci.* **11**:1748–1756.

Pfeffer, R. I., Afifi, A. A., and Chance, J. M., 1987, Prevalence of Alzheimer's disease in a retirement community, *Am. J. Epidemiol.* **125**:420–436.

Pike, C. J., Walencewicz, A. J., Glabe, C. G., and Cotman, C. W., 1991, Aggregation-related toxicity of synthetic β-amyloid protein in hippocampal cultures, *Eur. J. Pharmacol.* **207**:367–368.

Podlisny, M. B., Citron, M., Amarante, P., Sherrington, R., Xia, W., Zhang, J., Diehl, T., Levesque, G., Fraser, P., Haass, C., Koo, E. H. M., Seubert, P., St. George-Hyslop, P., Teplow, D. B., and Selkoe, D. J., 1997, Presenilin proteins undergo heterogeneous endoproteolysis between Thr_{291} and Ala_{299} and occur as stable N- and C-terminal fragments in normal and Alzheimer brain tissue, *Neurobiol. Dis.* **3**:325–337.

Poirier, J., Davignon, J., Bouthillier, D., Kogan, S., Bertrand, P., Gauthier, S., 1993, Apolipoprotein E polymorphism and Alzheimer's disease, *Lancet* **342**:697–699.

Price, D. L., and Sisodia, S. S., 1997, Mutant genes in familial Alzheimer's disease and transgenic models. *Annu. Rev. Neurosci.* in press.

Price, D. L., Kawas, C. H., and Sisodia, S. S., 1996, Aging of the brain and dementia of the Alzheimer's type, *in* "Principles of Neural Science" (E. R., Kandel, J. H., Schwartz, T. M., Jessell, eds.), Elsevier, New York.

Price, D. L., Sisodia, S. S., and Borchelt, D. R., 1998, Genetic neurodegenerative diseases: The human illness and transgenic models. *Science* **282:**1079–1083.

Price, D. L., Tanzi, R. E., Borchelt, D. R., and Sisodia, S. S., Alzheimer's disease: Genetic studies and transgenic models. *Annu. Rev. Genet.,* in press.

Probst, A., Brunnschweiler, H., Lautenschlager, C., and Ulrich, J., 1987, A special type of senile plaque, possibly an initial stage, *Acta Neuropathol.* **74:**133–141.

Probst, A., Langui, D., Ipsen, S., Robakis, N., and Ulrich, J., 1991, Deposition of β/A4 protein along neuronal plasma membranes in diffuse senile plaques, *Acta Neuropathol.* **83:**21–29.

Ratovitski, T., Slunt, H. H., Thinakaran, G., Price, D. L., Sisodia, S. S., and Borchelt, D. R., 1997, Endoproteolytic processing and stabilization of wild-type and mutant presenilin, *J. Biol. Chem.* **272:**24536–24541.

Rebeck, G. W., Reiter, J. S., Strickland, D. K., and Hyman, B. T., 1993, Apolipoprotein E in sporadic Alzheimer's disease: Allelic variation and receptor interactions, *Neuron* **11:**575–580.

Rogaev, E. I., Sherrington, R., Rogaeva, E. A., Levesque, G., Ikeda, M., Liang, Y., Chi, H., Lin, C., Holman, K., Tsuda, T., Mar, L., Sorbi, S., Nacmias, B., Piacentini, S., Amaducci, L., Chumakov, I., Cohen, D., Lannfelt, L., Fraser, P. E., Rommens, J. M., and St George-Hyslop, P. H., 1995, Familial Alzheimer's disease in kindreds with missense mutations in a gene on chromosome 1 related to the Alzheimer's disease type 3 gene, *Nature* **376:**775–778.

Roher, A., Wolfe, D., Palutke, M., and KuKurga, D., 1986, Purification, ultrastructure, and chemical analysis of Alzheimer disease amyloid plaque core protein, *Proc. Natl. Acad. Sci. USA* **83:**2662–2666.

Roses, A. D., 1994, Apolipoprotein E affects the rate of Alzheimer disease expression: β-Amyloid burden is a secondary consequence dependent on APOE genotype and duration of disease, *J. Neuropathol. Exp. Neurol.* **53:**429–437.

Roses, A. D., 1995, Apolipoprotein E genotyping in the differential diagnosis, not prediction, of Alzheimer's disease, *Ann. Neurol.* **38:**6–14.

Roses, A. D., 1997, Genetic testing for Alzheimer disease. Practical and ethical issues, *Arch. Neurol.* **54:**1226–1229.

Roses, A. D., Strittmatter, W. J., Pericak-Vance, M. A., Corder, E. H., Saunders, A. M., and Schmechel, D. E., 1994, Clinical application of apolipoprotein E genotyping to Alzheimer's disease, *Lancet* **343:**1564–1565.

Routtenberg, A., 1997, Measuring memory in a mouse model of Alzheimer's disease, *Science* **277:**839–840.

Saunders, A. M., Strittmatter, W. J., Schmechel, D., St. George-Hyslop, P. H., Pericak-Vance, M. A., Joo, S. H., Rosi, B. L., Gusella, J. F., Crapper-MacLachlan, D. R., Alberts, M. J., Hulette, C., Crain, B., Goldgaber, D., and Roses, A. D., 1993, Association of apolipoprotein E allele ϵ4 with late-onset familial and sporadic Alzheimer's disease, *Neurology* **43:**1467–1472.

Saito, K.-I., Elce, J. S., Hamos, J. E., and Nixon, R. A., 1993, Widespread activation of calcium-activated neutral proteinase (calpain) in the brain in Alzheimer disease: A potential molecular basis for neuronal degeneration, *Proc. Natl. Acad. Sci. USA* **90:**2628–2632.

Scheff, S. W., and Price, D. A., 1993, Synapse loss in the temporal lobe in Alzheimer's disease, *Ann. Neurol.* **33:**190–199.

Schellenberg, G. D., 1995, Progress in Alzheimer's disease genetics, *Curr. Opin. Neurol.* **8:**262–267.

Scherzinger, E., Lurz, R., Turmaine, M., Mangiarini, L., Hollenbach, B., Hasenbank, R., Bates, G. P., Davies, S. W., Lehrach, H., and Wanker, E. E., 1997, Huntingtin-encoded polyglutamine expansions form amyloid-like protein aggregates in vitro and in vivo, *Cell* **90:**549–558.

Scheuner, D., Eckman, C., Jensen, M., Song, X., Citron, M., Suzuki, N., Bird, T. D., Hardy, J., Hutton, M., Kukull, W., Larson, E., Levy-Lahad, E., Viitanen, M., Peskind, E., Poorkaj, P., Schellenberg, G., Tanzi, R., Wasco, W., Lannfelt, L., Selkoe, D., and Younkin, S., 1996, Secreted amyloid β-protein similar to that in the senile plaques of Alzheimer's disease is increased *in vivo* by the presenilin 1 and 2 and *APP* mutations linked to familial Alzheimer's disease, *Nat. Med.* **2:**864–870.

Schmechel, D. E., Saunders, A. M., Strittmatter, W. J., Crain, B. J., Hulette, C. M., Joo, S. H., Pericak-Vance, M. A., Goldgaber, D., and Roses, A. D., 1993, Increased amyloid β-peptide deposition in cerebral cortex as a consequence of apolipoprotein E genotype in late-onset Alzheimer's disease, *Proc. Natl. Acad. Sci. USA* **90:**9649–9653.

Shen, J., Bronson, R. T., Chen, D. F., Xia, W., Selkoe, D. J., and Tonegawa, S., 1997, Skeletal and CNS defects in *presenilin-1*-deficient mice, *Cell* **89:**629–639.

Sherrington, R., Rogaev, E. I., Liang, Y., Rogaeva, E. A., Levesque, G., Ikeda, M., Chi, H., Lin, C., Li, G., Holman, K., Tsuda, T., Mar, L., Foncin, J.-F., Bruni, A. C., Montesi, M. P., Sorbi, S., Rainero, I., Pinessi, L., Nee, L., Chumakov, I., Pollen, D., Brookes, A., Sanseau, P., Polinsky, R. J., Wasco, W., Da Silva, H. A. R., Haines, J. L., Pericak-Vance, M. A., Tanzi, R. E., Roses, A. D., Fraser, P. E., Rommens, J. M., and St George-Hyslop, P. H., 1995, Cloning of a gene bearing missense mutations in early-onset familial Alzheimer's disease, *Nature* **375:**754–760.

Shinkai, Y., Yoshimura, M., Ito, Y., Odaka, A., Suzuki, N., Yanagisawa, K., and Ihara, Y., 1995, Amyloid β-proteins 1–40 and 1–42 (43) in the soluble fraction of extra- and intracranial blood vessels, *Ann. Neurol.* **38:**421–428.

Sisodia, S. S., 1992, β-Amyloid precursor protein cleavage by a membrane-bound protease, *Proc. Natl. Acad. Sci. USA* **89:**6075–6079.

Sisodia, S. S., Koo, E. H., Beyreuther, K., Unterbeck, A., and Price, D. L., 1990, Evidence that β-amyloid protein in Alzheimer's disease is not derived by normal processing, *Science* **248:**492–495.

Slunt, H. H., Thinakaran, G., von Koch, C., Lo, A. C. Y., Tanzi, R. E., and Sisodia, S. S., 1994, Expression of a ubiquitous, cross-reactive homologue of the mouse β-amyloid precursor protein (APP), *J. Biol. Chem.* **269:**2637–2644.

Smale, G., Nichols, N. R., Brady, D. R., Finch, C. E., and Horton, W. E., Jr., 1995, Evidence for apoptotic cell death in Alzheimer's disease, *Exp. Neurol.* **133:**225–230.

St George-Hyslop, P. H., Haines, P., Rogaev, E., Mortilla, M., Vaula, G., Pericak-Vance, M., Foncin, J.-F., Montesi, M., Bruni, A., Sorbi, S., Rainero, I., Pinessi, L., Pollen, D., Polinsky, R., Nee, L., Kennedy, J., Macciardi, F., Rogaeva, E., Liang, Y., Alexandrova, N., Lukiw, W., Schlumpf, K., Tanzi, R., Tsuda, T., Farrer, L., Cantu, J.-M., Duara, R., Amaducci, L., Bergamini, L., Gusella, J., Roses, A., and Crapper McLachlan, D., 1992, Genetic evidence for a novel familial Alzheimer's disease locus on chromosome 14, *Nat. Genet.* **2:**330–334.

Strittmatter, W. J., Saunders, A. M., Schmechel, D., Pericak-Vance, M., Enghild, J., Salvesen, G. S., and Roses, A. D., 1993, Apolipoprotein E: High-avidity binding to β-amyloid and increased frequency of type 4 allele in late-onset familial Alzheimer disease, *Proc. Natl. Acad. Sci. USA* **90:**1977–1981.

Su, J. H., Anderson, A. J., Cummings, B. J., and Cotman, C. W., 1994, Immunohistochemical evidence for apoptosis in Alzheimer's disease, *Neuroreport* **5:**2529–2533.

Suzuki, N., Cheung, T. T., Cai, X.-D., Odaka, A., Otvos, L., Jr., Eckman, C., Golde, T. T., and Younkin, S. G., 1994, An increased percentage of long amyloid β protein secreted by familial amyloid β protein precursor (βAPP_{717}) mutants, *Science* **264:**1336–1340.

Sze, C.-I., Troncoso, J. C., Kawas, C. H., Mouton, P. R., Price, D. L., and Martin, L. J., 1997, Loss of the presynaptic vesicle protein synaptophysin in hippocampus correlates with early cognitive decline in aged humans, *J. Neuropathol. Exp. Neurol.* **56:**933–944.

Terry, R. D., Masliah, E., Salmon, D. P., Butters, N., DeTeresa, R., Hill, R., Hansen, L. A., and Katzman, R., 1991, Physical basis of cognitive alterations in Alzheimer's disease: Synapse loss is the major correlate of cognitive impairment, *Ann. Neurol.* **30:**572–580.

Thinakaran, G., Teplow, D. B., Siman, R., Greenberg, B., and Sisodia, S. S., 1996, Metabolism of the "Swedish" APP variant in Neuro2A (N2a) cells: Evidence that cleavage at the "β-secretase" site occurs in the Golgi apparatus, *J. Biol. Chem.* **271:**9390–9397.

Thinakaran, G., Borchelt, D. R., Lee, M. K., Slunt, H. H., Spitzer, L., Kim, G., Ratovitski, T., Davenport, F., Nordstedt, C., Seeger, M., Hardy, J., Levey, A. I., Gandy, S. E., Jenkins, N., Copeland, N., Price, D. L., and Sisodia, S. S., 1996, Endoproteolysis of presenilin 1 and accumulation of processed derivatives *in vivo, Neuron* **17:**181–190.

Thinakaran, G., Harris, C. L., Ratovitski, T., Davenport, F., Slunt, H. H., Price, D. L., Borchelt, D. R., and Sisodia, S. S., 1997, Evidence that levels of presenilins (PS1 and PS2) are coordinately regulated by competition for limiting cellular factors, *J. Biol. Chem.* **272:**28415–28422.

Troncoso, J. C., Sukhov, R. R., Kawas, C. H., and Koliatsos, V. E., 1996, *In situ* labeling of dying cortical neurons in normal aging and in Alzheimer's disease: correlations with senile plaques and disease progression, *J. Neuropathol. Exp. Neurol.* **55;**1134–1142.

Tsai, M.-S., Tangalos, E. G., Petersen, R. C., Smith, G. E., Schaid, D. J., Kokmen, E., Ivnik, R. J., and Thibodeau, S. N., 1994, Apolipoprotein E: Risk factor for Alzheimer disease, *Am. J. Hum. Genet.* **54;**643–649.

Van Broeckhoven, C., Haan, J., Bakker, E., Hardy, J. A., Van Hul, W., Wehnert, A., Vegter-Van der Vlis, M., and Roos, R. A. C., 1990, Amyloid β protein precursor gene and hereditary cerebral hemorrhage with amyloidosis (Dutch), *Science* **248:**1120–1122.

van Duijn, C. M., de Knijff, P., Wehnert, A., De Voecht, J., Bronzova, J. B., Havekes, L. M., Hofman, A., and Van Broeckhoven, C., 1995, The apolipoprotein E ε2 allele is associated with an increased risk of early-onset Alzheimer's disease and a reduced survival, *Ann. Neurol.* **37:**605–610.

Vogels, O. J. M., Broere, C. A. J., Ter Laak, H. J., Ten, Donkelaar, H. J., Nieuwenhuys, R., and Schultz, B. P. M., 1990, Cell loss and shrinkage in the nucleus basalis Meynert complex in Alzheimer's disease, *Neurobiol. Aging* **11:**3–13.

Wasco, W., Bupp, K., Magendantz, M., Gusella, J. F., Tanzi, R. E., and Solomon, F., 1992, Identification of a mouse brain cDNA that encodes a protein related to the Alzheimer disease-associated amyloid-beta-protein precursor, *Proc. Natl. Acad. Sci. USA* **89:**10758–10762.

Wasco, W., Gurubhagavatula, S., Paradis, M. D., Romano, D. M., Sisodia, S. S., Hyman, B. T., Neve, R. L., and Tanzi, R. E., 1993, Isolation and characterization of *APLP2* encoding a homologue of the Alzheimer's associated amyloid β protein precursor, *Nat. Genet.* **5:**95–99.

Weisgraber, K. H., Pitas, R. E., and Mahley, R. W., 1994, Lipoproteins, neurobiology, and Alzheimer's disease: Structure and function of apolipoprotein E, *Curr. Opin. Struct. Biol.* **4:**507–515.

Whitehouse, P. J., Price, D. L., Struble, R. G., Clark, A. W., Coyle, J. T., and DeLong, M. R., 1982, Alzheimer's disease and senile dementia: Loss of neurons in the basal forebrain, *Science* **215;**1237–1239.

Wisniewski, T., and Frangione, B., 1992, Apolipoprotein E. A pathological chaperone protein in patients with cerebral and systemic amyloid, *Neurosci. Lett.* **135:**235–238.

Wisniewski, H. M., and Terry, R. D., 1973, Reexamination of the pathogeneisis of the senile plaque *in* "Progress in Neuropathology," (H. M. Zimmerman, ed.), pp. 1–26, Grune & Stratton, New York.

Wisniewski, T., Ghiso, J., and Frangione, B., 1991, Peptides homologous to the amyloid protein of Alzheimer's disease containing a glutamine for glutamic acid substitution have accelerated amyloid fibril formation, *Biochem. Biophys. Res. Commun.* **179:**1247–1254.

Wong, P. C., Pardo, C. A., Borchelt, D. R., Lee, M. K., Copeland, N. G., Jenkins, N. A., Sisodia, S. S., Cleveland, D. W., and Price, D. L., 1995, An adverse property of a familial ALS-linked SOD1 mutation causes motor neuron disease characterized by vacuolar degeneration of mitochondria, *Neuron* **14;**1105–1116

Wong, P. C., Zheng, H., Chen, H., Becher, M. W., Sirinathsinghji, D. J. S., Trumbauer, M. E., Chen, H. Y., Price, D. L., Van der Ploeg, L. H. T., and Sisodia, S. S., 1997, Presenilin 1 is required for *Notch1* and *Dll1* expression in the paraxial mesoderm, *Nature* **387:**288–292.

Yankner, B. A., Dawes, L. R., Fisher, S., Villa-Komaroff, L., Oster-Granite, M. L., and Neve, R. L., 1989, Neurotoxicity of a fragment of the amyloid precursor associated with Alzheimer's disease, *Science* **245:**417–420.

Yoshikawa, K., Aizawa, T., and Hayashi, Y., 1992, Degeneration *in vitro* of post-mitotic neurons overexpressing the Alzheimer amyloid protein precursor, *Nature* **359:**64–67.

Zheng, H., Jiang, M.-H., Trumbauer, M. E., Sirinathsinghji, D. J. S., Hopkins, R., Smith, D. W., Heavens, R. P., Dawson, G. R., Boyce, S., Conner, M. W., Stevens, K. A., Slunt, H. H., Sisodia, S. S., Chen, H. Y., and Van der Ploeg, L. H. T., 1995, β-Amyloid precursor protein-deficient mice show reactive gliosis and decreased locomotor activity. *Cell* **81:**525–531.

CHAPTER 10

MODEL OF GENETIC SUSCEPTIBILITY TO LATE-ONSET ALZHEIMER'S DISEASE: MICE TRANSGENIC FOR HUMAN APOLIPOPROTEIN E ALLELES

DONALD E. SCHMECHEL, PU-TING XU, JOHN R. GILBERT, and ALLEN D. ROSES

Alzheimer's disease (AD) is a common neurodegenerative disorder in late life, affecting eventually up to 30% or more of individuals with a devastating and progressive syndrome of impairment of memory, cognition, and language. The

---

DONALD E. SCHMECHEL • Division of Neurology, Department of Medicine, Joseph and Kathleen Bryan Alzheimer's Disease Research Center, Durham, North Carolina 27705; Department of Neurobiology, Duke University Medical Center, Durham, North Carolina; and Durham Veterans Administration Medical Center, Durham, North Carolina. PU-TING XU, JOHN R. GILBERT, and ALLEN D. ROSES • Division of Neurology, Department of Medicine, Joseph and Kathleen Bryan Alzheimer's Disease Research Center, Durham, North Carolina 27705.

Mouse Models in the Study of Genetic Neurological Disorders, Volume 9 of *Advances in Neurochemistry,* edited by Popko, Kluwer Academic/Plenum Publishers, New York, 1999.

clinical diagnosis of AD is based on identification of adequate clinical evidence for dementia syndrome; exclusion of other etiologies of dementia, particularly reversible causes; and observation of progression over time. The pathological confirmation of AD is based on examination of vulnerable areas of cerebral cortex and identification of features such as neurofibrillary tangles and amyloid plaques in quantities in excess for age. These "classic" neuropathological features are accompanied by brain atrophy, neuronal loss, and abundant reactive astrogliosis and microgliosis. The profile of cortical vulnerability has been well defined for AD and prominently includes hippocampus, entorhinal cortex, and other association cortices. These common clinical and pathological features serve to group together what is now clearly a family of Alzheimer diseases. One part of this family can be defined by allelic effects of the AD susceptibility gene, apolipoprotein E (*APOE*).

In this chapter, we discuss background on the role of *APOE* in the normal and abnormal nervous system and the rationale for using mouse genetic models to examine the role of *APOE* as a susceptibility gene in late-onset AD.

1. GENETIC CLASSIFICATION OF AD

The definition of members of a family of Alzheimer diseases is based on genetic discoveries in the last 10–15 years that have identified subsets of AD according to age at onset and inheritance. The first group of familial AD is that of the rare autosomal dominant mutations including presenilin I, presenilin II, and amyloid precursor polypeptide (APP) (Roses, 1996). This group is distinguished by early age at onset (usually before 60 years of age) and is termed early-onset familial AD. The total involvement is roughly 120 families worldwide. The second group is represented by late-onset familial and sporadic AD occurring after age 60 and reflects in part the codominant influence of different alleles of the AD susceptibility gene (*APOE,* gene; apoE, protein; see later). Another susceptibility gene for late-onset AD has been linked to chromosome 12 (Pericak-Vance *et al.,* 1997). Other late-onset genes are being sought and may either interact with *APOE* or represent independent genetic factors for late-onset AD. In addition, mitochondrial mutations have been invoked as another source of genetic risk for AD, but are controversial at present. Table 1 presents a current proposed classification of AD based on genetic factors (Roses, 1996).

2. *APOE* SUSCEPTIBILITY GENE FOR AD

APOE is the most common genetic factor identified to date for the preponderance of AD cases that occur after age 60, accounting for up to 50% of all cases

TABLE 1. Genetic Classification of Alzheimer Diseases

Type	Chromosome	Gene
AD1: early-onset familial AD autosomal dominant mutations	21	*APP*
AD2: late-onset familial and sporadic AD susceptibility gene	19	*APOE* alleles (*E4* = high risk, *E2* = low risk)
AD3: early-onset familial AD autosomal dominant mutations	14	Presenilin I (*S182, STM1*)
AD4: early-onset familial AD (Volga-German and others) autosomal dominant mutations	1	Presenilin II (*E5.1, STM2*)
AD5: late-onset susceptibility gene	12	Unknown
Other late-onset susceptibility genes		

of AD (Roses, 1996). Apolipoprotein E is encoded by four exons, 3597 nucleotides in length, on the long arm of human chromosome 19. The mature form of apoE is a 34-kDa protein containing 299 amino acids coded for by a 1163-nucleotide mRNA (Mahley, 1988). There are three common alleles of *APOE* designated *E2, E3,* and *E4.* These three forms differ by one or two amino acids at the 112 and 158 position resulting in $cys_{112}cys_{158}$ (apoE2), $cys_{112}arg_{158}$ (apoE3), and $arg_{112}arg_{158}$ (apoE4) (Weisgraber *et al.,* 1982; Mahley, 1988; Weisgraber, 1994). The most common form, *E3,* has an allele frequency of 78% (Caucasian populations), and has a cysteine residue at position 112 and an arginine residue at position 158. The next most common form, *E4,* has an allele frequency of 15%, and differs from *E3* by presence of an arginine at 112. Finally, the least common form, *E2,* has an allele frequency of 7%, and differs from *E3* by presence of a cysteine residue at position 158 (Mahley, 1988). These three protein isoforms vary by a single unit of charge and can be separated and identified by isoelectric focusing (Utermann *et al.,* 1979). For most clinical and laboratory studies, a more reliable means of identification is polymerase chain reaction (PCR) analysis with appropriate primers of isolated DNA samples (Strittmatter *et al.,* 1993a).

Relatively little genetic variation of *APOE* gene occurs in human populations with rare mutants being recognized by their effect on lipid metabolism (Weisgraber, 1994). The one or two amino acid sequence difference between *E2, E3,* and *E4* alleles result in conformational and salt-bridging changes in apoE that change its binding to the low-density lipoprotein (LDL) receptor and are associated with changes in lipid metabolism and risk of heart disease (Mahley, 1988; Weisgraber, 1994). ApoE protein is highly conserved across mammalian species (mouse apoE has 80% or more amino acid homology with human apoE). In most animals including subhuman primates, the general form of apoE is "apoE4"-like based only on the 112/158 amino acid sites (Weisgraber *et al.,* 1982), except for rabbit, which is "apoE3"-like. Conformational and binding characteristics of rodent apoE are, however, quite different from human apoE4.

The discovery of *APOE* as an AD susceptibility gene resulted from attention to the discovery of chromosome 19 linkage of late-onset familial and sporadic AD (Pericak-Vance *et al.*, 1991) and the independent discovery that human apoE bound with high affinity to βA4 peptide *in vitro* (Strittmatter *et al.*, 1993a). Quick examination for *APOE* allele-specific differences in DNA samples from AD patients and controls from the Joseph and Kathleen Bryan Alzheimer's Disease Research Center demonstrated an increased frequency of *APOE4* allele in patients with late-onset familial and sporadic AD (Strittmatter *et al.*, 1993a; Saunders *et al.*, 1993a). This discovery has been confirmed widely by many groups across the world (e.g., Poirer *et al.*, 1993b; Rebeck *et al.*, 1993; reviewed in Roses, 1996). The positive effect of *E4* allele on age at onset distribution (average age 65 years) and protective effect of the *E2* allele on age at onset distribution (average age at onset of *E2/3* roughly 85 years) was demonstrated in a series of late-onset and sporadic families (Saunders *et al.*, 1993a,b; Corder *et al.*, 1993, 1994). The effect of *APOE4* allele may be censored by other disease effects of *E4* and death from other causes, and is also related to the varying allele frequency of *E4* in different ethnic groups. It is important to emphasize that *APOE* alleles are susceptibility factors influenced by other genetic and environmental factors. Thus, the above age at onset distributions do overlap, and the practical use of *APOE* genetic testing is at present limited to differential diagnosis of dementia syndromes, but not predictive testing for specific individuals (Saunders *et al.*, 1996).

The higher risk of *APOE4* allele associated with late-onset AD is accompanied by apoE-isoform specific differences in adverse outcome from environmental insult to the nervous system. This includes increased risk of AD after head injury, poorer cognitive outcome after thoracic surgery (CABG), and increased risk of death and morbidity after cerebral hemorrhage (Tardiff *et al.*, 1994; Alberts *et al.*, 1995; Mayeux *et al.*, 1995; Nicoll *et al.*, 1995; Roses and Saunders, 1995). Recent studies have supported that asymptomatic persons with one or more *APOE4* alleles may be distinguished on a population basis even at early ages by differences in language abilities, metabolic neuroimaging, or cognitive decline rates (Reed *et al.*, 1994; Small *et al.*, 1995).

3. APOE IN THE PERIPHERY

Initial studies of apoE focused on its role as a plasma protein involved in peripheral cholesterol metabolism and transport (Weisgraber *et al.*, 1982; Mahley, 1988; Weisgraber, 1994). ApoE is synthesized in many tissues and organs, including intestine, liver, kidney, and macrophages (Lin *et al.*, 1986;

Mahley, 1988). ApoE helps mediate cholesterol and lipid transport through a lipid binding domain and through other receptor binding domains that interact with LDL-receptor and a VLDL receptor. In addition, a more generalized receptor, the α_2-macroglobulin or *LDL-r*eceptor related *p*rotein (LRP), also recognizes apoE among other ligands. Via such receptors, apoE can be taken up by cells that apparently do not synthesize apoE such as pancreatic islet cells and others. Through hydrophobic domains and other regions such as the heparin binding domain, apoE can also bind to other proteins. This additional protein-binding role has led to the characterization of apoE as a chaperonin or more generalized transport protein (Mahley, 1988; Wisniewski and Frangione, 1991; Weisgraber, 1994). Recent evidence suggests that binding modulates the function of other proteins or provides a way of stabilizing/presenting them in the extracellular matrix. ApoE expression has been associated with specific immunological functions, particularly in macrophage reactions and the pathogenesis of atherosclerotic injury (Weisgraber *et al.*, 1982; Mahley, 1988; Weisgraber, 1994; Laskowitz, 1997a). In both cells that synthesize apoE and cells that uptake apoE, intracellular apoE may also subserve important roles in metabolism such as protein trafficking. Allelic differences are well described for cholesterol transport. The three isoforms differ in their affinity for the LDL receptor *in vitro,* with *E2* having a substantially lower affinity than *E3* and *E4.* Plasma lipoproteins containing *E4* are cleared more efficiently than those containing *E2* and *E3* (Weisgraber *et al.*, 1982). Isoform-specific differences in binding and other properties may depend on conformation and salt-bridging of conserved amino acid residues distant from the 112 and 158 cysteine/arginine pairs (Dong and Weisgraber, 1996).

4. APOE IN THE CENTRAL NERVOUS SYSTEM

ApoE is also synthesized in the nervous system and given the importance of lipid metabolism for the nervous system, early research efforts on apoE addressed the role of apoE in normal and injured nervous tissue. ApoE was demonstrated to be synthesized and secreted in the CNS by astrocytes (Boyles *et al.*, 1985; Ignatius *et al.*, 1986; Pitas *et al.*, 1987; Mahley, 1988; Namba *et al.*, 1991; Poirer *et al.*, 1991b). A role for apoE in reverse cholesterol transport in the central nervous system (CNS) was proposed (Mahley, 1988), which implied a role in processing lipids during repair of injury and in normal synaptic remodeling.

The periphery and nervous system form two separate compartments of apoE synthesis and metabolism. Cerebrospinal fluid from liver transplant patients shows that the normal central and peripheral nervous system within the blood–brain, blood–retinal, and blood–nerve barrier form a separate compartment and are not apparently accessible to peripherally generated apoE (Linton et al., 1991).

Investigation of apoE in nervous system function was initially driven by consideration of the role of apoE in reverse cholesterol transport, synaptic remodeling and repair, and tissue injury response (Skene and Shooter, 1983; Boyles et al., 1985; Muller et al., 1985; Ignatius et al., 1986; Snipes et al., 1986; Poirier et al., 1993a; Poirier, 1994). Indeed, increased levels of *APOE* mRNA were discovered during an unbiased analysis of hippocampal injury response (Poirier *et al.,* 1991a,b) and was observed in reactive astrocytes. Immunological studies in the CNS in *normal* rodents have demonstrated apoE in astrocytes, ependymal cells, choroid plexus cells, but not neurons (Lin *et al.,* 1986). Figure 1 displays immunocytochemical localization in mouse brain of apoE demonstrating immunoreactivity in astrocytes and related cells such as Bergmann radial glial cells of cerebellum. *In situ* hybridization methods demonstrate increased message in reactive astrocytes and support the apparent abundant message in normal astrocytes and support the apparent absence of apoE message in *normal* rodent neurons (e.g., Poirier et al., 1991a,b).

5. APOE IN THE PERIPHERAL NERVOUS SYSTEM

Injury to peripheral nerves results in the upregulation of apoE which is present in infiltrating macrophages and Schwann cells involved in the response to injury (Skene and Shooter, 1983; Boyles *et al.,* 1985; Mueller *et al.,* 1985; Ignatius *et al.,* 1986; Snipes *et al.,* 1986; Stoll and Mueller, 1986; Stoll *et al.,* 1989; Gelman *et al.,* 1991). Initial experiments on peripheral nerve regeneration demonstrated a role for apoE expression in reverse cholesterol transport in Schwann cells and particularly macrophages and non-neuronal cells (Mahley, 1988; Weisgraber, 1994). Participation of apoE in nerve regeneration may differ for peripheral compared to central nervous system (Stoll *et al.,* 1989).

6. APOE AND PATHOGENESIS OF AD—EXTRACELLULAR ROLE

The recent genetic and epidemiological studies demonstrating *APOE* as an AD susceptibility gene were independent of many earlier studies on immunopathological findings in AD and related neurodegenerative disorders. ApoE was described as one of many proteins associated with amyloid deposits in CNS disease, particularly AD (Diedrich *et al.,* 1991; Wisniewski and Frangione, 1991; Namba *et al.,* 1991), but also including such cerebral amyloidoses as scrapies and Creutzfeldt–Jakob disease or transmissible dementia (Diedrich *et al.,* 1991; Namba *et al.,* 1991).

Specific protein binding experiments then demonstrated that apoE specifi-

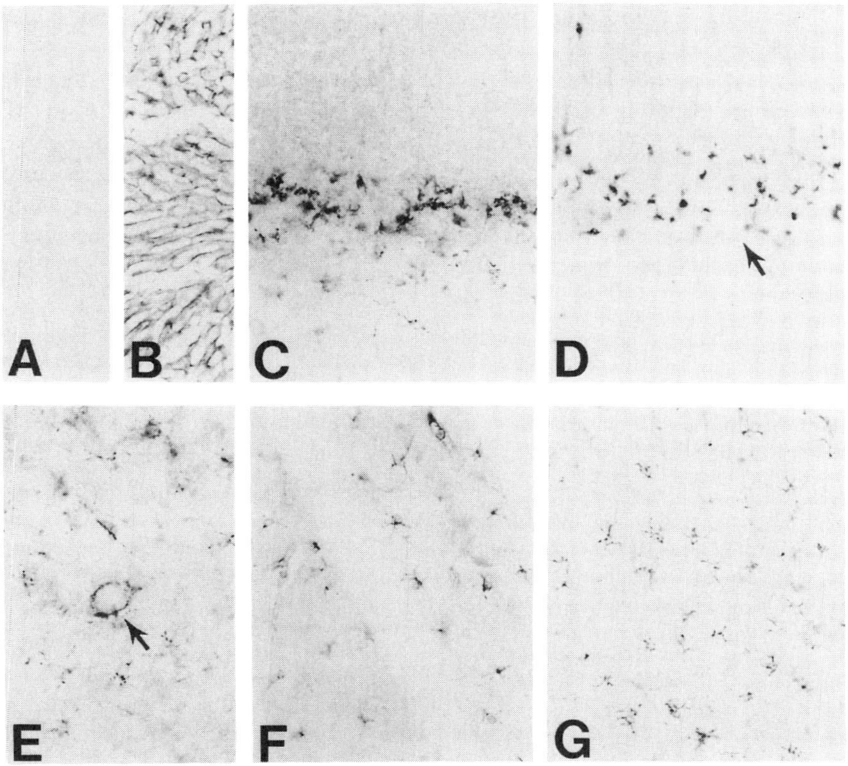

FIGURE 1. Immunocytochemical localization of apoE in mouse liver and brain. (A) Lack of staining of *APOE* knockout liver (ABC immunoperoxidase method). (B) Typical appearance of human apoE immunoreactivity in transgene and targeted replacement animals with staining of sinusoid walls as well as parenchymal cells (*APOE4-line 43*) similar to wild-type animal. (C) Bergmann glial cells of cerebellum and their proximal processes are strongly apoE-immunoreactive in wild-type animals without any detectable staining of Purkinje cells and granule cell neurons (wild-type animal). (D) Identical appearance of cerebellum in transgenic animal (*APOE4-line 81*). (E) Appearance of apoE immunoreactivity in inferior colliculus of brain-stem showing numerous glial cells and occasional glial net around large neurons (wild-type animal). (F) Appearance of apoE-immunoreactive glial cells in striatum with no staining of numerous striatal neurons (wild-type animal). (G) Appearance of apoE-immunoreactive glial cells in cerebral cortex of wild-type animal. Original magnification: A–G, 250×. (Adapted from Xu *et al.*, 1996.)

cally interacts with high affinity to one of the key proteins involved in amyloid CNS deposits in AD, the βA4 peptide derived from amyloid-precursor polypeptide (Strittmatter *et al.*, 1993a). Immunopathological studies confirmed the colocalization of apoE and βA4 in human brain (Strittmatter *et al.*, 1993a; Schmechel *et al.*, 1993), and demonstrated that this is also observed in nonhuman primates with central β-amyloidosis (Strittmatter *et al.*, 1993; Mufson *et al.*, 1994; Poduri *et al.*, 1994; Schmechel *et al.*, 1996a, 1996b).

Testing the possibility that *APOE* alleles might differentially affect the extracellular pathology of AD, we demonstrated that βA4-deposition in AD brains is highest in *APOE4* cases compared to *APOE2* or *APOE3* cases (Schmechel *et al.*, 1993). This finding correlates with the described *in vivo* differences in *APOE* allele interactions and Aβ fibrillogenesis (Strittmatter *et al.*, 1993b; Gallo *et al.*, 1994; Wisniewski *et al.*, 1994; Castano *et al.*, 1995) and has been confirmed by most other groups (e.g., Rebeck *et al.*, 1993, 1995; Hyman *et al.*, 1994; Gomez-Isla *et al.*, 1996). These data would support the viewpoint that one isoform-specific effect of *APOE* on the pathological features of AD could be mediated through the extracellular compartment in the brain with effects on plaque and vessel accumulation of amyloid (Wisniewski and Frangione, 1991; Selkoe, 1994; Sanan *et al.*, 1994). Recent articles have also raised the issue of effects of *APOE* alleles on Aβ-amyloid neuronal toxicity (Loo *et al.*, 1993; Wisniewski *et al.*, 1994; Sanan *et al.*, 1994; Selkoe, 1994; Anderson *et al.*, 1995; Cotman and Anderson, 1995; Yamatsuji *et al.*, 1996).

Another possibility for extracellular effects of apoE would be influencing growth factor metabolism through effects on extracellular matrix proteins and/or extracellular matrix "tethering" of growth factors through protein–protein interactions (Huang *et al.*, 1995; Gutman *et al.*, 1997). Compelling evidence from *in vivo* studies in cell culture support apoE isoform-specific effects on neurite outgrowth and alteration of extracellular matrix induction of cell extension. Dramatic effects can be demonstrated in cell culture of apoE isoform-specific effects on neuritic extension (Handelmann *et al.*, 1992; Huang *et al.*, 1995; Mahley *et al.*, 1996) with the possibility that both intracellular and extracellular effects may be involved.

7. APOE AND PATHOGENESIS OF AD—INTRACELLULAR ROLE

Cognitive decline in AD is most directly associated with neuronal loss and cytoskeletal abnormalities rather than amyloid burden. Another proposed role of apoE in pathogenesis of AD is influence of *APOE* alleles on intraneuronal metabolism through interaction with microtubule-associated proteins of both axons and dendrites (Strittmatter *et al.*, 1993b, 1994a,b; Huang *et al.*, 1994). An effect of *APOE* alleles on neuronal pathology could arise from extracellular interactions or remote effects. The early demonstration of apoE in neurons with neurofibrillary tangles also raised the possibility that apoE might affect neuronal metabolism from within the neuron either by uptake of glial-synthesized apoE by neurons or by endogenous synthesis (Han *et al.*, 1994a,b). If neurons prior to the formation of neurofibrillary tangles normally contain apoE in appropriate intracellular distribution, the observed *in vitro* effects of apoE isoforms on neuronal cytoskeletal proteins would be potential mechanisms of injury in AD (Han *et al.*, 1994a,b; Schmechel *et al.*, 1996a,b). *APOE* alleles have been proposed to affect the degree of cholinergic injury in Alzheimer's disease through mechanisms of altering synaptic plasticity (Poirier, 1994; Poirier *et al.*, 1996).

8. HUMAN PATTERN OF NEURONAL AND GLIAL LOCALIZATION OF APOE

A crucial question for evaluating the possible role of apoE alleles in intracellular mechanisms for AD pathogenesis is whether the intact nervous system normally contains neurons with apoE (Han et al., 1994a,b). In diseased human brain, it is now generally accepted that apoE is present in neurons with neurofibrillary tangles (Han *et al.*, 1994a,b; Benzing and Mufson, 1995; Metzger *et al.*, 1996). In our own and other studies in normal human controls, AD patients, and nonhuman primates, apoE-immunoreactive neurons without apparent cytoskeletal pathology are also observed particularly in hippocampus and cerebral cortex (Han *et al.*, 1994a,b; Schmechel *et al.*, 1996b). In rodent experiments with injury induced by trauma, ischemia, or excitotoxins, *APOE* mRNA and apoE immunoreactivity is seen in neurons in injured areas and might reflect neuronal synthesis or uptake of apoE (Hall *et al.*, 1995; Horsburgh and Nicoll, 1995; Kida *et al.*, 1995; Horsburgh *et al.*, 1997; Ong *et al.*, 1997). Rabbits on high cholesterol diets also demonstrate intraneuronal apoE (Sparks *et al.*, 1995). Studies on apoE receptors demonstrate that many neurons have LRP receptors on cell surface surrounding the soma and LDL receptors on distal processes (Moestrup *et al.*, 1992; Tooyama *et al.*, 1993; Hyman *et al.*, 1994; Rebeck *et al.*, 1995). Whether taken up by receptors or synthesized intraneuronally or both, intraneuronal apoE would presumably be able to influence neuronal function in the ways proposed based on *in vitro* observations and thus affect human neurodegenerative disease on an intracellular basis (Han *et al.*, 1994a,b; Schmechel *et al.*, 1996). Possible mechanisms would include interactions with cytoskeletal proteins such as tau or microtubule-associated protein-2C (MAP-2C). It is important to note that apoE mediation of neuronal injury might occur well before extracellular deposition of Aβ (Han *et al.*, 1994a,b; Schmechel *et al.*, 1996a,b; LaFerla *et al.*, 1997).

The full role of apoE in aging of the CNS and pathogenesis of AD may comprise both extracellular and intracellular effects. In nonhuman primate aging, apoE is found associated with amyloid plaques and deposits in susceptible aged animals (Mufson *et al.*, 1994; Poduri *et al.*, 1994; Schmechel *et al.*, 1996b) and also intraneuronally in normal young adult animals (Schmechel *et al.*, 1996b).

9. APOLIPOPROTEIN E AND OXIDATIVE INJURY

Apolipoprotein E participates in the macrophage response to oxidative injury of the intimal wall of vessels and to peripheral nerve injury (Mahley, 1988; Zhang *et al.,* 1992; Weisgraber, 1994; Bellosta, 1995; Boisvert *et al.,* 1995). Present evidence would suggest that the role of apoE in such injuries is multifactorial and might include both significant extracellular and intracellular actions (Linton *et al.,* 1995; Xu *et al.,* 1996; Schmechel *et al.,* 1996b). ApoE can be a target of oxidative damage and change during these injury responses both in the periphery and in the central nervous system (Palinski *et al.,* 1994; Huang *et al.,* 1995; Montine *et al.,* 1996). A recent report supports a direct isoform-specific modulation by apoE of oxidative injury provoked by Aβ-protein which may involve metal ions such as copper (Miyata and Smith, 1996). *APOE* knockouts are reported to have a higher rate of oxidative change in plasma and liver during aging (Hayek *et al.,* 1994; Palinski *et al.,* 1994). These results suggest that the particular effect of apoE in a given cellular oxidative injury response must be evaluated not only in terms of apoE isoform, but also posttranslational modification, oxidative change, or modification of apoE itself, and the effects of other ligands or associated proteins (Weisgraber, 1994; Miyata and Smith, 1996). In our study of nonhuman primate aging, we have pointed out the potential of apoE and APP/Aβ interactions and oxidative injury at iron-rich glial nets around many cortical neurons (Schmechel *et al.,* 1996a,b). Oxidative injury is a prominent factor to be considered in pathogenesis of AD and related disorders (Beal, 1995; Schmechel *et al.,* 1996a,b).

10. GENETIC MODELS OF *APOE* AS A SUSCEPTIBILITY GENE IN AD

To model the clearly demonstrated role of *APOE* alleles in age of onset distribution of late-onset AD, we need to consider a number of different factors. One factor is whether the allele-specific difference represents loss of a normal function of apoE or gain of an abnormal function. Another factor is the anatomical domain of this effect on function: extracellular or intracellular, neuronal or glial, and if intracellular in what compartment. Posttranslational changes through oxidation or other chemical changes (e.g., sialylation) may further complicate this analysis. Finally, the number of interactions of apoE with other proteins suggests that there may be more than one mechanism of action in AD.

The following genetic models of human *APOE* alleles are currently available: inactivated *APOE* gene ("*APOE* knockouts"), transgenic animals with genomic constructs of human *APOE* alleles including human regulatory sequences, transgenic animals with artificial regulatory sequences, and recently targeted replacement animals ("*APOE* knock-ins"). We analyze the current information for each of these models in the following sections (see also Table 2).

11. *APOE* GENE INACTIVATION MODEL

The differential effect of *APOE* alleles on age at onset of AD might be due to "loss of function" so that *APOE4* is less effective than *APOE3* or *APOE2* in maintaining normal integrity of the nervous system or in normal synaptic remodeling and repair. In this case, inactivation of the *APOE* gene in rodent models might be expected to show some effect on the central or peripheral nervous system or its response to injury.

In fact, the *APOE* gene can be inactivated or "knocked-out" in rodents with only moderate effect on reproductive fitness, general health, or life span. Although inactivation of the *APOE* gene is not lethal, the animals are hypercholesterolemic on normal diets (Piedrahita *et al.*, 1992; Zhang *et al.*, 1992; Plump and Breslow, 1995). These animals also have a greater susceptibility to atherosclerosis, which is enhanced on high-fat diets (Zhang *et al.*, 1992). Transfection of knockouts with *APOE* vectors or wild-type bone marrow transplants has demonstrated that a relatively small amount of apoE expression can lower cholesterol levels and reverse atherosclerotic lesions (Bellosta *et al.*, 1995; Boisvert *et al.*, 1995; Linton *et al.*, 1995).

The effect of *APOE* gene inactivation has been tested also in the peripheral and central nervous systems, both under normal conditions of aging and after injury. The reported effects have varied. Initial experiments on peripheral nerve regeneration after injury reported no difference between knockouts and wild-type controls (Popko *et al.*, 1993; Goodrum *et al.*, 1995). It is important to note in this context that the C57/BL6J background strain is already impaired with regard to rates of nerve regeneration compared to other inbred strains (Brown *et al.*, 1994). APOE knockouts are reported to have a decreased sensitivity to pain and abnormally blunted stress response (Gordon *et al.*, 1996a). More recent experiments on peripheral nerve report a significant change in small diameter, nonmyelinated axons with *APOE* knockouts having fewer and smaller fibers than control animals (Fullerton *et al.*, 1997).

In the CNS, *APOE* knockouts have been reported to have increased rates of synaptic loss with age compared to controls (Masliah *et al.*, 1995, 1997). Others have reported behavioral cholinergic deficits in *APOE* knockouts with restitution by pharmacological means (Gordon *et al.*, 1995, 1996b). These results have not been replicated in other laboratories. For example, a combined behavioral, anatomical, and biochemical study showed no effect of *APOE* gene inactivation

(Anderson and Higgins, 1997). In our laboratory, detailed morphological analysis of dendritic spines in the hippocampus of *APOE* knockouts likewise showed no difference with regard to wild-type control animals (Mann *et al.*, 1997). As a further example of varying results in *APOE* knockout models, an increase in phosphorylated tau in brain was reported (Genis *et al.*, 1995), but has not been confirmed by others (Mercken and Brion, 1995). These experimental results on *APOE* knockout models are summarized in Table 2.

With response to CNS injury, there is evidence that *APOE* gene inactivation may result in greater damage in animal models of cerebral ischemia. Larger stroke extent and greater mortality after filament induced transient middle cerebral artery occlusion is reported in *APOE* knockout animals (Laskowitz et al., 1997b). Closed head injury has also been reported to result in greater severity in *APOE* knockouts (Chen et al., 1997).

With regard to the influence of apoE on the expression of other AD-related genes, the recent report of decreased β-amyloidosis in the absence of a functional *APOE* gene suggests that the influence of apoE on nervous system function in rodent models may depend on context and the "humanization" of other relevant genes (Bales *et al.*, 1997).

12. CONSIDERATION OF GENETIC BACKGROUND IN ANALYSIS OF *APOE* "KNOCKOUTS" AND TRANSGENES

The varying results evident in the current literature on *APOE* genetic models (Table 2) may reflect different background strains of animals (Piedrahita *et al.*, 1992; Plump and Breslow, 1995) particularly given the chromosomal localization of the rodent *APOE* gene. The *APOE* gene in rodents is located on the short arm of chromosome 7 near the pink-eye dilution and also near a putative gene for brain weight and behavior (Belknap *et al.*, 1992; Lyon *et al.*, 1992). The observed very low rate of recombination in this region and the selection for inactivated *APOE* gene involved in generating these lines implies that back-crossing be carried out for longer periods than usual. This is needed to remove by recombination other genes close to the *APOE* locus that are present in the original stem-cell inbred mouse line. In practice, this means that the region adjacent to *APOE* locus in early backcrosses of *APOE* knockouts represents a portion of stem cell C129/ola chromosome 7 in midst of rest of chromosome 7 from backcross strain. Moreover, it is notable that this inbred strain used to derive the stem cells has low brain weight and impaired open field behavior ascribed in part to a locus in this exact region of chromosome 7 (Belknap *et al.*, 1992). This relative preservation of the stem cell surround on the short-arm of chromosome 7 probably affected the earliest, widely available line of 6th generation back-crossed ("N6," brown,

TABLE 2. Current Status of Genetic Models of APOE Effects on "AD" Pathologies

AD "components"	APOE condition	Finding	References
Behavioral deficit	*mAPOE* inactivation	Yes	Masliah *et al.*, 1995, 1997; Gordon *et al.*, 1995, 1998b Oitzl *et al.*, 1997; Mohlar *et al.*, 1997 (ovarectomy)
		No	Anderson and Higgins, 1997
	hAPOE alleles	Not completed	
Neuronal pathology	*mAPOE* inactivation	Yes	Masliah *et al.*, 1995, 1997
		No	Anderson and Higgins, 1997
	hAPOE alleles	Not completed	
β-Amyloidosis	*mAPOE* inactivation	No	
	+ presence APP mutations	Delayed	Bales *et al.*, 1997
	hAPOE alleles	No	
Neuronal loss	*mAPOE* inactivation	No	
	hAPOE alleles	Not completed	
Neurofibrils/ cytoskeletal pathology	*mAPOE* inactivation	Yes	Genis *et al.*, 1995
		No	Mercken and Brion, 1995
	hAPOE alleles	No	
Gliosis	*mAPOE* inactivation	No	
	hAPOE alleles	No	
Oxidative marker increase	*mAPOE* inactivation	Increased	Palinski *et al.*, 1994; Hayek *et al.*, 1994
	hAPOE alleles	Not completed	
Cholinergic deficit	*mAPOE* inactivation	Yes	Gordon *et al.*, 1995, 1997
		No	Anderson and Higgins, 1997
	hAPOE alleles	Not completed	
CNS Injury Model			
Ischemic injury	*mAPOE* inactivation	Greater infarct	Laskowitz *et al.*, 1997b
	hAPOE alleles	*E4* > *E3*	Sheng *et al.*, 1998
Closed head injury	*APOE* inactivation	Greater damage	Chen *et al.*, 1997
	APOE alleles	Not completed	

pink-eyed animals) *APOE* knockouts from Jackson Laboratories. More recently, further back-crossing to C57BL/6J to 10th generation ("N10," black, black-eyed animals) has presumably removed more of this surrounding, low-recombinant region of the *APOE* gene. This factor is a potential confound in terms of comparing the effects of *APOE* gene inactivation in different inbred strains and backgrounds, as the clear implication is that genetic variation near the *APOE* gene in various *APOE* knockout constructions would result in phenotypic differences according to strain. These phenotypic differences might also be subject to environmental modification.

13. *APOE* TRANSGENIC MODEL WITH HUMAN REGULATORY SEQUENCES

The approach in our laboratory for modeling human *APOE* allelic effects was to create animals transgenic for genomic constructs for each of the three human *APOE* alleles on a background of inactivation of the normal rodent *APOE* gene. Using standard molecular genetic methods, mice that were apoE deficient ("knockout") received large human genomic DNA fragment transgenes for each of the three common apoE alleles (Xu *et al.,* 1996). Genomic DNA fragments contained all *APOE* introns, both 5′ and 3′ human regulatory sequences and particularly downstream tissue expression elements (Simonet *et al.,* 1993). This strategy results in expression of human *APOE* alleles in the absence of any endogenous rodent *APOE* gene and with the possibility of gene regulation through human regulatory sequences. Such experiments assume that apoE protein will be reasonably biologically active in its interactions with mouse receptor molecules and other ligands, that endogenous mouse regulatory molecules will operate on human genomic DNA sequences and that position of insertion will not create non-specific effects.

APOE transgenes generated using this method show relatively normal patterns of expression of *APOE* mRNA and apoE protein in various tissues including brain (Xu *et al.,* 1996, 1998) (Figs. 2 and 3). This expression is associated with reduction of the elevated plasma cholesterol observed in the *APOE* knockout background animals to nearly normal levels (Xu *et al.,* 1996). On normal

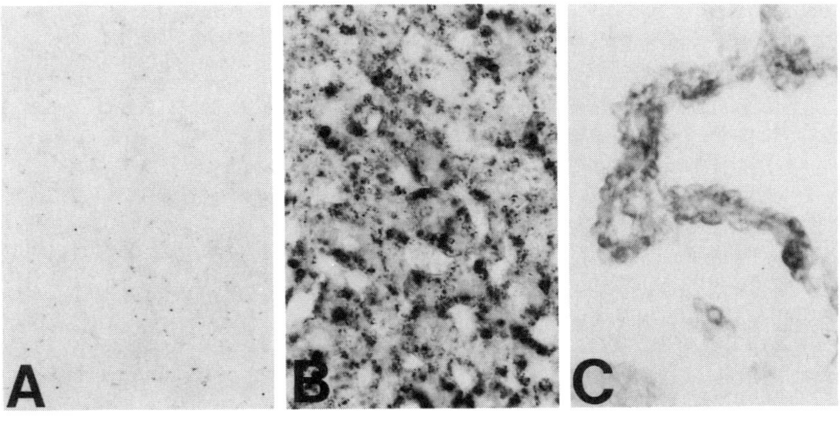

FIGURE 2. In situ hybridization for *APOE* mRNA in mouse liver and brain. *In situ* hybridization for *APOE* mRNA with lack of signal in *APOE* knockout liver (A) and signal in liver (B) and choroid plexus (C) of *APOE4* transgene (line 43, multiple copy). Original magnification: A–C, 400×.

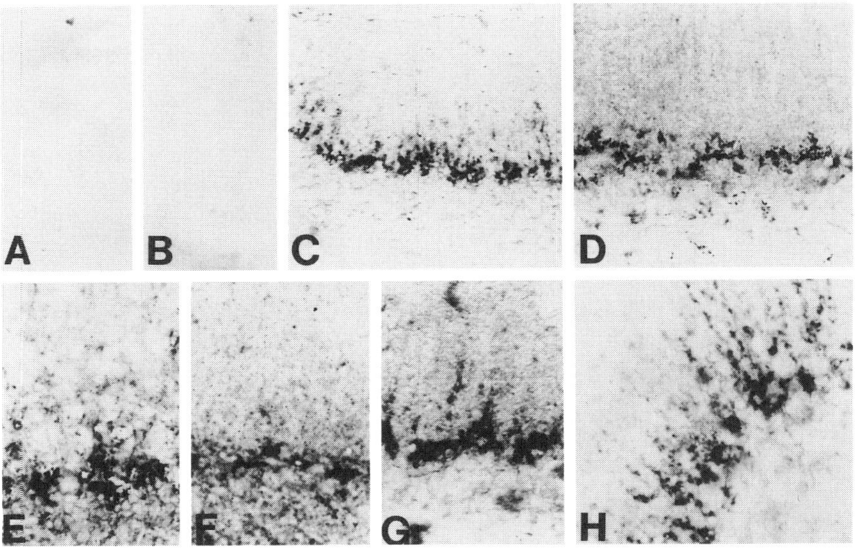

FIGURE 3. Pattern in mouse cerebellum of *APOE* transgenes. Sense (A) and antisense probes (B) for human *APOE* mRNA do not hybridize to wild-type (+/+) cerebellar cortex. In APOE4 (line 81) transgene, signal is located in line of Bergmann glial cells along Purkinje cell layer (C), is not present in nearby neurons, and resembles apoE immunolocalization (D). This pattern is observed in low- and high-copy animals: *APOE2*-line 205 (E), *APOE3*-line 437 (F), and *APOE4*-line 43 (G). Strong signal is observed in radial glial fibers especially in high copy *APOE4*-line 43 (H). Original magnifications, 250× (A–D), 400× (E–H). (From Xu *et al.*, 1998.)

rodent diets, these animals live up to 2 years without apparent physical illness and with good reproductive fitness.

The striking finding in our analysis of these *APOE* animals is the generation of human patterns of apoE cellular localization in the CNS (Xu *et al.*, 1996, 1998). The normal "rodent" pattern of localization of apoE protein and *APOE* mRNA is observed in astrocytes, ependymal cells, radial glial cells, and peripheral tissues in the transgenes (Figs. 2 and 3). Unlike normal rodents and similar to what is known of "human" pattern, apoE protein and *APOE* mRNA are also observed in a regionally specific pattern in neurons of cerebral cortex (Xu *et al.*, 1996, 1998) (Figs. 4 and 5).

The "human" pattern of neuronal and glial apoE localization is specific to animals containing the human regulatory sequences and contrasts with the normal "rodent" pattern found in animals with mouse regulatory sequences (Table 3; Fig. 6). These results suggest that full modeling of human genetic diseases of the nervous system may depend not only on expression of human protein sequences, but also on expression of human control sequences *cis* or *trans* to these regions.

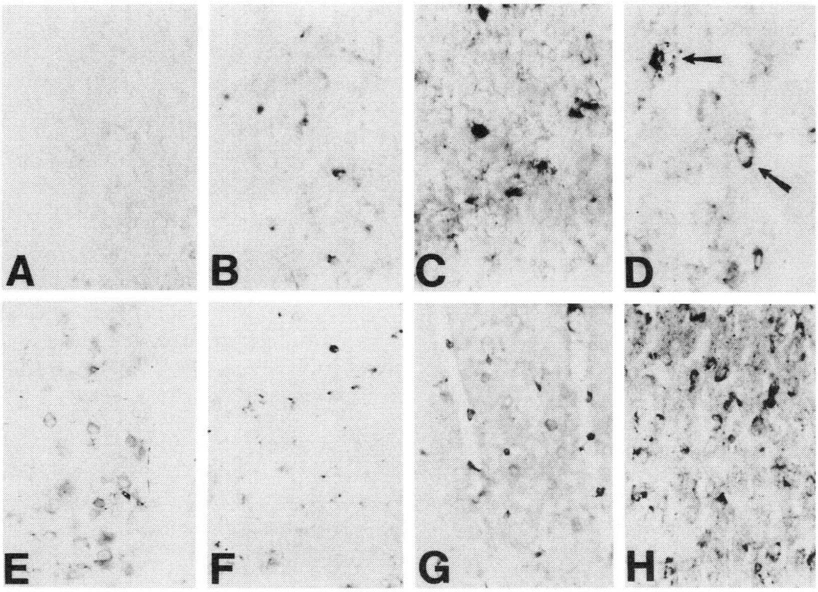

FIGURE 4. Pattern in mouse cerebral cortex of *APOE* transgenes. Absence of signal for human *APOE* mRNA in cerebral cortex of wild-type (+/+) animals (A). In transgenic lines, the strongest signal is observed in presumptive astrocytes illustrated for *APOE4*-line 81 (B) and *APOE4*-43 (C). In all transgenic lines, hybridization was also observed in larger cells consistent with pyramidal neurons shown in *APOE2*-line 205 (D and H) with presumptive glial cell (*horizontal arrow*) and pyramidal neuron (*diagonal arrow*). ApoE immunolocalization reveals both glial cells and neurons: *APOE4*-line 81 (E) and *APOE2*-line 205 (G). No neuronal localization is observed in wild-type animals (F) with antisera to either human apoE or rodent spoE. Both immunocytochemical (G) and *in situ* detection signal (H) are higher for glial cells compared to neurons. Original magnifications: 400× (A–D); 250× (E–H). (From Xu et al., 1998.)

The further implication is that the presence of apoE receptors on wild-type rodent neurons does not result in *detectable* apoE uptake under normal circumstances whether in the presence of normal rodent apoE or human apoE.

To test whether the presence of apoE intraneuronal immunoreactivity was due to *APOE* gene sequence differences between mouse and human, we compared the above *APOE* transgenic mice to other mice constructed using targeted replacement methodology so that the human gene was placed under mouse gene promoters (Sullivan *et al.,* 1997). These mice were available as homozygote human APOE3/3 and APOE2/2 animals with no rodent *APOE* gene. We were therefore able to compare transgenic mice expressing human *APOE* under the control of human or mouse regulatory sequences in absence of normal rodent *APOE* gene (Fig. 6). Because a high-cholesterol diet in rabbits is reported to

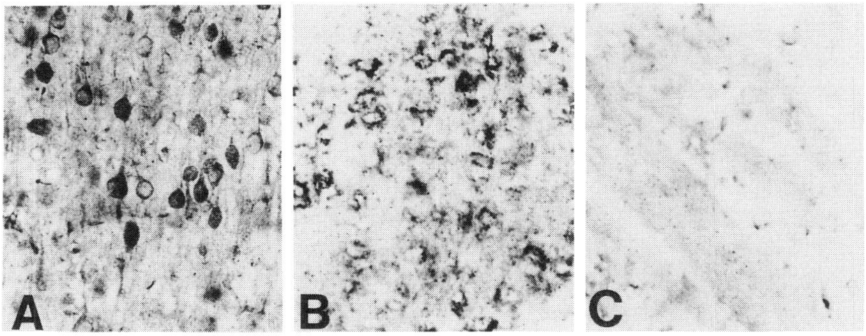

FIGURE 5. Regional specificity of *APOE* expression in transgenes. Regional specificity of *APOE* mRNA transcription in *APOE3*-line 437, a low copy line showing strong immunocytochemical (A) and *in situ* hybridization (B) signal for cerebral cortical neurons, but no appreciable hybridization signal for striatal neurons (C). Bar indicates 25 microns. Original magnification: A–C, 400×. (From Xu et al., 1998.)

result in appearance of neuronal apoE immunoreactivity (Sparks *et al.,* 1995), we also examined LDL-receptor knockout animals to control for effect of higher cholesterol and altered lipid metabolism in genetically modified animals compared to wild-type. LDL-receptor knockout animals also showed glial pattern of apoE immunoreactivity without neuronal staining (Table 3). These brain expres-

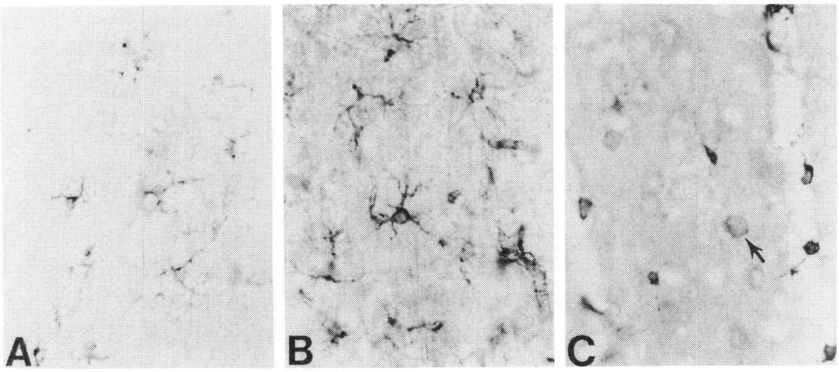

FIGURE 6. Comparison of cerebral cortex apoE immunolocalization in wild-type, *APOE* targeted replacement, and *APOE* transgenic animals. (A) ApoE-immunoreactive astrocytes in cerebral cortex of wild-type animal. (B) Intensely apoE-immunoreactive astrocytes in cerebral cortex of APOE targeted replacement animal (*APOE2/2*). (C) ApoE immunoreactivity of both glial cells and selected cortical neurons in cerebral cortex of APOE transgenic animal (*APOE2-line 205*). Original magnification: A, B, C 520×.

TABLE 3. ApoE Immunoreactive Patterns for Different Mouse Strains

Animal	*rAPOE*	*hAPOE*	glial IR	neuronal IR	cis-promoter
Transgene	(copies)	(copies)	0 to ++++	0 to +++	(*hAPOE*)
APOE2-205	0	8	++	++	+
APOE3-437	0	2	+	+++	+
APOE4-81	0	2	++	++	+
APOE4-43	0	6	++	++	+
LDLK-KO	2	0	+++	Rare	0
APOE-KO	0	0	0	0	0
Replacement:					
APOE2	0	2	+++	0	0
APOE3	0	2	+++	0	0
APOE4	ND	ND	ND	ND	ND
Wild-type					
C57Bl/6j	2	0	+++	Rare	0
C129/j	2	0	+++	Rare	0
CD-1(ICR)	2	0	+++	Rare	0

hAPOE and *rAPOE* refer to human and rodent *APOE* respectively; IR refers to immunoreactivity; *-##* after *APOE* transgene designation indicates founder line (see Xu *et al.*, 1996); *LDLR-KO* indicates LDL receptor knockout animals; *APOE-KO* indicates *APOE* knockout animals; ND indicates not determined.

sion data are consistent with earlier results demonstrating that the regulatory regions necessary for hepatic expression are located at some distance from the *APOE* gene, between the *APOC1* gene and *APOC1'* pseudogene approximately 18 kb downstream from the *APOE* promoter and 9 kb downstream from the *APOC1* promoter (Simonet *et al.,* 1993; Xu *et al.,* 1996).

Experiments are currently underway to test these animals for age-related pathology (Table 2). The demonstration of intraneuronal synthesis of apoE (Xu *et al.,* 1998) in a regionally specific pattern to cerebral cortex points out that models of AD pathogenesis must take into account intraneuronal mechanisms of injury. It may well be necessary to introduce through cross-breeding other human genes such as APP or tau into such models to produce full extent of AD pathology. Although models of APP mutations are important (e.g., Hsiao *et al.,* 1995, 1996), it is important to realize that most human cases of late-onset AD occur in the setting of normal APP, presenilin and tau genes with the influence of *APOE* alleles and/or other genes to be determined. We have recent evidence that apoE localization within neurons can occur in cell body region, axon, and synaptic terminals for selected classes of cortical neurons (Fig. 7). Other recent results support axonal transport of apoE in the central nervous system (Amaratunga *et al.,* 1996). Thus, *APOE* transgenic animals with human regulatory sequences may permit modeling the full interaction of apoE within neurons that may be pertinent to AD pathogenesis (Han *et al.,* 1994a,b; Xu *et al.,* 1996, 1998).

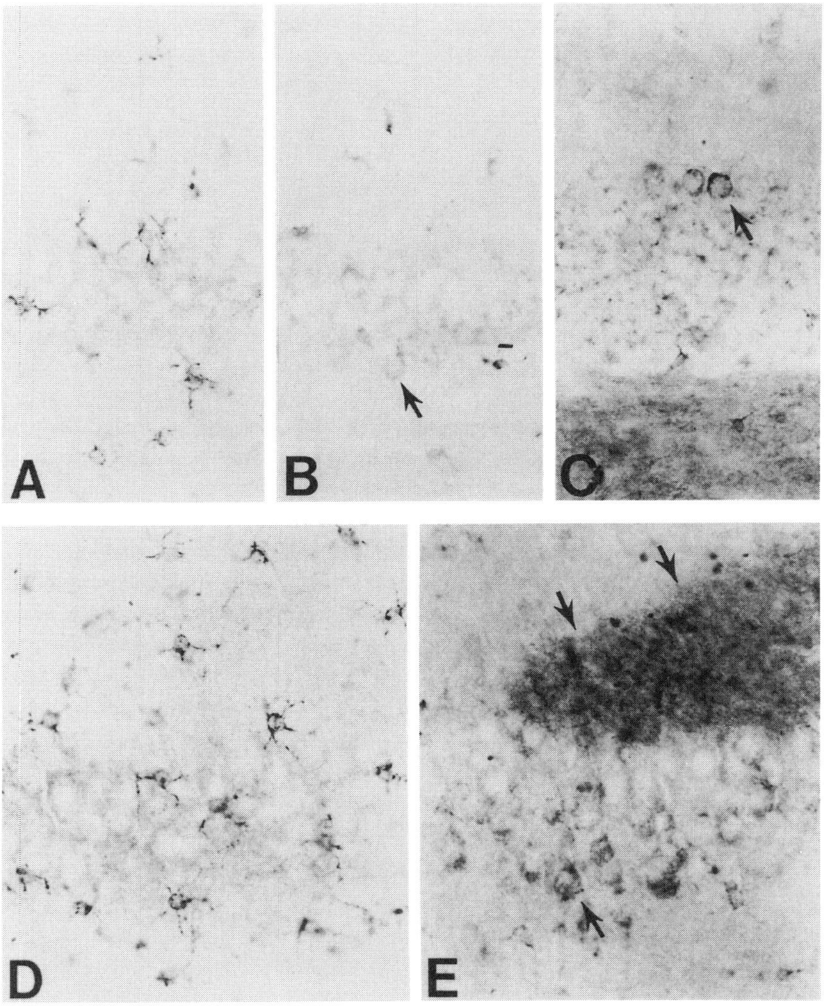

FIGURE 7. Selective patterns of apoE neuronal immunoreactivity in *APOE* transgene hippocampus. (A) ApoE immunoreactivity of astrocytes in hippocampal CA1-2 layer with unstained neurons (wild-type animal). (B) ApoE immunoreactivity in same layer of transgene showing less glial immunoreactivity and some faintly stained neurons (*arrow*) just at the threshold of positivity (*APOE4-line 81*). (C) Two apoE immunoreactive neurons in dentate granule cell layer of *APOE4-line 81* animal with most neurons nonreactive. (D) ApoE immunoreactivity of astrocytes in CA3 layer of wild-type animal with lack of staining of neurons and mossy fiber layer. Original magnification: A–E, 520×. (E) Selected apoE-immunoreactive neurons in CA3 layer of *APOE4-line 81* animal with strongly stained mossy fiber layer above (see text).

14. *APOE* TRANSGENIC MODEL WITH NONHUMAN REGULATORY SEQUENCES

A number of groups have created *APOE* transgenetic animals without human regulatory sequences. These animals show correction of lipid status, and the pattern of nervous system expression in accordance with their particular artificial promoter sequences (Bowman *et al.,* 1996). They afford a potential way to model expression of human *APOE* alleles in defined cell classes or with controllable promoters.

15. *APOE* TARGETED REPLACEMENT ANIMALS

Recently, targeted replacement *APOE* animals have been created that place the human *APOE* gene under control of normal mouse regulatory sequences (Sullivan *et al.,* 1997). Homozygous animals for each human *APOE* allele allow testing the effects of human apoE expression under mouse regulatory control. These animals demonstrate correction of lipid status under normal diet conditions, and mimic some of the human lipid disorders under high-fat conditions. In the nervous system, apoE cellular localization resembles normal "rodent" pattern with glial localization and not "human pattern" observed in transgenes with human regulatory sequences (Table 3, Fig. 6).

16. CURRENT MODELS FOR THE EFFECT OF *APOE* ALLELES ON SUSCEPTIBILITY TO AD

Human *APOE* transgenes made with constructs that contain the downstream hepatic control region generally correct cholesterol with apoE expression in liver and other tissues (Bowman *et al.,* 1996; Xu *et al.,* 1996). Homologous replacement animals likewise have normal serum cholesterol and partially correct their abnormal lipid status with expression of the human *APOE* gene (Sullivan *et al.,* 1997). This suggests at least partial recognition of human apoE by the various mouse apoE receptor proteins.

Both transgenic and targeted replacement *APOE* mice with normal human or mouse regulatory sequences and no endogenous rodent *APOE* expression offer important tools for study of human AD susceptibility gene *APOE.* They offer models of human apoE expression in human pattern (genomic APOE transgenes) and in rodent pattern (targeted replacement) animals. Such genetic models are contrasted to models with gene constructs using engineered promoters that result in an artificial pattern of expression without normal human or rodent regulatory control sequences. Creation of metabolically relevant mouse models of AD neuropathology for late-onset sporadic and familial AD will need to take into account differences in brain expression of apoE isoforms. The important issues are not only allelic coding sequence differences, but also species-specific differences in expression pattern apparently due to regulatory control regions.

Further experiments are necessary to identify the responsible control regions for neuronal expression of apoE and to resolve the issue of cytoplasmic compared to vesicular compartmentation of apoE (Xu *et al.,* 1996). The importance of isoform-specific 3′-untranslated sequences in sorting mRNAs to different cytoplasmic compartments has been previously demonstrated for the mRNAs encoding two actin isoforms, β-cytoplasmic and α-cardiac. Each mRNA can occupy different cytoplasmic compartments within the same cell (Kislauskis *et al.,* 1993). Kislauskis *et al.* demonstrated that each 3′UTR was "necessary and sufficient to direct reporter transcripts to its appropriate compartment." The regulatory regions necessary for hepatic expression of *APOE* mRNA are found downstream from the *APOE* gene, between the *APOC1* gene and *APOC1′* pseudogene approximately 18 kb downstream from the *APOE* promoter and 9 kb downstream from the *APOC1* promoter (Simonet *et al.,* 1993). It is not yet clear whether 5′ or 3′UTR or both UTRs may also regulate species differences in brain-specific expression of apoE immunoreactivity. The species difference is not apparently related to the coding sequence differences between mouse and human *APOE* (Xu *et al.,* 1996, 1998). Creation of *APOE* transgenic mice using only human coding sequences and an artificial promoter leads to mice that do not demonstrate intraneuronal apoE immunoreactivity (Bowman *et al.,* 1996) and are not useful to study the unique aspects of apoE metabolism that occur in humans and some nonhuman primates. The implications of these data for widespread attempts to create transgenic models of disease are obvious: expressing a gene product in its appropriate cellular distribution may be critical to modeling the pathogenic process leading to the human disease. Successful transgenic models in which the expected disease process is reproduced may result from appropriate tissue expression, rather than simply expressing or overexpressing a particular gene. Transgenic strategies involving only coding regions with artificial promoters for *APOE* gene may be insufficient to model the common sporadic and familial forms of human AD. The failure of many attempts to model the full neuropathological features of AD using APP mutation transgenes, particularly with respect to neuronal degeneration, may be explained by differences in apoE metabolism and apoE/Aβ interactions in rodents compared to humans. Current evidence in humans (Han *et al.,* 1994a, 1994b; LaFerla et al., 1997) and primates (Schmechel *et al.,* 1996a,b) suggest that neuronal injury and intraneuronal apoE interactions occur well before extracellular deposition of Aβ. Interaction of apoE with soluble forms of Aβ and interaction with other AD-related proteins relevant to neuronal and/or glial apoptosis may be significant (Anderson *et al.,* 1995;

Cotman and Anderson, 1995; Miyata and Smith, 1996; Gearing *et al.*, 1996; Yamatsuji *et al.*, 1996; Ishii *et al.*, 1997).

17. SUMMARY/CONCLUSIONS

Apolipoprotein E (*APOE*, gene; apoE, protein) is a susceptibility gene for late-onset Alzheimer's Disease (AD) with the three common alleles, *E2, E3* and *E4* influencing risk and age of onset distribution. *APOE* transgenic mice have been created with artificial promoters, human regulatory sequences (large genomic inserts), and targeted replacement relying largely on endogenous mouse regulatory sequences. We have compared the expression pattern of apoE in brain for the latter two approaches using specific immunocytochemistry and *in situ* hybridization. The results demonstrate that mouse regulatory sequences are associated with high glial expression of apoE in brain and human regulatory sequences with "human-like" pattern of neuronal and glial expression in cerebral cortex. The human pattern of apoE expression is observed only in mice with human regulatory sequences of the *APOE* gene and not in targeted replacement animals that lack these human sequences. *In situ* hybridization suggests that this difference is due to a regionally specific pattern of endogenous synthesis of apoE in selected classes of cerebral cortical neurons. These data demonstrate that it will be important to consider both allelic differences in the peptide coding sequence and specific influences of regulatory sequences when constructing models to study the role of apoE in the pathogenesis of AD. Finally, the role of *APOE* alleles in human susceptibility to late onset familial and sporadic AD occurs in the setting of normal APP, presenilin, and tau genes and will need to be modeled accordingly.

ACKNOWLEDGMENTS

This publication was made possible by NIA grant 5P5O AGO5128 and NIEMS IAA YOI-ES-40290.

REFERENCES

Alberts, M., Graffagnino, C., McClenny, C., Saunders, A., and Roses, A., 1995, ApoE genotype and survival from intracerebral hemorrhage, *Lancet* **346:**575.

Amaratunga, A., Abraham, C. R., Edwards, R. B., Sandell, J. H., Schreiber, B. M., and Fine, R. E., 1996, Apolipoprotein E is synthesized in the retina by Muller glial cells, secreted into the vitreous, and rapidly transported into the optic nerve by retinal ganglion cells, *J. Biol. Chem.* **271:**5628–5632.

Anderson, A. J., Pike, C. J., and Cotman, C. W., 1995, Differential induction of immediate early proteins in cultured neurons by beta-amyloid (A-beta): association of c-jun with A-beta induced apoptosis, *J. Neurochem.* **65:**1482–1497.

Anderson, R., and Higgins, G. A., 1997, Absence of central cholinergic deficits in APOE knockout mice, *Psychopharm.* **132:**135–144.

Bales, K. R., Verina, T., Dodel, R. C., Du, Y., Altstiel, L., Bender, M., Hyslop, P., Johnstone, E. M., Little, S. P., Cummins, D. J., Piccardo, P., Ghetti, B., and Paul, S. M., 1997, Lack of apolipoprotein E dramatically reduces amyloid beta-peptide deposition, *Nat. Genet.* **17:**263–264.

Beal, M. F., 1995, Aging, energy, and oxidative stress in neurodegenerative diseases, *Ann. Neurol.* **38:**357–366.

Belknap, J. K., Phillips, T. J., and O'Toole, L. A., 1992, Quantitative trait loci associated with brain weight in the BXD/Ty recombinant inbred mouse strains, *Brain Res. Bull.* **29:**337–344.

Bellosta, S., Mahley, R. W., Sanan, D. A., Murata, J., Newland, D. L., Taylor, J. M., and Pitas, R. E., 1995, Macrophage-specific expression of human apolipoprotein E reduces atherosclerosis in hypercholesterolemic apolipoprotein E-null mice, *J. Clin. Invest.* **96:**2170–2179.

Benzing, W. C., and Mufson, E. J., 1995, Apolipoprotein E immunoreactivity within neurofibrillary tangles: relationship to Tau and PHF in Alzheimer's disease, *Exp. Neurol.* **132:**162–171.

Boisvert, W. A., Spangenberg, J., and Curtiss, L. K., 1995, Treatment of severe hypercholesterolemia in apolipoprotein E-deficient mice by bone marrow transplantation, *J. Clin. Invest.* **96:**1118–1124.

Bowman, B. H., Yang, F., Buchanan, J. M., Adrain, G. S., Martinez, A. O., Jansen, L., Zhao, M., Atherton, S. L., and Hixson, J. E., 1996, Human APOE protein localized in brains of transgenic mice, *Neurosci. Lett.,* **219:**57–59.

Boyles, J. K., Pitas, R. E., Wilson, E., Mahley, R. W., and Taylor, J. M., 1985, Apolipoprotein E associated with astrocytic glia of the central nervous system and with nonmyelinating glia of peripheral nervous system, *J. Clin. Invest.* **76:**1501–1513.

Brown, M. C., Perry, V. H., Hunt, S. P., and Lapper, S. R., 1994, Further studies on motor and sensory nerve regeneration in mice with delayed Wallerian degeneration, *Eur. J. Neurosci.* **6:**420–428.

Castano, E. M., Prelli, F., Wisniewski, T., Golabeck, A., Kumar, R. A., Soto, C., and Frangione, B., 1995, Fibrillogenesis in Alzheimer's disease of amyloid beta peptides and apolipoprotein E, *Biochem. J.* **306:**599–604.

Chen, Y., Lomnitski, L., Michaelson, D. M., and Shohami, E., 1997, Motor and cognitive deficits in apolipoprotein E-deficient mice after closed head injury. *Neurosci.* **80:**1255–1262.

Corder, E. H., Saunders, A. M., Strittmatter, W. J., Schmechel, D. E., Gaskell, P. C., Small, G. W., Roses, A. D., Haines, J. L., and Pericak-Vance, M. A., 1993, Gene dose of apolipoprotein E type 4 and the risk of Alzheimer's disease in late-onset families, *Science* **261:**921–923.

Corder, E. H., Saunders, A. M., Risch, N. J., Strittmatter, W. J., Schmechel, D. E., Gaskell, P. C., Rimmler, J. B., Locke, P. A., Conneally, P. M., Schmader, K. E., Small, G. W., Roses, A. D., Haines, J. L., and Pericak-Vance, M. A., 1994, Protective effect of apolipoprotein E type 2 allele for late onset Alzheimer disease, *Nature Genet.* **7:**180–184.

Cotman, C. W., and Anderson, A. J., 1995, A potential role for apoptosis in neurodegeneration and Alzheimer's disease, *Molec. Neurobiol.* **10:**19–45.

Diedrich, J. F., Minnigan, H., Carp, R. I., Whitaker, J. N., Race, R., Frey, W. II, and Hasse, A. T., 1991, Neuropathological changes in scrapie and Alzheimer's disease are associated with increased expression of apolipoprotein E and cathepsin D in astrocytes, *J. Virol.* **65:**4759–4768.

Dong, L. M., and Weisgraber, K. H., 1996, Human apolipoprotein E4 domain interaction: arginine 61 and glutamic acid 255 interact to direct the preference for very low density lipoproteins, *J. Biol. Chem.* **271:**19053–19057.

Fullerton, S. M., Strittmatter, W. J., Montine, T. J., and Matthew, W. D., 1997, Altered morphology of the sciatic nerve in apoE deficient mice. *Soc. Neurosci. Abstr.* **23:**572.

Gallo, G., Wisniewski, T., Choi-Miura, N. H., Ghiso, J., and Frangione, B., 1994, Potential role of apolipoprotein E in fibrillogenesis, *Amer. J. Pathol.* **145:**526–530.

Gearing, M., Mori, H., and Mirra, S. S., 1996, Abeta-peptide length and apolipoprotein E genotype in Alzheimer's Disease, *Ann. Neurol.* **39:**395–399.

Gelman, B. B., Goodrum, J. F., and Bouldin, 1991, Macrophage apolipoprotein synthesis and endoneurial distribution as a response to segmental demyelination, *J. Neuropathol. Exp. Neurol.* **50:**383–407.

Genis, I., Gordon, I., Sehayek, E., and Michaelson, D. M., 1995, Phosphorylation of tau in apolipoprotein E-deficient mice, *Neurosci. Letters* **199:**5–8.

Gomez-Isla, T., West, H. L., Rebeck, G. W., Harr, S. D., Growdon, J. H., Lacascio, J. J., Perls, T. T., Lipsitz, L. A., and Hyman, B. T., 1996, Clinical and pathological correlates of apolipoprotein E epsilon 4 in Alzheimer's disease, *Ann. Neurol.* **39:**62–70.

Goodrum, J. F., Bouldin, T. W., Zhang, S. H., Maeda, N., and Popko, B., 1995, Nerve regeneration and cholesterol reutilization occur in the absence of apolipoproteins E and A-I in mice, *J. Neurochem.* **64:**408–416.

Gordon, I., Grauer, E., Genis, I., Sehayek, E., and Michaelson, D. M., 1995, Memory deficits and cholinergic impairments in apolipoprotein E-deficient mice, *Neurosci. Letters* **199:**1–4.

Gordon, I., Ben-Eliyahu, S., Rosenne, E., Sehayek, E., and Michaelson, D. M., 1996a, Derangement of stress response of apolipoprotein E-deficient mice, *Neurosci. Letters* **206:**212–214.

Gordon, I., Genis, I., Grauer, E., Sehayek, E., and Michaelson, D. M., 1996b, Biochemical and cognitive studies of apolipoprotein-E-deficient mice, *Mol. Chem. Neuropath.* **28:**97–103.

Gutman, C. R., Weisgraber, K. H., Strittmatter, W. J., and Matthew, W. D., 1997, Apolipoprotein E binds to and potentiates the biological activity of CNTF, *J. Neurosci.* **17:**6114–6121.

Hall, E. D., Oostveen, J. A., Dunn, E., and Carter, D. B., 1995, Increased amyloid protein precursor and apolipoprotein E immunoreactivity in the selectively vulnerable hippocampus following transient forebrain ischemia in gerbil, *Exp. Neurol.* **135:**17–27.

Han, S-H., Hulette, C., Saunders, A., Einstein, G., Pericak-Vance, M., Strittmatter, W., Roses, A. and Schmechel, D., 1994a, Apolipoprotein E is present in hippocampal neurons without neurofibrillary tangles in Alzheimer's disease and in age-matched controls, *Exp. Neurol.* **128:**13–26.

Han, S-H., Einstein, G., Weisgraber, K., Strittmatter, W., Saunders, A., Pericak-Vance, M., Roses, A., and Schmechel, D., 1994b, Apolipoprotein E is localized to the cytoplasm of human cortical neurons: a light and electron microscopic study, *J. Neuropath. Exp. Neurol.* **53:**535–544.

Handelmann, G. E., Boyles, J. K., Weisgraber, K. H., Mahley, R. W., and Pitas, R. E., 1992, Effects of apolipoprotein E, β-very low density lipoproteins, and cholesterol on the extension of neurites by rabbit dorsal root ganglion neurons *in vitro, J. Lipid Res.* **33:**1677–1688.

Hayek, T., Oiknine, J., Brook, J. G., and Aviram, M., 1994, Increased plasma and lipoprotein lipid peroxidation in apo E-deficient mice, *Biochem. Biophys. Res. Comm.* **201:**1567–1574.

Horsburgh, K., and Nicoll, J., 1995, Selective alterations in the cellular distribution of apolipoprotein E immunoreactivity following transient cerebral ischaemia, *Alz. Res.* **1:**169–176.

Horsburgh, K., Fitzpatrick, M., Nilsen, M., and Nicoll, J. A., 1997, Marked alterations in the cellular localization and levels of apolipoprotein E following acute subdural hematoma in rat, *Brain Res.* **763:**103–110.

Hsiao, K. K., Borchelt, D. R., Olsen, K., Johannsdottir, R., Kitt, C., Yunis, W., Xu, S., Eckman, C., Younkin, S., Price, D., Iadecola, C., Clark, H. B., and Carlson, G., 1995, Age-related CNS disorder and early death in transgenic FVB/N mice overexpressing Alzheimer amyloid precursor proteins, *Neuron* **15:**1203–1218.

Hsiao, K., Chapman, P., Nilsen, S., Eckman, C., Harigaya, Y., Younkin, S., Yang, F., and Cole, G., 1996, Correlative memory deficits, Aβ elevation, and amyloid plaques in transgenic mice, *Science* **274:**99–102.

Huang, D. Y., Goedert, M., Jakes, Weisgraber, K., Garner, C., Saunders, A., Pericak-Vance, M. A., Schmechel, D., Roses, A., and Strittmatter, W., 1994, Isoform-specific interactions of apolipoprotein E with the microtubule associated protein MAP2c: Implications for Alzheimer's disease, *Neurosci. Lett.;* **182**:55–58.

Huang, D. Y., Weisgraber, D. H., Strittmatter, W. J., and Matthew, W. D., 1995, Interaction of apolipoprotein E with laminin increases neuronal adhesion and alters neurite morphology, *Exp. Neurol.* **136**:251–157.

Hyman, B. T., Strickland, D., and Rebeck, G. W., 1994, Alpha 2-macroglobulin receptor/low density lipoprotein receptor-related protein: relationship to apolipoprotein E and role in Alzheimer disease senile plaques, *Ann. NY Acad. Sci.* **737**:88–95.

Ignatius, M. J., Gebicke-Harter, P. J., Skene, J. H. P., Schilling, J. W., Weisgraber, K. H., Mahley, R. W., and Shooter, E. M., 1986, Expression of apolipoprotein E during nerve degeneration and regeneration, *Proc. Natl. Acad. Sci. USA* **83**:1125–1129.

Ishii, K., Tamaoka, A., Mizusawa, H., Shoji, S., Ohtake, T., Fraser, P. E., Takahashi, H., Tsuji, S., Gearing, M., Mizutani, T., Yamada, S., Kato, M., St. George-Hyslop, P. H., Mirra, S. S., and Mori, H., 1997, Abeta1-40 but not Abeta1-42 levels in cortex correlate with apolipoprotein E epsilon4 allele dosage in sporadic Alzheimer's disease, Ann. Neurol. **39**:395–399.

Kida, E., Pluta, R., Lossinsky, A. S., Golabek, A. A., Choi-Miura, N. H., Wisniewski, H. M., Mossakowski, M. J., 1995, Complete cerebral ischemia with short-term survival induced by cardiac arrest. II. Extracellular and intracellular accumulation of apolipoproteins E and J in the brain, *Brain Res.* **674**:341–346.

Kislauskis, E. H., Li, Z., Singer, R. H., and Taneja, K. L., 1993, Isoform-specific 3′-untranslated sequences sort alpha-cardiac and beta-cytoplasmic actin messenger RNAs to different cytoplasmic compartments, *J. Cell Biol.* **123**:165–172.

LaFerla, F. M., Troncoso, J. C., Strickland, D. K., Kawas, C. H., and Jay, G., 1997, Neuronal cell death in Alzheimer's disease correlates with apoE uptake and intracellular Abeta stabilization, *J. Clin. Invest.* **100**:310–320.

Laskowitz, D. T., Goel, S., Bennett, E. R., and Matthew, W. D., 1997a, Apolipoprotein E suppresses glial cell secretion of TNFα, *J. Neuroimmunol.* **76**:70–74.

Laskowitz, D. T., Sheng, H., Bart, R. D., Joyner, K. A., Roses, A. D., and Warner, D. A., 1997b. Apolipoprotein E-deficient mice have increased susceptibility to focal cerebral ischemia, *J. Cereb. Blood Flow Metab.* **17**:753–758.

Lin, C-T., Xu, Y., Wu, J-Y., and Chan, L., 1986, Immunoreactive apolipoprotein E is a widely distributed cellular protein. Immunohistochemical localization of apolipoprotein E in baboon tissues, *J. Clin. Invest.* **78**:947–958.

Linton, M. F., Gish, R., Hubl, S. T., Butler, E., Esquivel, C., Bry, W. I., Boyles, J. K., Wardell, M. R., and Young, S. G., 1991, Phenotypes of apolipoprotein B and apolipoprotein E after liver transplantation, *J. Clin. Invest.* **88**:270–281.

Linton, M. F., Atkinson, J. B., and Fazio, S., 1995, Prevention of atherosclerosis in apolipoprotein E-deficient mice by bone marrow transplantation, *Science* **267**:1034–1037.

Loo, D. T., Copani, A., Pike, C. J., Whittemore, E. R., Walencewicz, A. J., and Cotman, C. W., 1993, Apoptosis is induced by beta-amyloid in cultured CNS neurons, *Proc. Natl. Acad. Sci. USA* **90**:7951–7955.

Lyon, M. F., King, T. R., Gondo, Y., Gardner, J. M., Nakatsu, Y., Eicher, E. M., and Brilliant, M. H., 1992, Genetic and molecular analysis of recessive alleles at the pink-eyed dilution (p) locus of the mouse, *Proc. Natl. Acad. Sci. (U.S.A.)* **89**:6968–6972.

Mahley, R. W., 1988, Apolipoprotein E: cholesterol transport protein with expanding role in cell biology, *Science* **240**:622–630.

Mahley, R. W., Nathan, B. P., Bellosta, S., and Pitas, R. E., 1996, Apolipoprotein E: structure, function and possible roles in modulating neurite extension and cytoskeletal activity. p49–58. In *Apolipoprotein E and Alzheimer's Disease,* eds. Roses, A., Weisgraber, K., and Christen Y. Fondation IPSEN, Springer-Verlag.

Mann, S., Varner, J., Xu, P.-T., Saunders, A. M., Roses, A. D., Gilbert, J. R., Schmechel, D. E., and

Einstein, G., 1997, Comparison of morphology of hippocampal dentate granule cells in APOE KO and human E4 transgenic mice, *Soc. Neurosci. Abstr.* **23**:828.

Masliah, E., Mallory, M., Ge, N., Alford, M., Veinbergs, I., and Roses, A. D., 1995, Neurodegeneration in the central nervous system of apoE-deficient mice, *Exp. Neurol.* **136**:107–122.

Masliah, E., Samuel, W., Veinbergs, I., Mallory, M., Mante, M., and Saitoh, T., 1997, Neurodegeneration and cognitive impairment in apoE-deficient mice is ameliorated by infusion of recombinant apoE, *Brain Res.*, **751**:307–314.

Mayeux, R., Ottman, R., Maestra, G., Ngai, C., Tang, M. X., Ginsberg, H., Chun, M., Tycko, B., and Shelanski, M., 1995, Synergistic effects of traumatic head injury and apolipoprotein-epsilon 4 in patients with Alzheimer's disease, *Neurol.* **45**:555–557.

Mercken, L., and Brion, J. P., 1995, Phosphorylation of tau protein is not affected in mice lacking apolipoprotein E, *Neuroreport* **6**:2381–2384.

Metzger, R. E., LaDu, M. J., Dan, J. B., Getz, G. S., Frail, D. E., and Falduto, M. T., 1996, Neurons of the human frontal cortex display apolipoprotein E immunoreactivity: implications for Alzheimer's disease, *J. Neuropath. Exp. Neurol.* **55**:372–380.

Miyata, M., and Smith, J. D., 1996, Apolipoprotein E allele-specific antioxidant activity and effects on cytotoxicity by oxidative insults and β-amyloid peptides, *Nature Gen.* **14**:55–61.

Moestrup, S. K., Gliemann, J., and Pallesen, G., 1992, Distribution of the α_2-macroglobulin receptor/low density lipoprotein receptor-related protein in human tissues, *Cell Tissue Res.* **269**:375–382.

Mohlar, E. C., Einstein, G., and Williams, C. L., 1997, Spatial memory is impaired in female apolipoprotein E-deficient mice, *Soc. Neurosci. Abstr.* **23**:564.

Montine, T. J., Huang, D. Y., Valentine, W. M., Amarnath, V., Saunders, A., Weisgraber, K. H., Graham, D. G., and Strittmatter, W. J., 1996, Crosslinking of apolipoprotein E by products of lipid peroxidation, J. Neuropath. Exp. Neurol. **55**:202–210.

Muller, H. W., Gebick-Harter, P. J., Hangen, D. H., and Shooter, E. M., 1985, A specific 37,000-dalton protein that accumulates in regenerating but not in non-regenerating mammalian nerves, *Science* **228**:499–501.

Mufson, E. J., Benzing, W. C., Cole, G. M., Wang, H., Emerich, D. F., Sladek, J. R., Morrison, J. H., and Kordower, J. H., 1994, Apolipoprotein E-immunoreactivity in aged rhesus monkey cortex: colocalization with amyloid plaques, *Neurobiol. Aging* **15**:621–627, 1994.

Namba, Y. M., Tomonaga, H., Kawasaki, et al., 1991, Apolipoprotein E immunoreactivity in cerebral amyloid deposits and neurofibrillary tangles in Alzheimer's disease and kuru plaque amyloid in Creutzfeldt-Jacob disease, *Brain Res.* **541**:163–166.

Nicoll, J. A., Roberts, G. W., and Graham, D. I., 1995, Apolipoprotein E epsilon 4 allele is associated with deposition of amyloid beta-protein following head injury, *Nature Med.* **1**:135–137.

Oitzl, M. S., Mulder, M., Lucassen, P. J., Havekes, L. M., Grootendorst, J., and de Kloet, E. R., 1997, Severe learning deficits in apolipoprotein E-knockout mice in a water maze task, *Brain Res.* **752**:189–196.

Ong, W. Y., He, Y., Suresh, S., and Patel, S. C., 1997, Differential expression of apolipoprotein D and apolipoprotein E in the kainic acid-lesioned rat hippocampus, *Neurosci.* **79**:359–367.

Palinski, W., Ord, V. A., Plump, A. S., Breslow, J. L., Steinberg, D., and Witztum, J. L., 1994, ApoE-deficient mice are a model of lipoprotein oxidation in atherogenesis: demonstration of oxidation-specific epitopes in lesions and high titers of antibodies to malondialdehyde lysine in serum, *Arterioscler. Thromb.* **14**:605–616.

Pericak-Vance, M. A., Bebout, J. L., Gaskell, P. C., Jr., Yamaoka, L. H., Hung, W.-Y., Albert, M. J., Walker, A. P., Bartlett, R. J., Haynes, C. A., Welsh, K. A., Earl, N. L., Heyman, A., Clark, C. M., and Roses, A. D., 1991, Linkage studies in familial Alzheimer's Disease, *Amer. J. Hum. Genet.* **48**:1034–1050.

Pericak-Vance, M. A., Bass, M. P., Yamaoka, L. H., Gaskell, P. C., Scott, W. K., Terwedow, H. A., Menold, M. M., Conneally, P. M., Small, G. W., Vance, J. M., Saunders, A. M., Roses, A. D., and Haines, J. L., 1997, Complete genomic screen in late-onset familial Alzheimer's Disease: evidence for a new locus on chromosome 12, *J.A.M.A.* **278:**1237–1241.

Piedrahita, J. A., Zhang, S. H., Hagaman, J. R., Oliver, P. M., and Maeda, N., 1992, Generation of mice carrying a mutant apolipoprotein E gene inactivated by gene targeting in embryonic stem cells, *Proc. Natl. Acad. Sci. USA* **89:**4471–4475.

Pitas, R. E., Boyles, J. K., Lee, S. H., Foss, D., and Mahley, R. M., 1987, Astrocytes synthesize apolipoprotein E and metabolize apolipoprotein E-containing lipoproteins, *Biochem. Biophys. Acta* **917:**148–161.

Plump, A. S., and Breslow, J. L., 1995, Apolipoprotein E and the apolipoprotein E-deficient mouse, *Ann. Rev. Nutr.* **15:**495–518.

Poduri, A., Gearing, M., Rebeck, G. W., Mirra, S. S., Tigges, J., and Hyman, B. T., 1994, Apolipoprotein E4 and beta amyloid in senile plaques and cerebral blood vessels of aged rhesus monkeys, *Amer. J. Pathol.* **144:**1183–1187.

Poirier, J., 1994. Apolipoprotein E in animal models of CNS injury and in Alzheimer's disease, *TINS* **17:**525–530.

Poirier, J., Hess, M., May, P. C., and Finch, C. E., 1991a, Cloning of hippocampal poly (A) RNA sequences that increase after entorhinal cortex lesion in adult rat, *Mol. Brain Res.* **9:**191–195.

Poirier, J., Hess, M., May, P. C., and Finch, C. E., 1991b, Astrocytic apolipoprotein E mRNA and GFAP mRNA in hippocampus after entorhinal cortex lesioning, *Mol. Brain Res.* **11:**97–106.

Poirier, J., Baccichet, A., Dea, D., and Gauthier, S., 1993a, Cholesterol synthesis and lipoprotein reuptake during synaptic remodelling in hippocampus in adult rats, *Neurosci.* **55:**81–90.

Poirier, J., Davignon, J., Bouthillier, D., Kogan, S., Bertrand, P., and Gauthier, S., 1993b, Apolipoprotein E polymorphism and Alzheimer's disease, *Lancet* **342:**697–699.

Poirier, J., Delisle, M.-C., Quirion, R., Aubert, I., Rocheford, J., Rousse, I., Gracon, S., Farlow, M., and Gauthier, S., 1996, Apolipoprotein E4, cholinergic integrity, synaptic plasticity, and Alzheimer's Disease. p20–28. In *Apolipoprotein E and Alzheimer's Disease,* eds. Roses, A., Weisgraber, K., And Christen Y. Fondation IPSEN, Springer-Verlag.

Popko, B., Goodrum, J. F., Bouldin, T. W., Zhang, S. H., and Maeda, N., 1993, Nerve regeneration occurs in the absence of apolipoprotein E in mice, *J. Neurochem.* **60:**1155–1158.

Rebeck, G. W., Reiter, J. S., Strickland, D. K., and Hyman, B. T., 1993, Apolipoprotein E in sporadic Alzheimer's disease: allelic variation and receptor interactions, *Neuron* **11:**575–580.

Rebeck, G. W., Harr, S. D., Strickland, D. K., and Human, B. T., 1995, Multiple, diverse senile plaque associated proteins are ligands of an apolipoprotein E receptor, the alpha 2-macroglobulin receptor/low-density-lipoprotein receptor related protein, *Ann. Neurol.* **37:**211–217.

Reed, T., Carmelli, D., Swan, G. E., Breitner, J. C., Welsh, K. A., Jarvik, G. P., Deeb, S., and Auwerx, J., 1994, Lower cognitive performance in normal older adult male twins carrying the apolipoprotein E epsilon 4 allele, *Arch. Neurol.* **51:**1189–1192.

Roses, A. D., 1996, Apolipoprotein E alleles as risk factors in Alzheimer's Disease, *Annu. Rev. Med.,* **47:**387–400.

Roses, A., and Saunders, A., 1995, Head injury, amyloid β and Alzheimer's disease, *Nat. Med.* **1:**603–604.

Sanan, D. A., Weisgraber, K. H., Russell, S. J., Mahley, R. W., Huang, D., Saunders, A., Schmechel, D., Wisniewski, T., Frangione, B., Roses, A. D., et al., 1994, Apolipoprotein E associates with beta-amyloid peptide of Alzheimer's disease to form novel monofibrils: apolipoprotein E4 assembles more efficiently than apolipoprotein E3, *J. Clin. Invest.* **94:**860–869.

Saunders, A. M., Strittmatter, W. J., Schmechel, D., St. George-Hyslop, P. H., Pericak-Vance, M. A., Joo, S. H., Rosi, B. L., Gusella, J. F., Crapper-MacLachlan, D. R., Albert, M. J., Hulette, C., Crain, B., Goldgaber, D., and Roses, A. D., 1993a, Association of apolipoprotein E allele ε4 with late-onset familial and sporadic Alzheimer's disease, *Neurology* **43:**1467–1472.

Saunders, A. M., Schmader, K., Breitner, J. C., Benson, M. D., Brown, W. T., Goldfarb, L., Goldgaber, D., Manwaring, M. G., Szymanski, M. H., McCown, N., Dole, K. C., Schmechel, D. E., Strittmatter, W. J., Pericak-Vance, M. A., and Roses, A. D., 1993b, Apolipoprotein E ϵ4 allele distributions in late-onset Alzheimer's disease and in other amyloid-forming diseases, *Lancet* **342:**710–711.

Saunders, A. M., Hulette, C., Welsh-Boehmer, K. A., Schmechel, D. E., Crain, B., Burke, J. R., Alberts, M. J., Strittmatter, W. J., Breitner, J. C., and Rosenberg, C., 1996, Specificity, sensitivity, and predictive value of apolipoprotein-E genotyping for sporadic Alzheimer's Disease, *Lancet* **348:**90–93.

Schmechel, D. E., Saunders, A. M., Strittmatter, W. J., Crain, B. J., Hulette, C. M., Joo, S. H., Pericak-Vance, M. A., Goldgaber, D., and Roses, A. D., 1993, Increased amyloid β-peptide deposition in cerebral cortex as a consequence of apolipoprotein E genotype in late-onset Alzheimer's disease, *Proc. Natl. Acad. Sci. USA* **90:**9649–9653.

Schmechel, D. E., Burkhart, D. S., Ange, R. L., and Izard, M. K., 1996a, Cholinergic axonal dystrophy and mitochondrial pathology in prosimian primataes, *Exp. Neurol.,* **142:**111–127.

Schmechel, D., Tiller, O., Tong, P., McSwain, M., Han, S-H, Ange, R., Burkhart, D., and Izard, M., 1996b. Pattern of Apolipoprotein E immunoreactivity during brain aging. p27–48. In *Apolipoprotein E and Alzheimer's Disease,* eds. Roses, A., Weisgraber, K., And Christen Y. Fondation IPSEN, Springer-Verlag.

Selkoe, D. J., 1994, Cell biology of the amyloid beta-protein precursor and the mechanism of Alzheimer's disease, *Ann. Rev. Cell Biol.* **10:**373–403.

Sheng, H., Laskowitz, D. L., Bennett, E., Schmechel, D. E., Bart, R. D., Saunders, A. M., Pearlstein, R. D., Roses, A. D., and Warner, D. S., 1998, Apolipoprotein E isoform-specific differences in outcome from focal ischemia in transgenic mice, *J. Cereb. Blood Flow Metab.,* in press.

Simonet, W. S., Bucay, N., Lauer, S. J., and Taylor, J. M., 1993, A far-downstream hepatocyte-specific control region directs expression of the linked human apolipoprotein E and C1 genes in transgenic mice, *J. Biol. Chem.* **268:**8221–8229.

Skene, J. H. P., and Shooter, E. M., 1983, Denervated sheath cells secrete a new protein after nerve injury, *Proc. Natl. Acad. Sci. USA* **80:**4169–4173.

Small, G. W., Mazziotta, J. C., Collins, M. T., Baxter, L. R., Phelps, M. E., Mandelkern, M. A., Kaplan, A., La Rue, A., Adamson, C. F., Chang, L., Guze, B. H., Corder, E. H., Saunders, A. M., Haines, J. L., Pericak-Vance, M. A., and Roses, A. D., 1995, Apolipoprotein E type 4 allele and cerebral glucose metabolism in relatives at risk for familial Alzheimer disease, *J.A.M.A.* **273:**942–947.

Snipes, G. J., McGuire, C. B., Norden, J. J., and Freeman, J. A., 1986, Nerve injury stimulates the secretion of apolipoprotein E by nonneuronal cells, *Proc. Natl. Acad. Sci. USA* **83:**1130–1134.

Sparks, D. L., Liu, H., Gross, D. R., and Scheff, S. W., 1995, Increased density of cortical apolipoprotein E immunoreactive neurons in rabbit brain after dietary administration of cholesterol, *Neurosci. Lett.* **187:**142–144, 1995.

Stoll, G., and Mueller, H. W., 1986, Macrophages in the peripheral nervous system and astroglia in the central nervous system of rat commonly express apolipoprotein E during development but differ in their response to injury. *Neurosci. Letters* **72:**233–238.

Stoll, G., Mueller, H. W., Trapp, B. D., and Griffin, J. W., 1989, Oligodendrocytes but not astrocytes express apolipoprotein E after injury of rat optic nerve. *Glia* **2:**170–176.

Strittmatter, W. J., Saunders, A. M., Schmechel, D., Pericak-Vance, M., Enghild, J., Salvesen, G. S., and Roses, A. D., 1993a, Apolipoprotein E: High affinity binding to β-amyloid and increased frequency of type 4 allele in familial Alzheimer's disease, *Proc. Natl. Acad. Sci. USA* **90:**1977–1981.

Strittmatter, W. J., Weisgraber, W. J., Huang, D., Dong, L., Salvesen, G. S., Pericak-Vance, M., Schmechel, D., Saunders, A. M., Goldgaber, D., and Roses, A. D., 1993b, Binding of human apolipoprotein E to βA4 peptide: Isoform-specific effects and implications for late-onset Alzheimer's disease, *Proc. Natl. Acad. Sci. USA* **90:**8098–8102.

Strittmatter, W. J., Weisgraber, K. H., Goedert, M., Saunders, A. M., Huang, D., Corder, E. H., Dong, L., Jakes, R., Albert, M. J., Gilbert, J. R., Han, S., Hulette, C., Einstein, G., Schmechel, D. E., Pericak-Vance, M. A., and Roses, A. D., 1994a, Hypothesis: Microtubule instability and paired helical filament formation in the Alzheimer's disease brain are related to apolipoprotein E genotype, *Exper. Neurol.* **125:**163–171.

Strittmatter, W. J., Saunders, A. M., Goedert, M., Weisgraber, K. H., Dong, L., Jakes, R., Huang, D. Y., Pericak-Vance, M., Schmechel, D., and Roses, A. D., 1994b, Isoform-specific interactions of apolipoprotein E with microtubule-associated protein Tau: implications for Alzheimer's disease, *Proc. Natl. Acad. Sci. USA* **91:**11183–11186.

Sullivan, P. M., Mezdour, H., Aratani, Y., Knouff, C., Najib, J., Reddick, R. L., Quarfordt, S. H., and Maeda, N., 1997, Targeted replacement of the mouse apolipoprotein E gene with the common human apoE3 allele enhances diet-induced hypercholesterolemia and atherosclerosis, *J. Biol. Chem.* **272:**17972–17980.

Tardiff, B., Newman, M., Saunders, A., Roses, A. et al., 1994, Apolipoprotein E allele frequency in patients with cognitive defects following cardiopulmonary bypass, *Circulation* **90:**I–201.

Tooyama, I., Kawamata, T., Akiyama, H., Moestrupp, S. K., Gliemann, J., and McGeer, P. L., 1993, Immunohistochemical study of α_2-macroglobulin receptor in Alzheimer and control postmortem human brain, *Mol. Chem. Neuropath.* **18:**153–159.

Uterman, G., Vogelberg, K. H., Steinmetz, A., Schoenborn, W., Pruin, N., Jaeschke, M., Hees, M., and Canzler, H., 1979, Polymorphism of apolipoprotein E, *Clin. Genet.* **15:**37–62.

Weisgraber, K. H., 1994, Apolipoprotein E: structure-function relationships, *Adv. Protein Chem.* **45:**249–302.

Weisgraber, K. H., Innerarity, T. L., and Mahley, R. W., 1982, Abnormal lipoprotein receptor-binding activity of the human E apoprotein due to cysteine-arginine interchange at a single site, *J. Biol. Chem.* **257:**2518–2521.

Wisniewski, T., and Frangione, B., 1991, Apolipoprotein E: a pathological chaperone protein in patients with cerebral and systemic amyloid, *Neurosci. Lett.* **135:**235–238.

Wisniewski, T., Castano, E. M., Golabek, A., Vogel, T., and Frangione, B., 1994, Acceleration of Alzheimer's fibril formation by apolipoprotein E *in vitro, Amer. J. Pathol.* **145:**1030–1035.

Xu, P.-T., Schmechel, D. E., Rothrock-Christian, T., Burkhart, D. S., Qui, H. L., Popko, B., Sullivan, P., Maeda, N., Saunders, A. M., Roses, A. D., and Gilbert, J. R., 1996, Human apolipoprotein E2, E3, and E4 isoform specific transgenic mice: human-like pattern of neuronal and glial immunoreactivity in central nervous system not observed in wild type mice, *Neurobiol. Dis.,* **3:**229–235.

Xu, P.-T., Gilbert, J. R., Qui, H.-L., Rothrock-Christian, T., Settles, D. L., Roses, A. D., and Schmechel, D. E., 1998, Regionally specific neuronal expression of human *APOE* gene in transgenic mice, *Neurosci. Letters,* **246:**65–68.

Yamatsuji, T., Matsui, T., Okamoto, T., Komatsuzaki, K., Takeda, S., Fukumoto, H., Iwatsubo, T., Suzuki, N., Asami-Odaka, A., Ireland, S., Kinane, T. B., Giambarella, U., and Nishimoto, I., 1996, G protein-mediated neuronal DNA fragmentation by familial Alzheimer's disease-associated mutants of APP, *Science* **272:**1349–1352.

Zhang, S. H., Reddick, R. L., Piedrahita, J. A., and Maeda, N., 1992, Spontaneous hypercholesterolemia and arterial lesions in mice lacking apolipoprotein E, *Science* **258:**468–471.

CHAPTER 11

LYSOSOMAL DISORDERS

KUNIHIKO SUZUKI, MARIE T. VANIER, and KINUKO SUZUKI

1. INTRODUCTION

Based on biochemical evidence, de Duve first defined the concept of the lysosome as the subcellular organelle responsible for physiological turnover of cellular constituents. It contains catabolic enzymes that are generally glycoproteins themselves and have characteristically very low pH optima for their function and other associated metabolic machineries. Defective activity of any of these enzymes results in a block in the digestive process of the cellular materials essential for normal function. The term *inborn lysosomal disease* was then proposed by Hers to categorize a group of genetic diseases that satisfied two major criteria: (1) an acidic hydrolase normally localized in the lysosome is genetically defective and (2) consequently, the substrate of the defective enzyme accumulates abnormally within pathologically altered secondary lysosomes (Hers, 1966).

KUNIHIKO SUZUKI • Neuroscience Center and Departments of Neurology and Psychiatry, University of North Carolina School of Medicine, Chapel Hill, North Carolina 27599. MARIE T. VANIER • INSERM-CNRS U189, Lyon-Sud School of Medicine and Fondation Gillet-Mérieux, Lyon-Sud Hospital, F-69921 Oullins Cedex, France. KINUKO SUZUKI • Neuroscience Center and Department of Pathology & Laboratory Medicine, University of North Carolina School of Medicine, Chapel Hill, North Carolina 27599.

Mouse Models in the Study of Genetic Neurological Disorders, Volume 9 of *Advances in Neurochemistry,* edited by Popko, Kluwer Academic/Plenum Publishers, New York, 1999.

These disorders are often referred to as "storage diseases" because the abnormal "storage" of the undigested substrates is often the most conspicuous clinical and pathological manifestation. The original concept of Hers has inevitably been expanded and modified since its inception to include not only lysosomal enzymes but also other noncatalytic lysosomal proteins and abnormalities of more general lysosomal functions. Traditionally, they are subcategorized according to the nature of the affected substrates, such as sphingolipidosis, mucopolysaccharidosis, glycoprotein storage disease, etc., although any given disease may encompass more than one category owing to the overlapping substrate specificities of the affected enzyme. In recent years, most of the genes encoding lysosomal enzymes have been cloned and disease-causing mutations characterized (Suzuki and Vanier, 1998).

The general rarity of genetic lysosomal disorders and the obvious ethical considerations limit experimentation, even of therapeutic nature, on human patients. Genetically equivalent animal models are useful in overcoming the limitations. Although genetic lysosomal disorders occur naturally among many mammalian species, relatively few spontaneous mouse models are known. In contrast, the combination of the homologous recombination and embryonic stem cell technology now allows artificial inactivation and other manipulations of any genes in the mouse genome once it is cloned. With this technology, almost all of the known human sphingolipidoses and some other lysosomal disorders have been duplicated in the mouse. It is important to note that this technology also allows generation of genetic diseases that are either unknown or statistically highly unlikely to occur in humans.

2. NATURALLY OCCURRING MOUSE MODELS

Three naturally occurring mouse models of human genetic lysosomal diseases are known. Two of them, the *twitcher* (galactosylceramidase deficiency) and the β-glucuronidase deficiency, satisfy the classic concept of the lysosomal disease, whereas the Niemann–Pick type C mouse fits a more expanded categorization.

2.1. *Twitcher* Mouse

The *twitcher* mouse is a genetically equivalent model of one of the classic human genetic leukodystrophies, globoid cell leukodystrophy (Krabbe disease) caused by deficiency of galactosylceramidase (Suzuki *et al.*, 1995a). The typical globoid cell leukodystrophy (Krabbe disease) is a rapidly progressive, invariably fatal disease of infants. It is transmitted as an autosomal recessive trait. The human galactosylceramidase gene has been mapped to chromosome 14 (Cannizzaro *et al.*, 1994). The disease usually begins between 3 and 6 months of age with ambiguous symptoms, such as irritability or hypersensitivity to external stimuli, but soon progresses to severe mental and motor deterioration. Patients rarely survive the second year. Clinical manifestations are limited to the nervous system, with prominent long-tract signs. Hypertonicity with hyperactive reflexes is present in the early stages, but patients later become flaccid and hypotonic. Blindness and deafness are common. Peripheral neuropathy is almost always detectable. The clinical picture of the classic infantile form is relatively uniform but atypical or late-onset forms of the disease with highly variable clinical manifestations are being uncovered with increasing frequency. The presence of numerous multinucleated globoid cells, almost total loss of myelin and oligodendroglia, and extensive astrocytic gliosis in the white matter are the morphological basis for diagnosis. The globoid cells are hematogenous macrophages that contain undigested galactosylceramide. Segmental demyelination, axonal degeneration, fibrosis, and histiocytic infiltration are common in the peripheral nervous system (PNS). The clinical and pathological phenotypes of the late-onset forms are generally milder and more slowly progressive. One of the unique biochemical features of Krabbe disease is that, despite the genetic block in its immediate catabolic pathway, the amount of galactosylceramide in the brain is much lower than in controls (Svennerholm *et al.*, 1980). Instead, an abnormal metabolite, galactosylsphingosine (psychosine), is dramatically increased over the essentially undetectable levels in normal brain (Svennerholm *et al.*, 1980). The human galactosylceramidase gene was cloned relatively recently (Chen *et al.*, 1993) but a large number of disease-causing mutations have already been described including the combined point mutation and a large deletion that underlie the classic infantile phenotype of the Nordic origin (Rafi *et al.*, 1995).

The *twitcher* mouse was identified as an enzymatically authentic mouse model of human Krabbe disease in 1980 (Kobayashi *et al.*, 1980; Suzuki and Suzuki, 1995). Affected mice become generally less active than their littermates by 15–20 days. Early clinical signs include failure to gain weight and generalized tremulousness. At the terminal stage around 40 days, their body weight is about one third that of normal mice. Progressive muscle weakness and eventual paralysis develop, particularly conspicuous in the hindlimbs and the neck muscles. The life span of inbred *twi/twi* with the C57BL background rarely extends beyond 40 days. Motor functions generally deteriorate rapidly after 20 days. Pathological changes are almost entirely limited to the nervous system. In *twitcher* mice younger than 15 days without noticeable neurological symptoms, no neuropathological changes are apparent on the light microscopic level, although inclusions in oligodendrocytes can be detected on the ultrastructural level as early as 5 days. Demyelination progresses in an orderly fashion such that myelin degeneration begins at 10–20 days after commencement of myelination in any of the given nerve fiber tracts. The characteristic pathology in the central nervous system (CNS) is demyelination, accompanied by degeneration of oligodendrocytes and infiltration of macrophages ("globoid cells"), containing periodic acid-Schiff (PAS)-positive materials. These pathological features are essentially identical with those of human Krabbe disease. Crystalloid or slender tubular inclusions, typical in human galactosylceramidase deficiency (GLD), are present largely in the macrophages and the oligodendrocytes. Peripheral nerves are abnormally thick, translucent, and firm and can easily be distinguished from normal nerves by visual inspection. Segmental demyelination with some evidence of remyelination is conspicuous. The Schwann cells of myelinated fibers frequently contain the characteristic GLD inclusions. Similar to the human disease, there is no abnormal accumulation of galactosylceramide in the brain. However, unlike in the human disease, there is a very large accumulation of galactosylceramide in the kidney (Ida *et al.,* 1982; Igisu *et al.,* 1983). More importantly, there is a progressive accumulation of psychosine in the brain (Igisu and Suzuki, 1984). The finding on psychosine had been postulated by what is known as "the psychosine hypothesis" for the pathogenesis of human and animal globoid cell leukodystrophy (Miyatake and Suzuki, 1972; Suzuki, 1997). The mouse gene encoding galactosylceramidase was cloned and the mutation underlying the *twitcher* mutant has been identified as a nonsense mutation in the middle of the enzyme protein (Sakai *et al.,* 1994). In most respects the *twitcher* mouse is an excellent model of human Krabbe disease and it has been extensively utilized for studies of the pathogenesis and treatment of this disorder (see later).

2.2. β-Glucuronidase Deficiency

Genetic β-glucuronidase deficiency in humans causes autosomal recessive mucopolysaccharidosis VII (MPS VII, Sly disease) (Neufeld and Muenzer, 1995). Patients with typical MPS VII generally show mild to severe mental retardation, coarse facies, organomegaly, skeletal abnormalities, short stature, inguinal or umbilical hernia, and corneal clouding. The clinical manifestations and course vary from severe infantile form with hydrops foetalis and dysostosis multiplex to older and milder forms. Some patients with late-onset forms may have essentially normal intelligence.

The β-glucuronidase deficiency (gus^{mps} $-/-$) in the mouse is, in most respects, an excellent model of the human disease (Birkenmeier *et al.,* 1989; Vogler *et al.,* 1990). Affected mice appear normal at birth but the characteristic dysmorphic features become evident by 20 days. They are smaller and have shorter and stubby limbs and a shorter and thicker tail than normal littermates. They have shorter nasal bones and other skeletal abnormalities with shortened and broadened long bones and ribs. They are more sluggish than their littermates and after several weeks, they develop hobbled gait suggestive of joint dysfunction. Visceromegaly is not evident. Numerous granulocytic inclusions are present in the nucleated cells in the peripheral blood and in many solid tissues. Distended lysosomes containing fibrillar and granular materials are abundant in all reticuloendothelial systems as well as in many other tissues including brain. Affected mice have a generally reduced life span. Although females are fertile, affected males are sterile. The mutation responsible for the mouse disease is a single-base deletion in exon 10 in the β-glucuronidase gene that results in a termination codon downstream soon after the deletion. This frameshift appears to affect the stability of the mRNA, resulting in the steady-state level of mRNA at less than 0.5% of the normal level (Sands and Birkenmeier, 1993). The fact that the [*gus*] complex is one of the best characterized genetic loci in the mouse makes this model particularly interesting for studies of the gene regulation, expression, pathogenesis, and treatment trials. This mutant has been the subject for many studies aimed at treating genetic lysosomal diseases (see later).

2.3. Niemann–Pick Type C Mouse

It is important to note that despite the eponym, Niemann–Pick type C (NPC) disease has nothing in common with Niemann–Pick type A and B disease (due to genetic deficiency of lysosomal acid sphingomyelinase; see later), other than superficial resemblance in the phenotype. Niemann–Pick type C is now recognized as a disorder of intracellular trafficking and homeostasis of endocytosed low-density lipoprotein (LDL)-cholesterol, with lysosomal storage of unesterified cholesterol (Pentchev *et al.,* 1995; Vanier and Suzuki, 1996). Age at presentation in human patients may vary from the perinatal period to adult age. A transient neonatal cholestatic icterus with hepatosplenomegaly occurs in about 50% of the cases. Typical signs and symptoms in patients with the most common late-infantile and juvenile forms are splenomegaly, cerebellar ataxia, cataplexia, and vertical supranuclear gaze palsy. Other neurological manifestations include seizures, progressive dementia, dystonia, dysphagia, and occasional psychiatric disturbances. The majority of cases exhibit characteristic vertical supranuclear gaze palsy. Foam cells or "sea-blue histiocytes" are commonly present in many tissues including liver and spleen. Neuronal storage, meganeurites, and axonal spheroids are found throughout the CNS. Abundant neurofibrillary tangles similar to those seen in Alzheimer disease and senile dementia are present in most cases with a slowly progressive course (Suzuki *et al.,* 1995b). Major (95%) and minor (5%) genetic complementation groups are recognized (Vanier *et al.,* 1996), which result in similar clinical, pathological, and biochemical phenotypes. The gene responsible for the major complementation group (*NPC1*) has recently been cloned and several disease-causing mutations identified (Carstea *et al.,* 1997; Vanier and Suzuki, 1998). The *NPC1* gene contains a sterol-sensing domain homologous to that in some of the known cholesterol homeostatic genes, and also shows homology to the *patched* gene. Its precise function is not yet elucidated.

Two mouse models (BALB/c npc^{nih} and C57BL/KsJ *spm*) are known (Miyawaki *et al.*, 1982; Morris *et al.*, 1982; Ohno *et al.*, 1992; Shio *et al.*, 1982). The C57BL/KsJ *spm* mutant initially described as a "sphingomyelinosis" mouse has since been correctly and unambiguously reclassified as an NPC model. Both strains of mice show similar clinicopathological features. The neurological disease starts at approx. 35–50 days, with tremor of the back legs and ataxic gait. Mice die between 70 and 120 days. Foamy storage cells are conspicuous in the visceral organs, and the brain shows marked neuronal storage and numerous axonal spheroids. Rare argyrophilic fibers are detected in storage neurons in the cerebral cortex and brain stem in the mouse model but no paired helical filaments have been reported for any of the animal models. Pictures of neuroaxonal dystrophy become more numerous with age, at variance with human NPC, in which neuroaxonal dystrophy is more numerous in young patients. Glutamic acid decarboxylase (GAD) immunoreactivity was demonstrated on these dystrophic axonal spheroids. The lipid storage pattern in various tissues is essentially similar to that in human patients. Liver and spleen show a progressive accumulation of unesterified cholesterol, two phospholipids (sphingomyelin and bis(monoacylglycero)phosphate), glycolipids (primarily glucosyl- and lactosylceramide), and free sphingoid bases. However, in the brain, levels of free cholesterol and phospholipids are not increased, but there are striking increases of glucosyl- and lactosylceramides, as well as of GM3 and GM2 gangliosides. Reflecting loss of myelin, all myelin lipids including galactosylceramide and sulfatide are decreased in the brain (Pentchev *et al.*, 1980; Weintraub *et al.*, 1985; Goldin *et al.*, 1992). Differences include a much larger increase of unesterified cholesterol in non-neural tissues in the mouse, as well as a significant decrease of sphingomyelinase and glucocerebrosidase activities in the same tissues. Unlike other naturally occurring mice models, the NPC mice were essential for elucidation of the genetic defect in the human counterpart of the disease. Demonstration of an impaired cholesterol homeostatic reaction as a response to LDL in the BALB/c npc^{nih} model (Pentchev *et al.*, 1984) was the seminal observation that led to similar and more refined investigations in human patients. Similarly, early mapping of the *spm* gene to chromosome 18 (Sakai *et al.*, 1991) and restoration of normal phenotype in mouse 3T3 cells by transfection of human chromosome 18 (Kurimasa *et al.*, 1993) greatly facilitated linkage of human *NPC1* to chromosome 18 (Carstea *et al.*, 1993), and thus positional cloning of the gene. Crossbreeding (Yamamoto *et al.*, 1994) and cell hybridization experiments (Akaboshi *et al.*, 1997) have shown that the *spm* and npc^{nih} mutations are allelic and belong to the same complementation group as human *NPC1*. Recently, the BALB/c npc^{nih} was also localized to murine chromosome 18 in a region syntenic to the human *NPC1* locus (Erickson *et al.*, 1997). An integrated human–mouse positional candidate approach pursued in parallel to the cloning of the human *NPC1* gene has allowed isolation of the corresponding murine gene. The human and the murine genes show 85% sequence identity. The *npc*^{nih} mutation has been characterized (Loftus *et al.,* 1997) but the *spm* mutation is not yet identified.

3. ARTIFICIALLY GENERATED MOUSE MODELS

3.1. Models of Known Human Lysosomal Disorders

All lysosomal disorders to be described in the following are inherited as autosomal recessive traits except Fabry disease, which is X-linked. A recent review article of induced mouse models of genetic lysosomal disorders with emphasis on their pathology is available (Suzuki *et al.,* 1998).

3.1.1. Sphingolipidosis

3.1.1.1. Gaucher Disease (Glucosylceramidase Deficiency) Gaucher disease in humans is caused by deficient activity of glucosylceramidase (glucocerebrosidase), resulting in massive accumulation of glucocerebroside in cells of the reticuloendothelial system. These storage cells, Gaucher cells, found throughout the visceral tissues, and hepatosplenomegaly are the most conspicuous clinical features in human patients. According to the traditional classification, type I is a non-neuronopathic form affecting mostly adults, type II is a severe infantile neuronopathic form, and type III is phenotypically intermediate. In addition, a rapidly progressive fulminating phenotype in neonates, a subtype of type II Gaucher disease, was recognized more recently.

In 1992, Tybulewicz and co-workers (Tybulewicz *et al.,* 1992) generated a murine model through gene targeting. The murine glucocerebrosidase gene was disrupted by the neomycin-resistance (*Neo*) gene that replaced the entire intron 9 and portions of exons 9 and 10, which encode part of the active site of the enzyme. The glucosylceramidase activity in these homozygous mutant mice was less than 4% of control. Increased amounts of glucocerebroside were demonstrated in the liver, lung, brain, and bone marrow. The homozygous mutant mice, however, were akinetic with irregular respiration and poor feeding, and died within 24 h of birth with rapidly progressive cyanosis. In more than 100 affected mice studied, no typical Gaucher cells were observed in any of the tissues that showed increased glucocerebroside. The stored lipid, which had the same physical appearance as the inclusions found in cells of Gaucher patients, was identified by electron microscopy in the macrophages in the liver, spleen, and bone marrow. However, the extent of glucocerebroside accumulation in these organs did not appear to be sufficient to explain the rapid death within 1 day of birth. In the

CNS, no Gaucher cells were observed although tubular Gaucher-type inclusions were identified by electron microscopy in microglia, neurons of the red nucleus and vestibular nucleus, and sensory and motor neurons (Willemsen *et al.,* 1995). These mice did not show hepatosplenomegaly. The most significant pathology was in the skin. The skin showed abnormally prominent rugation and hyperkeratosis. These features were nearly identical to those seen in human patients with the severe neonatal subtype of type II Gaucher disease ("collodion" babies, or "Gaucher infants having congenital ichthyosis" or "cellophane like" skin) (Sidransky *et al.,* 1992). Hydrolysis of glucocerebroside and consequent accumulation of ceramide normally occurs during the maturational process in the skin structure to establish a permeability barrier. The absence of glucosylceramidase may inhibit normal production of ceramide and thus affect functional skin integrity. During the final stages of normal epidermal differentiation, extrusion of lamellar body contents (at the stratum granulosum/stratum corneum interface) is followed sequentially by unfurling, elongation, and processing into a mature lamellar bilayer unit structure. However, in the mutant mice, ultrastructural abnormalities including the persistence of incompletely processed lamellar body-derived content, indicating perturbed maturation, was demonstrated throughout the striatum corneum interstices. Furthermore, these mice demonstrated markedly elevated transepidermal water loss as well as altered permeability barrier function as shown by an electron-dense tracer study (Holleran *et al.,* 1994). Such changes were not detected in the skin of type I or III Gaucher patients (Sidransky *et al.,* 1996). Thus, phenotypically the mouse model with complete inactivation of the glucosylceramidase gene resembles the rare and the most severe neonatal subtype of type II Gaucher disease (Sidransky *et al.,* 1992).

Mouse models of point mutations have been generated very recently to mimic human Gaucher disease types II and III (Liu *et al.,* 1998). These authors used single insertion mutagenesis technology to introduce point mutations into the glucosylceramidase gene, *L444P* (type 3), and a combined *L444P* and *A456P* (type 2, *RecNicI* mutation). The mouse with the double mutation *L444P* and *A456P* gave little enzyme activity and an accumulation of glucosylceramide in the liver and brain. The mouse with only *L444P* gave higher residual glucosylceramidase activity and no accumulation of the substrate in the liver or brain. However, in contrast to human patients with these types of Gaucher disease, both mice with the point mutations died within 48 h of birth apparently due to an abnormal epidermal permeability barrier as in the mice with complete gene inactivation. This conspicuous difference between the mouse and human disease may well be due to the fact that mice are born at a much earlier stage of development than humans.

3.1.1.2. GM2-Gangliosidosis GM2-gangliosidoses are caused by genetic defect in degradation of GM2-ganglioside. Three distinct lysosomal proteins, the β-hexosaminidase α- and β-subunits and the GM2-activator protein are required for normal degradation of GM2-ganglioside *in vivo.* The two subunits of β-hexosaminidase form three forms of dimers, hexosaminidase A (αβ), hexosaminidase B (ββ), and hexosaminidase S (αα). The gene encoding the β-hexosaminisase α subunit (*HEXA*) is located on chromosome 15, while the β subunit (*HEXB*) and GM2-activator (*GM2A*) genes are on chromosome 5. Thus, abnormalities in these three genes cause three genetically distinct forms of GM2-gangliosidoses: Tay-Sachs disease (hexosaminidase A and S deficiency), Sandhoff disease (hexosaminidase A and B deficiency), and AB variant (GM2-activator deficiency). In all forms, massive accumulation of GM2-ganglioside and other glycolipids in neuronal lysosomes occurs. These glycolipids are stored as multilamellar structures known as membranous cytoploasmic bodies (MCBs) (Terry and Korey, 1960; Terry and Weiss, 1963). In Sandhoff disease, the additional absence of β-hexosaminidase B results in accumulation of other substrates, such as GA2 (asialo-GM2-ganglioside), globoside, oligosaccharides, and glycosaminoglycans in the visceral organs. The infantile forms are rapidly progressive, resulting in death in early childhood. The late-onset forms, which usually retain low but detectable level of enzyme activity, progress more slowly with a wide spectrum of clinical phenotypes (Gravel *et al.,* 1995; Suzuki and Suzuki, 1996).

Both cDNAs for the mouse α and β subunits have 84% and 75% identity with their human counterparts in their predicted protein sequences (Bapat *et al.,* 1988; Bellachioma *et al.,* 1993; Yamanaka *et al.,* 1994b). Through targeted disruption of the mouse *Hexa, Hexb,* and *Gm2a* genes (corresponding to human *HEXA, HEXB,* and *GM2A* genes) by homologous recombination, mouse models of all genetic forms of human GM2-gangliosidoses have been generated independently in a few laboratories (Cohen-Tannoudji *et al.,* 1995; Liu *et al.,* 1997; Phaneuf *et al.,* 1996; Sango *et al.,* 1995; Taniike *et al.,* 1995; Yamanaka *et al.,* 1994a).

3.1.1.3. Tay–Sachs Disease (β-Hexosaminidase α Subunit Deficiency) A mouse model of β-hexosaminidase α subunit deficiency was generated by disrupting the mouse *Hexa* gene with the neomycin-resistant gene in exon 8. Affected mice with the *Hexa* null mutation are profoundly deficient in β-hexosaminidase A activity with an accumulation of GM2-ganglioside in the brain. Storage neurons contain MCBs. No histological abnormalities were noted in the liver, spleen, and other visceral organs on the light microscopic level, although modest lysosomal storage was described in the hepatocytes on the ultrastructural level by one group (Cohen-Tannoudji *et al.,* 1995). In these respects, *Hexa−/−* mice are similar biochemically and pathologically to Tay–Sachs disease in humans. However, there are many features in the mouse model that are distinctly different from the human disease. First, these *Hexa−/−* mice are clinically normal, fertile, and have a normal life span. The accumulation of GM2-ganglioside is progressive but its degree is much milder than that seen in the brain of human patients with infantile Tay–Sachs disease, reaching only 15% of total brain ganglioside at 6 months (Yamanaka *et al.,* 1994a). The localization of storage neurons is limited to certain regions in the brain such as cerebral cortex, amygdala, piriform cortex, hippocampus, and hypothalamus, in contrast to the infantile human disease in which GM2-ganglioside accumulation is widespread in all neurons. Notably absent is storage in the Purkinje and granular cells in the cerebellar folia and in neurons in the spinal cord. In the cerebral cortex, large pyramidal neurons show more storage than other cortical neurons. The storage materials stain positively with anti-GM2-ganglioside antibody (Taniike *et al.,* 1995). Identical distribution of storage neurons and GM2-ganglioside immunoreactivity was reported in the *Hexa*−/− mice generated by Phaneuf *et al.* (1996). However, in the mice described by Cohen-Tannoudji and co-workers, no storage was noted in the hippocampus at 150 days (Cohen-Tannoudji *et al.,* 1995). The mild accumulation of GM2-ganglioside, the limited neuronal storage, and the apparently normal clinical phenotype can be explained by a difference in the GM2-ganglioside degradation pathway between human and mouse. In humans GM2-ganglioside cannot easily be degraded to GA2 by sialidase, while the mouse sialidase is active toward GM2-ganglioside. Because GA2 can be further degraded by β-hexosaminidase B, which is not deficient in the *Hexa*−/− mice, there is an effective bypass for degradation of GM2-ganglioside in the Tay–Sachs mice (see next section for further clarification).

3.1.1.4. Sandhoff Disease (β-Hexosaminidase β Subunit Deficiency) The *Hexb*−/− mice were generated by disruption of exon 13 of the *Hexb* gene with the neomycin-resistant gene (Sango *et al.,* 1995). Liver extracts from *Hexb*−/− mice are deficient in both β-hexosaminidase A and B activity. The "residual activity" at about 2% of the wild-type is likely to be due to β-hexosaminidase S (αα). The *Hexb*−/− mice appear normal at birth. However, progressive worsening of motor coordination and balance can be detected by the rotorod test starting at approx. 12 weeks. Generalized locomotion in a Digiscan open field and passive avoidance learning are not significantly different from wild-type controls at this stage of development. The first overt signs of severe motor dysfunction begin at approx. 3 months. The neurological signs include gait abnormalities with spastic movements starting from hindlimbs and progressing to forelimbs. By 4–4.5 months, muscles of the hindlimbs become atrophic, and the mice are no longer able to effectively take food or water. Both male and female *Hexb*−/− mice are fertile, however. Thus, unlike Tay–Sachs and Sandhoff diseases in humans, where the clinical course is nearly identical, the corresponding mouse models, *Hexa*−/− and *Hexb*−/− show quite different clinical phenotypes. Extensive neuronal storage is present throughout the cerebrum, cerebellum, brain stem, spinal cord, trigeminal and dorsal root ganglia, retina, and myenteric plexus. Unlike *Hexa*−/− mice, abundant storage is noted in the Purkinje and granular cell neurons as well as in many cells in the molecular layer of the cerebellum and in the spinal neurons. These storage materials are best demonstrated with the PAS stain on frozen sections. In addition to neurons, PAS-positive storage materials are present in small round cells in the cerebrum and cerebellum. The storage materials in these small cells, but not those in neurons, stain positive with Alcian blue and colloidal iron consistent with their being glycosaminoglycans. These cells also stain positively with a lectin Ricinus communis agglutinin (RCA-1), indicative of their being macrophage/microglia. Similarly PAS and Alcian blue/colloidal iron positive cells are noted in the hepatic sinusoids. Epithelial cells of the proximal tubules in the kidney also contain materials of similar staining characteristics. These histological features are identical to those in human Sandhoff disease. The cerebral white matter is diffusely hypomyelinated compared with the *Hexa−/−* or wild-type mice. White matter degeneration with axonal spheroids is recognized in the spinal cord of older mice. Clinical as well as pathologic phenotypes of *Hexb−/−* generated by Phaneuf and co-workers (1996) are essentially identical to those generated by Sango and co-workers (1995). In the brains of *Hexb−/−* mice, a massive and progressive accumulation of both GM2-ganglioside and GA2 is readily demonstrated from early stages of development. Thus, *Hexb−/−* mice are very similar to human Sandhoff disease in clinical, pathological, and biochemical phenotype. As indicated in the preceding section, the presence of an alternate pathway for GM2-ganglioside degradation in mice can explain these differences between the two mutants in this species. In both human Tay–Sachs patients and *Hexa−/−* mice, the conversion of GM2 to GM3 is blocked by the absence of β-hexosaminidase A. However, in mice the alternate degradative pathway, conversion of GM2 to GA2 by sialidase, is present. In the *Hexa−/−* mice, GA2 can then be degraded by the genetically intact β-hexosaminidase B, explaining their limited GM2-ganglioside storage and the lack of clinical symptoms. In the *Hexb−/−* mice, however, both GM2 and GA2 cannot be further degraded, leading to extensive neuronal storage of both GA2 and GM2-ganglioside and a severe clinical phenotype similar to infantile GM2-gangliosidosis in humans. An analogous alternate pathway through sialidase exists in the mouse for degradation of GM1-ganglioside (see later).

3.1.1.5. GM2-Gangliosidosis AB Variant (GM2 Activator Protein Deficiency) The *Gm2a* gene was inactivated by a deletion of a portion of exon 3 and all of the coding region of exon 4 (Liu *et al.,* 1997). *Gm2a−/−* mice grow normally and are fertile. Although displaying a normal life span, they exhibit subtle neurological dysfunction. When tested beginning at 13 and ending at 33 weeks, overall rotorod performance was significantly impaired. Passive-avoidance task as a test for learning and memory suggested possible memory deficit. Similar to human patients with GM2-activator deficiency, the *Gm2a−/−* mice had normal levels of β-hexosaminidase activity in liver extracts when tested with artificial substrates. Analysis at 4 months demonstrated an accumulation of GM2-ganglioside in the brains of *Gm2a−/−* mice at a level comparable to the *Hexa−/−* mice but less than the *Hexb−/−* mice. In *Hexb−/−* mice, both GM2-ganglioside and GA2 accumulate in the brain, whereas only GM2-ganglioside accumulates in the brains of *Hexa−/−* mice at about one fourth the level of that in the *Hexb−/−* mice. In *Gm2a−/−* mice, in addition to GM2-ganglioside, a slight accumulation of GA2 is detectable. The distribution of storage neurons is generally very similar to that of *Hexa−/−* mice with the exception of the cerebellum, where storage materials are present in the Purkinje and granular cells and also in some cells in the molecular layer. Very little storage is present in these cells in the cerebellum of *Hexa−/−* mice even at age 18 months. The significant storage in the cerebellum in *Gm2a−/−*, but not in *Hexa−/−* mice, can explain the impaired motor coordination in the former. The minimal GA2 storage in the *Gm2a−/−* mice indicates that the hexosaminidase-mediated degradation of GA2 can proceed at least to some extent, if not optimally, in the absence of the activator protein.

3.1.1.6. β-Hexosaminidase α and β Subunit Double Deficiency Mice in which both *Hexa* and *Hexb* genes are inactivated have been produced by cross-breeding double heterozygotes (*Hexa+/− Hexb+/−*) or by mating *Hexa−/−* and *Hexb+/−* mice (Sango *et al.,* 1996). The "double knockout" (*Hexa−/−Hexb−/−*) mice are totally deficient in all forms of lysosomal β-hexosaminidase including β-hexosaminidase S present in the *Hexb−/−* mice. The equivalent genetic condition is not known in humans. Unexpectedly, these mice showed clinical, pathological, and biochemical features of a severe mucopolysaccharidosis. Affected mice are indistinguishable from their littermates at birth but they can be distinguished by their smaller size and physical dysmorphism by 4–5 weeks. Their heads are shorter and the snouts broader. Their feet are thick and broad with flexion contracture of the digits. Corneal opacity is present. Occasional seizure-like activities and unresponsiveness to sudden loud noise suggestive of deafness are observed. A radiographic study showed kyphosis, an abnormally shaped rib cage, and short and thickened long bones. Their life span was only 1–4 months. There is a massive accumulation of GM2-ganglioside and GA2 to a degree comparable to the level of *Hexb−/−* mouse brain. They excrete into urine large amounts of glycosaminoglycans consisting of *N*-acetylgalactosamine and iduronic acid consistent with dermatan sulfate. However, the activity of enzymes involved in the classic degradation pathway of glycosaminoglycans such as α-*N*-acetylglucosaminidase, α-L-iduronidase, β-glucuronidase, and β-galactosidase were generally higher than in wild-type control mice.

Pathology is similar to that of other mucopolysaccharidoses (Suzuki *et al.,* 1997). It consists of storage of glycosaminoglycans in macrophages, chondrocytes, splenic sinusoidal cells, and in some neurons (see description in section describing mucopolysaccharidoses models later). Macrophages containing PAS, Alcian blue, and colloidal iron-positive materials, consistent with glycosaminoglycans, are present in the bone marrow, skin, and adipose and connective tissues in visceral organs. Similar cells are also present in the myocardium, heart valves, lung, liver, spleen, and kidney. In the CNS, extensive neuronal storage, similar to that seen in *Hexb−/−* mice, is present throughout. In addition, there are numerous small cells (probably macrophages) with distended cytoplasm containing colloidal iron and Alcian blue-positive materials in the brain, leptomeninges, and choroid plexus stroma. These cells stain strongly positive with RCA-1. The white matter is diffusely hypomyelinated with increased numbers of glial fibrillary acidic protein (GFAP)-positive astrocytes and RCA-1-positive microglia/macrophages. At the terminal stage, nerve fiber tracts in the white matter contain many axonal spheroids and scattered myelin figures indicative of myelin degeneration. In the PNS, neuronal storage is seen in the trigeminal and dorsal root ganglia, and in neurons in the myenteric plexus and visceral autonomic ganglia. Macrophages similar to those found in the brain are also very conspicuous in the peripheral nerves, although nerve fibers appear well preserved. These morphological features together with biochemical data indicate that β-hexosaminidase is a crucial enzyme in some steps of degradation of some glycosaminoglycans. Although a few colloidal iron-containing cells are found in the brains of *Hexb−/−* mice, lack of significant glycosaminoglycan storage in the mice with single gene defect and in human patients with Tay–Sachs and Sandhoff diseases suggests possible functional redundancy in the β-hexosaminidase system. These total hexosaminidase-deficient mice are not a model of a known lysosomal storage disease in humans but are a valuable tool for studying the role of the β-hexosaminidase in the degradation of glycosaminoglycans and in understanding the pathological consequences of excessive accumulation of glycosaminoglycans.

3.1.1.7. Metachromatic Leukodystrophy (Arylsulfatase A Deficiency) Metachromatic leukodystrophy (MLD) is caused by deficiency of arylsulfatase A (ASA) (Kolodny and Fluharty, 1995). Deficiency of ASA results in an accumulation of the substrate, sulfatide, in various organs including the brain. Sulfatide is a major myelin lipid and the perturbed sulfatide metabolism causes diffuse demyelination of the white matter and peripheral nerves. Late infantile, juvenile, and adult types are known clinically. The late infantile MLD is the most common type with rapidly progressive neurological symptoms such as psychomotor deterioration with blindness, seizures, quadriparesis, and peripheral neuropathy, usually manifesting in the second year of life. Patients with this type usually die within a few years. Juvenile and adult types are more slowly progressive and neurological symptoms appear between age 4 and 12 years or between mid-teens and the seventh decade, respectively. Clinical manifestations of the late-onset forms are widely variable. Behavioral disturbance, dementia, and progressive peripheral neuropathy are often the major presenting signs. The *ASA* gene is located near the end of the long arm of chromosome 22 and more than 30 disease-causing mutations have been identified in human patients (Gieselmann *et al.,* 1994). The pathology of MLD in the nervous system is primarily a diffuse demyelination in association with deposits of the characteristic metachromatic materials (sulfatide) in the brain, kidney, gallbladder, liver, pancreas, and various other visceral organs. In the brain, metachromatic materials accumulate in oligodendrocytes, and also in neurons in certain regions including dentate nucleus of the cerebellum, some brain stem nuclei, and the spinal anterior horn (Peng and Suzuki, 1987). The storage materials stain brown with acidic cresyl violet in frozen sections. Prismatic and tuffstone inclusions are the characteristic ultrastructural features of these metachromatic deposits (Gregoire *et al.,* 1966).

The mouse model of MLD was generated by targeted disruption of the arylsulfatase A gene (Hess *et al.,* 1996). ASA mRNA is undetectable in the *ASA−/−* mice. No degradation of sulfatide can be detected in cultured fibroblasts in the sulfatide loading test. Thus, biochemically the *ASA−/−* mice resemble the severe late infantile form of human MLD. However, *ASA−/−* mice show a much milder clinical phenotype. Affected mice develop normally for several months without overt clinical signs. At 12–14 months of age, statistically significant but subtle abnormalities can be detected in the neuromuscular function and behavior when they are evaluated by the walking pattern, rotorod test, and the Morris water maze. The most notable neurological sign in *ASA−/−* mice is total absence of auditory brain stem evoked potentials. In the second year of life, *ASA−/−* mice develop low-frequency tremors of the head. Sulfatide storage is observed histologically in the kidney, gall bladder, and bile duct at 6–11 months. However, the storage is not as widespread as that seen in human patients and there is no evidence of storage in hepatocytes, adrenal glands, and skeletal muscle. In the brains of 11-month-old *ASA−/−* mice, storage of metachromatic material (sulfatide) was noted in the white matter (corpus callosum, hippocampal fimbria, internal capsule, and optic nerve) but demyelination is not apparent. Ultrastructurally, lamellar deposits, herringbone- and tuffstone-like materials, similar to those found in human MLD, can be demonstrated in astrocytes, oligodendrocytes, microglia, and Schwann cells. The myelin sheath has a normal ultrastructure. However, there is a statistically significant reduction of the cross-sectional area of myelinated axons in the corpus callosum and optic nerve. Astrogliosis is present at 1 year and activation of microglia becomes conspicuous with increasing age. Neuronal storage is present in several nuclei of the brain stem, diencephalon, spinal cord, and cerebellum. Dramatic pathology is seen in the inner ear of *ASA−/−* mice. The number of acoustic ganglion cells and myelinated nerve fibers is greatly reduced. The Schwann cells in the acoustic and vestibular ganglia show marked sulfatide storage. Storage is also evident in the Schwann cells in the peripheral nerve but demyelination is not apparent. Thus, the basic pattern of pathological process in *ASA−/−* mice is reminiscent of human MLD but the process is slower and milder.

3.1.1.8. Fabry Disease (α-Galactosidase A Deficiency) Fabry disease is an X-linked disorder caused by a deficiency of α-galactosidase A (α-Gal A) (Desnick *et al.,* 1995). Neutral glycosphingolipids with terminal α-linked galactosyl moieties—predominantly globotriaosylceramide—accumulate in the endothelial, perithelial, and smooth muscle cells of blood vessels and many other cell types in the liver, heart, spleen, kidney, and also in the plasma of the patients. Classic clinical manifestations include pain and paresthesia in the extremities, angiokeratoma of the skin and mucous membranes, and hypohidrosis during childhood or adolescence. Corneal and lenticular opacity may be present. With increasing age, severe renal failure due to progressive glycosphingolipid accumulation in blood vessels results in hypertension and uremia. The hemizygous male patients usually die of renal, cardiac, or cerebrovascular complications. A late-onset milder form with a primary cardiac involvement ("cardiac variant") is known. Heterozygous females may have an attenuated form of the disease, although they are usually asymptomatic. The gene encording α-galactosidase A is localized to Xq22. The characteristic pathology is widespread deposits of crystalline glycosphingolipids, which show birefringence with characteristic "Maltese crosses" under polarized light. The lipid deposits were also reported in neurons in certain regions in the brain and spinal cord (Sung, 1979; DeVeber *et al.,* 1992).

The mouse model of Fabry disease has recently been generated by targeted gene disruption (Wang *et al.,* 1996; Ohshima *et al.,* 1997). α-Galactosidase A activity is undetectable in liver homogenates from *α-Gal A−/0* mice. These mutant mice are clinically normal at 10–14 weeks of age. No obvious histological lesions are present in the kidney, liver, heart, spleen, lung, and brain on sections stained with hematoxylin and eosin. However, lipid inclusions consisting of concentric lamellar structures can be demonstrated by electron microscopy in the renal tubular epithelium. They appear similar to those seen in human patients with Fabry disease. With fluorescent-labeled *Griffonia (Bandeiraea) simplicifolia* lectin which selectively binds to α-D-galactosyl residues, significant accumulation of compounds containing α-D-galactosyl residues was demonstrated in the kidneys of 10-week-old mutant mice. Skin fibroblasts from *α-Gal−/0* mouse embryos also displayed significant accumulation of α-galactosidase A substrates. Significant accumulation of globotriaosylceramide is present in the liver and kidney. Thus, the mouse model of Fabry disease closely resembles the human disease. A new model of Fabry disease with a specific point mutation has also been recently developed (Shimmoto *et al.,* 1997).

3.1.1.9. Niemann–Pick Type A and B (Acid Sphingomyelinase Deficiency) Niemann–Pick disease is caused by deficient activity of the lysosomal acid sphingomyelinase (Schuchman and Desnick, 1995; Vanier and Suzuki, 1996). Two clinical types, A and B, are known. It should not be confused with Niemann–Pick type C disease (see earlier), which is, despite the unfortunate common eponym, unrelated to Niemann–Pick A and B disease. Type A patients have severe neurovisceral storage and usually die by 3 years of age. Type B patients have similarly severe hepatosplenomegaly but little or no neurological involvement and often survive until adulthood. The underlying mechanism for such a phenotypic difference is not well understood. The acid sphingomyelinase gene locus is on chromosome 11 and many mutations that cause type A or type B disease have been identified. Sphingomyelin and cholesterol accumulate within the cells of the reticuloendothelial system, causing massive hepatosplenomegaly. In addition, diffuse neuronal storage is a conspicuous neuropathological finding in the type A patients. Consistent with the pathological finding, abnormal accumulation of sphingomyelin in the brain appears to occur only in patients with the type A disease (Kamoshita *et al.,* 1969).

Recently two laboratories generated murine models of Niemann–Pick disease by disrupting acid sphingomyelinase gene with insertion of the neomycin-resistant gene into exon 2 (Horinouchi *et al.,* 1995) and exon 3 (Otterbach and Stoffel, 1995), respectively. While the locations of the gene disruption was in adjacent exons, the two laboratories reported significantly different findings in the phenotype of the respective mutant mice.

The acid sphingomyelinase null mutant (*asmase*−/−) mice generated by Otterbach and Stoffel (Otterbach and Stoffel, 1995) were normal until 8–10 weeks of age when fine tremor of the entire body and severe intention tremor developed. Their gait became increasingly ataxic and displayed tottering with zigzag movement characteristic of cerebellar dysfunction. Around 60 days of age, hepatosplenomegaly became apparent and the affected mice died around 4 months of age after a period of severe dyspnea. In the total protein extracts of liver, spleen, and brain, no acid sphingomyelinase activity was measurable. Northern blot analysis of total RNA of liver, spleen, and brain detected no acid sphingomyelinase mRNA. Neutral sphingomyelinase activity in the brain, however, was normal. Accumulation of sphingomyelin was present in the liver and spleen and to a lesser extent in the brain. The size of the liver and spleen of the affected mice increased about 1.5 and 2 times in weight, respectively, in the terminal stage. Massive accumulation of lipid material was histologically evident in the Kupffer cells and in macrophages in the bone marrow, spleen, and lung. The brain weight and size appeared similar to that of age-matched controls. Neurons in the CNS were frequently swollen with pale vacuolation. The cerebral cortical structures were slightly disturbed. The most striking feature in the brain was loss of Purkinje cells within 60–90 days after birth. The cerebellar molecular cell layer became very thin in the final stage of the disease at days 210 and 225 (Kuemmel *et al.,* 1997). Thus, this murine model is phenotypically closely similar to Niemann–Pick disease type A in humans.

The acid sphingomyelinase knockout mice generated by Horinouchi and coworkers (Horinouchi *et al.,* 1995) showed ataxia, mild tremor, and feeding difficulty around 8 weeks but survived longer and were fertile. Death occurred much later than in the affected mice described by Otterbach and Stoffel at between 6 and 8 months of age. Unlike the mice of Otterbach and Stoffel, which showed no significant brain atrophy, the brain of the mice generated by Horinouchi *et al.* were atrophic and less than half the weight and volume of control mice. A remarkable loss of Purkinje cells and general atrophy of the cerebellum and midbrain were similar in both models. Hepatosplenomegaly, which was a striking feature in the mice generated by Otterbach and Stoffel, was not evident in these mice, although lipid-laden foam cells were found in most major organs, particularly in bone marrow and spleen. Multilamellar inclusions were demonstrated in the neurons in both mutant mice by electron microscopy. The reason for such phenotypic differences between these acid sphingomyelinase deficient mouse lines is not clear.

3.1.1.10. GM1-Gangliosidosis (Acid β-Galactosidase Deficiency) GM1-gangliosidosis is a progressive neurological disease caused by a genetic deficiency of lysosomal acid β-galactosidase (Suzuki *et al.,* 1995c; Suzuki and Suzuki, 1996). The clinical course is most rapidly progressive in the infantile form, usually associated with visceromegaly and mucopolysaccharidosis-like dysmorphic features. Patients with the infantile form usually succumb to the disease within a few years. In the later-onset types, the clinical course is more prolonged and visceral involvement is less common. The adult type is the most chronic form and often manifests itself as a movement disorder. Radiologic evidence of bony abnormalities is frequently present. Morquio B disease is allelic to GM1-gangliosidosis also caused by β-galactosidase deficiency and occurs at various ages. Unlike the cerebral forms of the disease mentioned, Morquio B disease is free of neurological symptoms. The main clinical symptoms are related to severe bony deformities. Patients with intermediate clinical forms have also been described. The infantile type shows diffuse and extensive neuronal storage of GM1-ganglioside and, to a much lesser extent, its asialo derivative, GA1. Oligosaccharides with terminal β-galactose residues derived from keratan sulfate and glycoproteins are the primary storage materials in the visceral organs (Suzuki, 1968) and are excreted into the urine. In the chronic adult type, neuronal storage tends to be more localized in the basal ganglia and cerebellum (Suzuki, 1991). The human acid β-galactosidase cDNA and the gene have been cloned and many disease-causing mutations have been identified (Suzuki *et al.,* 1995c).

Recently a mouse model lacking the functional β-galactosidase gene (β-*gal*−/−) has been generated by homologous recombination and embryonic stem cell technology in two laboratories. The β-*gal* gene was inactivated by introduction of the neomycin-resistant gene into exon 6 (Hahn *et al.,* 1997) or exon 15 (Matsuda *et al.,* 1997a,b). The β-*gal*−/− mice are fertile and appear phenotypically normal until about 4–5 months of age, when generalized tremor, ataxia, and abnormal gait become apparent. No β-galactosidase mRNA is detectable by Northern blot analysis of kidney, brain, and liver. β-Galactosidase activity in the kidney, brain, liver, and spleen as measured with a fluorescent substrate is markedly decreased (ranging from 1% in the spleen to 4% in the brain) compared with wild-type littermates. Degradation of the natural substrate, GM1-ganglioside, is not detectable at all. By thin layer chromatography, markedly increased GM1-ganglioside and GA1 are noted in the brains of β-*gal*−/− mice. In the brains of human patients with infantile GM1-gangliosidosis, a similar degree of GM1-ganglioside accumulation occurs but accumulation of GA1 is only slight. Thus, marked accumulation of GA1 in this mouse model suggests that the murine sialidase is more active toward GM1-ganglioside than the corresponding human enzyme, a difference between human and mouse analogous to GM2-ganglioside degradation (see earlier). In β-*gal*−/− mice, the total amounts of brain ganglioside sialic acid and GM1-ganglioside increase dramatically from 3 weeks to 3.5 months. Histologically, neuronal storage is already widespread as early as 3 weeks of age. By 5 weeks, PAS-positive storage materials are present in almost all neurons in the cerebrum, cerebellum, brain stem, spinal cord, and the spinal dorsal root ganglia. Diffuse neuronal storage in these mice is similar to the pattern of neuronal storage noted in the human infantile GM1-gangliosidosis. Ultrastructural features of neuronal storage materials are also similar to those of human infantile GM1-gangliosidosis. Unlike the human disease, β-*gal*−/− mice show no hepatosplenomegaly and no storage materials are detected in visceral organs even at 3.5 months. Biochemical analysis shows only a minimal storage of oligosaccharides in the liver and a low level of abnormal urinary oligosaccharides. The β-*gal*−/− mice reported by Matsuda and co-workers (Matsuda *et al.,* 1997a) were also apparently healthy for the first 4 months. Subsequently, horizontal movement became slower and rearing or vertical climbing became less frequent. Definite gait disturbance was noted by 6–8 months of age. They described that β-*gal*−/− mice showed unusual postures with all four limbs flexed when hung vertically with the tail held upward. Spastic diplegia progressed and the mice died of extreme emaciation at 7–10 months of age because of difficulty feeding. They described ballooning of neurons in various areas of the CNS and a 20- to 30-fold increase of GM1-ganglioside in the brains of these mice. β-Galactosidase activity in β-*gal*−/− fibroblasts was reported to be 0–1% of control value. Thus, both of these mouse models are similar and appear to be excellent models of the infantile type of human GM1-gangliosidosis as far as the CNS involvement is concerned. Minimal involvement of visceral organs in the mouse model may suggest differences in the metabolism of keratan sulfate between humans and mice.

3.1.1.11. Total Sphingolipid Activator Protein (SAP, Saposin) Deficiency Sphingolipid activator proteins are small nonenzymatic lysosomal glycoproteins that function as essential cofactors for physiological degradation of sphingolipids with relatively short hydrophilic head groups (Sandhoff *et al.,* 1995). Two genes encode all established sphingolipid activator proteins. One is the GM2-activator protein localized on chromosome 5 that stimulates degradation of GM2-ganglioside and asialo GM2-ganglioside (GA2) (see GM2-gangliosidosis section earlier) (Burg *et al.,* 1983). The other gene is on chromosome 10 coding for the SAP precursor (prosaposin), which is processed to the four homologous activator proteins, sap-A, sap-B, sap-C, and sap-D. Genetic deficiencies of *sap*-B (sulfatide activator) (Mehl and Jatzkewitz, 1964) or sap-C (glucosylceramide activator) (Ho and O'Brien, 1971) result in human disorders mimicking metachromatic leukodystrophy and Gaucher disease, respectively. In 1989, Harzer and co-workers reported two sibs in a consanguineous family, affected by a rapidly fatal disorder (Harzer *et al.,* 1989). None of the *saps* were detected in the tissue from these sibs (Bradová *et al.,* 1993) and they were later found to be homoallelic for a mutation in the initiation codon of the *SAP* precursor gene (Schnabel *et al.,* 1992). Shortly after birth, the first patient in the family showed hyperkinetic or cloniform motor abnormalities and generalized clonic seizures. The patient had fasciculation of tongue and periauricular muscle, spontaneous Babinski sign, and an exaggerated Moro reflex. Massive hepatosplenomegaly was present and storage macrophages resembling Gaucher cells were observed in a bone marrow smear at 5 weeks. Enzyme studies with leukocytes and cultured fibroblasts showed deficiency of galactosylceramidase and partial deficiency of glucosylceramidase activity. Ultrastructure of liver, nerve, and skin biopsies revealed the presence of vesicular inclusions and membranous bodies suggestive of a lysosomal storage disease. An *in situ* test with cultured fibroblasts demonstrated a defect in ceramide catabolism similar to Farber disease. Neuroimaging indicated an atrophic brain with hydrocephalus. Postmorten examination was not performed, however. The second patient, an aborted female sib of the first patient, showed a similar defect in the ceramide catabolism and had contracted hand joints consistent with a Farber-like disease. The SAP precursor and mature *saps* were completely absent in this fetus who had generalized accumulation of multiple sphingolipids (ceramide, glucosylceramide, sulfatide, lactosylceramide, digalactosylceramide, GM1-ganglioside) with the exception of sphingomyelin in the brain and systemic organs.

The mouse SAP gene was disrupted by insertion of the neomycin-resistant gene within exon 3 to generate a *SAP−/−* mouse line (Fujita *et al.,* 1996). SAP mRNA was completely absent in the tissue of *SAP−/−* mice. Of the first 126 offspring from heterozygous mating, 36 (26%) wild-type, 68 (55%) heterozygotes, and 16 (13%) homozygotes were obtained. Five of the nine mice that died within a day or two of birth were *SAP−/−* mice. These results suggest strongly that a disproportionate percentage of *SAP−/−* mice die *in utero* or during the perinatal period. Those *SAP−/−* mice that survive the neonatal period appear healthy and grow normally until about 18–20 days, although they were slightly smaller in size than their normal littermates. Tremulousness of the head and mild weakness/ataxia are the initial symptoms, appearing at approx. 20 days of age. These symptoms progress rapidly and *SAP−/−* mice usually die at 35± days in an emaciated condition. The brain and visceral organs are of normal size except for the kidneys, which are smaller than in controls. The neuropathology of *SAP−/−* mice is that of combined neuronal storage and leukodystrophy. There is an extensive neuronal storage throughout the cerebrum, cerebellum, brain stem, spinal cord, and retinal ganglion cells. The storage materials stain brilliantly with PAS on frozen sections. Many axonal spheroids and macrophages/microglia containing storage materials are also conspicuous. The white matter is hypomyelinated but myelin ovoids suggestive of myelin degeneration are also noted in the corpus callosum, internal capsule, and spinal white matter. Similar changes of myelin are present in the spinal roots and trigeminal and sciatic nerves. No metachromatic or sudanophilic materials are found in frozen sections of the cerebrum. In the liver, clusters of histiocytes with abundant eosinophilic cytoplasm are present in the hepatic sinusoids. Similar cells are also seen in the spleen and lymph nodes. No significant abnormalities are detected in the kidney despite the smaller size. The inclusions in neurons are pleomorphic; some consist of concentric or short lamellar structures combined with an electron-dense granular structure and others are electron-dense granular structures surrounded by a single membrane. These inclusions resemble those of cultured skin fibroblasts from the patients with total SAP deficiency. The inclusions within the cells in the hepatic sinusoids also consist of membranous and vesicular structures and are more complex than neuronal inclusions (Oya *et al.,* 1998). Biochemically the most conspicuous specific abnormality is a major accumulation of lactosylceramide, which appears to account for the strong PAS staining on frozen sections. The total ganglioside sialic acid is increased significantly with large relative increases in the monosialogangliosides, GM1, GM2, and GM3. However, on the wet weight or protein basis, many of the disialo- and trisialogangliosides are also significantly increased, a condition known so far only in an unusual genetic disorder occurring in the emu (Bermudez *et al.,* 1995; Freischütz *et al.,* 1997). Similar to the brain, lactosylceramide is conspicuously increased in the liver and kidney. In addition, ceramide, glucosylceramide, globotriaosylceramide, and globoside are all significantly increased in the liver. Galactosylceramide and sulfatide appear to be increased in the kidney. Thus, accumulation of varieties of sphingolipids noted in the patients with total SAP deficiency is also present in the *SAP−/−* mice. As noted in human patients, activities of glucosylceramidase and galactosylceramidase are also decreased in *SAP−/−* mice. The results of the serine-loading tests on cultured fibroblasts were remarkably similar to the results obtained with the human patient fibroblasts. Therefore, *SAP−/−* mice closely mimic biochemical changes of the known human patients with total SAP deficiency. Neuropathology cannot be compared, however, as postmortem investigation was not carried out on the original patient with SAP deficiency.

3.1.2. Mucopolysaccharidosis

3.1.2.1. Hurler Disease (MPS I) (α-L-Iduronidase Deficiency) MPS I, considered to be the prototype of MPS, is caused by a deficiency of α-L-iduronidase (Neufeld and Muenzer, 1995). The spectrum of the clinical phenotype in MPS I ranges from severe mental retardation with hepatosplenomegaly, dysostosis multiplex, corneal clouding, cardiac involvement, and death in early childhood to milder symptoms consisting of corneal clouding, hearing loss, and mild visceral involvement with normal intelligence and life span. Hurler syndrome (MPS IH) represents the most common and severe form and Scheie syndrome (MSP IS) represents the milder form. A large number of disease-causing mutations have been identified in human cases of MPS I and there is a good genotype/phenotype correlation (Scott *et al.,* 1995).

The targeted disruption of murine *Idua* gene resulted in mice with total lack of *Idua* mRNA and total deficiency of α-L-iduronidase activity in the liver, kidney, brain, and tail clippings (Clarke *et al.,* 1997). Urinary glycosaminoglycan secretion was markedly increased (twofold at 4 weeks and fivefold at 15 weeks). *Idua−/−* mice were apparently normal clinically during the first 2–3 weeks. At 4 weeks, however, characteristic facial abnormalities of MPS I (broadness of the face with a loss of the fine tapered snout) and broadened and thickened digits became apparent. Prominent dysostosis multiplex was noted radiologically at 15 weeks of age. However, weight gain of *Idua−/−* and control mice were similar without any increase in mortality in the former during the first 20 weeks of life, and *Idua−/−* mice were fertile. No hepatosplenomegaly was detected. The pathology was that of diffuse lysosomal storage disease involving cells of the reticuloendothelial system, including the Kupffer cells, splenic sinusoidal lining cells, pulmonary macrophages, chondrocytes, and glial cells. Cytoplasmic vacuolation indicative of lysosomal storage was present in these cells as early as 4 weeks. Cytoplasmic vacuolation was not present in neurons at 4 weeks but appeared at 8 weeks. Thus, biochemically and pathologically the *Idua−/−* mice closely resemble the severe forms of MPS I in humans except for the relatively mild clinical phenotype.

3.1.2.2. Maroteaux–Lamy Disease (Arylsulfatase B Deficiency) Maroteaux–Lamy syndrome or mucopolysaccharidosis VI (MPS VI) is caused by deficiency of lysosomal enzyme, arylsulfatase B (ASB; *N*-acetylgalactosamine-4-sulfatase) (Neufeld and Muenzer, 1995). The patients show dysmorphic skeletal abnormalities (dysostosis multiplex with macrocephaly, deformities of the chest and vertebral bodies, pelvic bones, and dysplasia of long bones) similar to MPS I but mental development is normal. Corneal clouding, hernia, thickened skin, hepatosplenomegaly, and cardiac involvement with valvular dysfunction are common. Multiple mutations in the arylsulfatase B gene have been described in MPS VI patients.

The arylsulfatase B deficient (*Asl-s*−/−) mice were generated by disruption of ASB gene (*Asl-s*) (Evers *et al.*, 1996). In the tissue homogenates of these mice, no arylsulfatase B activity was detected. At birth and during the first postnatal week, *Asl-s*−/− mice could not be distinguished from littermate controls. Facial dysmorphism, shortened limbs and coarse paws became apparent around 4 weeks of age. Symptoms progressed and at age 9–12 months, the body weight of *Asl-s*−/− mice was about 15% less than that of controls. Urinary excretion of glucosaminoglycans was increased. Coarse granular inclusions equivalent to the Alder–Reilly bodies in human MPSVI were observed in almost all polymorphonuclear leukocytes and in about 50% of lymphocytes. The *Asl-s*−/− mice were fertile and no increase in mortality was noted up to 15 months. In addition to bony abnormalities, the striking radiographic abnormality in *Asl-s*−/− mice was persisting growth plates in long bones and tail vertebrae at 4 and even 8 months of age when growth plates had completely disappeared in control mice. Abnormalities in bone and cartilage with ballooned vacuolated chondrocytes were observed in the vertebrae and skull of the newborn *Asl-s*−/− mice. Accumulation of glycosaminoglycans was widespread in interstitial fibroblast-like cells and macrophages in all tissues investigated histologically. However, parenchymal cells only rarely contained storage materials. With exceptions of leptomeninges and choroid plexus stroma in older mice, no storage of glycosaminoglycans was observed in the brains of *Asl-s*−/− mice. All these pathological features observed in this mouse model are comparable with those in human disease.

3.1.3. Other Lysosomal Diseases

Mouse models of a few other genetic lysosomal disorders have been generated. The following three can be considered primarily glycoprotein disorders. Although the nature of the genetic defects in two of them, galactosialidosis and Schindler disease, suggests possible involvement of sphingolipids, solid evidence is lacking.

3.1.3.1. Galactosialidosis (β-Galactosidase/Sialidase Protective Protein Deficiency) Galactosialidosis is caused by mutations in the gene encording a 32/20-kDa protective dimeric protein with cathepsin A-like activity (protective protein/cathepsin A; PPCA). This protein forms a complex with two lysosomal enzymes, β-galactosidase and sialidase (*N*-acetyl-α-neuraminidase). Deficiency of PPCA secondarily causes a combined deficiency of β-galactosidase and sialidase (Wenger *et al.,* 1978; d'Azzo *et al.,* 1995). Three phenotypic subtypes—early infantile, late infantile, and juvenile/adult types—are known. The early infantile type is associated with fetal hydrops, edema, ascites, visceromegaly, skeletal dysplasia, and death within the first year. The late infantile type is characterized by hepatosplenomegaly, growth retardation, cardiac involvement, and relatively mild neurologic signs. The juvenile/adult form appears to be the most common phenotype, in particular in Japan. Patients develop slowly progressive neurological deterioration with mental retardation, myoclonus, ataxia, macular cherry-red spot, angiokeratoma, and facial and/or skeletal dysmorphia (d'Azzo *et al.,* 1995). Malfunction of the protective protein leads to intralysosomal proteolysis of the enzymes that results in an accumulation of sialyloligosaccharides in lysosomes and excessive excretion of sialyloligosaccharides in urine. Multiple cytoplasmic vacuoles are present in varieties of cell types in the central, peripheral, and autonomic nervous system, retina, liver, kidney, skin, and leukocytes (Yamano *et al.,* 1985; Nordborg *et al.,* 1997). Membranous lamellar inclusions have been described in neurons (Yamano *et al.,* 1985; Oyanagi *et al.,* 1991; Nordborg *et al.,* 1997).

A murine model of galactosialidosis was generated by insertion of the *hygro* cassette into exon 2 of the mouse PPCA gene (Zhou *et al.,* 1995). The mice carrying the null mutation at the PPCA locus (*PPCA*−/−) are viable and fertile. They appear normal during early development except for apparent flattening of the face. However, they usually weigh by 25–40% less than the wild-type or heterozygous mice. With age, progressive and diffuse cutaneous edema accompanied by ataxia and tremor becomes apparent. By 10 months of age, the characteristic broad face, rough coat, and extensive swelling of subcutaneous tissues, limbs, and eyelids are conspicuous. These features closely resemble human patients with galactosialidosis. *PPCA*−/− mice usually die by 12 months of age. Cathepsin A activity is absent or markedly reduced in *PPCA*−/− mice, from being undetectable in bone marrow to 8% of normal in the liver. Urinary excretion of sialyloligosaccharides increases progressively, reaching levels 25 times higher than that of controls at 6–8 months. Activity of α-neuraminidase is severely reduced, especially in fibroblasts and kidney. β-Galactosidase levels varied considerably, however. In human patients, PPCA deficiency results in the combined deficiency of both neuraminidase and β-galactosidase but apparently this was not the case in the murine model, suggesting that, in the mouse, β-galactosidase may be less dependent for its stability and activity on the complex formation with PPCA than α-neuraminidase. Similar to the human infantile form of galactosialidosis, hepatosplenomegaly is present in *PPCA*−/− mice. Prominent and progressive vacuolation is observed in 30% of lymphocytes in the peripheral blood, macrophages in the skin, and connective tissues around visceral organs, renal tubular epithelium, parietal and visceral epithelium, endothelium, and mesangial cells. Vacuolation may also be found in the Kupffer cells, hepatocytes, and splenic macrophages. In the brain, vacuolated cells similar to the macrophages in the visceral organs are found often in the perineuronal or perivascular areas. The choroid plexus epithelial cells are conspicuously vacuolated. These vacuoles contain PAS-positive storage materials. Neuronal storage is more pronounced in the entorhinal cortex and hippocampus than in the somatosensory cortex. Neuronal storage is also noted in the trigeminal and spinal dorsal root ganglia. In the peripheral nerve, many vacuolated cells are scattered among well-myelinated fibers. Pathological features of *PPCA*−/− mice closely resemble the most severe forms of galactosialidosis in humans, although the mouse model survives longer.

3.1.3.2. Aspartylglycosaminuria (Glycosylasparaginase Deficiency) Aspartylglycosaminuria is the most common disorder of glycoprotein degradation in Finland (Thomas and Beaudet, 1995). Only rare isolated cases are reported elsewhere. It is caused by a deficiency of the lysosomal enzyme, glycosylasparginase (*Aga*) resulting in tissue accumulation and urinary excretion of aspartylglucosamine. The clinical course is that of a slowly progressive psychomotor retardation. The patients are relatively healthy for the first year of life except for recurrent infection, diarrhea, and hernia. In later childhood, head circumference and stature are decreased and hepatomegaly may be present. With age, sagging skin folds and coarsening of the facies become progressively obvious. Mental retardation is detected between the ages of 6 and 15. Behavioral abnormalities and skeletal dysplasia have been described. Patients usually die in the third to fifth decade of pulmonary infection. Only very limited reports on the pathology are available. The most notable finding is vacuolated cytoplasm of various cells with variable PAS-positive staining.

Through targeted disruption of the mouse *Aga* gene in embryonic stem cells, a mouse line that completely lacks glycosylasparaginase activity (*Aga*−/−) has been generated (Kaartinen *et al.*, 1996). In *Aga*−/− mice, the natural substrate aspartylglucosamine accumulates in the brain and liver and is excreted in the urine. The mice do not exhibit an overt clinical phenotype and are fertile. At 5 months, however, *Aga*−/− mice show gradual loss of motor coordination and balance. Massively dilated urinary bladder (neurogenic?) is seen at 10 months. As noted in human patients, vacuolation of cells in the CNS and visceral organs are the most prominent pathology of *Aga*−/− mice. The Kupffer cells and the cells in the renal glomeruli and proximal renal tubules are also severely affected. In the brain, vacuolation is present in neurons and in endothelial and glial cells. In addition, axonal spheroids are present in the nucleus gracilis in the medulla and spinal cord. Thus, manifestations in *Aga*−/− mice appear to be clinically and pathologically similar to the human disease.

*3.1.3.3. Schindler Disease (*N*-Acetyl α-Galactosaminidase Deficiency)* Schindler disease in humans is due to genetic deficiency of *N*-acetyl α-galactosamindase, also known as α-galactosidase B. Type I disease manifests itself as infantile neuroaxonal dystrophy. Patients with type I disease undergo a rapidly progressive neurodegenerative course starting at 1–2 years with severe psychomotor retardation, cortical blindness, and myoclonic seizures. Type II patients have much later onset and only mild mental deterioration. Although a mouse model was generated relatively early by gene targeting, only a meeting abstract that announced the successful production of the line has been published (Wang *et al.*, 1993). It is understood that the model is either phenotype-free or of a relatively mild phenotype. More detailed accounts of this mouse model would be welcome.

3.2. Lysosomal Disorders Not Known in Humans

The technology of homologous recombination in embryonic stem cells allows inactivation of any genes in the murine genome once they are cloned. Many induced murine mutants of lysosomal enzymes or associated proteins are being generated including those that have no known equivalent human diseases. Simultaneous genetic deficiency of the β-hexosaminidase α- and β-subunits already described is not known in humans. Others that belong to this category include lysosomal acid phosphatase deficiency (Saftig *et al.*, 1997) and deficiency of cation-dependent and cation-independent mannose 6-phosphate receptors and their double knockouts (Köster *et al.*, 1993; Ludwig *et al.*, 1993; Lau *et al.*, 1994).

In a recent meeting of the European Study Group for Lysosomal Disorders, successful generation of several new mouse mutants with genetic deficiency of lysosome-associated enzymes/proteins was reported with preliminary characterization of their clinical and biochemical phenotypes. The inactivated genes included *LAMP I, LAMP II, LIMP I,* cathepsin B, cathepsin L, and tartrate-resistant acid phosphatase (abstracts for the 11th meeting of the European Study Group for Lysosomal Disorders, Bad Deutsch-Altenburg, Austria, September 18–21, 1997). Full characterization of these mutants should become available in the literature in the near future. None of these induced mutants have known counterparts in humans.

Even without known human counterparts, these mutants are important because they can shed new light on the structure and function of the lysosome in relation to the general cellular physiology and on the interactions among lysosomal proteins. It is also conceivable that some of them may in fact anticipate discoveries of equivalent human genetic disorders, as in β-mannosidosis which was first discovered in goat and subsequently in man.

4. UTILITY OF EXPERIMENTAL ANIMAL MODELS

4.1. For Studies of Pathogenetic Mechanisms

Despite the recent dizzying advances in studies of genetic lysosomal diseases on the nucleic acid level, the molecular sequences of the pathogenesis of these disorders, particularly those of brain dysfunction, are still poorly understood. Mouse models are well suited for exploration of the pathogenetic mechanisms of genetic lysosomal disorders. This will be illustrated by representative examples.

4.1.1. β-Hexosaminidase Deficiencies

As described previously, cross-breeding of the β-hexosaminidase α- and β-subunit-deficient mouse lines produced a double knockout with an unexpected phenotype. Very few mucopolysaccharidosis-like features that are conspicuous in the double-knockout mice are present in either of the single-gene defects. This finding pointed to the likely functional redundancy among the three β-hexosaminidase isozymes, A, B, and S. Both β-hexosaminidase A and GM2-activator-deficient mice gave relatively mild phenotype. The current interpretation of the interactions among the hexosaminidase A, B, sialidase, and GM2 activator can be explored by generating mice simultaneously deficient in both β-hexosaminidase α-subunit and GM2-activator. Utilizing the β-hexosaminidase-deficient mice, Huang *et al.* recently reported that the neuronal death in these disorders follows the apoptotic process (Huang *et al.,* 1997).

4.1.2. Interaction with an Immune System Gene

In another cross-breeding experiment, the interaction of a major histocompatibility complex gene in the pathogenesis of Krabbe disease was explored using the *twitcher* mutant (Matsushima *et al.,* 1994). It had been noted in the *twitcher* mutant that the major histocompatibility complex (MHC) class II (Ia) might have a specific involvement in the pathogenesis of this genetic demyelinating disease. Mice doubly deficient in both galactosylceramidase (*twitcher*) and the MHC class II molecules were generated by cross-breeding the *twitcher* line with a MHC class II knockout mouse line. In these mice, clinical symptoms and histopathology of the cerebrum and the brain stem/cerebellar region were milder than those in *twitcher* mice with the MHC-II positive background but there was no noticeable improvement in the pathology of the spinal cord. Preliminary analysis of psychosine level also showed less accumulation than in the *twitcher* mice with the normal MHC background, consistent with the lesser degree of pathology in these mice. Although this experiment specifically explored the influence of the *MHC II* gene on the course of the *twitcher* mutant, it also points to the more general question of the influence of different genetic background on the course of genetic lysosomal disease. Because humans are hopelessly complex genetically, mouse models are ideally suited for such studies.

4.1.3. Development of Psychosine Hypothesis

The psychosine hypothesis was proposed initially on the basis of enzymology to explain the unique pathological and biochemical features of human Krabbe disease (Miyatake and Suzuki, 1972), and its accumulation in the brains of human patients was first demonstrated eight years later (Vanier and Svennerholm, 1976). It was in the *twitcher* mouse brain where the rapidly progressive nature of the accumulation was first established (Igisu and Suzuki, 1984). The brain psychosine level appears to be a sensitive parameter for assessment of the efficacy of therapeutic trials of this disorder using the *twitcher* mouse as a model (see later).

4.2. For Therapeutic Trials

One of the major utilities of mouse models of human lysosomal disorders is as a tool for therapeutic trials. The spontaneous or induced murine models of human lysosomal diseases are being used with increasing frequency as the vehicle to test various approaches for treatment of these generally devastating disorders. Genetically authentic mouse models can overcome many of the scientific disadvantages in working with human patients. Often overlooked but in fact very important is that these mice can be bred to a homogeneous genetic background, which can never be achieved in the human population. Consequences of complex interactions among different genes in these disorders are being recognized in increasing frequency (e.g., Matsushima *et al.,* 1994). Then, one can do well-controlled experiments with sufficient numbers of normal, carrier, and affected animals with relative ease. Experimental manipulations that are not allowed in human patients can be done with mice, including germline genetic manipulation. Because of the longer availability, the spontaneous mutants have been utilized more frequently for therapeutic attempts but more extensive use of induced mouse models for gene therapy attempts can be predicted in the near future. The results will be "cleaner" and the conclusions more reliable than human patient studies. There is no room for anecdotal cases which so often clutter clinical studies and obscure the true logical conclusions or lack thereof. For strictly scientific inquiries into how Nature works, the mouse models provide a much better vehicle. On the other hand, if the ultimate aim is to understand and treat human diseases, it is equally important that we always keep in mind the simple and obvious dictum: "Mouse is not human." Even though the mouse models are useful and even though some other models in larger animals including monkeys (Luzi *et al.,* 1997) can fill the gap to some extent, studies on human patients are indispensable for pragmatic investigations. Animal experiments and human studies are complementary and both are needed.

4.2.1. Enzyme Replacement

The enzyme replacement therapy has been highly successful in human patients with chronic non-neuropathic form of Gaucher disease. Patient trials for enzyme replacement for Fabry disease is being contemplated. However, relatively few attempts have been made for replacing the deficient enzymes using the mouse models. The naturally occurring β-glucuronidase deficient mouse has been used extensively for both enzyme replacement or bone transplantation alone and combined enzyme replacement and bone marrow transplantation (Sands *et al.*, 1993, 1994, 1997a; Vogler *et al.*, 1993, 1996). They found that these treatments effectively reduced the abnormal storage and pathology in systemic tissues and to some extent also in the brain. One important conclusion was that the treatment should be initiated as soon as possible after birth for maximum effectiveness. Enzyme replacement was reported in the induced mouse model of Fabry disease with encouraging results (Ioannou *et al.*, 1996).

4.2.2. Transgenic Treatment

Treating genetic disorders by introducing normal genes into the germline cells is not an approach permitted for human patients. However, as a biological experiment, this approach provides an ideal means to supplement the defective genes with the normal genes. Studies of their consequences in animal models will be invaluable for designing the strategy for other therapeutic approaches. Kyle *et al.* reported a highly effective result of a transgenic therapy of the β-glucuronidase-deficient mouse (Kyle *et al.*, 1990). Matsumoto *et al.* introduced the human galactosylceramidase gene into the *twitcher* mouse as a transgene driven by the mouse myelin basic protein promoter (Matsumoto *et al.*, 1997). Although they failed to achieve the aimed-at robust expression of transgene, the results suggested that a relatively small increment in the activity of galactosylceramidase might be sufficient for effective treatment of the disease. The most significant changes were observed in the level of brain psychosine which was reduced to approximately half that in the brain of untreated *twitcher* mice at 35 days despite the nearly negligible expression of the introduced exogenous gene.

4.2.3. Bone Marrow Transplantation

In addition to the β-glucuronidase-deficient mouse discussed previously, the *twitcher* mutant has been subjected to the bone marrow transplantation treatment (e.g., Yeager *et al.*, 1998; Hoogerbrugge *et al.*, 1988a,b; Ichioka *et al.*, 1987; Suzuki *et al.*, 1988). Transplantation of normal donor bone marrow cells at about 7–8 days prolonged the life span of affected *twitcher* mice by a factor of 3, up to 100 days. The enzyme levels were increased substantially in all organs including the brain, and donor-derived cells were demonstrated within the CNS. In later stages, frequent pictures of remyelination could be demonstrated by electron microscopy. Zhou and co-workers (Zhou *et al.*, 1995) reported successful treatment of the galactosialidosis mouse model with transplantation of the bone marrow cells. They used bone marrow cells from transgenic mice that overexpressed human protective protein precursor gene in the erythroid lineage cells under the control of the β-globin promotor and the locus control region of the β-globin gene (Grosveld *et al.*, 1987). They achieved dramatic improvement in the clinical phenotype and systemic pathology. However, no significant improvement could be observed in the CNS. The sphingomyelinase-deficient (Niemann–Pick A and B) mouse has been used to evaluate various parameters that affect the outcome of bone marrow transplantation (Miranda *et al.*, 1997). Bone marrow transplantation into the *Hexb*−/− (Sandhoff disease) gave a surprising and encouraging correction of the clinical phenotype including neurological manifestations (Norflus *et al.*, 1998). However, despite the neurological improvement, pathological, enzymatic, and biochemical corrections were limited to systemic organs and no detectable effects were observed in the brain, except the presence of β-hexosaminidase-positive cells.

4.2.4. Gene Therapy

Effective gene therapy in the narrow sense, correction of the gene defects by introduction and expression of normal genes into somatic cells of the whole animal, is the ultimate goal for treatment of human lysosomal disorders. The mouse models are suitable for this approach. While many laboratories are working with their favorite models with their favorite vectors, the field is largely in the preliminary stage. Cultured cell lines and/or reporter genes instead of the native genes that are defective in the respective diseases are being used to establish the optimal conditions for experiments using the whole animal (e.g., Sands *et al.*, 1997b). *In vivo* experiments are still of limited scale. It is likely that we are at the verge of an explosive growth of use of these animal models for gene therapy experiments. Correct cellular and subcellular targeting of the introduced exogenous gene will be crucial for successful outcome of the treatment. The unique features of the brain must also be considered because the nervous system is severely affected in many of the genetic lysosomal disorders. Some of the characteristics of the brain include the existence of the blood–brain barrier, the highly complex and temporally tightly regulated developmental process of the brain, the multiple cell types with their unique functions (neurons, astrocytes, oligodendroglia, microglia, blood vessels), and the largely postmitotic nature of the neural cells.

4.2.5. Treatment with Inhibitors of the Substrate Synthesis

Because the lysosomal disorders are diseases of genetic blocks in the catabolic steps of the affected substrates, the idea of administering metabolic inhibitors to reduce synthesis of the substrates to slow down the abnormal accumulation has been suggested in the past 20 years by Radin, Lev, and other investigators. This approach has a few fundamental logical weaknesses compared to the more conventional approach of increasing the defective catabolic capacity. It is logically impossible to ensure complete absence of side effects, as it is not possible to test all possibilities. Also, rodents may not be a sensitive enough experimental animal for testing side effects in human patients. Furthermore, the nature of this approach is such that the procedure may be able to slow down the process of abnormal accumulation of the substrates but not to reduce the already accumulated materials, unless there is meaningful residual activity of the defective enzymes or a minor alternate metabolic pathway through which the accumulating substrates can be degraded to some extent. This approach was applied recently to the *Hexa−/−* mice (Tay–Sachs disease) with *N*-butyldeoxynojirimycin, which is a potent inhibitor of glucosylceramide synthesis (Platt *et al.*, 1997). They described encouraging slowdown of GM2-ganglioside accumulation in the brain, although comparison of single neurons from treated and untreated mice was inadequate as demonstration of the claimed improved pathology. Assessment of clinical improvement was not possible because this knockout model is clincally normal. Undoubtedly, this approach will be further tested in various mouse models.

5. ON THE HORIZON

As described, genetic diseases corresponding to nearly all human sphingolipidoses, three mucopolysaccharidoses and aspartylglycosaminuria exist either spontaneously or have been duplicated artificially in mice. Undoubtedly mouse models of other human lysosomal disorders will be generated in the near future. Technologies now exist to allow introduction of more subtle gene abnormalities than mere complete inactivation, including tissue/organ and developmental stage-specific expression, expression that can be switched on and off by external manipulation, and point mutations. Examples of mouse models of point mutations have been mentioned previously for Gaucher and Fabry diseases. These recent methodologies will add more varieties to available mouse models. However, many of the already generated mutants have not been extensively explored for their potential as models of human lysosomal diseases beyond the basic characterization. Genetic manipulation by cross-breeding among these mutants or with other mutant mice are being applied with increasing freqnency. These double-knockout mice should bring insight into the interactions among related or seemingly unrelated genes and also into the pathogenetic mechanism of these disorders unattainable with other means as has already been demonstrated by the double-knockout of the α- and β-subunits of β-hexosaminidase.

Acknowledgments

The work from the authors' laboratories was supported in part by research grants RO1-NS24289 and RO1-NS28997, and a Mental Retardation Research

Center Core Grant, P30-HD03110, to Ku.S., RO1-24453 to Ki.S. from the USPHS, and a research grant from the ELA to M. T. V.

REFERENCES

Akaboshi, S., Yano, T., Miyawaki, S., Ohno, K., and Takeshita, K., 1997, A C57BL/KsJ mouse model of Niemann–Pick disease (spm) belongs to the same complementation group as the major childhood type of Niemann–Pick disease type C, *Hum. Genet.* **99:**350–353.

Bapat, B., Ethier, M., Neote, K., Mahuran, D., and Gravel, R. A., 1988, Cloning and sequence analysis of a cDNA encoding the β-subunit of mouse β-hexosaminidase, *FEBS Lett.* **237:**191–195.

Bellachioma, G., Stirling, J. L., Orlacchio, A., and Beccari, T., 1993, Cloning and sequence analysis of a cDNA clone coding for the mouse GM2 activator protein, *Biochem. J.* **294:**227–230.

Bermudez, A. J., Johnson, G. C., Vanier, M. T., Schröder, M., Suzuki, K., Stogsdill, P. L., Johnson, G. S., O'Brien, D., Moore, C. P., and Fry, W. W., 1995, Gangliosidosis in emus (Dromaius novaehollandiae), *Avian. Dis.* **39:**292–303.

Birkenmeier, E. H., Davisson, M. T., Beamer, W. G., Ganschow, R. E., Vogler, C. A., Gwynn, B., Lyford, K. A., Maltais, L. M., and Wawrzyniak, C. J., 1989, Murine mucopolysaccharidosis type VII. Characterization of a mouse with β-glucuronidase deficiency, *J. Clin. Invest.* **83:**1258–1266.

Bradová, V., Smid, F., Ulrich-Bott, B., Roggendorf, W., Paton, B. C., and Harzer, K., 1993, Prosaposin deficiency: Further characterization of the sphingolipid activator protein-deficient sibs. Multiple glycolipid elevations (including lactosylceramidosis), partial enzyme deficiencies and ultrastructure of the skin in this generalized sphingolipid storage disease, *Hum. Genet.* **92:**143–152.

Burg, J., Banerjee, A., Conzelmann, E., and Sandhoff, K., 1983, Activating proteins for ganglioside GM2 degradation by beta-hexosaminidase isoenzymes in tissue extracts from different species, *Hoppe Seylers Z. Physiol. Chem.* **364:**821–829.

Cannizzaro, L., A., Chen, Y. Q., Rafi, M. A., and Wenger, D. A., 1994, Regional mapping of the human galactocerebrosidase gene (GALC) to 14q31 by in situ hybridization, *Cytogenet. Cell Genet.* **66:**244–245.

Carstea, E. D., Polymeropoulos, M. H., Parker, C. C., Detera-Wadleigh, S. D., O'Neill, R. R., Patterson, M. C., Goldin, E., Xiao, H., Straub, R. E., Vanier, M. T., Brady, R. O., and Pentchev, P. G., 1993, Linkage of Niemann–Pick disease type C to human chromosome 18, *Proc. Natl. Acad. Sci USA* **90:**2002–2004.

Carstea, E. D., Morris, J. A., Coleman, K. G., Loftus, S. K., Zhang, D., Cummings, C., Gu, J., Rosenfeld, M. A., Pavan, W. J., Krizman, D. B., Nagle, J., Polymeropoulos, M. H., Sturley, S. L., Ioannou, Y. A., Higgins, M. E., Comly, M., Cooney, A., Brown, A., Kaneski, C. R., Blanchette-Mackie, E. J., Dwyer, N. K., Neufeld, E. B., Chang, T. Y., Liscum, L., Strauss, J. F., III, Ohno, K., Zeigler, M., Carmi, R., Sokol, J., Markie, D., O'Neill, R. R., van Diggelen, O. P., Elleder, M., Patterson, M. C., Brady, R. O., Vanier, M. T., Pentchev, P. G., and Tagle, D. A., 1997, Niemann–Pick C1 disease gene: Homology to mediators of cholesterol homeostasis, *Science* **277:**228–231.

Chen, Y. Q., Rafi, M. A., De Gala, G., and Wenger, D. A., 1993, Cloning and expression of cDNA encoding human galactocerebrosidase, the enzyme deficient in globoid cell leukodystrophy, *Hum. Mol. Genet.* **2:**1841–1845.

Clarke, L. A., Russell, C. S., Pownall, S., Warrington, C. L., Borowski, A., Dimmick, J. E., Toone, J., and Jirik, F. R., 1997, Murine mucopolysaccharidosis type I: Targeted disruption of the murine α-L-iduronidase gene, *Hum. Mol. Genet.* **6**:503–511.

Cohen-Tannoudji, M., Marchand, P., Akli, S., Sheardown, S. A., Puech, J.-P., Kress, C., Gressens, P., Nassogne, M.-C., Beccari, T., Muggleton-Harris, A. L., Evrard, P., Stirling, J. L. Poenaru, L., and Babinet, C., 1995, Disruption of murine *Hexa* gene leads to enzymatic deficiency and to neuronal lysosomal storage, similar to that observed in Tay–Sachs disease, *Mamm. Genome* **6**:844–849.

d'Azzo, A., Andrea, G., Strisciuglio, P., and Galjaard, H., 1995, Galactosialidosis, *in* "The Metabolic and Molecular Basis of Inherited Disease" (C. R. Scriver, A. L. Beaudet, W. S. Sly, and D. Valle, eds.), pp. 2825–2838, McGraw-Hill, New York.

Desnick, R. J., Ioannou, Y. A., and Eng, C. M., 1995, α-Galactosidase A deficiency: Fabry Diseasse *in* "The Metabolic and Molecular Basis of Inherited Disease" (C. R. Scriver, A. L. Beaudet, W. S. Sly, and D. Valle, eds.), pp. 2741–2784, McGraw-Hill, New York.

DeVeber, G. A., Schwarting, G. A., Kolodny, E. H., and Kowall, N. W., 1992, Fabry disease: Immunocytochemical characterization of neuronal involvement, *Ann. Neurol.* **31**:409–415.

Erickson, R. P., Aviles, R. A., Zhang, J., Kozloski, M. A., Garver, W. S., and Heidenreich, R. A., 1997, High-resolution mapping of the *spm* (Niemann–Pick type C) locus on mouse chromosome 18, *Mamm. Genome* **8**:355–356.

Evers, M., Saftig, P., Schmidt, P., Hafner, A., McLoghlin, D. B., Schmahl, W., Hess, B., von Figura, K., and Peters, C., 1996, Targeted disruption of the arylsulfatase B gene results in mice resembling the phenotype of mucopolysaccharidosis VI, *Proc. Natl. Acad. Sci. USA* **93**:8214–8219.

Freischütz, B., Tokuda, A., Ariga, T., Bermudez, A. J., and Yu, R. K., 1997, Unusual gangliosides in emu (*Dromaius novaehollandiae*) *J. Neurochem.* **68**:2070–2078.

Fujita, N., Suzuki, K., Vanier, M. T., Popko, B., Maeda, N., Klein, A., Henseler, M., Sandhoff, K., and Nakayasu, H., 1996. Targeted disruption of the mouse sphingolipid activator protein gene: A complex phenotype, including severe leukodystrophy and wide-spread storage of multiple sphingolipids, *Hum. Mol. Genet.* **5**:711–725.

Gieselmann, V., Zlotogora, J., Harris, A., Wenger, D. A., and Morris, C. P., 1994, Molecular genetics of metachromatic leukodystrophy, *Hum. Mutat.* **4**:233–242.

Goldin, E., Roff, C. F., Miller, S. P. F., Rodriguez-Lafrasse, C., Vanier, M. T., Brady, R. O., and Pentchev, P. G., 1992, Type C Niemann–Pick disease: A murine model of the lysosomal cholesterol lipidosis accumulates sphingosine and sphinganine in liver, *Biochim. Biophys. Acta* **1127**:303–311.

Gravel, R. A., Clarke, J. T. R., Kaback, M. M., Mahuran, D., Sandhoff, K., and Suzuki, K., 1995, The GM2 gangliosidosis *in* "The Metabolic and Molecular Basis of Inherited Disease" (C. R. Scriver, A. L. Beaudet, W. S. Sly, and D. Valle, eds.), pp. 2839–2879, McGraw-Hill, New York.

Gregoire, A., Perier, O., and Dustin, P., 1966, Metachromatic leukodystrophy, an electron microscopic study, *J. Neuropath. Exp. Neurol.* **25**:617–636.

Grosveld, F., van Assendelft, G. B., Greaves, D. R., and Kollias, G., 1987, Position-independent, high-level expression of the human β-globin gene in transgenic mice, *Cell* **51**:975–985.

Hahn, C. N., Martin, M. D., Schröder, M., Vanier, M. T., Hara, Y., Suzuki, K., and d'Azzo, A., 1997, Generalized CNS disease and massive G_{M1}-ganglioside accumulation in mice defective in lysosomal acid β-galactosidase, *Hum. Mol. Genet.* **6**:205–211.

Harzer, K., Paton, B. C., Poulos, A., Kustermann-Kuhn, B., Roggendorf, W., Grisar, T., and Popp, M., 1989, Sphingolipid activator deficiency in a 16-week-old atypical Gaucher disease patient and his fetal sibling: Biochemical signs of combined sphingolipidoses, *Eur. J. Pediatr.* **149**:31–39.

Hers, H. G., 1966, Inborn lysosomal disease, *Gastroenterology* **48**:625–633.

Hess, B., Saftig, P., Hartmann, D., Coenen, R., Lüllmann-Rauch, R., Goebel, H. H., Evers, M., von Figura, K., D'Hooge, R., Nagels, G., De Deyn, P., Peters, C., and Gieselmann, V., 1996,

Phenotype of arylsulfatase A-deficient mice: Relationship to human metachromatic leukodystrophy, *Proc. Natl. Acad. Sci. USA* **93:**14821–14826.

Ho, M. W., and O'Brien, J. S., 1971, Gaucher's disease: Deficiency of "acid" β-glucosidase and reconstitution of enzyme activity in vitro, *Proc. Natl. Acad. Sci USA* **68:**2810–2813.

Holleran, W. M., Ginns, E. I., Menon, G. K., Grundmann, J.-U., Fartasch, M., McKinney, C. E., Elias, P. M., and Sidransky, E., 1994, Consequences of β-glucocerebrosidase deficiency in epidermis. Ultrastructure and permeability barrier alterations in Gaucher disease, *J. Clin. Invest.* **93:**1756–1764.

Hoogerbrugge, P. M., Poorthuis, B. J. H. M., Romme, A. E., van de Kamp, J. J. P., Wagemaker, G., and van Bekkum, D. W., 1988a, Effect of bone marrow transplantation on enzyme levels and clinical course in the neurologically affected twitcher mouse, *J. Clin. Invest.* **81:**1790–1794.

Hoogerbrugge, P. M., Suzuki, K., Suzuki, K., Poorthuis, B. J. H. M., Kobayashi, T., Wagemaker, G., and van Bekkum, D. W., 1988b, Donor derived cells in the central nervous system of twitcher mice after bone marrow transplantation, *Science* **239:**1035–1038.

Horinouchi, K., Erlich, S., Perl, D. P., Ferlinz, K., Bisgaier, C. L., Sandhoff, K., Desnick, R. J., Stewart, C. L., and Schuchman, E. H., 1995, Acid sphingomyelinase deficient mice: A model of types A and B Niemann–Pick disease, *Nat. Genet.* **10:**288–293.

Huang, J.-Q., Trasler, J. M., Igdoura, S., Michaud, J., Hanai, N., and Gravel, R. A., 1997, Apoptotic cell death in mouse models of GM2 gangliosidosis and observations on human Tay–Sachs and Sandhoff diseases, *Hum. Mol. Genet.* **6:**1879–1885.

Ichioka, T., Kishimoto, Y., Brennan, S., Santos, G. W., and Yeager, A. M., 1987, Hematopoietic cell transplantation in murine globoid cell leukodystrophy (the twitcher mouse): Effects on levels of galactosylceramidase, psychosine, and galactocerebrosides, *Proc. Natl. Acad. Sci. USA* **84:**4259–4263.

Ida, H., Umezawa, F., Kasai, E., Eto, Y., and Maekawa, K., 1982, An accumulation of galactocerebroside in kidney from mouse globoid cell leukodystrophy (twitcher), *Biochem. Biophys. Res. Commun.* **109:**634–638.

Igisu, H., and Suzuki, K., 1984, Progressive accumulation of toxic metabolite in a genetic leukodystrophy, *Science* **224:**753–755.

Igisu, H., Takahashi, H., Suzuki, K., and Suzuki, K., 1983, Abnormal accumulation of galactosylceramide in the kidney of twitcher mouse, *Biochem. Biophys. Res. Commun.* **110:**940–944.

Ioannou, Y. A., Zeidner, K. M., Friedman, B., and Desnick, R. J., 1996, Fabry disease: Enzyme replacement therapy in α-galactosidase A deficient mice, *Am. J. Hum. Genet.* **59** (Suppl.):A15 (abstract no. 72).

Kaartinen, V., Mononen, I., Voncken, J. W., Noronkoski, T., Gonzalez-Gomez, I., Heisterkamp, N., and Groffen, J., 1996, A mouse model for the human lysosomal disease aspartylglycosaminuria, *Nat. Med.* **2:**1375–1378.

Kamoshita, S., Aaron, A. M., Suzuki, K., and Suzuki, K., 1969, Infantile Niemann–Pick disease: A chemical study with isolation and characterization of membranous cytoplasmic bodies and myelin, *Am. J. Dis. Child.* **117:**379–394.

Kobayashi, T., Yamanaka, T., Jacobs, J. M., Teixeira, F., and Suzuki, K., 1980, The twitcher mouse: An enzymatically authentic model of human globoid cell leukodystrophy (Krabbe disease), *Brain Res.* **202:**479–483.

Kolodny, E. H., and Fluharty, A. L., 1995. Metachromatic leukodystrophy and multiple sulfatase deficiency: Sulfatide lipidosis *in* "The Metabolic and Molecular Basis of Inherited Disease" (C. R. Scriver, A. L. Beaudet, W. S. Sly, and D. Valle, eds.), pp. 2693–2739, McGraw-Hill, New York.

Köster, A., Saftig, P., Matzner, U., von Figura, K., Peters, C., and Pohlmann, R., 1993, Targeted disruption of othe M(r) 46,000 mannose 6-phosphate receptor gene in mice results in misrouting of lysosomal proteins, *EMBO J.* **12:**5219–5223.

Kuemmel, T. A., Schroeder, R., and Stoffel, W., 1997, Light and electron microscopic analysis of the central and peripheral nervous system of acid sphingomyelinase-deficient mice resulting from gene targeting, *J. Neuropathol. Exp. Neurol.* **56:**171–179.

Kurimasa, A., Ohno, K., and Oshimura, M., 1993, Restoration of the cholesterol metabolism in 3T3 cell lines derived from the sphingomyelinosis mouse (spm/spm) by transfer of a human chromosome 18, *Hum. Genet.* **92:**157–162.

Kyle, J. W., Birkenmeier, E. H., Gwynn, B., Vogler, C., Hoppe, P. C., Hoffmann, J. W., and Sly, W. S., 1990, Correction of murine mucopolysaccharidosis VII by a human β-glucuronidase transgene, *Proc. Natl. Acad. Sci. USA* **87:**3914–3918.

Lau, M. M., Stewart, C. E., Liu, Z., Bhatt, H., Rotwein, P., and Stewart, C. L., 1994, Loss of imprinted IGF2/cation-independent mannose 6-phosphate receptor results in fetal overgrowth and perinatal lethality, *Genes Dev.* **8:**2953–2963.

Liu, Y., Hoffmann, A., Grinberg, A., Westphal, H., McDonald, M. P., Miller, K. M., Crawley, J. N., Sandhoff, K., Suzuki, K., and Proia, R. L., 1997, Mouse model of GM2 activator deficiency manifests cerebellar pathology and motor impairment, *Proc. Natl. Acad Sci. USA* **94:**8138–8143.

Liu, Y., Suzuki, K., Reed, J., Grinberg, A., Westphal, H., Hoffmann, A., Döring, T., Sandhoff, K., and Proia, R. L., 1998, Mice with type 2 and 3 Gaucher disease point mutations generated by a single insertion mutagenesis procedure (SIMP), *Proc. Natl. Acad. Sci. USA* **95:**2503–2508.

Loftus, S. K., Morris, J. A., Carstea, E. D., Gu, J. Z., Cummings, C., Brown, A., Ellison, J., Ohno, K., Rosenfeld, M. A., Tagle, D. A., Pentchev, P. G., and Pavan, W. J., 1997, Murine model of Niemann–Pick C disease: Mutation in a cholesterol homeostasis gene, *Science* **277:**232–235.

Ludwig, T., Ovitt, C. E., Bauer, U., Hollinshead, M., Remmler, J., Lobel, P., Rüther, U., and Hoflack, B., 1993, Targeted disruption of the mouse cation-dependent mannose 6-phosphate receptor results in partial missorting of multiple lysosomal enzymes, *EMBO J.* **12:**5225–5235.

Luzi, P., Rafi, M. A., Victoria, T., Baskin, G. B., and Wenger, D. A., 1997, Characterization of the rhesus monkey galactocerebrosidase (GALC) cDNA and gene and identification of the mutation causing globoid cell leukodystrophy (Krabbe disease) in this primate, *Genomics* **42:**319–324.

Matsuda, J., Suzuki, O., Oshima, A., Ogura, A., Naiki, M.L, and Suzuki, Y., 1997a, Neurological manifestations of knockout mice with β-galactosidase deficiency, *Brain Dev.* **19:**19–20.

Matsuda, J., Suzuki, O., Oshima, A., Ogura, A., Noguchi, Y., Yamamoto, Y., Asano, T., Takimoto, K., Sukegawa, K., Suzuki, Y., and Naiki, M., 1997b, β-Galactosidase-deficient mouse as an animal model for G_{M1}-gangliosidosis, *Glycoconjugate J.* **14:**729–736.

Matsumoto, A., Vanier, M. T., Oya, Y., Kelly, D., Popko, B., Wenger, D. A., Suzuki, K., and Suzuki, K., 1997, Minimal increment in galactosylceramidase expression is sufficient for significant phenotypic improvement in twitcher mouse, *Dev. Brain Dysfunction,* **10:**142–154.

Matsushima, G. K., Taniike, M., Glimcher, L. H., Grusby, M. J., Frelinger, J. A., Suzuki, K., and Ting, J. P. Y., 1994, Absence of MHC class II molecules reduces CNS demyelination, microglial/macrophage infiltration, and twitching in murine globoid cell leukodystrophy, *Cell.* **78:**645–656.

Mehl, E., and Jatzkewitz, H., 1964, Eine Cerebrosidsulfatase aus Schweineniere, *Hoppe Seylers Z. Physiol. Chem.* **339:**260–276.

Miranda, S. R., Erlich, S., Visser, J. W., Gatt, S., Dagan, A., Friedrich, V. L., Jr., and Schuchman, E. H., 1997, Bone marrow transplantation in acid sphingomyelinase-deficient mice: Engraftment and cell migration into the brain as a function of radiation, age, and phenotype, *Blood* **90:**444–452.

Miyatake, T., and Suzuki, K., 1972, Globoid cell leukodystrophy: Additional deficiency of psychosine galactosidase, *Biochem. Biophys. Res. Commun.* **48:**538–543.

Miyawaki, S., Mitsuoka, S., Sakiyama, T., and Kitagawa, T., 1982, Sphingomyelinosis, a new mutation in the mouse. A model of Niemann–Pick disease in humans, *J. Hered.* **73:**257–263.

Morris, M. D., Bhuvaneswaren, C., Shio, H., and Fowler, S., 1982, Lysosomal lipid storage disorder in NCTR-BALB c mice. I. Description of the disease and genetics, *Am. J. Pathol.* **108:**140–149.

Neufeld, E. F., and Muenzer, J., 1995, The mucopolysaccharidoses *in* "The Metabolic and Molecular Basis of Inherited Disease" (C. R. Scriver, A. L., Beaudet, W. S. Sly, and D. Valle, eds.), pp. 2465–2494, McGraw-Hill, New York.

Nordborg, C., Kyllerman, M., Conradi, N.L, and Månsson, J. E., 1997, Early-infantile galactosialidosis with multiple brain infarctions: Morphological, neuropathological and neurochemical findings, *Acta Neuropathol. (Berl.).* **93:**24–33.

Norflus, F., Tifft, C. J., McDonald, M. P., Goldstein, G., Crawley, J. N., Hoffmann, A., Sandhoff, K., Suzuki, K., and Proia, R. L., 1998, Bone marrow transplantation prolongs life span and ameliorates neurologic manifestations in Sandhoff disease mice, *J. Clin. Invest,* **101:**1881–1888.

Ohno, K., Nanba, E., Miyawaki, S., Sakiyama, T., Kitagawa, T., and Takeshita, K., 1992, A cell line derived from sphingomyelinosis mouse shows alterations in intracellular cholesterol metabolism similar to those in type C Niemann–Pick disease, *Cell Struct. Funct.* **17:**229–235.

Ohshima, T., Murray, G. J., Swaim, W. D., Longenecker, G., Quirk, J. M., Cardarelli, C. O., Sugimoto, Y., Pastan, I., Gottesman, M. M., Brady, R. O., and Kulkarni, A. B., 1997, α-Galactosidase A deficient mice: A model of Fabry disease, *Proc. Natl. Acad. Sci. USA* **94:**2540–2544.

Otterbach, B., and Stoffel, W., 1995, Acid sphingomyelinase-deficient mice mimic the neurovisceral form of human lysosomal storage disease (Niemann–Pick disease), *Cell* **81:**1053–1061.

Oya, Y., Nakayasu, H., Fujita, N., Suzuki, K., and Suzuki, K., 1998, Pathological study of mice with total deficiency of sphingolipid activator proteins (SAP knockout mice), *Acta Neuropathol. (Berl.).* **96:**29–40.

Oyanagi, K., Ohama, E., Miyashita, K., Yoshino, H., Miyatake, T., Yamazaki, M., and Ikuta, F., 1991, Galactosialidosis: Neuropathological findings in a case of the late infantile type, *Acta Neuropathol. (Berl).* **82:**331–339.

Peng, L., and Suzuki, K., 1987, Ultrastructural study of neurons in metachromatic leukodystrophy, *Clin. Neuropathol.* **6:**224–230.

Pentchev, P. G., Gal, A. E., Booth, A. D., Omodeo-Sale, F., Fouks, J., Neumeyer, B. A., Quirk, J. M., Dawson, G., and Brady, R. O., 1980, A lysosomal storage disorder in mice characterized by a dual deficiency of sphingomyelinase and glucocerebrosidase, *Biochim. Biophys. Acta* **619:**669–679.

Pentchev, P. G., Boothe, A. D., Kruth, H. S., Weintraub, H., Stivers, J., and Brady, R. O., 1984, A genetic storage disorder in BALB C mice with metabolic block in esterification of exogenous cholesterol, *J. Biol. Chem.* **259:**5784–5791.

Pentchev, P. G., Vanier, M. T., Suzuki, K., and Patterson, M. C., 1995, Niemann–Pick disease type C: A cellular cholesterol lipidosis *in* "The Metabolic and Molecular Basis of Inherited Disease" (C. R. Scriver, A. L. Beaudet, W. S. Sly, and D. Valle, eds.), pp. 2625–2639, McGraw-Hill, New York.

Phaneuf, D., Wakamatsu, N., Huang, J. Q., Borowski, A., Peterson, A. C., Fortunato, S. R., Ritter, G., Igdoura, S. A., Morales, C. R., Benoit, G., Akerman, B. R., Leclerc, D., Hanai, N., Marth, J. D., Trasler, J. M., and Gravel, R. A., 1996, Dramatically different phenotypes in mouse models of human Tay–Sachs and Sandhoff diseases, *Hum. Mol. Genet.* **5:**1–14.

Platt, F. M., Neises, G. R., Reinkensmeier, G., Townsend, M. J., Perry, V. H., Proia, R. L., Winchester, B., Dwek, R. A., and Butters, T. D., 1997, Prevention of lysosomal storage in Tay–Sachs mice treated with N-butyldeoxynojirimycin, *Science* **276:**428–431.

Rafi, M. A., Luzi, P., Chen, Y. Q., and Wenger, D. A., 1995, A large deletion together with a point mutation in the GALC gene is a common mutant allele in patients with infantile Krabbe disease, *Hum. Mol. Genet.* **4:**1285–1289.

Saftig, P., Hartmann, D., Lüllmann-Rauch, R., Wolff, J., Evers, M., Köster, A., Hetman, M., von Figura, K., and Peters, C., 1997, Mice deficient in lysosomal acid phosphatase develop lysosomal storage in the kidney and central nervous system, *J. Biol. Chem.* **272:**18628–18635.

Sakai, N., Inui, K., Fujii, N., Fukushima, H., Nishimoto, J., Yanagihara, I., Isegawa, Y., Iwamatsu, A., and Okada, S., 1994, Krabbe disease: Isolation and characterization of a full-length cDNA for human galactocerebrosidase, *Biochem. Biophys. Res. Commun.* **198:**485–491.

Sakai, Y., Miyawaki, S., Shimizu, A., Ohno, K., and Watanabe, T., 1991, A molecular genetic linkage map of mouse chromosome 18, including spm, Grl-1, Fim-2; c-fms, and Mbp, *Biochem Genet.* **29:**103–113.

Sandhoff, K., Harzer, K., and Fürst, W., 1995. Sphingolipid activator proteins *in* "The Metabolic and Molecular Basis of Inherited Disease" (C. R. Scriver, A. L. Beaudet, W. S. Sly, and D. Valle, eds.), pp. 2427–2441, McGraw-Hill, New York.

Sands, M. S., and Birkenmeier, E. H., 1993, A single-base-pair deletion in the β-glucuronidase gene accounts for the phenotype of murine mucopolysaccharidosis type VII, *Proc. Natl. Acad. Sci. USA* **90:**6567–6571.

Sands, M. S., Barker, J. E., Vogler, C., Levy, B., Gwynn, B., Galvin, N., Sly, W. S., and Birkenmeier, E., 1993, Treatment of murine mucopolysaccharidosis type VII by syngeneic bone marrow transplantation in neonates, *Lab. Invest.* **68:**676–686.

Sands, M. S., Vogler, C., Kyle, J. W., Grubb, J. H., Levy, B., Galvin, N., Sly, W. S., and Birkenmeier, E. H., 1994, Enzyme replacement therapy for murine mucopolysaccharidosis type VII, *J. Clin. Invest.* **93:**2324–2331.

Sands, M. S., Vogler, C., Torrey, A., Levy, B., Gwynn, B., Grubb, J., Sly, W. S., and Birkenmeier, E. H., 1997a, Murine mucopolysaccharidosis type VII: Long term therapeutic effects of enzyme replacement and enzyme replacement followed by bone marrow transplantation, *J. Clin. Invest.* **99:**1596–1605.

Sands, M. S., Wolfe, J. H., Birkenmeier, E. H., Barker, J. E., Vogler, C., Sly, W. S., Okuyama, T., Freeman, B., Nicholes, A., Muzyczka, N., Chang, P. L., and Axelrod, H. R., 1997b, Gene therapy for murine mucopolysaccharidosis type VII, *Neuromusc. Disord.* **7:**352–360.

Sango, K., Yamanaka, S., Hoffmann, A., Okuda, Y., Grinberg, A., Westphal, H.L, McDonald, M. P., Crawley, J. N., Sandhoff, K., Suzuki, K., and Proia, R. L., 1995, Mouse models of Tay–Sachs and Sandhoff diseases differ in neurologic phenotype and ganglioside metabolism, *Nat. Genet.* **11:**170–176.

Sango, K., McDonald, M. P., Crawley, J. N., Mack, M. L., Tifft, C. J., Skop, E., Starr, C. M., Hoffmann, A., Sandhoff, K., Suzuki, K., and Proia, R. L., 1996, Mice lacking both subunits of lysosomal β-hexosaminidase display gangliosidosis and mucopolysaccharidosis, *Nat. Genet.* **14:**348–352.

Schnabel, D., Schröder, M., Fürst, W., Klein, A., Hurwitz, R., Zenk, T., Weber, J., Harzer, K., Paton, B. C., Poulos, A., Suzuki, K., and Sandhoff, K., 1992, Simultaneous deficiency of sphingolipid activator proteins 1 and 2 is caused by a mutation in the initiation codon of their common gene, *J. Biol. Chem.* **267:**3312–3315.

Schuchman, E. H., and Desnick, R. J., 1995, Niemann–Pick disease type A and B: Acid sphingomyelinase deficiencies *in* "The Metabolic and Molecular Basis of Inherited Disease" (C. R. Scriver, A. L. Beaudet, W. S. Sly, and D. Valle, eds.), pp. 2601–2624, McGraw-Hill, New York.

Scott, H. S., Bunge, S., Gal, A., Clarke, L. A., Morris, C. P., and Hopwood, J. J., 1995, Molecular genetics of mucopolysaccharidosis type I: Diagnostic, clinical and biological implications, *Hum. Mutat.* **6:**288–302.

Shimmoto, M., Kase, R., Itoh, K., Utsumi, K., Ishii, S., Taya, C., Yonekawa, H., and Sakuraba, H., 1997, Generation and characterization of transgenic mice expressing a human mutant α-galactosidase with an R301Q substitution causing a variant form of Fabry disease, *FEBS Lett.* **417:**89–91.

Shio, H., Fowler, S., Bhuveneswaren, C., and Morris, M. D., 1982, Lysosomal lipid storage disorder in NCTR-BALB c mice II. Morphologic and cytochemical studies, *Am. J. Pathol.* **108:**150–159.

Sidransky, E., Sherer, D. M., and Ginns, E. I., 1992, Gaucher disease in the neonate: A distinct

Gaucher phenotype is analogous to a mouse model created by targeted disruption of the glucocerebrosidase gene, *Pediatr. Res.* **32**:494–498.

Sidransky, E., Fartasch, M., Lee, R. E., Metlay, L. A., Abella, S., Zimran, A., Gao, W., Elias, P. M., Ginns, E. I., and Holleran, W. M., 1996, Epidermal abnormalities may distinguish type 2 from type 1 and type 3 of Gaucher disease, *Pediatr. Res.* **39**:134–141.

Sung, J. H., 1979, Autonomic neurons affected by lipid storage in the spinal cord in Fabry's disease: Distribution of autonomic neurons in the sacral cord, *J. Neuropathol. Exp. Neurol.* **38**:87–98.

Suzuki, K., 1968, Cerebral GM1-gangliosidosis: Chemical pathology of visceral organs, *Science.* **159**:1471–1472.

Suzuki, K., 1991, Neuropathology of late onset gangliosidoses. A review, *Dev. Neurosci.* **13**:205–210.

Suzuki, K., 1997, Twenty-five years of the psychosine hypothesis: A personal perspective of its history and present status, *Neurochem. Res.* **23**:251–259.

Suzuki, K., and Suzuki, K., 1995, The twitcher mouse: A model for Krabbe disease and for experimental therapies, *Brain Pathol.* **5**:249–258.

Suzuki, K., and Suzuki, K., 1996, The gangliosidoses, *in* "Neurodystrophies and Neurolipidoses" (H. W. Moser, ed.), pp. 247–280, Elsevier, Amsterdam.

Suzuki, K., and Vanier, M. T., 1999, Lysosomal and peroxisomal diseases, *in* "Basic Neurochemistry, 6th ed.," (G. Siegel, B. W. Agranoff, R. W. Albers, S. Fischer, and M. Uhler, eds.), pp. 821–839, Lippincott-Raven, New York.

Suzuki, K., Hoogerbrugge, P. M., Poorthuis, B. J. H. M., Romme, A. E., van de Kamp, J. J. P., Wagemaker, G., van Bekkum, D. W., and Suzuki, K., 1988, The twitcher mouse: Central nervous system pathology after bone marrow transplantation (BMT), *Lab. Invest.* **58**:302–309.

Suzuki, K., Suzuki, Y., and Suzuki, K., 1995a, Galactosylceramide lipidosis: Globoid cell leukodystrophy (Krabbe disease) *in* "Metabolic and Molecular Basis of Inherited Disease" (C. R. Scriver, A. L. Beaudet, W. S. Sly, and D. Valle, eds.), pp. 2671–2692, McGraw-Hill, New York.

Suzuki, K., Parker, C. C., Pentchev, P. G., Katz, D., Ghetti, B., D'Agostino, A. N., and Carstea, E. D., 1995b, Neurofibrillary tangles in Niemann–Pick disease type C, *Acta Neuropathol. (Berl).* **89**:227–238.

Suzuki, Y., Sakuraba, H., and Oshima, A., 1995c, β-Galactosidase deficiency (β-galactosidosis): GM1 gangliosidosis and Morquio B disease *in* "The Metabolic and Molecular Basis of Inherited Disease" (C. R. Scriver, A. L. Beaudet, W. S. Sly, and D. Valle, eds.), pp. 2785–2823, McGraw-Hill, New York.

Suzuki, K., Sango, K., Proia, R. L., and Langaman, C., 1997, Mice deficient in all forms of lysosomal beta-hexosaminidase show mucopolysaccharidosis-like pathology, *J. Neuropathol. Exp. Neurol.* **56**:693–703.

Suzuki, K., Proia, R. L., and Suzuki, K., 1998, Mouse models of human lysosomal diseases, *Brain Pathol.* **8**:195–215.

Svennerholm, L., Vanier, M.-T., and Månsson, J.-E., 1980, Krabbe disease: A galactosylsphingosine (psychosine) lipidosis, *J. Lipid Res.* **21**:53–64.

Taniike, M., Yamanaka, S., Proia, R. L., Langaman, C., Bone-Turrentine, T., and Suzuki, K., 1995, Neuropathology of mice with targeted disruption of Hexa gene, a model of Tay–Sachs disease. *Acta Neuropathol. (Berl).* **89**:296–304.

Terry, R. D., and Korey, S. R., 1960, Membranous cytoplasmic granules in infantile amaurotic idiocy, *Nature* **188**:1000–1002.

Terry, R. D., and Weiss, M., 1963, Studies in Tay–Sachs disease II. Ultrastructure of cerebrum, *J. Neuropathol. Exp. Neurol.* **22**:18–55.

Thomas, G. H., and Beaudet, A. L., 1995, Disorders of glycoprotein degradation and structure: α-mannosidosis, β-mannosidosis, fucosidosis, sialidosis, aspartylglucosaminuria, and carbohydrate-deficient glycoprotein syndrome *in* "The Metabolic and Molecular Basis of Inherited

Disease" (C. R. Scriver, A. L. Beaudet, W. S. Sly, and D. Valle, eds.), pp. 2529–2561, McGraw-Hill, New York.

Tybulewicz, V. L. J., Tremblay, M. L., LaMarca, M. E., Willemsen, R., Stubblefield, B. K., Winfield, S., Zablocka, B., Sidransky, E., Martin, B. M., Huang, S. P., Mintzer, K. A., Westphal, H., Mulligan, R. C., and Ginns, E. I., 1992, Animal model of Gaucher's disease from targeted disruption of the mouse glucocerebrosidase gene, *Nature* **357:**407–410.

Vanier, M. T., and Suzuki, K., 1996, Niemann–Pick diseases *in* "Neurodystrophies and Neurolipidoses, Handbook of Clinical Neurology, Vol. 22" (H. W. Moser, ed.), pp. 133–162, Elsevier, Amsterdam.

Vanier, M. T., and Suzuki, K., 1998, Recent advances in elucidating Niemann–Pick C disease, *Brain Pathol.* **8:**163–174.

Vanier, M. T., and Svennerholm, L., 1976, Chemical pathology of Krabbe disease: The occurrence of psychosine and other neutral sphingolipids. *Adv. Exp. Med. Biol.* **68:**115–126.

Vanier, M. T., Duthel, S., Rodriguez-Lafrasse, C., Pentchev, P., and Carstea, E. D., 1996, Genetic heterogeneity in Niemann–Pick C disease: A study using somatic cell hybridization and linkage analysis, *Am. J. Hum. Genet.* **58:**118–125.

Vogler, C., Birkenmeier, E. H., Sly, W. S., Levy, B., Pegors, C., Kyle, J. W., and Beamer, W. G., 1990, A murine model of mucopolysaccharidosis-VII—gross and microscopic findings in β-glucuronidase-deficient mice, *Am. J. Pathol.* **136:**207–217.

Vogler, C., Sands, M., Higgins, A., Levy, B., Grubb, J., Birkenmeier, E. H., and Sly, W. S., 1993, Enzyme replacement with recombinant β-glucuronidase in the newborn mucopolysaccharidosis type VII mouse, *Pediatr. Res.* **34:**837–840.

Vogler, C., Sands, M. S., Levy, B., Galvin, N., Birkenmeier, E. H., and Sly, W. S., 1996, Enzyme replacement with recombinant β-glucuronidase in murine mucopolysaccharidosis type VII: Impact of therapy during the first six weeks of life on subsequent lysosomal storage, growth, and survival, *Pediatr. Res.* **39:**1050–1054.

Wang, A. M., Stewart, C. L., and Desnick, R. J., 1993, α-N-acetylgalactosaminidase: Characterization of the murine cDNA and genomic sequences and generation of the mice by targeted gene disruption, *Am. J. Human Genet.* **53:**A99.

Wang, A. M., Ioannou, Y. A., Zeidner, K. M., Gotlib, R. W., Dikman, S., Stewart, C. L., and Desnick, R. J., 1996, Fabry disease: Generation of a mouse model with α-galactosidase A deficiency, *Am. J. Hum. Genet.* **59** (Suppl.):A208 (abstract no. 1192).

Weintraub, H., Abramovici, A., Sandbank, U., Pentchev, P. G., Brady, R. O., Sekine, A., Suzuki, A., and Sela, B., 1985, Neurological mutation characterized by dysmyelination in NCTR-Balb C mouse with lysosomal lipid storage disease, *J. Neurochem.* **45:**665–672.

Wenger, D. A., Tarby, T. J., and Wharton, C., 1978, Macular cherry-red spots and myoclonus with dementia: Coexistent neuraminidase and beta-galactosidase deficiencies, *Biochem. Biophys. Res. Commun.* **82:**589–595.

Willemsen, R., Tybulewicz, E., Sidransky, E., Eliason, W. K., Martin, B. M., LaMarca, M. E., Reuser, A. J. J., Tremblay, M., Westphal, H., Mulligan, R. C., and Ginns, E. I., 1995, A biochemical and ultrastructural evaluation of the type 2 Gaucher mouse, *Mol. Chem. Neuropathol.* **24:**179–192.

Yamamoto, T., Tokoro, T., and Eto, Y., 1994, The attenuated elevation of cytoplasmic calcium concentration following the uptake of low density lipoprotein in type C Niemann–Pick fibroblasts, *Biochem. Biophys. Res. Commun.* **198:**438–444.

Yamanaka, S., Johnson, M. D., Grinberg, A., Westphal, H., Crawley, J. N., Taniike, M., Suzuki, K., and Proia, R. L., 1994a, Targeted disruption of the Hexa gene results in mice with biochemical and pathologic features of Tay–Sachs disease, *Proc. Natl. Acad. Sci. USA* **91:**9975–9979.

Yamanaka, S., Johnson, O. N., Norflus, F., Boles, D. J., and Proia, R. L., 1994b, Structure and expression of the mouse β-hexosaminidase genes, HEXA and HEXB, *Genomics* **21:**588–596.

Yamano, T., Shimada, M., Sugino, H., Dezawa, T., Koike, M., Okada, S., and Yabuuchi, H., 1985,

Ultrastructural study on a severe infantile sialidosis (β-galactosidase-α-neuraminidase deficiency), *Neuropediatrics* **16:**109–112.

Yeager, A. M., Brennan, S., Tiffany, C., Moser, H. W., and Santos, G. W., 1984, Prolonged survival and remyelination after hematopoietic cell transplantation in the twitcher mouse, *Science* **225:**1053–1054.

Yeager, A. M., Shinn, C., Shinohara, M., and Pardoll, D. M., 1993, Hematopoietic cell transplantation in the twitcher mouse: The effects of pretransplant conditioning with graded doses of busulfan, *Transplantation* **56:**185–190.

Zhou, X. Y., Morreau, H., Rottier, R., Davis, D., Bonten, E., Gillemans, N., Wenger, D., Grosveld, F. G., Doherty, P., Suzuki, K., Grosveld, G. C., and d'Azzo, A., 1995, Mouse model for the lysosomal disorder galactosialidosis and correction of the phenotype with overexpressing erythroid precursor cells, *Genes Dev.* **9:**2623–2634.

CHAPTER 12

NEUROLOGICAL IMPLICATIONS OF THE GENETIC MOUSE MODELS FOR HUMAN PHENYLKETONURIA AND HYPERPHENYLALANINEMIA

J. DAVID MCDONALD

1. INTRODUCTION

The inability to catabolize dietary phenylalanine (PHE) was first described by the Norwegian scientist and physician Asbørn Fölling (Fölling, 1934). In this seminal work, Dr. Fölling made a preliminary description of the biochemical basis of the defect that would come to be known as phenylketonuria (PKU), its apparent

J. DAVID MCDONALD • Department of Biological Sciences, Wichita State University, Wichita, Kansas 67260.

Mouse Models in the Study of Genetic Neurological Disorders, Volume 9 of *Advances in Neurochemistry,* edited by Popko, Kluwer Academic/Plenum Publishers, New York, 1999.

heritability, and its most obvious neurological manifestations. In the more than 60 years that have elpased since this breakthrough, an abundant amount of research has been undertaken to extend and complete this initial characterization. (For a recent review, see Scriver *et al.,* 1994.)

With time, the biochemical process by which excess dietary PHE is catabolized became more fully defined. This definition was begun by Jervis (1947) and Udenfriend and Cooper (1952) and was more fully delineated by Mitoma (1956) and Kaufman (1959, 1963). Through the efforts of these researchers, it gradually became well established that the enzyme PHE hydroxylase catalyzed the rate-limiting step in PHE catabolism. Shortly after the initial descriptions of the biochemical basis of PKU began emerging, a treatment was devised based on dietary PHE limitation that led to the amelioration of the most severe effects of untreated PKU (Bickel *et al.,* 1954; Armstrong and Tyler, 1955; Woolf *et al.,* 1955). Ultimately, the genetic definition of PKU was brought to a very high level when Kwok *et al.* (1985) cloned and determined the sequence of the human PHE hydroxylase gene (i.e., the gene that comprises the PKU locus). These cloned sequences were then used to identify the precise chromosomal location of the human PKU locus on chromosome 12 in band region q22–q24.1 (Lidsky *et al.,* 1985).

Ultimately, it was discovered that the disease PKU is separable into two distinct syndromes: Neonatal PKU and maternal PKU. Neonatal PKU is that disease process that arises after the birth of a child who is homozygous for the PKU mutation. During gestation, the embryo and fetus are protected from the harmful effects of hyperphenylalaninemia by maternal metabolism. After parturition, however, the neonate must begin to metabolize his own diet and the metabolic insufficiency of PKU becomes apparent and, without treatment, this leads to a range of neurological defects. The most readily demonstrable neurological defect associated with PKU is a PHE-dependent and irreversible impairment of cognitive development, a condition that can be prevented by strict dietary PHE control. Untreated PKU patients have also been reported to have an unusual gait, unusual standing and sitting postures, and a tendency toward epilepsy (Paine, 1957). Further, lack of dietary PHE control in adult PKU patients has been associated with neuropsychological dysfunction caused by neurotransmitter depletion (Krause *et al.,* 1985, 1986). Maternal PKU, first noted in 1956 by Dent (1957) and expanded by Mabry (1963), is a syndrome of the disease that occurs when a PKU mother who is not in dietary compliance becomes pregnant. The severe metabolic imbalance in the mother creates a toxic developmental environment in her uterus and the fetus developing therein is at very high risk for birth defects or fetal death. Here, the factors that predispose to birth defects are entirely maternal and, in contrast to neonatal PKU, the fetal genotype does not seem to be a factor in the development of birth defects. In this developmental situation, the gestating fetus is typically exposed to nearly twice the level of hyperphenylalaninemia as the mother because this metabolite is concentrated across the placental barrier into the fetal compartment (Levy, 1984). In an international survey, 92% of the offspring of untreated PKU mothers are mentally-retarded or exhibit subnormal cognitive development and 73% are born microcephalic (Lenke and Levy, 1980).

2. EARLIER EFFORTS TO PRODUCE MOUSE PKU MODELS

In an attempt to more thoroughly examine the relationship between dietary PHE intake and PKU pathology, there has long been a keenly felt need for pertinent animal models of the disease. As with the history of the characterization of the genetic and biochemical bases for PKU described previously, the pursuit of animal models for human PKU is long and storied. The first PKU models to appear made use of the fact that PHE hydroxylase is produced at relatively low levels in newborn (Waisman, 1962) and albino (Perez and Schmidt, 1963) rats. These models were inaccurate paradigms for PKU because, while exhibiting the elevated blood PHE levels (hyperphenylalaninemia) characteristic of PKU, they also exhibited elevated levels of blood tyrosine which is decidedly uncharacteristic of PKU. Because of this inaccuracy, these PHE overadministration models soon gave way to other models that made use of chemical inhibitors of the PHE hydroxylase enzyme. The most popular of these chemical inhibitors is *p*-chlorophenylalanine, a compound that irreversibly inhibits PHE hydroxylase *in vivo* (Guroff, 1969; Chang *et al.*, 1979). In the early 1960s there was a brief excitement owing to the supposed identification of a genetic mouse model for human PKU (Coleman, 1960; Rauch and Yost, 1963). Soon, however, it was revealed that the few apparent similarities that suggested this as a model for PKU came from different causes and that mutants did not exhibit the chronically elevated blood PHE levels needed for a pertinent animal model for human PKU (Zannoni *et al.*, 1966). Although some pertinent pathology was produced by these early modeling efforts, comparatively little of lasting significance was learned from either the overadministration models or from the albino mouse. The enzyme inhibition models have been considerably more useful due to the ease of administration and the extent of metabolic manipulation. Here a normal animal can be converted rather quickly into the chronic hyperphenylalaninemia state needed to produce pertinent modeling of human PKU and this state can be easily modulated to adjust the degree of hyperphenylalaninemia. As recently as 1994, some fruitful research has been accomplished with chemically induced animal models (Diamond *et al.*, 1994). In this study, learning and behavioral effects were studied among rats that had been exposed to moderately elevated levels of PHE during gestation and early postnatal development. Animals exposed both pre- and postnatally, and those exposed only postnatally showed significant cognitive deficits compared to untreated or mock-treated controls.

There is, however, a common thread in these early models that leads to notable weaknesses. These models are of a type termed phenocopies; they have been manipulated to a condition that resembles a heritable phenotype but for a different reason than the one that leads to the heritable modeled disease. Because of the difference in the cause of the disease phenotype, such a strategy often introduces doubts about accuracy of disease modeling. One example of this comes from the chemically induced models described earlier. In this case, it has been amply demonstrated that numerous unrelated and confounding side effects are induced by some of the chemical inhibitors (Koe and Weissman, 1966; Tagliamonte *et al.*, 1969; Gal *et al.*, 1970; Prichard and Guroff, 1971; Brown and Fawke, 1972; Hyppa *et al.*, 1972; McFarlain and Bloom, 1972).

3. THE GENETIC PKU MOUSE PROGRAM

To avoid the shortcomings of previous models, we chose to produce a mouse model for PKU that was as similar as possible to the human heritable condition. For this task we used the highly efficient mutagen ethylnitrosourea to mutagenize the mouse germline (reviewed in McDonald *et al.*, 1990a,b; McDonald, 1995). The first mutant that this mutagenesis program yielded, *HPH-1*, exhibits a deficiency in GTP-cyclohydrolase I (GTPCH-I) (Bode *et al.*, 1988; McDonald *et al.*, 1988; McDonald and Bode, 1988). GTPCH-I catalyzes the rate-limiting step in the formation of the pteridine cofactor tetrahydrobiopterin (BH_4), a cofactor absolutely required for PHE hydroxylase activity. BH_4 is required not only for the catabolism of PHE, but also for the synthesis of catecholamine neurotransmitters, and for the catalytic function of nitric oxide synthase.

HPH-1 mutants display a very striking deficiency of GTPCH-I activity early in life but gain some activity later in life. While deficiency of GTPCH-I activity has numerous and severe effects in humans (Niederweiser *et al.*, 1984; Dhondt *et al.*, 1985; Naylor *et al.*, 1987; Ichinose *et al.*, 1994), the phenotype of HPH-1 mutants is far less severe. It appears that this is due to the incomplete deficiency of GTPCH-I activity among mutants (Jacobson and Manos, 1989; Cha *et al.*, 1991). Nevertheless, HPH-1 mutants have been useful for unraveling some of the many roles of tetrahydrobiopterin in brain chemistry.

Although the HPH-1 line is not truly a PKU model but rather a model for BH_4-dependent hyperphenylalaninemia, considerable work has been done to characterize the neurochemical defects that occur in them. Hyland and Bola (1989) found that, although HPH-1 mutants develop some ($\approx$50%) ability to produce BH_4 by adulthood, concentrations of 3,4-dihydroxyphenylacetic acid and 5-hydroxylindoleacetic acid were significantly reduced for all age classes. This result was taken to indicate impaired turnover of dopamine and serotonin as a consequence of chronic BH_4 deficiency.

Brand *et al.* (1995) used the *HPH-1* mutants to study the effect of chronic BH_4 deficiency on brain nitric oxide synthase (bNOS). They found that, in the absence of exogenous BH_4, there was a significant reduction in bNOS activity and that reduced BH_4 led to a reduced binding affinity for arginine. They concluded that less binding of BH_4 to bNOS occurs during the chronic BH_4 deficiency state of *HPH-1* mutants, that this reduced BH_4 binding leads to a reduction of arginine binding, and that reduced arginine binding may lead to a new rate-limiting step in the function of bNOS in *HPH-1* mutants and in similar human patients.

Brand *et al.* (1996) again used the *HPH-1* mutant as a model of chronic BH_4 deficiency to study the neurochemical effects of treatments with different BH_4 analogues. They found that one level of BH_4 administration (100 μmol/kg) led to a transient increase in brain serotonin and dopamine. The BH_4 analogue 6-methyltetrahydropterin was found to cross the blood–brain barrier to a great extent than BH_4 but was also metabolized faster, thus obviating any advantage. Further, BH_4 linked to a lipid carrier was ineffective in stimulating an increase in the brain neurotransmitters. Finally, higher levels of BH_4 (i.e., 1000 μmol/kg) were found to be toxic. The study concluded that these types of treatments were ineffective as possible treatments for dopamine and serotonin deficiency due to BH_4 deficiency.

Further work on the mutagenesis project resulted in the successful production of three distinct mutations in the mouse PHE hydroxylase gene; mutations originally named *hph-5.0, hph-5.1,* and *hph-5.2* but later renamed Pah^{enu1}, Pah^{enu2}, and Pah^{enu3} respectively (McDonald *et al.*, 1990c; Shedlovsky *et al.*, 1993; McDonald and Charlton, 1997). PAH^{enu1} mutants are mildly affected. They do not exhibit hyperphenylalaninemia under normal holding conditions but only when exposed to extraneous PHE. In contrast, PAH^{enu2} and PAH^{enu3} are severely deficient in PAH activity and exhibit many of the phenotypic manifestations of classical human PKU; they are chronically hyperphenylalaninemic under normal holding conditions and are hypopigmented and runted. Further, female mutants exhibit some of the same pregnancy complications seen in human maternal PKU (Shedlovsky *et al.*, 1993; McDonald *et al.*, 1997). These mouse classical PKU models have so far proven useful for a number of non-neurological investigations such as gene therapy (Liu *et al.*, 1992; Cristiano *et al.*, 1993; Eisensmith and Woo, 1994; Fang *et al.*, 1994), enzyme replacement therapy (Safos and Chang, 1995), and maternal PKU-related defects in cardiovascular development (McDonald *et al.*, 1997). Nevertheless, the focus of this chapter is to discuss the potential utility of the model for studying the neurological defects that result from PKU.

Hommes and Moss (1992) found that myelin turnover is increased in weanling age *PAH^enu1^* mutants made hyperphenylalaninemic to a level of 2.5 mM by exposure to PHE. The half-life of fast myelin was reduced from the normal 15 days to 4.5 days among such animals. The half-life of the slow myelin compoinent was also reduced by hyperphenylalaninemia, but to a lesser extent. Animals at an intermediate level of hyperphenylalaninemia showed the same overall phenomenon but to a proportionally reduced degree. Later, Hommes (1993) and Hommes (1994) found that, in PAH^{enu1} mutants made hyperphenylalaninemic by adding PHE to their drinking water, there was a gradual decrease in the number of binding sites for the muscarinic acetylcholine receptor. These findings in a genetic mouse model confirmed the validity of previous similar findings in rats made hyperphenylalaninemic with PAH inhibitors. The effect of hyperphenylalaninemia on the number of binding sites for muscarinic acetylcholine was not uniform throughout the brain: there was no change in the putamen; there was a gradual decrease in the external layer of the olfactory bulb in addition to the parietal, occipital, the cingulate areas of the cerebral cortex, and the CA_1 and CA_2 layers of the hippocampus; and there was a decrease preceded by a transient increase in the frontal area of the cerebral cortex and also in the caudate nucleus.

With regard to the issue of PHE-dependent hypomyelination, there is some discrepancy in the published record for PAH^{enu2} mutants. Using magnetic resonance imaging to near microscopic resolution, Kornguth *et al.* (1994) found no evidence for hypomyelination or other structural abnormalities in the brains of hyperphenylalaninemic PAH^{enu2} mutants compared with normal controls. In contrast, Dyer *et al.* (1996) reported the neuropathological findings of hypomyelination and gliosis in the PAH^{enu2} mutants; findings very similar to the pathological findings typically seen in untreated adult human PKU patients (Allen and Kirk, 1992).

Thus far very little has been done to characterize behavior or learning in these PKU models. Nevertheless, owing to the high level of development of mouse behavior testing and the apparent similarity of the neuropathology reported by some groups, it seems likely that the behavioral testing will be successful at revealing deficits. There are, however, some areas of learning and behavior that may present problems for disease modeling. It may be very difficult to accurately model the behavioral problems reported by some clinicians among PKU children due to the possible lack of the highly sophisticated brain functions in the laboratory mouse compared to humans. Further, untreated PKU in the mouse model has been observed to result in some degree of physical impairment (e.g., developmental delay, runting, lethargy, uncoordination) and these may partially confound the accurate detection of some learning and behavioral endpoints. Lastly, it is possible that the premature parturition of mice compared to humans may introduce important differences in the developmental disturbances in mouse and human. Mice are born at the developmental stage that approximates the end of the second trimester of human development. During the weeks that follow, both humans and mice exhibit a huge burst of brain development that occurs *in utero* for humans and *ex utero* for the mouse. Mice may be spared important exposures to teratogens by this developmental fact.

4. FUTURE STUDIES ENABLED BY THE GENETIC MOUSE MODEL FOR HUMAN PKU

By producing a mouse model for human PKU, at least three shortcomings of studies involving humans have been surmounted. Human studies often suffer from the lack of important *control groups* owing to ethical considerations. In addition, the data from such studies also are subject to contamination from *dietary noncompliance* and the *genetic heterogeneity* of the study population. The diet of these mouse mutants can be precisely controlled and data can easily be obtained from all desired test and control groups. Moreover, because the mutations are in an inbred genetic background, the test and control animals are genetically similar in every respect except for the presence of the PKU mutations. Therefore these models can be used to examine both varying levels and types of dietary constituents with the goal of rigorously examining the neurological effects in exposed animals. Further, mutant females can be fed diets of different compositions to examine the effect on gestational outcome. Such studies may involve inducing the disease neuropathology with a goal of explaining the etiology or they may take the form of examining the palliative effect of defined diets.

The induced mutations can be placed on different genetic backgrounds to examine the effect of genetic background on the modification of the development of neurological effects. If strong modifying effects are discovered from one or a few genes, they can be mapped to high resolution and cloned from map position as has recently been done for other modifying genes (MacPhee *et al.,* 1995; Rozmahel *et al.,* 1996). This would extend the base of knowledge about the precise etiology of the disease caused by such mutations.

REFERENCES

Allen, I. V., and Kirk, J., 1992, "Greenfield's Neuropathology," 5th ed. (J. H. Adams and L. W. Duchen, eds.), pp. 447–520, Oxford University Press, New York.

Armstrong, M. D., and Tyler, F. H., 1955, Studies on phenylketonuria. I. Restricted phenylalanine intake in phenylketonuria, *J. Clin. Invest.* **34**:565–580.

Bickel, H., Gerrard, J., and Hickmans, E. M., 1954, The influence of phenylalanine intake on the chemistry and behavior of a phenylketonuria child, *Acta Pediatr. Scand.* **43**:64–77.

Bode, V., McDonald, J. D., Guenet, J-L., and Simon, D., 1988, *hph-1*: A mouse mutant with hereditary hyperphenylalaninemia induced by ethylnitrosourea mutagenesis, *Genetics* **118:**299–305.

Brand, M. P., Heales, S. J. R., and Clark, J. B., 1995, Tetrahydrobiopterin deficiency and brain nitric oxide synthase in the HPH-1 mouse, *J. Inher. Metab. Dis.* **18:**33–39.

Brand, M. P., Hyland, K., and Heales, S. J. R., 1996, Neurochemical effects following peripheral administration of tetrahydrobiopterin derivatives to the *hph-1* mouse, *J. Neurochem.* **66:**1150–1156.

Brown, P. S., and Fawke, L., 1972, Effects of reserpine, *p*-chlorophenylalanine, α-methyltyrosine, thyroxamine or methallibure on pituitary FSH in male rats, *J. Reprod. Fert.* **28:**167–175.

Cha, K. A., Jacobson, K. B., and Yim, J. J., 1991, Isolation and characterization of GTP cyclohydrolase I from mouse liver. Comparison of normal and the HPH-1 mutant, *J. Biol. Chem.* **266:**12294–12300.

Chang, N., Kaufman, S., and Milstein, S., 1979, The mechanism of irreversible inhibition of rat liver phenylalanine hydroxylase due to treatment with *p*-chlorophenylalanine, *J. Biol. Chem.* **254:**2665–2668.

Coleman, D. L., 1960, Phenylalanine hydroxylase activity in dilute and nondilute strains of mice, *Arch. Biochem. Biophys.* **91:**300–306.

Cristiano, R. J., Smith, L. C., and Woo, S. L. C., 1993, Hepatic gene therapy: Adenovirus enhancement of receptor-mediated gene delivery and expression in primary hepatocytes. *Proc. Natl. Acad. Sci. USA* **90:**2122–2126.

Dent, C. E., 1957, Relation of biochemical abnormality to development of mental defect in phenylketonuria, *in* "Etiological Factor in Mental Retardation: Report of 23rd Ross Pediatric Research Conference," Ross Laboratories, Columbus, OH.

Dhondt, J.-L., Farriaux, J.-P., Boudha, A., Largillière, C., Ringel, J., Roger, M.-M., and Leeming, R. J., 1985, Neonatal hyperphenylalaninemia presumably caused by guanosine triphosphate-cyclohydrolase deficiency, *J. Pediatr.* **106:**954–956.

Diamond, A., Ciaramitaro, V., Donner, E., Djali, S., and Robinson, M. R., 1994, An animal model for early-treated PKU, *J. Neurosci.* **14:**3072–3082.

Dyer, C. A., Kendler, A., Philibotte, T., Gardiner, P., Cruz, J., and Levy, H. L., 1996, Evidence for central nervous system glial cell plasticity ikn phenylketonuria, *J. Neuropathol. Exp. Neurol.* **55:**795–814.

Eisensmith, R. C., and Woo, S. L. C., 1994, Gene therapy for PKU, *Acta Pediatr.* **83** (Suppl. 407):124–129.

Fang, B., Eisensmith, R. C., Li, X. H. C., Finegold, M. J., Shedlovsky, A., Dove, W. F., and Woo, S. L. C., 1994, Gene therapy for phenylketonuria: Phenotypic correction in a genetically deficient mouse model by adenovirus-mediated hepatic gene transfer, *Gene Ther.* **1:**247–254.

Fölling, A., 1934, Uber ausscheidung von phenylbrenztraubensaure in den harn als stoflwechselanomalis in verbindung mit imbezilliat, *Hoppe Seyler's Z. Physiol. Chem.* **227:**169–176.

Gal, E. M., Roggeveen, A. E., and Millard, S. A., 1970, DL-[2-^{14}C]*p*-Chlorophenylalanine as an inhibitor of tryptophan 5-hydroxylase, *J. Neurochem.* **17:**1221–1235.

Guroff, G., 1969, Irreversible *in vivo* inhibition of rat liver phenylalanine hydroxylase by *p*-chlorophenylalanine, *Arch. Biochem. Biophys.* **134:**610–611.

Hommes, F. A., 1993, The effect of hyperphenylalaninemia on the muscarinic acetylcholine receptor in the HPH-5 mouse brain, *J. Inher. Metab. Dis.* **16:**962–974.

Hommes, F. A., 1994, Loss of neurotransmitter receptors by hyperphenylalaninemia in the HPH-5 mouse brain, *Acta Paediatrica* **83:**120–121.

Hommes, F. A., and Moss, L., 1992, Myelin turnover in hyperphenylalaninemia. A reevaluation with the HPH-5 mouse, *J. Inher. Metab. Dis.* **15:**243–251.

Hyland, K., and Bola, F., 1989, Tetrahydrobiopterin and biogenic amine status of the HPH-1 mouse model, *Biol. Chem. Hoppe Seyler* **370:**387.

Hyppa, M., Lampinen, P., and Lehtinen, P., 1972, Alteration in the sexual behavior of male and female rats after neonatal administration of *p*-chlorophenylalanine, *Psychopharmacology* **25:**152–161.

Ichinose, H., Ohye, T., Takahashi, E., Seki, N., Hori, T., Segawa, M., Nomura, Y., Endo, K., Tanaka, H., Tsuji, S., Fujita, K., and Nagatsu, T., 1994, Hereditary progressive dystonia with marked diurnal fluctuation caused by mutations in the GTP cyclohydrolase I gene, *Nat. Genet.* **8:**236–242.

Jacobson, K. B., and Manos, R. E., 1989, Effects of sepiapterin and 6-acetyldihydrohomopterin on the guanosine triphosphate cyclohydrolase I of mouse, rat and the fruit-fly *Drosophila, Biochem. J.* **260:**135.

Jervis, G. A., 1947, Studies on phenylpyruvic oligophrenia. The position of the metabolic error, *J. Biol. Chem.* **169:**651–656.

Kaufman, S., 1959, Studies on the mechanism of the enzymatic conversion of phenylalanine to tyrosine, *J. Biol. Chem.* **234:**2677–2682.

Kaufman, S., 1963, The structure of the phenylalanine-hydroxylation cofactor, *Proc. Natl. Acad. Sci. USA* **50:**1085–1093.

Koe, B.K., and Weissman, A., 1966, *p*-Chlorophenylalanine: A specific inhibitor of brain serotonin, *J. Pharmacol. Exp. Ther.* **154:**499–516.

Kornguth, S., Anderson, M., Markley, J. L., and Shedlovsky, A., 1994, Near-microscopic magnetic resonance imaging of the brains of phenylalanine hydroxylase-deficient mice, normal littermates, and of normal BALB/c mice at 9.4 tesla, *Neuroimage* **1:**220–229.

Krause, W., Halminski, M., McDonald, L., Demure, P., Salvo, R., Freides, D., and Elsas, L., 1985, Biochemical and neuropsychological effects of elevated plasma phenylalanine in patients with treated phenylketonuria, *J. Clin. Invest.* **75:**40–48.

Krause, W., Epstein, C., Averbrook, A., Demure, P., and Elsas, L., 1986, Phenylalanine alters the mean power frequency of electroencephalograms and plasma L-DOPA in treated patients with phenylketonuria, *Pediatr. Res.* **20:**1112–1116.

Kwok, S. C. M., Ledley, F. D., DiLella, A. G., Robson, K. J. H., and Woo, S. L. C., 1985, Nucleotide sequence of a full-length complementary DNA clone and amino acid sequence of human phenylalanine hydroxylase, *Biochemistry* **24:**556–561.

Lenke, R., and Levy, 1980, Maternal phenylketonuria and hyperphenylalaninemia: An international survey of the outcome of untreated and treated pregnancies, *N. Engl. J. Med.* **303:**1202–1208.

Levy, H. L., 1984, Lack of fetal effect on blood phenylalanine concentration in maternal phenylketonuria, *J. Pediatr.* **104:**245–247.

Lidsky, A. S., Law, M. L., Morse, H. G.L, Kao, F. T., Rabin, M., Ruddle, F., and Woo, S. L. C., 1985, Regional mapping of the human phenylalanine hydroxylase gene and the phenylketonuria locus on chromosome 12, *Proc. Natl. Acad. Sci. USA* **82:**6221–6225.

Liu, T. J., Kay, M. A., Darlington, G., and Woo, S. L. C., 1992, Reconstitution of enzymatic activity in hepatocytes of phenylalanine hydroxylase-deficient mice, *Somat. Cell Mol. Genet.* **18:**89–96.

Mabry, C. C., Denniston, J. C., Nelson, T. L., and Son, C. D., 1963, Maternal phenylketonuria: A cause of mental retardation in children without metabolic defect, *N. Engl. J. Med.* **269:**1404–1408.

MacPhee, M., Chepenik, K. P., Liddell, R. A., Nelson, K. K., Siracusa, L. D., and Buchberg, A. M., 1995, The secretory phospholipase A2 gene is a candidate for the *Mom-1* locus, a major modifier of APO^{min}-induced intestinal neoplasia, *Cell* **81:**957–966.

McDonald, J. D., 1995, The PKU mouse project: Its history, potential, and implications, *Acta Paediatr.* **84:**122–123.

McDonald, J. D., and Bode, V., 1988, Hyperphenylalaninemia in the *hph-1* mouse mutant, *Pediatr. Res.* **23:**63–67.

McDonald, J. D., and Charlton, C. K., 1997, Characterization of mouse phenylalanine hydroxylase mutations and their correlation with genotype/phenotype relationships in human phenylketonuria, *Genomics* **39:**402–405.

McDonald, J. D., Cotton, R., Jennings, I., Ledley, F., Woo, S., and Bode, V., 1988, The biochemical defect in the *hph-1* mouse mutant is a defect in GTP-cyclohydrolase activity, *J. Neurochem.* **50:**655–657.

McDonald, J. D., Shedlovsky, A., and Dove, W. F., 1990a, Investigating inborn errors of phenylalanine metabolism by efficient mutagenesis of the mouse germline, *in* "Banbury Report 34, Biology of Mammalian Germ Cell Mutagenesis" (J. W. Allen, B. A. Bridges, M. F. Lyon, M. J. Moses, and L. B. Russell, eds.), pp. 259–270, Cold Spring Harbor Laboratory Press, Cold Spring Harbor, New York.

McDonald, J. D., Shedlovsky, A., and Dove, W. F., 1990b, The use of ethylnitrosourea to produce mouse models for human phenylketonuria and hyperphenylalaninemia, *in* "Progress in Clinical and Biological Research, Mutation and the Environment Part C: Somatic and Heritable Mutation, Adduction and Epidemiology," Vol. 340 (M. Mendelsohn and R. Albertini, eds.), pp. 407–413, Wiley-Liss, New York.

McDonald, J. D., Bode, V. C., Dove, W. F., and Shedlovsky, A., 1990c, $Pah^{hph\text{-}5}$: A mouse mutant deficient in phenylalanine hydroxylase, *Proc. Natl. Acad. Sci. USA* **87:**1965–1967.

McDonald, J. D., Kirby, M., Gaillis, L., and Dyer, C., 1997, Cardiovascular defects among the progeny of mouse phenylketonuria females, *Pediatr. Res.* **42:**103–107.

McFarlain, R. A., and Bloom, J. M., 1972, The effects of *p*-chlorophenylalanine on brain serotonin, food intake, and U-maze behavior, *Psychopharmacology* **27:**85–92.

Mitoma, C., 1956, Studies on partially purified phenylalanine hydroxylase, *Arch. Biochem. Biophys.* **60:**476–484.

Naylor, E. W., Ennis, D., Davidson, A. G. F., Wong, L. T. K., Applegarth, D. A., and Niederweiser, A., 1987, Guanosine triphosphate cyclohydrolase I deficiency: Early diagnosis by routine urine pteridine screening, *Pediatrics* **79:**374–378.

Niederweiser, A., Blau, N., Wang, M., Joller, P., Atarés, M., and Cardesa-Garcia, J., 1984, GTP cyclohydrolase I deficiency, a new enzyme defect causing hyperphenylalaninemia with deficiency, a new enzyme defect causing hyperphenylalaninemia with neopterin, biopterin, dopamine, and serotonin deficiencies and muscular hypotonia, *Eur. J. Pediatr.* **141:**208–214.

Paine, R. S., 1957, The variability in manifestations of untreated patients with phenylketonuria (phenylpyruvic aciduria), *Pediatrics* **20:**290–302.

Perez, V. J., and Schmidt, H., 1963, Experimental phenylketonuria: Effects of excessive amounts of L-phenylalanine upon maze learning ability of the albino rat, *Am. Psychol.* **18:**430–436.

Prichard, J. W., and Guroff, G., 1971, Increased cerebral excitability caused by *p*-chlorophenylalanine in young rats, *J. Neurochem.* **18:**153–160.

Rauch, H., and Yost, M. T., 1963, Phenylalanine metabolism in dilute-lethal mice, *Genetics* **48:**1487–1495.

Rozmahel, R., Wilschansk, M., Matin, A., Plyte, S., Oliver, M., Auerbach, W., Moore, A., Forstner, J., Durie, P., Nadeau, J., Bear, C., and Tsui, L-C., 1996, Modulation of disease severity in cystic fibrosis transmembrane conductance regulator deficient mice by a secondary genetic factor, *Nat. Genet.* **12:**280–283.

Safos, S., and Chang, T. M., 1995, Enzyme replacement therapy in ENU2 phenylketonuric mice using oral microencapsulated phenylalanine ammonia-lyase: A preliminary report, *Artif. Cell, Blood Sub. Immobil. Biotechnol.* **23:**681–692.

Scriver, C. R., Eisensmith, R. C., Woo, S. L. C., and Kaufman, S., 1994, The hyperphenylalaninemias of man and mouse, *Annu. Rev. Genet.* **28:**141–165.

Shedlovsky, A., McDonald, J. D., Symula, D., and Dove, W. F., 1993, Mouse models for human phenylketonuria, *Genetics* **134:**1205–1210.

Tagliamonte, A., Tagliamonte, P., Gessa, G. L, and Brodie, B. B., 1969, Compulsive sexual activity induced by *p*-chlorophenylalanine in normal and and pinealectomized rats, *Science* **166:**1433–1435.

Undenfriend, S., and Cooper, J. R., 1952, The enzymatic conversion of phenylalanine to tyrosine, *J. Biol. Chem.* **194:**503–511.

Waisman, H. A., 1962, Role of phenylalanine excess in experimental phenylketonuria, *in* "Ultrastructure and Metabolism of the Nervous System," Vol. 50 (R. Korey, ed.), pp. 32–56, Williams & Wilkins, Baltimore.

Woolf, L. I., Griffiths, R., and Moncrieff, A., 1955, Treatment of phenylketonuria with a diet low in phenylalanine, *Br. Med. J.* **1:**57–64.

Zannoni, V. G., Weber, W. W., Van Allen, P., Rubin, A., Bernstein, R., and La Du, N., 1966, Phenylalanine metabolism and "phenylketonuria" in dilute-lethal mice, *Genetics* **54:**1391–1399.

CHAPTER 13

MOUSE MODELS OF DOWN SYNDROME

MURIEL T. DAVISSON
and ALBERTO C. S. COSTA

1. INTRODUCTION

Down syndrome (DS) is a complex condition caused by a triplicate state (trisomy) of all or part of human chromosome 21 (HSA21). Trisomy 21 (Ts21) is the most frequently observed aneuploidy among liveborn infants (Hook, 1981; Hassold and Jacobs, 1984; Hayes and Batshaw, 1993) and accounts for 90–95% of trisomies in newborns (Sankaranarayanan, 1979). Molecular genetics studies have traced the origin of the extra chromosome in Ts21 as maternal in 95% of the cases, with nondisjunction in meiosis I the error about 80% of the time (Abruzzo and Hassold, 1995). The only predisposing factor that has been established for the occurrence of trisomies, particularly Ts21, is advanced maternal age, a phenomenon that still awaits a mechanistic explanation (Gaulden, 1992; Abruzzo and Hassold, 1995; Lamb *et al.,* 1996). It should be noted, however that, in absolute terms, the importance of the maternal age effect in increasing the overall

---

MURIEL T. DAVISSON and ALBERTO C. S. COSTA • The Jackson Laboratory, Bar Harbor, Maine 04609.

Mouse Models in the Study of Genetic Neurological Disorders, Volume 9 of *Advances in Neurochemistry,* edited by Popko, Kluwer Academic/Plenum Publishers, New York, 1999.

incidence of DS is offset by the higher birth rate in the younger female population, which explains why about 80% of all infants with DS are born to mothers younger than 35 years (Hook *et al.,* 1983).

DS is a significant health problem for the human population because of the constellation of associated congenital defects and postnatal disorders. The estimated number of people with DS in the United States is 250,000, and this figure is expected to increase significantly over the next decade due to projected increased life expectancy of people with DS (Baird and Sadovnick, 1987; Rasore-Quartino and Cominetti, 1995). This populational trend is a direct consequence of improvements in general health care of individuals with DS. An expansion of our basic understanding of this condition will be needed, however, before one can envision any rational pharmaco/genetic therapeutic intervention in such key areas as neurodevelopmental deficits and neurodegenerative processes.

The most frequently cited examples of neuropsychological and neurological components of DS are various degrees of mental retardation increased incidence of seizure disorders, hypotonia, strabismus, nystagmus, deficits in audiovestibular function, and a neuropathology indistinguishable from Alzheimer's disease (Coyle *et al.,* 1986, 1988). In addition, individuals with DS present with variable frequency and intensity several other features, including short stature, congenital cardiac and gastrointestinal anomalies, skeletal and facial dysmorphisms, immune and hematopoietic system deficits, and increased risk of leukemia (Epstein, 1986; Korenberg *et al.,* 1994).

The correlation of various DS features with trisomy for specific HSA21 regions, so-called phenotypic mapping of HSA21, has led to associations between specific segments of HSA21 and specific DS features (Korenberg *et al.,* 1994; Korenberg, 1995). Although molecular techniques have defined the gene content and expression levels in tissues from these individuals (Korenberg, 1993), the phenotypic mapping of DS still lacks precise molecular resolution. This approach is further limited by the low frequency of segmentally trisomic individuals in the human population and the phenotypic heterogeneity of persons with full Ts21. In fact, a central question in the DS field is why the clinical features of DS vary so greatly in type and severity from one individual to another. This variability is probably a consequence of the combined influence of heterogeneous genetic backgrounds and modulation of the phenotype by developmental or environmental variables. Still, the investigation of the relationship between the dosage of individual genes and the expression of the various clinical features of DS is a necessary first step if we are to understand this genetic condition in molecular terms.

It has long been recognized that single gene mutations in mice provide valuable experimental models of human inherited diseases. Mice share many biochemical and physiological characteristics with humans; they can serve as our surrogates for experiments not practically or ethically permissible in humans.

Their small size, short gestation and life span, and ease of genetic manipulation makes them an ideal experimental system. Finally, the extensive knowledge of the mouse genome and comparative mapping with the human genome allows us to move easily between mice and human genomes to identify and characterize genes and their function. On the other hand, we must always keep in mind some caveats associated with use of the mouse central nervous system (CNS) to model our own. For example, the neocortex is not as important to the mouse as it is to humans; in more primitive mammals, the function of the paleocortex and the archicortex may differ from that in primates; the range of cognitive abilities in mice is much more limited than in humans, and the stages of neurodevelopment do not perfectly map between species. As long as we understand such limitations, mice provide the best existing whole animal model for studying the cognitive and neurobiological deficits associated with many human conditions that compromise the CNS.

Since Gropp (1975) first proposed mouse models for aneuploidy, they have become widely accepted. Polani and Adinolfi (1980) subsequently proposed Ts16 as a model for DS. Comparative genome mapping has revealed the conservation of many HSA21 genes in mouse Chromosome 16 (MMU16) and led to the use of the trisomy 16 (Ts16) mouse as a model for DS (Epstein *et al.,* 1985; Reeves *et al.,* 1987). Additional mouse models have subsequently been developed by the creation of single- and multigene duplications of HSA21 genes, segmental trisomy for the HSA21-homologous region of MMU16, single transgenes, and multigene YAC transgenes.

The particular features of each of the various mouse models for DS are reviewed in this chapter. Introductory information is provided on (1) the current status of comparative chromosomal mapping between HSA21 and MMU16, 10, and 17 and (2) the history of the development of mouse model systems. The remainder of the chapter describes specific neurological features of DS and the degree to which DS is modeled by each of the mouse systems: Ts16, Ts16↔diploid (Ts16↔2*n*) chimeras, Ts16 cell cultures and transplants, segmental Ts16 mice (Ts65Dn and Ts1Cje), and single- or multigene transgenic mice.

2. LOCATION OF HUMAN CHROMOSOME 21 GENES IN THE MOUSE GENOME

Homologs of HSA21 genes are located in three mouse chromosomal segments: MMU16 (~36 cM), MMU10 (~8 cM), and MMU17 (~1 cM). At least 27 homologous genes have been identified to date on HSA21 and MMU16, seven on MMU10, and five on MMU17 (Table 1). Because the genes are closely linked in each of these mouse chromosomal segments and appear to be in the same order as in HSA21, it is expected that much, if not all, of each segment is

TABLE 1. Human Chr 21 Genes Conserved in the Mouse Genome

Human Chr location	Human symbol	Mouse Chr	cM	Symbol	Gene name
21	PCP4[a]	16	68.00	*Pcp4*	Purkinje cell protein 4
21q11	*D21S13E*	16	47.00	*D16H21S13*	DNA segment, Chr 16, human D21S13
21q11	*D21S16*	16	47.00	*D16H21S16*	DNA segment, Chr 16, human D21S16
21q11	*D21S52*	16	47.00	*D16H21S52*	DNA segment, Chr 16, human D21S52
21q11.1	*STCH*	16	~47.00	*Stch*	Stress 70 protein chaperone, microsome-associated, 60kD (Chr 16, Reeves and Kao, 5/27/97)
21q21.21	*NCAM2*	16	56	*Ncam2*	Neural cell adhesion molecule 2
21q21–q22.1	*GABPA*	16	65.00	*Gabpa*	GA repeat binding protein, α
21q21.2	*APP*	16	56.00	*App*	Amyloid β (A4) precursor protein
21q22.1	*IFNAR1*	16	63.00	*Ifnar*	Interferon (α and β) receptor
21q22.1	*IFNGR2*	16	63.00	*Ifngr2*	Interferon γ receptor 2
21q22.1	*GART*	16	63.00	*Gart*	Phosphoribosylglycinamide formyltransferase
21q22.1	*KCNJ6*	16	68.00	*Kcnj6*	Potassium inwardly rectifying channel, subfamily J, member 6
21q22.1	*SOD1*	16	61.00	*Sod1*	Superoxide dismutase 1, soluble
21q22.1–q22.1	*TIAM1*	16	61.80	*Tiam1*	T-cell lymphoma invasion and metastasis 1
21q22.1–q22.2	*CBFA2*	16	62.20	*Cbfa2*	Core binding factor α2
21q22.1–q22.2	*IL10RB*	16	61.00	*Il10rb*	Interleukin 10 receptor, beta
21q22.1–q22.2	*GRIK1*	16	58.00	*Grik1*	Glutamate receptor, ionotropic, kainate 1
21q22.1–q22.2	*SON*	16	64.00	*Son*	Son cell proliferation protein

21q22.2	*ERG*	16	69.00	*Erg*	Avian erythroblastosis virus E-26 (v-ets) oncogene related
21q22.2	*SIM*	16	69.00	*Sim2*	Single-minded 2
21q22.3	*CBR*	16	67.00	*Cbr*	Carbonyl reductase
21q22.2	*DYRK*	16	67.60	*Dyrk*	Dual-specificity tyrosine-(Y)-phosphorylation regulated kinase
21q22.3	*ETS2*	16	69.00	*Ets2*	E26 avian leukemia oncogene 2, 3′ domain
21q22.3	*HMG14*	16	70.00	*Hmg14*	High-mobility group protein 14
21q22.3	*KCNE1*	16	64.40	*Kcne1*	Potassium channel Kvs1
21q22.3	*MX1*	16	70.00	*Mx1*	Myxovirus (influenza virus) resistance 1
21q22.2	*MX2*	16	70.00	*Mx2*	Myxovirus (influenza virus) resistance 2
21q22.3	*D21S56*	17	17.20	*D17H21S56*	DNA segment, Chr 17, human D21S56
21q22.3	*CBS*	17	17.40	*Cbs*	Cystathionine β-synthase
21q22.3	*CRYAA*	17	17.40	*Crya1*	α-Crystallin 1
21q22.3	*TFF3*	17	17.00	*Tff3*	Trefoil factor 3, intestinal
21q22.3	*PKNOX1*	17	syntenic	*Pknox1*	Pbx/knotted 1 homeobox
21q22.3	*ITGB2*	10	41.50	*Itgb2*	Integrin β2
21q22.3	*PFKL*	10	41.50	*Pfkl*	Phosphofructokinase, liver, B-type
21q22.3	*COL6A1*	10	41.00	*Col6a1*	Procollagen, type VI, α1
21q22.3	*COL6A2*	10	41.00	*Col6a2*	Procollagen, type VI, α2
21q22.3	*COL18A1*	10	41.00	*Col18a1*	Procollagen, type XVIII, α1
21q22.3	*S100B*	10	40.50	*S100b*	S100 protein, β-polypeptide, neural
21q22.3	*CSTB*	10	43	*Cstb*	Cystatin B

From MGB, 1998.
[a]Human symbol not yet approved.

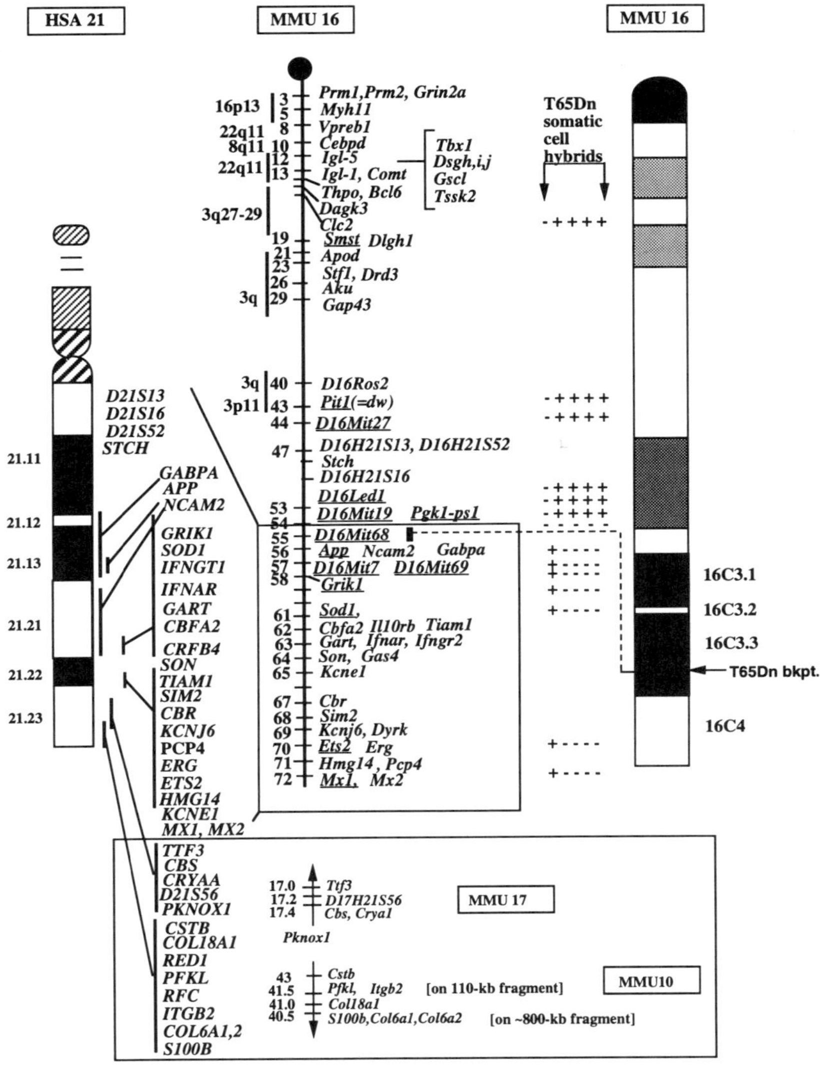

FIGURE 1. Comparative maps of the segments conserved between HSA21 and MMU10, 16, and 17, showing the segment present in Ts65Dn. The + and − signs under the heading T65Dn hybrids indicate the presence or absence in five somatic cell hybrids of the loci underlined on the mouse linkage map at the left. The cytological breakpoint of T65Dn is indicated by an arrow. Arrows on the Chrs 10 and 17 segments indicate the direction of the centromere. Map locations for mouse genes are from MGD (1997), Wei *et al.* (1996; *Gabpa*). Mariano *et al.* (1996; *Ifngr2*). Selected G-band identifications are given to the left of HSA21 and to the right of MMU16. *PCP4 is not yet an approved human symbol.

conserved intact. The majority of the long arm of HSA21 is conserved in MMU16, extending from the centromere to the myxovirus resistance (*MX1*, *MX2*) genes. Genes in the cytological band 21q22.3 distal to the *MX* genes are conserved in MMU17 and MMU10, in that order. To date no homologs of HSA21 genes have been located on any other mouse chromosomes (Fig. 1 and Table 1; MGD, 1998).

3. HISTORY OF MOUSE TRISOMY MODELS FOR DS

As early as 1975, Gropp proposed mouse models for human aneuploidy (Gropp, 1975). Using mice with Robertsonian (Rb) chromosomes from wild populations of *Mus domesticus,* Gropp and colleagues developed a breeding scheme that would produce high frequencies of mouse embryos with specific whole chromosome trisomies. In brief, when mice doubly heterozygous for two Rbs with a chromosome arm in common (homobrachial) are mated to normal laboratory mice, 25–40% of the conceptuses are trisomic for the shared chromosome (Gropp *et al.,* 1975). In 1980, Ts16 was proposed as a model for DS based on similarities in pathology (Polani and Adinolfi, 1980). Subsequently, comparative mapping showed that many HSA21 genes are conserved in MMU16 (Fig. 1; Epstein *et al.,* 1985; Reeves *et al.,* 1987; MGD, 1998), and Ts16 mice have been broadly used as a model system to investigate various aspects of DS traceable to early developmental defects (Epstein *et al.,* 1985, 1990; Gearhart *et al.,* 1986a). Ts16 embryos have endocardial cushion-derived heart anomalies (Gui *et al.,* 1996; Webb *et al.,* 1996); neuropathology (Lacey-Casem and Oster-Granite, 1994), and delayed neural development (Haydar *et al.,* 1996) somewhat similar to defects observed in individuals with DS. Ts16 mice cannot be used to model postnatal aspects of DS, such as mental deficits, however, because they do not survive past birth. These mice are also trisomic for many genes the human counterparts of which are located in other human chromosomes (Fig. 1).

Mouse aggregation chimeras between Ts16 and diploid embryos (Ts16↔2n), transplants of Ts16 cells into normal mice, and cultured Ts16 cells have been used to circumvent the survival problem. Aggregation chimeras are made by combining Ts16 and 2n embryonic cells at the inner cell mass stage or earlier (Gearhart *et al.,* 1986b). The genetic content of the resulting chimeric mouse cannot be predetermined. Although Ts16 cells are present in adult, Ts16↔2n chimeric mice, the Ts16 cell content differs from mouse to mouse and tissue to tissue reducing the reproducibility of results (Cox *et al.,* 1984; Gearhart *et al.,* 1986b). Cultured neurons have provided valuable insights into electrophysiological and biochemical aspects of Ts16 brains, but are obviously limited with respect to whole animal studies.

Davisson *et al.* (1990, 1993) created a segmentally trisomic mouse called Ts65Dn [full symbol Ts(17^{16})65Dn] that is trisomic for most of the region of mouse MMU16 that contains HSA21-homologous genes and very little else (Davisson *et al.,* 1990, 1993; Reeves *et al.,* 1995). First, reciprocal translocations were produced in germ cells of male mice by cesium irradiation of the testes and progeny from these animals were screened for small translocation chromosomes containing the noncentromeric end of MMU16. Then, mice carrying such translocations were bred to determine which would produce mice that survived to adulthood and carried the small translocation chromosome in addition to a normal chromosomal complement ($2n = 41$,Ts) (Davisson *et al.,* 1990). Ts65Dn mice produced in this way are trisomic for the MMU16 segment from the β amyloid precursor protein (*App*) gene to the myxovirus susceptibility (*Mx1*) gene and probably the telomere (Fig. 1; Reeves *et al.,* 1995). Cytological analysis indicates that the extra chromosome contains the entire distal end of MMU16 attached to the centromere and no more than 5% of MMU17 (Fig. 1; Lambert *et al.*).

Because mice with trisomy for this DS-homologous region survive to adulthood, Ts65Dn has opened possibilities of DS-related studies not feasible with complete Ts16 mice because of the *in utero* death. These include studies of aging, learning and memory, neurological abnormalities, cancer development, and postnatal immunologic deficts. Several research groups have demonstrated phenotypes in Ts65Dn mice that are similar to some DS features. Most exciting are neural/cognitive deficits, because the occurrence of some degree of mental retardation is one of the most invariant features of DS. Ts65Dn mice exhibit spatial learning deficits as assessed in a Morris water maze (Escorihuela *et al.,* 1995; Coussons-Read and Crnic, 1996; Reeves *et al.,* 1995; Holtzman *et al.,* 1996a; Klein *et al.,* 1996). Individuals with DS have developmental delay in growth and certain sensorimotor skills (Epstein, 1986); Ts65Dn mice are often smaller than their euploid littermates (Davisson *et al.,* 1993; Hotzman *et al.,* 1996a) and show a delay in achieving sensorimotor milestones and several adultlike reflexes (Holtzman *et al.,* 1996a). Recently, a second segmental trisomy, Ts(16C-ter)1Cje, has been created (Sago *et al.,* 1998). Ts1Cje mice also survive to adulthood. They have been less well characterized so far than Ts65Dn (see section 4.3.2.).

Since the 1980s, when technology for making transgenic animals became available, several research groups have made transgenic mice carrying genes that map to HSA21. Transgenic mice enable investigators to isolate the major effects of individual genes. More recently, technology has evolved to enable the insertion of large fragments of DNA potentially containing multiple genes (Smith *et al.,* 1995) and the creation of multigene duplications using the *Cre-loxP* system (Ramirez-Solis *et al.,* 1995). These multigene transgenic mice provide a bridge between aneuploid and single-gene transgenic mouse models. Comparison of the abnormalities in aneuploid and transgenic mice will enable the determination of single-gene and gene-interactive effects.

4. MENTAL RETARDATION AND NEUROLOGICAL DEFICITS

4.1. Down Syndrome

Mental retardation and hypotonia are arguably the most common features in persons with DS. The majority of individuals with DS function in the mild to moderate range of mental retardation, whereas some individuals function in the borderline to low average range, and others are severely mentally retarded. One should note, however, that mental retardation is a ubiquitous feature in the clinical presentation of chromosomal imbalances, which suggests the existence of a nonspecific component to the mental retardation seen in DS. The available behavioral data suggest that the mental retardation associated with DS also may have some specific features, such as a decline in developmental rate as the children age and a disproportionately delayed language development (mainly in the area of syntax acquisition) (Nadel, 1995; Wishart, 1995). The careful identification of these specific learning deficits is of extreme importance, because they represent the component of mental retardation in the DS population that is most likely correlated with the increased expression of single genes located in HSA21.

Hypotonia in persons with DS may or may not be accompanied by hyporeflexia and reduced muscular strength. Hypotonia manifests itself either at or shortly after birth. Typically it is of moderate intensity in infants and starts to improve after 1 year of age. The origin of the hypotonia seen in DS is generally thought to be in the central nervous system (CNS) (Shumway-Cook and Woollacott, 1985), even though there has been at least one attempt to correlate it with a neuromuscular deficit (Landing and Shankle, 1982). Just as with mental retardation, however, it is difficult to attribute specific characteristics to the hypotonia seen in children with DS, given that hypotonia is a feature of several other chromosomal disorders, the most commonly cited being Prader–Willi syndrome, *cri du chat* syndrome, trisomy 13, ring chromosome 14, ring chromosome 22, translocation of the long arm of chromosome 10, and (in one third of cases) individuals with XXXXY karyotype (OMIM, 1998).

The frequency of seizure disorders in persons with DS (5–10%) is about 10 times greater than is seen in the general population (Haslam, 1995), but still 2–10 times lower than in the combined pool of individuals with mental retardation of all etiologies (Stafstrom *et al.*, 1991). The distribution of occurrence of seizures by age seems to follow a bimodal pattern, with peaks occurring in infancy and in the fourth or fifth decade of life. Besides abnormalities in brain development and function, the increased incidence of seizures also can be traced to problems associated with cardiac defects and infections (Stafstrom *et al.*, 1991; Haslam, 1995).

There are three remarkable features of seizures in the DS population: (1) Infantile spasms are the most common type of seizures seen in infancy but, whereas the occurrence of the same type of epilepsy in the general population has an ominous prognosis in terms of pharmacological control and cognitive outcome, infantile spasms in children with DS usually have a better prognosis both in terms of seizure control and cognitive outcome (Haslam, 1995). In the late 1960s and early 1970s, infantile spasms were reportedly produced by the use of 5-hydroxytryptophan, which was used to control hypotonia in children with DS (Coleman, 1971). (2) There seems to be a disproportionately high frequency of reflex seizures in patients with DS and epilepsy (20%) compared to populations of unselected patients with epilepsy (1%) (Guerrini *et al.,* 1990). Reflex seizures are seizures that are initiated by a triggering stimulus and occur in association with spontaneous seizures. (3) Older persons with DS who develop Alzheimer-type dementia usually have associated seizures. This is particularly true for patients with severe dementia and DS, who almost always have some type of associated seizure activity (Pueschel *et al.,* 1991).

Even though virtually all individuals with DS will develop Alzheimer-like neuropathological findings by the end of their third decade of life, the actual frequency of clinical dementia in this population is an issue far from being settled. There is little controversy over the findings that overall incidence of clinical dementia in the DS population is higher and that onset age is lower than in the general population. Earlier studies, however, seem to have overestimated the relative number of older individuals with DS and clinical dementia by failing to consider other psychiatric disorders, such as depressive disorder, hearing problems, and hypothyroidism in the differential diagnosis (Haslam, 1995; Nelson *et al.,* 1995).

Other reported neuropathological findings in individuals with DS include decreased brain weight; frequently reduced ratio of hindbrain to cerebral volume; hypoplasia of cerebellar middle lobe; anomalies of sulcation of the cerebral cortex, including shallow sulci and fewer secondary sulci; reduction in neuronal number and density in cochlear nuclei, cerebellum, and granular layers of the cerebral cortex; reduction in cholinergic neurons in the basal forebrain, and decreases in cholinergic markers greater than explained by reduced cell numbers; reduction in noradrenergic neurons in the locus coeruleus; reduction in overall cell number in the hippocampus; abnormally positioned cell clusters in cerebral and cerebellar cortices; dendritic atrophy in the visual cortex; delayed myelination; and altered neuronal plasticity and astrocyte reactivity (Becker *et al.,* 1986; Coyle *et al.,* 1986).

4.2. Trisomy 16 (Ts16) in Mice

Overall features of Ts16 mouse fetuses include smaller size, transient edema around 14–17 days gestation (e14–e17), and cellular hypoplasia in several organ systems (reviewed in Lacey-Casem and Oster-Granite, 1994; Holtzman *et al.,*

1996b). The CNS of Ts16 fetuses has many features that resemble those seen in individuals with DS. Grossly, the Ts16 brain is smaller in size than the brains of normal littermates (Gearhart *et al.,* 1986a; Haydar *et al.,* 1996). Growth of the cerebral cortex is delayed during midgestation (e13–e16), but the thickness and number of layers of the pallium are not different from controls by e18 (Haydar *et al.,* 1996). Nevertheless, the overall dimensions of the brain remain smaller. There is a 50–60% reduction in the size of the cerebellum, thalamus, and colliculi of the Ts16 brain (Kornguth *et al.,* 1986; Kornguth and Bersu, 1988). Analysis of the cerebral hemispheres and diencephalon/brain stem regions in embryos at days 15–18 of gestation (e15–e18) showed that wet weights and protein content of both regions were significantly lower than in normal littermates (Singer *et al.,* 1984). Because the placenta is small and has reduced vascularity, Kornguth *et al.* (1986) proposed that placental insufficiency alone might account for the cellular deficits in Ts16 fetuses. However, Ts16 mice have unique neuropathologic features not present in Ts19 mice, which also have placental insufficiency (Lacey-Casem and Oster-Granite, 1994).

As in individuals with DS, cell numbers and density in the CNS are reduced in Ts16 fetuses and various postsynaptic neurotransmitter systems are altered (Coyle *et al.,* 1988; reviewed in Lacey-Casem and Oster Granite, 1994; Holtzman *et al.,* 1996b). However, the reduced numbers of cholinergic neurons may result from different etiologies in DS and Ts16. At birth, individuals with DS have normal levels of cholinergic markers (Kish *et al.,* 1989); basal forebrain cholinergic neurons, at least, degenerate later (Yates *et al.,* 1983). In Ts16 fetuses there is a 30–40% reduction in the number of cholinergic neurons in the septal region and the basal forebrain and a loss of dopaminergic cells in the pons (Kiss *et al.,* 1989), although the pattern of neuronal placement and overall brain organization are not substantially different from controls (Kiss *et al.,* 1989; Sweeney *et al.,* 1989). The reduced neuronal number in Ts16, however, appears to result from early cessation of neurogenesis rather than neurodegeneration. Although generation of basal forebrain neurons begins at about the same time as in normal littermates, cessation of neurogenesis occurs prematurely in a rostral–caudal gradient, leading to 40% fewer putative cholinergic and 13% fewer noncholinergic neurons in the basal forebrain (Sweeney *et al.,* 1989). Because Ts16 mice do not survive past birth, it is impossible to assess whether postnatal neuronal degeneration might also occur. When fetal basal forebrain tissue was transplated into young adult hippocampus, selective atrophy of cholinergic neurons occurred over time in Ts16-derived tissue but not in control tissue (Holtzman *et al.,* 1992a). Ts16 neuronal survival rate can be improved in culture by the addition of nerve growth factor (Kim and Hammond, 1995). Given the well known role of nerve growth factor in protecting cholinergic neurons from apoptotic cell death (see, e.g., Cortazzo *et al.,* 1996), this finding in Ts16 mice may be associated with the observation by Buscigho and Yankner (1995) that cultured neurons from human fetuses with DS have a higher rate of apoptosis than control neurons derived from euploid fetuses.

The reduced neuronal number is associated with neurotransmitter deficits in the CNS, especially in cholinergic neurons (Kiss *et al.,* 1989; Lacey-Casem and Oster Granite, 1994). Between e15 and e18, regional reductions in norepinephrine and dopamine are noted in the cerebral hemispheres and diencephalon/brain stem. Serotonin content and choline acetyltransferase activity are reduced in the diencephalon/brain stem but not in the cerebral hemispheres; γ-aminobutyric acid-ergic (GABAergic) neurons are not affected (Singer *et al.,* 1984). Ts16 cultures have an increased number of somatostatin-immunoreactive neurons (Caserta *et al.,* 1990). Although the gene for preprosomatostatin (*Smst*) maps to MMU16 and SMST protein is elevated in Ts16 tissues (O'Hara *et al.,* 1989), the human *SMST* gene maps to HSA3 and is, therefore, unlikely to have a role in DS neuropathology. Neurotransmitter system alterations also are present in the Ts16 peripheral nervous system (Koistinaho *et al.,* 1991).

Both DS and Ts16 neurons appear to have altered cytoskeletal structure and membrane function. Cultured Ts16 neurons have shorter, less branched neurites and smaller cell bodies than do those from controls (Caserta *et al.,* 1990; Corsi and Coyle, 1991). At the ultrastructural level, Ts16 neuronal microtubules are curved whereas in controls they are straight (Plioplys and Bedford, 1989), and microtubule-associated proteins MAP2 (gene symbol *Mtap2*) and tau (gene symbol *Mtapt*) are reduced (Lacey-Casem and Oster Granite, 1994). Because *Mtap2* and *Mtapt* map to MMU11 and MMU1, respectively, alterations in these two proteins must be secondary to MMU16 triplication. Cellular membranes from Ts16 mice appear fragile as they are difficult to fix for electron microscopic studies (Lacey-Casem and Oster Granite, 1994). The ability of Ts16 chromaffin cells to internalize norepinephrine is inhibited (Koistinaho *et al.,* 1991). Such alterations in the membranes of Ts16 cells could alter cellular adhesion or migration.

In terms of active electrical membrane properties, it has been observed that the time course of action potentials is similarly altered in cultured dorsal root ganglion (DRG) cells from both human embryos with DS (Nieminen *et al.,* 1988; Caviedes *et al.,* 1990), and Ts16 mice (Orozco *et al.,* 1987; Ault *et al.,* 1989). Trisomic neurons from both origins show increased depolarization and repolarization rates and, consequently, decreased spike duration compared to control euploid neurons. The chromosome-specific nature of these alterations is supported by the fact that they are not observed in neurons from Ts19 mice (Caviedes *et al.,* 1990). Also, specificity in terms of anatomical origin of the neurons is suggested by the finding that cultured hippocampal neurons from the Ts16 mouse actually depolarized more slowly than in euploid controls due to reduced inward sodium currents (Galdzicki *et al.,* 1993). It is interesting to note that the DRG is the site where glutamate receptor 5 (GluR5) subunits are thought to be a major component of functional kainate receptors and are expressed in relatively high levels (Sommer *et al.*, 1992). GluR5 (gene symbols: human *GRIK1*, mouse *Grik1*) has been mapped to both HSA21 and MMU16 (GDB, 1998; MGD, 1998). One can speculate that the overexpression of these glutamate receptors in the DRG might produce local changes in expression levels and/or properties of voltage-gated channels in these neurons. This could provide an explanation for the discrepancies between the results obtained from DRG and hippocampal neurons (the hippocampus expresses GluR5 receptor subunits very poorly—Wisden and Seeburg, 1993).

The sugar reduction product *myo*-inositol is elevated in cerebrospinal fluid from DS individuals and in the brains of Ts16 fetuses (Shetty *et al.*, 1996). A sodium/*myo*-inositol contransporter gene (*SLC5A3*) has been mapped to HSA21 (GDB, 1998) and overexpression may cause increased transport of *myo*-inositol into cells. Because metabolically appropriate levels of intracellular *myo*-inositol are essential for synthesis of membrane inositol phospholipids and related cellular signaling pathways, the elevation of *myo*-inositol may contribute to membrane anomalies found in DS and Ts16.

Amyloid β precursor protein (APP) is elevated about 1.5-fold in Ts16 fetuses although the expression patterns appear normal (Holtzman *et al.*, 1992b). APP may be involved in the early-onset Alzheimer's-like dementia and neuropathology that has been observed in individuals with DS. The role of APP in neuropathology in mice can be more effectively studied in adult mice, such as *APP* transgenic mice. The gene for Cu/Zn superoxide dismutase (SOD1) also is overexpressed about 1.5-fold in Ts16 (Holtzman *et al.*, 1992b). For further discussion on both these genes, see section that follows on transgenic mice.

Limited studies in Ts16↔2n chimeras are consistent with expectations from developmental studies in Ts16 fetuses. Chimeras usually lived a normal life span and Ts16 cells survived in most of the tissues examined. An exception was that no trisomic cells were observed in the heart septa, derived from endocardial cushions, which are delayed in development in Ts16 fetuses. Behavioral assessment at 1 year of age showed hyperactivity and stereotypic behavior, such as repetitive grooming. The measurement of presynaptic markers for cholinergic, GABAergic, and catecholaminergic neurotransmitter systems in the brain revealed increased activity of choline acetyltransferase and tyrosine hydroxylase in the cortex, decreased activity in the striatum, and no change in the hippocampus or cerebellum (Gearhart *et al.*, 1986b). Glutamic acid decarboxylase was unchanged in all four regions examined. Again, chimeras have limited value because one cannot control the location, distribution, or proportion of Ts16 cells in different tissues within or between chimeras.

Data from Ts16 brain tissue transplanted into the CNS of diploid mice also are consistent with data from Ts16 fetuses (Höhmann *et al.*, 1990; Holtzman *et al.*, 1992a). Six months after Ts16 basal forebrain neurons were transplanted into diploid mice, cholinergic but not noncholinergic neurons were atrophic (Holtzman *et al.*, 1992a). Unlike Ts16 fetuses, however, the non-HSA21 gene homolog *Smst* was not overexpressed in the transplants. Hippocampal grafts from Ts16 showed neuropathology after 12 months, including axonal degeneration and membrane alterations (Richards *et al.*, 1990). Other results of chimera and transplantation studies are reviewed in Lacey-Casem and Oster-Granite (1994).

In conclusion, Ts16 fetuses show some of the neurologic phenotypes commonly seen in human fetuses with Ts21. In certain instances, there is a very good match between the human and the mouse phenotypes, as is the case for the membrane active electrical properties of cultured neurons from the DRG. In other cases, the phenotype is greatly exaggerated in Ts16 compared to human fetuses with DS, e.g., the overall dimensions of the brain are significantly more reduced in Ts16 than in human fetuses with DS when compared to euploid controls. The true value of the Ts16 mouse as a model of neurodevelopment and neuropathology in DS derives from the fact that, as in the human condition, the genetic material of an entire chromosome is triplicated. Extra caution should be taken in the interpretation of experimental results, however, because potentially neurologically important genes in MMU16 are not mapped to HSA21 but elsewhere in the human genome. Examples of this are the NMDA receptor subunit NMDA2A (gene symbol, *Grin2A*—homologous to human *GRIN2A* on HSA16), dopamine receptor 3 (*Drd3*—homologous to human *DRD3* on HSA3), and chloride channel 2 (*Clc2*). The overexpression of the *Clc2* gene is probably implicated in the recent finding of an increased chloride conductance molecular mechanism for hypotonia in DS. A more important example, though, can be found in the fact that Ts65Dn mice (see later) do not present any congenital heart defects, whereas nearly 100% of Ts16 fetuses do. Because Ts65Dn mice are trisomic for only the segment of MMU16 homologous to HSA21, this suggests that the cardiac phenotype, originally thought to model DS, actually has no relationship with the overexpression of mouse genes homologous to HSA21. In fact, because of the homology between the proximal portion of MMU16 and part of the region 22q11 of HSA22, it is tempting to speculate that the heart defect seen in Ts16 actually may have the same mechanism as the defect seen in complete human Ts22 (Bacino *et al.*, 1995) and cat eye syndrome (McDermid *et al.*, 1986).

4.3. Segmental Ts16

4.3.1. Ts65Dn

Segmental trisomy mice Ts65Dn provide, for the first time, an aneuploid model for systematic and reproducible postnatal studies of systems affected by DS. Although Ts65Dn mice have been available for a relatively short time and many studies remain to be done, research on these mice so far suggests they have many neurobiologic features observed in people with DS.

Most Ts65Dn mice lag behind their littermates in overall growth and are smaller throughout life (Davisson *et al.,* 1993; Holtzman *et al.,* 1996a). About 20% of Ts65Dn mice develop hydrocephalus during their lifetimes (Davisson *et al.,* 1993). This can be compared with the finding that some cerebrospinal fluid filled spaces, such as the suprasellar cistern, are significantly enlarged in adults with DS; no evidence of age-related ventriculomegaly in the brain of persons with DS has been found, however—reviewed in Raz *et al.* (1995). Some Ts65Dn mice exhibit tremors (Davisson *et al.,* 1993; Escorihuela *et al.,* 1995) and seizures have been seen in a few older Ts65Dn mice (Costa and Davisson, unpublished). It will be interesting to determine whether these seizures have some of the characteristics of the seizures noted in older people with DS and, also, if they are accompanied by a further decrease in spatial learning. This could potentially provide a model of DS dementia that is independent from the formation plaques and tangles in the CNS. The Ts65Dn brain appears grossly and neuroanatomically normal (Davisson, unpublished; Holtzman *et al.,* 1996a) and is not smaller than the brains of normal euploid littermates relative to body size (Holtzman *et al.,* 1996a). The volumes of the hippocampus, cortex, striatum, and cerebellum are not different from euploid littermates over ages ranging from 2 days after birth to adults (Holtzman *et al.,* 1996a).

In spite of this lack of gross neuropathological findings, Ts65Dn mice have learning deficits and behavioral features compatible with those observed in humans with DS. In two different tests measuring spatial learning and memory, Ts65Dn mice perform more poorly than their euploid littermates. In the Morris water maze paradigm, they learn more slowly and they never reach the same level of competency, nor do they have the same level of recall during the second round of testing (Escorihuela *et al.,* 1995; Reeves *et al.,* 1995; Holtzman *et al.,* 1996a; Klein *et al.,* 1996). In the radial arm maze, they also perform more poorly than controls, although they learn the task eventually (Demas *et al.,* 1996; Demas *et al.,* 1998). The observation that in two instances Ts65Dn mice showed impaired long-term memory in the radial arm maze task after a 50-day interval, but returned to choice accuracy levels similar to controls after one trial, led one group to suggest that these mice have a defect in memory retrieval rather than storage (Demas *et al.,* 1996). Impaired performance cannot be explained by sensorimotor deficits that might affect these tasks—deficits such as visual acuity, equilibrium, or muscle strength. Although Ts65Dn mice are delayed in achieving certain physical and sensorimotor milestones, as tested by several reflexes, they reach the same level of maturation as littermates by weaning age (3–4 weeks of age), which occurs prior to the ages when testing has been done (Holtzman *et al.,* 1996a). Ts65Dn mice performed similar to controls in tests for visual placing, balance, prehensile reflex and traction on a horizontal bar, motor coordination, and swimming ability (Escorihuela *et al.,* 1995; Holtzman *et al.,* 1996a; Klein *et al.,* 1996), as well as in olfaction, orienting reactions, and climbing (Klein *et al.,* 1996). Thus, the results of the spatial learning and memory testing define a HSA21-homologous region of MMU16 that, when trisomic, causes measurable and reproducible learning deficits.

Ts65Dn mice show spontaneous locomotor hyperactivity in both the horizontal and vertical planes of an open field during the dark phase of a 24-h cycle and in the elevated plus maze (Escorihuela *et al.,* 1995; Reeves *et al.,* 1995; Coussons-Read and Crnic, 1996; Holtzman *et al.,* 1996a). It should be noted that there were two discrepancies among these studies. Unlike the studies by other investigators, Demas *et al.* (1996) found no increase in activity or exploratory behavior in the elevated plus arm maze. Coussons-Read and Crnic (1996) found hyperactivity in both light and dark phases of open field testing, while others noted increased activity only in the dark phase, when mice are normally more active (Reeves *et al.,* 1995; Escorihuela *et al.,* 1996). Although the reasons are not clear, these discrepancies are probably due to differences in testing protocols. Hyperactivity also has been observed in Ts16↔2n chimeras (Gearhart *et al.,* 1986b). Hyperactivity in both Ts65Dn and Ts16↔2n chimeric mice can be seen as indicative of a genetic basis for this phenomenon and that the genes are in the HSA21-homologous region of MMU 16. However, hyperactivity has been noted in rodents with other CNS lesions or neurodegenerative disorders (Sanberg *et al.,* 1987) and may be a nonspecific marker of CNS abnormality. A parallel situation is found in the human DS population, in which hyperactive behavior is more frequent than in the general population, but has the same frequency as in the population of individuals with mental retardation of other etiologies (Myers and Pueschel, 1991; Lovell and Reiss, 1993).

Ts65Dn mice tested in an open field paradigm or in an open–closed-arm maze are less cautious than littermates (Reeves *et al.,* 1995; Coussons-Read and Crnic, 1996). In the open field, they spend more time in the center of the field than control euploid littermates and in the maze they spend equal amounts of time in open and closed arms, whereas control mice spend more time in the closed, darkened arms of the maze (Coussons-Read and Crnic, 1996; Demas *et al.,* 1998). Ts65Dn mice either fail to process cues that signal danger to control euploid mice or do not respond to them. Given a choice of open versus protected fields, Ts65Dn mice do not discriminate between the open and protected areas and show inappropriate sudden movements into areas normal mice would avoid. Although hyperactivity and lack of caution or behavior inhibition suggest a prefrontal cortex functional deficit, this hypothesis is inconsistent with the results of two independent studies in which Ts65Dn mice were not impaired in learning or remembering a passive avoidance task (Coussons-Read and Crnic, 1996; Holtzman *et al.,* 1996a). Since both studies noted a high degree of variability between mice in these tests, it is possible that the defect is too subtle to be detected by these passive avoidance tasks.

Ts65Dn mice frequently show stereotypic behavior, such as repeated jumping in one place (Davisson and Crnic, unpublished; Escorihuela *et al.*, 1995). Note that stereotypic behavior is a general clinical feature of individuals with severe to profound mental retardation mental, including those with DS (Lovell and Reiss, 1993).

An important finding is that Ts65Dn mice show at least some of the neuronal changes seen in individuals with DS. Basal forebrain cholinergic neurons (BFCNs), which supply the major cholinergic input to the hippocampus and neocortex, show age-related degeneration in Ts65Dn brains. Although no difference from controls was observed in BFCN numbers in the medial septal nucleus in the first 3–4 weeks after birth, Ts65Dn mice have 30% fewer BFCNs than controls by 6 months of age and 40% fewer by 20 months of age (Holtzman *et al.*, 1996a). This is in contrast to Ts16 fetuses, which appear to have a deficit in neurogenesis, as discussed in the previous section.

Glial anomalies in DS, including astrocytic hypertrophy and increased astrocyte numbers, also are found in Ts65Dn mice (Holtzman *et al.*, 1996a). Overexpression of the S100β neural peptide (*S100b*) gene has been associated with astrocytosis (Reeves *et al.*, 1994) and is seen in DS. Because the *S100b* gene is not included in the Ts65Dn segment (it maps to MMU10), the astrocytosis in Ts65Dn mice may not have the same genetic basis as that seen in *S100b* transgenic mice (see next section). Either astrocytosis and gliosis can result from different insults to the CNS or, possibly, genes in the Ts65Dn segment may regulate S100β expression. Analysis of S100β expression levels in Ts65Dn mice remains to be done.

Cu/Zn superoxide dismutase (SOD1) mRNA and protein are elevated about 1.5-fold in livers (Reeves *et al.*, 1995) and brain (Holtzman *et al.*, 1996a) and amyloid precursor protein (APP) message and protein are elevated two-fold in cerebral cortex and hippocampus (Reeves *et al.*, 1995) and in Northern analysis of overall brain (Holtzman *et al.*, 1996a) in Ts65Dn mice. The possible roles of these two proteins are discussed in the next section on transgenic mice. Interestingly, brain expression of the apolipoprotein E (*Apoe*) gene on Chr 7 also appears to be elevated, suggesting a gene or genes in the Ts65Dn segment regulates *Apoe* expression (Holtzman *et al.*, 1996a). Despite elevated APP, up to 21 months of age Ts65Dn mice do not show abnormal deposits or plaques containing Aβ4 or neurofibrilliary tangles characteristic of AD-like pathology.

Recent neurochemical studies have shown that Ts65Dn mice have lower basal production of cAMP in the cerebral cortex and hippocampus compared to control animals (Dierssen *et al.*, 1996; Dierssen *et al.*, 1997). This change seems to be region-specific, because no differences were observed in the cerebellum

(Dierssen *et al.,* 1997). In another set of experiments, these authors stimulated the production of cAMP with the β-adrenergic agonist isoprenaline and adenylate cyclase activator forskolin. Again, the maximum responses of cAMP to both stimuli were significantly and specifically lower in the cerebral cortex and hippocampus of Ts65Dn mice when compared to control animals. The slope of the response curve to isoprenaline was markedly depressed, whereas the slope of response to forskolin was similar to what was observed in control animals. This led the authors to conclude that Ts65Dn mice display severe regional deficiencies in β-adrenergic neurotransmission (Dierssen *et al.,* 1997).

Ts65Dn mice do not exactly mimic DS genetically because they are not trisomic for genes in the most distal segment of HSA21q22.3-ter, which are conserved in MMU10 and 17. Clearly, many genes in this region are likely to play an important role in DS. Breeding studies are in progress to produce mice carrying transgenes with some of these MMU10 and 17 genes and the Ts65Dn segment; however, the viability of mice with increasing loads of chromosomal material remains to be determined (Davisson, unpublished).

4.3.2. Ts1Cje

One other segmental trisomy for the HSA21-homologous region of MMU16 has been reported (Sago *et al.,* 1998). During creation of a mouse carrying a *Sod1* targeted mutation, the noncentromeric end of MMU16 was translocated to MMU12. Ts1Cje [full symbol: Ts(16C-ter)1Cje] contains *Sod1* and *Ets2* but not *App.* The *Sod1* gene was probably disrupted by the chromosomal breakpoint because mice with the extra segment do not have increased expression of *Sod1.* Ts1Cje mice exhibit learning deficits in the Morris water maze, although they were not as severely impaired as Ts65Dn mice. Unlike Ts65Dn mice, they were able to perform as well as the controls in the cued or non-spatial (visible platform) part of the test and only showed moderate to severe impairment in the hidden platform, probe and reverse learning parts of the test. In addition, the age-dependent degeneration of BFCN, which is observed in Ts65Dn mice and individuals with DS, was absent in Ts1Cje mice (Sago *et al.,* 1998). Ts1Cje is important because it will provide mice trisomic for a subset of the genes triplicated in Ts65Dn. Further comparison of the phenotypes of Ts65Dn and Ts1Cje mice will help to assess which parts of the MMU16 segment are associated with which features of the abnormal phenotype. Other research groups are making multigenic transgenic mice that will subdivide the segment into even smaller segments. Some of these are discussed in the next section. Nevertheless, segmental trisomic mice are the only currently available aneuploid models of DS for postnatal studies, such as correlation of cognitive deficits and behavioral anomalies with pathologic or functional changes in the nervous system.

4.4. Transgenic Mice

Transgenic mice overexpressing specific genes can be used to assess the effects of individual genes on HSA21. Transgenic mice have been made with the HSA21 genes (or their mouse homologs) for *APP, S100B, SOD1, PFKL,* and *ETS2*. The neurologic phenotypes of these are described briefly in the following paragraphs.

The *APP* gene has been implicated as one of at least four genes with a role in Alzheimer's disease (AD) (OMIM, 1998), because some familial forms of AD are genetically linked to the *APP* gene and its processed product Aβ4 is present in plaques observed in autopsied brains from AD patients. Because many individuals with DS have AD-like dementia and pathology, *APP* may be involved in these features of DS as well. Aβ4 deposits form in the human brain during the aging process, but an immunohistochemical study of 50 individuals with DS and 127 control euploid aging subjects showed that Aβ4 deposition began 50 years earlier in the DS subjects (Rumble *et al.*, 1989). Attempts in several laboratories to generate a mouse model for AD by creating mice overexpressing the normal *APP* gene have been unsuccessful. The lack of AD-like plaques and neurofibrillary tangles in *APP* transgenic mice and Ts65Dn mice suggests that either *APP* overexpression alone is insufficient for this AD-like pathology or mice do not develop this type of pathology. Overexpression of mutant forms of human *APP*, however, has now been shown to cause some AD-like pathologic features in mice, including extracellular Aβ4 deposits, neuritic plaques, synaptic loss, astrocytosis, and microgliosis (Games *et al.*, 1995; Hsiao *et al.*, 1996). In behavioral testing, mice carrying one of these transgenes showed deficits in spatial reference learning and memory in the Y-maze spatial alteration task and in the Morris water maze at 9–10 months of age (Hsaio *et al.*, 1996). Mice transgenic for the mouse *App* gene with mutations specifically in the α-secretase site in the Aβ4 region were aggressive, had seizures, and died earlier than controls. The severity of these symptoms was correlated with the expression level and the number of copies of the transgene in three independent transgenic sublines. The behavioral anomalies were associated with neuropathology, including necrosis, apoptosis, and astrogliosis in the cortex, hippocampus, and other brain regions (Moechars *et al.*, 1996). Interestingly, deficiency of APP in mice homozygous for a targeted mutation is associated with reactive gliosis and decreased locomotor activity and forelimb strength (Zheng *et al.*, 1995). These observations, together with the aggressive behavior reported in Ts65Dn and Ts↔2n chimeric mice, suggest that the *App* gene may be involved in the hyperactivity and, possibly, the seizure activity of individuals with DS.

S100B is one of a family of small acidic calcium binding proteins. It is expressed primarily in the nervous system by astrocytes and Schwann cells.

Overexpression by 10–100-fold of the human *S100B* gene in mice has been reported to have no obvious pathologic effect (Friend *et al.,* 1992). Two- to seven-fold overexpression of the mouse *S100b* gene, however, resulted in astrocytosis and increased axonal proliferation in the hippocampus and especially the dentate gyrus in gene-dose-dependent levels (Reeves *et al.,* 1994). Otherwise, the brains of *S100b* transgenic mice appear of normal size and have no abnormal gross pathology. Elevated S100β protein, astrocytosis, and neurite sprouting occur in brains from DS individuals, but it is not clear whether the elevated S100β causes or is the result of astrocytosis. These observations in transgenic mice give some support to the former explanation.

SOD1 is a key enzyme in the metabolism of oxygen free radicals. Because it is thought to play a role in normal aging, it may be involved in the premature aging thought to be associated with DS. It has been hypothesized that increased SOD1 activity might elevate hydrogen peroxide within brain cells and this, in turn, could alter peroxidation of membrane lipids (Brooksbank and Balazs, 1984; Anneren and Epstein, 1987). This phenomenon has been observed in cells from DS patients, but has been difficult to demonstrate in Ts16 mice (Anneren and Epstein, 1987) and has not been studied in Ts65Dn. On the other hand, oxyradicals generated by microglia cells (CNS-specific macrophages) could lead to neuropathologic changes. When cultured Ts16 microglia cells were stimulated with opsonized zymosan or phorbol myristate acetate, they produced significantly higher levels of superoxide per milligram of protein than littermate controls (Colton *et al.,* 1990). In studies to determine the ability of neurons to handle oxidative stress, cultures of cerebellar neurons from *SOD1* transgenic mice with 3.6 times the normal levels of SOD1 activity were more resistant to damage induced by hydrogen peroxide or arachidonic acid (Epstein *et al.,* 1990). *SOD1* transgenic mice also appear to be more resistant to biochemical-induced neurotoxic damage associated with increased free radicals (Beal *et al.,* 1995).

SOD1 overexpression affects the neurotransmitter system and neuromuscular junctions as well. The uptake of neurotransmitters by the chromaffin granulate transport mechanism is impaired in transfected rat cells with elevated levels of human SOD1 (Elroy-Stein and Groner, 1988). Neuromuscular junctions in the tongues of transgenic mice overexpressing *SOD1* showed terminal axon degeneration and motor end plate pathology, including reduced synaptic contact, reminiscent of what is seen in DS. In addition, tongues from *SOD1* transgenic mice had increased concentrations of calcium and copper, which could result from elevated SOD1 activity (Avraham *et al.,* 1988). These mice, like many individuals with DS, also have diminished serotonin uptake in platelets. The same phenomenon could affect serotonin in the brain. Overexpression of human *SOD1* in transgenic mice also caused a decrease in the mossy fiber projection area in the hippocampus (Barkats *et al.,* 1993). In another transgenic mouse, overexpression of *SOD1* leads to cell death, which could account for the apparent loss of neurons in brains of individuals with DS (Groner *et al.*, 1994). The *SOD1* gene has been shown to be mutated in cases of familial amyotrophic lateral sclerosis (Rosen *et al.*, 1993). Mice transgenic for this mutated *SOD1* gene undergo vacuolar neuron degeneration in both the CNS and spinal cord and Wallerian type degeneration of anterior roots and peripheral nerves. Although mice transgenic for normal human *SOD1* do not exhibit clinical signs, they do develop mild vacuolar changes in the anterior horns of the spinal cord and in motor axons (Dal Canto and Gurney, 1995). Thus, *SOD1* overexpression is likely to play a role in some features of the neuropathology in individuals with DS.

Phosphofructokinase (PFK) is a key glycolytic enzyme that catalyzes a major rate-limiting step in glycolysis. Three subunits encoded by three different genes associate at random to form the active tetrameric molecule. The subunit composition is regulated by tissue specific expression of the genes. Overexpression of the liver form (*PFKL*) in a transfected rat cell line led to a 40% increase in the rate of glycolysis (Elson *et al.*, 1992). In *PFKL* transgenic mice, overexpression of *PFKL* occurred in fetal but not adult tissues (Elson *et al.*, 1994). Thus, if PFKL plays a role in DS, it is likely to do so during development. Activated membrane-bound protein kinase C and mRNA levels of the three *Jun* cellular oncogenes (*Jun, Junb, Junc*) were secondarily elevated in these cells. These are all genes that might be expected to alter the expression of other genes, leading to widespread alteration in gene expression.

The human homolog of the 3′ domain of the E26 avian leukemia oncogene (*ETS2*) is important for thymic development and appears to play a role in some neoplasias. *ETS2* is overexpressed in DS cells. Although no neurologic abnormalities have been reported in mice transgenic for the mouse *Ets2* gene, these mice have hypoplasia of skull, facial, and vertebral bones and various DS-like skeletal anomalies in bones derived from intracartilaginous ossification, including shortened snouts, abnormally shaped heads, shorter necks, and kyphosis. Similar skeletal anomalies also were seen in Ts16 fetuses (Sumarsono *et al.*, 1996). *Ets2* transgenic mice also have thymic abnormalities with histological pathology similar to that in thymuses from DS individuals. Ts65Dn have not been analyzed for skeletal anomalies as yet, but their skull and facial structure appears subtly different from those of their littermate controls (Davisson, unpublished). Curiously, Ts65Dn mice do not have the extreme skeletal malformations seen in *Ets2* transgenic mice although the expression level of *Ets2* in the transgenic mice is ~1.5-fold, not greater than what might be expected in Ts65Dn mice.

Finally, several groups are making transgenic mice that carry multigene segments from HSA21 or MMU16. Smith *et al.* (1995) have made a panel of multigenic transgenic mice with YAC and P1 segments of DNA that cover 2 Mb of HSA21q22.2 between the carbonyl reductase (*CBR*) and Purkinje cell protein 4 (mouse gene symbol *Pcp4*). Learning and memory deficits have been reported in mice from some sublines in this panel. The deficits are specific to the YAC transgene carried, suggesting that single- or multigene transgenics can be used to associate specific types of learning and memory to specific genes on HSA21 (Rubin, 1996). One of these transgenic segments contains the gene for dual specificity tyrosine phosphorylation-regulated kinase (DYRK), which is the human homolog of the *Drosophila* minibrain gene (Smith *et al.,* 1996). The minibrain gene appears to play a role in early neurogenesis in *Drosophila* and is expressed in the adult mouse brain in the cortex, piriform cortex, cerebellum, hippocampus, hypothalamus, and olfactory bulb, all regions of the brain that show abnormalities in individuals with DS (Guimera *et al.,* 1996). Whether the minibrain homolog is the gene in the transgenic segment that causes the learning and behavioral anomalies in the transgenic mice remains to be determined.

4.5. Genetic Background Effects

Genetic background can have a strong effect on the phenotype caused by aneuploidies and transgenes. Differences in genetic background have been shown to affect the incidence and the survival time of Ts16 fetuses (Epstein and Vekemans, 1990). The degree of severity of one of the *APP* transgenes has been shown to vary from mild deficits in learning and near normal life span on the original FVB/N genetic background on which it was created (Hsaio *et al.,* 1996) to a more severe pathology and significantly reduced life span on other inbred backgrounds (Carlson *et al.,* 1996). The phenotypes of single gene targeted mutations that have been made to study behavior and cognition also have been shown to depend on genetic background (Gerlai, 1996; Wehner and Silva, 1996). The effect of multiple genes on specific types of behavior was demonstrated by the ability to breed selectively for extremes in open-field activity of mice (DeFries *et al.,* 1978). These behavioral differences were subsequently correlated to changes in the size of the intrainfrapyramidal mossy fiber projection, suggesting that the projection influences or mediates exploratory activity in mice (Hausheer–Zarmakupi *et al.,* 1996). The Ts65Dn segmental trisomy is currently maintained by repeated backcrossing of trisomic females to C57BL/6JEi × C3H/HeSnJ (B6EiC3H) F1 hybrid males to maintain breeding vigor. Although the litter size is reduced by nearly half in comparison to control hybrid mice, Ts65Dn females still produce litters of six to seven and about 26% of the progeny are trisomic. Attempts to backcross the Ts65Dn chromosome onto inbred genetic backgrounds (C57BL/6J, C57BL/6JEi and a C3H congenic strain) have resulted in litter sizes of one to three and failure to recover Ts65Dn mice after two to three generations of backcrossing. Thus, recovery of Ts65Dn mice for analysis may depend on maintaining a segregating background. However, the segregating background complicates analysis because each locus at which B6Ei and C3H differ can be either heterozygous or homozygous for the allele from either parental strain. For example, C3H mice are homozygous for a retinal degeneration mutation (*Pdeb*^{rd1}) and all Ts65Dn mice must be tested for vision or genotyped at the *Pdeb*^{rd1} gene before they can be used for behavioral studies. (A new B6EiC3H F1 hybrid is being created using a C3H congenic that has the wild-type allele of *Pdeb,* but Ts65Dn on the new background must be compared phenotypically with Ts65Dn on the present background before investigators can switch to using the new background.) In one sense, this segregating genetic background is of value because it mirrors the heterogeneous genetic background of human beings with DS. On the other hand, some experimental variability may be due to genetic background rather than the trisomic segment per se and some studies may require large sample sizes to be statistically significant. For robust effects, the segregating genetic background is unlikely to be a serious problem. For example, the behavioral studies have clearly been reproducible with relatively small sample sizes in at least five laboratories (Reeves *et al.,* 1995, Coussens-Read and Crnic, 1996; Escourihila *et al.,* 1996; Holtzman *et al.,* 1996a; Klein *et al.,* 1996). Many of the transgenic mouse models for DS are being made on a 129 × B6 genetic background, which will give rise to stocks differing in genetic content of B6 and 129 genes unless proper matings are made beginning with the original founder animal. Thus, it is critical to use standard genetic backgrounds for genetically altered mice created to model DS.

5. SUMMARY

Common neurological and neuropsychological features in mammalian trisomies may reflect similar underlying mechanisms affecting developmental processes in the presence of excessive chromosomal material or the limited repertoire of devices available to the CNS to respond to various developmental perturbations. For example, reductions in cell numbers precede mental retardation in several human conditions (Lacey-Casem and Oster-Granite, 1994) and may result in the smaller brain size in individuals with DS. Delay in development of the CNS may lead to improper or inadequate connections in the adult brain. Reduced neuronal numbers or inappropriate or inadequate synaptic connections in the hippocampus could underlie cognitive defects and behavioral anomalies seen in persons with DS. Much of this remains speculation because of the lack of studies that connect observed CNS anomalies with postnatal cognition and behavior. Some investigators believe that an overall disruption of developmental homeostasis leads to DS features (Shapiro, 1983). Most investigators, however, favor the hypothesis that DS is caused by the effects of gene dosage and the interaction of abnormally expressed genes. Given that generalized features are common to several mouse trisomies and to many human aneuploidies, it is likely that the observed phenotypes of both persons with DS and aneuploid mouse models for DS are produced by a combination of specific and generalized components. Generalized features are as important as gene-specific anomalies, but the latter may be more amenable to drug or gene therapy. Transgenic mice are valuable for determining the effects of overexpression of individual genes. Analyzing gene interactions and testing therapies will require experimental models with triplication of as much of the DS-critical region of HSA21 as possible. Although it is clear that Ts65Dn mice do not have all the features of DS, they are the most robust and genetically sound aneuploid mouse model currently available to reproducibly study behavioral and mental deficits associated with human Ts21-DS.

ACKNOWLEDGMENTS

The authors thank Drs. Eva M. Eicher and Roderick T. Bronson for helpful suggestions on the manuscript. Unpublished data from the Davisson research group was collected by Ms. Cecilia Schmidt, Ms. Ellen C. Akeson, and Dr. Roderick T. Bronson. Dr. Davisson's support for study of Ts65Dn mice includes NIH-NICHD Grants HD24605 and HD17449, and contract HD73265. Dr. Costa was the recipient of the NDSS's 1996 Science Scholar Award and has received partial support from the Enoch Gelbard Foundation.

REFERENCES

Abruzzo, M. A., and Hassold, T. J., 1995, Etiology of nondisjunction in humans, *Environ. Mol. Mutagen.* 25 (Suppl): **26:**38–47.

Anneren, K. G., and Epstein, C. J., 1987, Lipid peroxidation and superoxide dismutase-1 and glutathione peroxidase activities in trisomy 16 fetal mice and human trisomy 21 fibroblasts, *Pediatr. Res.* **21:**88–92.

Ault, B., Caviedes, P., and Rapoport, S. I., 1989, Neurophysiological abnormalities in cultured dorsal root ganglion neurons from the trisomy-16 mouse fetus, a model for Down syndrome, *Brain Res.* **485:**165–170.

Avraham, K. B., Schickler, M., Sapoznikov, D., Yarom, R., and Groner, Y., 1988, Down's syndrome: Abnormal neuromuscular junction in tongue of transgenic mice with elevated levels of Cu/Zn superoxide dismutase, *Cell* **54:**823–829.

Bacino, C. A., Schreck, R., Fischel-Ghodsian, N., Pepkowitz, S., Prezant, T. R., and Graham, J. M., Jr., 1995, Clinical and molecular studies in full trisomy 22: Further delineation of the phenotype and review of the literature, *Am. J. Med. Genet.* **56:**359–365.

Baird, P. A., and Sadovnick, A. D., 1987, life expectancy in Down syndrome, *J. Pediatr.* **110:**849–854.

Bambrick, L. L., Golovina, V. A., Blaustein, M. P., Yarowsky, P. J., and Krueger, B. K., 1997, Abnormal calcium homeostasis in astrocytes from the trisomy 16 mouse, *Glia* **19:**352–358.

Barkats, M., Bertholet, J.-Y., Venault, P., Ceballos-Picot, I., Nicole, A., Phillips, J., Moutier, R., Roubertoux, P., Sinet, P.-M., and Cohen-Salmon, C., 1993, Hippocampal mossy fiber changes in mice transgenic for the human copper-zinc superoxide dismutase gene, *Neurosci. Lett.* **160:**24–28.

Beal, M. F., Ferrante, R. J., Henshaw, R., Matthews, R. T., Chan, P. H., Kowall, N. W., Epstein, C. J., and Schultz, J. B., 1995, 3-Nitropropionic acid neurotoxicity is attenuated in copper/zinc superoxide dismutase transgenic mice, *J. Neurochem.* **65:**919–922.

Becker, L. E., Armstrong, D. L., and Chan, F., 1986, Dendritic atrophy in children with Down's syndrome, *Ann. Neurol.* **20:**520–526.

Brooksbank, B., and Balazs, R., 1984, Superoxide dismutase, glutathione peroxidase and lipoperoxidation in Down's syndrome fetal brain, *Brain Res. Dev. Brain Res.* **16:**37–44.

Buscigho, J., and Yankner, B. A., 1995, Apoptosis and increased generation of reactive oxygen species in Down's syndrome neurons in vitro, *Nature* **378:**776–779.

Carlson, G. A., Dake, A., Danielson, V., Davis, A., Perry, B., Turner, S., Borchelt, D. R., Younkin, S. G., and Hsaio, K. K., 1996, Modifiers of the phenotypes induced by Alzheimer amyloid precursor protein (APP), *in* "10th International Mouse Genome Conference," Paris, 7–10 October 1996.

Caserta, M. T., Corsi, P., Oster-Granite, M. L., and Coyle, J. T., 1990, Increased number of somatostatin-immunoreactive neurons in primary cultures of trisomy 16 mouse neocortex, *Brain Res. Mol. Brain Res.* **7:**269–272.

Caviedes, P., Ault, B., and Rapoport, S. I., 1990, Electrical membrane properties of cultured dorsal root ganglion neurons from trisomy 19 mouse fetuses: A comparison with the trisomy 16 mouse fetus, a model for Down syndrome, *Brain Res.* **511:**169–172.

Coleman, M., 1971, Infantile spasms associated with 5-hydroxytryptophan administration in patients with Down's syndrome, *Neurology* **21:**911–919.

Colton, C. A., Yao, J., Gilbert, D., and Oster-Granite, M. L., 1990, Enhanced production of superoxide anion by microglia from trisomy 16 mice, *Brain Res.* **519:**236–242.

Corsi, P., and Coyle, J. T., 1991, Nerve growth factor corrects developmental impairments of basal forebrain cholinergic neurons in the trisomy 16 mouse, *Proc. Natl. Acad. Sci. USA* **88:**1793–1797.

Cortazzo, M. H., Kassis, E. S., Sproul, K. A., and Schor, N. F., 1996, Nerve growth factor (NGF)-mediated protection of neural crest cells from antimitotic agent-induced apoptosis: the role of the low-affinity NGF receptor, *J. Neurosci.* **16:**3895–3899.

Coussons-Read, M. E., and Crnic, L. S., 1996, Behavioral assessment of the Ts65Dn mouse, a model for Down syndrome: Altered behavior in the elevated plus maze and open field, *Behav. Genet.* **26:**7–13.

Cox, D. R., Smith, S. A., Epstein, L. B., and Epstein, C. J., 1984, Mouse trisomy 16 as an animal model of human trisomy 21 (Down syndrome): Production of viable trisomy 16↔diploid mouse chimeras, *Dev. Biol.* **101:**416–424.

Coyle, J. T., Oster-Granite, M. L., and Gearhart, J. D., 1986, The neurobiologic consequences of Down syndrome, *Brain Res. Bull.* **16:**773–787.

Coyle, J. T., Oster-Granite, M. L., Reeves, R. H., and Gearhart, J. D., 1988, Down syndrome, Alzheimer's disease and the trisomy 16 mouse, *Trends Neurosci.* **11:**390–394.

Dal Canto, M. C., and Gurney, M. E., 1995, Neuropathological changes in two lines of mice carrying a transgene for mutant human Cu,Zn SOD, and in mice overexpressing wild type human SOD: A model of familial amyotrophic lateral sclerosis (FALS), *Brain Res.* **676:**25–40.

Davisson, M. T., Schmidt, C., and Akeson, E. C., 1990, Segmental trisomy of murine chromosome 16: A new model system for studying Down Syndrome, *in* "Molecular Genetics of Chromosome 21 and Down Syndrome. Progress in Clinical and Biological Research," Vol. 360 (D. Patterson, and C. J. Epstein, eds.), pp. 263–280, John Wiley & Sons, New York.

Davisson, M. T., Schmidt, C., Reeves, R. H., Irving, N. G., Akeson, E. C., Harris, B. S., and Bronson, R. T., 1993, Segmental trisomy as a model for Down syndrome, *in* "The Phenotypic Mapping of Down Syndrome and Other Aneuploid Conditions. Progress in Clinical and Biological Research," Vol. 384 (C. J. Epstein, ed.), pp. 117–133, Wiley-Liss, New York.

DeFries, J. C., Gervais, M. C., and Thomas, E. A., 1978, Response to 30 generations of selection for open-field activity in laboratory mice, *Behav. Genet.* **8:**3–13.

Demas, G. E., Nelson, R. J., Krueger, B. K., and Yarowsky, P. J., 1996, Spatial memory deficits in segmental trisomic Ts65Dn mice, *Behav. Brain Res.* **82:**85–97.

Demas, G. E., Nelson, R. J., Krueger, B. K., and Yarowsky, P. J., 1998, Impaired spatial working and reference memory in segmental trisomy (Ts65Dn) mice, *Behav. Brain Res.* **90:**199–201.

Dierssen, M., Vallina, I. F., Baamonde, C., Lumbreras, M. A., Martinez-Cue, C., Calatayud, S. G., and Florez, J., 1996, Impaired cyclic AMP production in the hippocampus of a Down syndrome murine model, *Dev. Brain Res.* **95:**122–124.

Dierssen, M., Vallina, I. F., Baamonde, C., Garcia-Calatayud, S., Lumbreras, M. A., and Florez, J., 1997, Alterations of central noradrenergic transmission in Ts65Dn mouse, a model for Down syndrome, *Brain Res.* **749:**238–244.

Elroy-Stein, O., and Groner, Y., 1988, Impaired neurotransmitter uptake in PC12 cells overexpressing human Cu/Zn-superoxide dimutase [sic]—implication for gene dosage effects in Down syndrome, *Cell* **52:**259–267.

Elson, A., Bernstein, Y., Degani, H., Levanon, D., Ben-Hur, H., and Groner, Y., 1992, Gene dosage and Down's syndrome: Metabolic and enzymatic changes in PC12 cells overexpressing transfected human liver-type phosphofructokinase, *Somat. Cell Mol. Genet.* **18:**143–161.

Elson, A., Levanon, D., Weiss, Y., and Groner, Y., 1994, Overexpression of liver-type phosphofructokinase (PFKL) in transgenic-PFKL mice: Implication for gene dosage in trisomy 21, *Biochem. J.* **299:**409–415.

Epstein, C. J., 1986, "Consequences of Chromosome Imbalance: Principles, Mechanisms, and Models," Cambridge University Press, New York.

Epstein, C. J., Cox, D. R., and Epstein, L. B., 1985, Mouse trisomy 16: An animal model for human trisomy 21 (Down syndrome), *Ann. NY Acad. Sci.* **450:**157–168.

Epstein, D. J., and Vekemans, M., 1990, Genetic control of the survival of murine trisomy 16 fetuses, *Teratology* **42:**571–580.

Epstein, C. J., Berger, C. N., Carlson, E. J., Chan, P. H., and Huang, T. T., 1990, Models for Down Syndrome: Chromosome 21-specific genes in mice, *in* "Molecular Genetics of Chromosome 21 and Down Syndrome. Progress in Clinical and Biological Research," Vol. 360 (D. Patterson, and C. J. Epstein, eds.), pp. 215–232, John Wiley & Sons, New York.

Escorihuela, R. M., Fernández-Teruel, A., Vallina, I. F., Baamonde, C., Lumbreras, M. A., Dierssen, M., Tobeña, A., and Flórez, J., 1995, A behavioral assessment of Ts65Dn mice: A putative Down syndrome model, *Neurosci. Lett.* **199:**143–146.

Friend, W. C., Clapoff, S., Landry, C., Becker, L. E., O'Hanlon, D., Allore, R. J., Brown, I. R., Marks, A., Roder, J., and Dunn, R. J., 1992, Cell-specific expression of high levels of human S100β in transgenic mouse brain is dependent on gene dosage, *J. Neurosci.* **12:**4337–4346.

Galdzicki, Z., Coan, E., and Rapoport, S. I., 1993, Cultured hippocampal neurons from trisomy 16 mouse, a model for Down's syndrome, have an abnormal action potential due to a reduced inward sodium current, *Brain Res.* **604:**69–78.

Games, D., Adams, D., Alessandrini, R., Barbour, R., Berthelette, P., Blackwell, C., Carr, T., Clemens, J., Donaldson, T., Gillespie, F., Guido, T., Hagoplan, S., Johnson-Wood, K., Khan, K., Lee, M., Leibowitz, P., Lieburger, I., Little, S., Masliah, E., McConlogue, L., Montoya-Zavala, M., Mucke, L., Paganini, L., Penniman, E., Power, M., Schenk, D., Seubert, P., Snyder, B., Soriano, F., Tan, H., Vitale, J., Wadworth, S., Wolozin, B., and Zhao, J., 1995, Alzheimer-type neuropathology in transgenic mice overexpressing V717F β-amyloid precursor protein, *Nature* **373:**523–527.

Gaulden, M. E., 1992, Maternal age effect: The enigma of Down syndrome and other trisomic conditions, *Mutat. Res.* **296:**69–88.

GDB(TM) Genome Database [database online], 1998, Baltimore (Maryland, USA): Johns Hopkins University, URL: http://gdbwww.gdb.org.

Gearhart, J. D., Davisson, M. T., and Oster-Granite, M. L., 1986a, Autosomal aneuploidy in mice: Generation and developmental consequences, *Brain Res. Bull.* **16:**789–801.

Gearhart, J. D., Singer, H. S., Moran, T. H., Tiemeyer, M., Oster-Granite, M. L., and Coyle, J. T., 1986b, Neurochemical and behavioral studies of trisomy 16↔normal mouse chimera, *Brain Res. Bull.* **16:**815–824.

Gerlai, R., 1996, Gene-targeting studies of mammalian behavior: Is it the mutation or the background genotype? *Trends Neurosci.* **19:**177–181.

Groner, Y., Elroy-Stein, O., Avraham, K. B., Schickler, M., Knobler, H., Minc-Golomb, D., Bar-Peled, O., Yarom, R., and Rotshenker, S., 1994, Cell damage by excess CuZnSOD and Down's syndrome, *Biomed. Pharmacother.* **48:**231–240.

Gropp, A., 1975, Chromosomal animal models of human disease. Fetal trisomy and developmental failure, *in* "Teratology" (C. L. Berry, and D. E. Poswillo, eds.), pp. 17–33, Springer, Berlin.

Gropp, A., Kolbus, U., and Giers, D., 1975, Systematic approach to the study of trisomy in the mouse. II. *Cytogenet. Cell Genet.* **14:**42–62.

Guerrini, R., Genton, P., Bureau, M., Dravet, C., and Roger, J., 1990, Reflex seizures are frequent in patients with Down syndrome and epilepsy, *Epilepsia* **31:**406–417.

Gui, Y. H., Linask, K. K., Khowsathit, P., and Huhta, J. C., 1996, Doppler echocardiography of normal and abnormal embryonic mouse heart, *Pediatr. Res.* **40:**633–642.

Guimera, J., Casas, C., Pucharcos, C., Solans, A., Domenech, A., Planas, A. M., Ashley, J., Lovett, M., Estivill, X., and Pritchard, M. A., 1996, A human homologue of *Drosophila minibrain* (*MNB*) is expressed in the neuronal regions affected in Down syndrome and maps to the critical region, *Hum. Mol. Genet.* **5:**1305–1310.

Haslam, R. H. A., 1995, Neurological problems associated with Down syndrome, *in* "Down Syndrome: Living and Learning in the Community" (L. Nadel, and D. Rosenthal, eds.), pp. 107–114, Wiley-Liss, New York.

Hassold, T. J., and Jacobs, P. A., 1984, Trisomy in man, *Annu. Rev. Genet.* **18:**69–97.

Hausheer-Zarmakupi, Z., Wolfer, D. P., Leisinger-Trigona, M. C., and Lipp, H. P., 1996, Selective breeding for extremes in open-field activity of mice entails a differentiation of hippocampal mossy fibers, *Behav. Genet.* **26:**167–176.

Haydar, T. F., Blue, M. E., Molliver, M. E., Krueger, B. K., and Yarowsky, P. J., 1996, Consequences of trisomy 16 for mouse brain development: Corticogenesis in a model of Down syndrome, *J. Neurosci.* **16:**6175–6182.

Hayes, A., and Batshaw, M. L., 1993, Down syndrome, *Pediatr. Clin. North Am.* **40:**523–535.

Höhmann, C. F., Capone, G., Oster-Granite, M. L., and Coyle, J. T., 1990, Transplantation of brain tissue from murine trisomy 16 into euploid hosts: Effects of gene imbalance on brain development, *Prog. Brain Res.* **82:**203–214.

Holtzman, D. M., Li, Y., DeArmond, S. J., McKinley, M. P., Gage, F. H., Epstein, C. J., and Mobley, W. C., 1992a, Mouse model of neurodegeneration: Atrophy of basal forebrain cholinergic neurons in trisomy 16 transplants, *Proc. Natl. Acad. Sci. USA* **89:**1383–1387.

Holtzman, D. M., Bayney, R. M., Li, Y. W., Khosrovi, H., Berger, C. N., Epstein, C. J., and Mobley, W. C., 1992b, Dysregulation of gene expression in mouse trisomy 16, an animal model of Down syndrome, *EMBO J.* **11:**619–627.

Holtzman, D. M., Santucci, D., Kilbridge, J., Chua-Couzens, J., Fontana, D. J., Daniels, S. E., Johnson, R. M., Chen, K., Sun, Y., Carlson, E., Alleva, E., Epstein, C. J., and Mobley, W. C., 1996a, Developmental abnormalities and age-related neurodegeneration in a mouse model of Down syndrome, *Proc. Natl. Acad. Sci. USA* **93:**13333–13338.

Holtzman, D. M., Epstein, C. J., and Mobley, W. C., 1996b, The human trisomy 21 brain: Insights from mouse models of Down syndrome, *Ment. Retard. Dev. Dis. Res. Rev.* **2:**66–72.

Hook, E. B., 1981, Down syndrome: Its frequency in human populations and some factors pertinent to variation in rates, *in* "Trisomy 21 (Down Syndrome)" (F. F. de la Cruz and P. S. Gerald, eds.), pp. 3–68, University Park Press, Baltimore.

Hook, E. B., Cross, P. K., and Schreinemachers, D. M., 1983, Chromosomal abnormality rates at amniocentesis and in live-born infants, *JAMA* **249**:2034–2038.

Hsiao, K., Chapman, P., Nilsen, S., Eckman, C., Harigaya, Y., Younkin, S., Yang, F., and Cole, G., 1996, Correlative memory deficits, Aβ elevation, and amyloid plaques in transgenic mice, *Science* **274**:99–102.

Kim, J. H., and Hammond, D. N., 1995, Septal cell lines derived from the trisomy 16 mouse: Generation, characterization, and response to NGF, *Brain Res.* **671**:299–304.

Kish, S., Karlinsky, H., Becker, L., Gilbert, J., Rebbetoy, M., Chang, L. J., DiStefano, L., and Hornykiewicz, O., 1989, Down's syndrome individuals begin life with normal levels of brain cholinergic markers, *J. Neurochem.* **52**:1183–1187.

Kiss, J., Schlumpf, M., and Balázs, R., 1989, Selective retardation of the development of the basal forebrain cholinergic and pontine catecholaminergic nuclei in the brain of trisomy 16 mouse, an animal model of Down's syndrome, *Brain Res. Dev. Brain Res.* **50**:251–264.

Klein, S. L., Kriegsfeld, L. J., Hairston, J. E., Rau, V., Nelson, R. J., and Yarowsky, P. J., 1996, Characterization of sensorimotor performance, reproductive and aggressive behaviors in segmental trisomic 16 (Ts65Dn) mice, *Physiol. Behav.* **60**:1159–1164.

Koistinaho, J., Hervonen, A., Winking H., and Rapoport, S. I., 1991, Histochemically demonstrable catecholamines in the sympathetic nervous system of trisomy 16 and normal fetal mice, *Mech. Ageing Dev.* **57**:101–110.

Korenberg, J. R., 1993, Toward a molecular understanding of Down Syndrome, *in* "The Phenotypic Mapping of Down Syndrome and Other Aneuploid Conditions. Progress in Clinical and Biological Research," Vol. 384 (C. J. Epstein, ed.), pp. 87–116, Wiley-Liss, New York.

Korenberg, J. R., 1995, Mental modelling, *Nat. Genet.* **11**:109–111.

Korenberg, J. R., Chen, X. N., Schipper, R., Sun, Z., Gonsky, R., Gerwehr, S., Carpenter, N., Daumer, C., Dignan, P., Disteche, C., Graham, J. M., Jr., Hugdins, L., McGillivray, B., Miyazaki, K., Ogasawara, N., Park, J. P., Pagon, R., Pueschel, S., Sack, G., Say, B., Schuffenhauer, S., Soukup, S., and Yamanka, T., 1994, Down syndrome phenotypes: The consequences of chromosomal imbalance, *Proc. Natl. Acad. Sci. USA* **91**:4997–5001.

Kornguth, S., and Bersu, E., 1988, Cerebellar and collicular development in murine trisomy 16 animals: Morphological changes, monocytic infiltration and increased expression of MHC $H\text{-}2K_k$, *Brain Dysfunct.* **1**:255–271.

Kornguth, S. E., Bersu, E. T., Auerbach, R., Sobkowicz, H. M., Schutta, H. A., and Scott, G. L., 1986, Trisomy 16 mice: Neural, morphological and immunological studies, *Ann. NY Acad. Sci.* **477**:160–178.

Lacey-Casem, M. L., and Oster-Granite, M. L., 1994, The neuropathology of the trisomy 16 mouse, *Crit. Rev. Neurobiol.* **8**:293–322.

Lamb, N. E., Freeman, S. B., Savage-Austin, A., Pettay, D., Taft, L., Hersey, J., Gu, Y., Shen, J., Saker, D., May, K. M., Avramopoulos, D., Petersen, M. B., Hallberg, A., Mikkelsen, M., Hassold, T. J., and Sherman, S. L., 1996, Susceptible chiasmate configurations of chromosome 21 predispose to non-disjunction in both maternal meiosis I and meiosis II, *Nat. Genet.* **14**:400–405.

Lambert, J. P., Davisson, M. T., and Narayanswami, S., 1996, Localization of the translocation breakpoint in a mouse model for Down syndrome, *Mamm. Genome* submitted.

Landing, B. H., and Shankle, W. R., 1982, Reduced number of skeletal muscle fiber nuclei in Down syndrome: Speculation on a "shut off" role of chromosome 21 in control of DNA and nuclear replication rates, possibly via determination of cell surface area per nucleus, *Birth Defects: Original Article Series* **18**:81–87.

Lovell, R. W., and Reiss, A. L., 1993, Dual diagnoses. Psychiatric disorders in developmental disabilities, *Pediatr. Clin. North Am.* **40**:579–592.

Mariano, T. M., Muthukumaran, G., Donnelly, R. J., Wang, N., Adamson, M. C., and Kozak, C. A., 1996, Genetic mapping of the gene for the mouse interferon-γ receptor signaling subunit to the distal end of Chromosome 16, *Mamm. Genome* **7**:321–322.

McDermid, H. E., Duncan, A. M. V., Brasch, K. R., Holden, J. J., Magenis, E., Sheehy, R., Burn, J., Kardon, N., Noel, B., Schinzel, A., Teshima, I., and White, B. N., 1986, Characterization of the supernumerary chromosome in cat eye syndrome, *Science* **232**:646–648.

MGD (Mouse Genome Database), 1998, Mouse Genome Informatics, The Jackson Laboratory, Bar Harbor, Maine. World Wide Web URL:http://www.informatics.jax.org/.

Moechars, D., Lorent, K., De Strooper, B., Dewachter, I., and Van Leuven, F., 1996, Expression in brain of amyloid precursor protein mutated in the alpha-secretase site causes disturbed behavior, neuronal degeneration and premature death in transgenic mice, *EMBO J.* **15**:1265–1274.

Myers, B. A., and Pueschel, S. M., 1991, Psychiatric disorders in persons with Down syndrome, *J. Nervous Ment. Dis.* **179**:609–613.

Nadel, L., 1995, Neural and cognitive development in Down syndrome, *in* "Down Syndrome: Living and Learning in the Community" (L. Nadel and D. Rosenthal, eds.), pp. 107–114, Wiley-Liss, New York.

Nelson, L., Lott, I., Touchette, P., Satz, P., and D'Elia, L., 1995, Detection of Alzheimer disease in individuals with Down syndrome, *Am. J. Ment. Retard.* **99**:616–622.

Nieminen, K., Suarez-Isla, B. A., and Rapoport, S. I., 1988, Electrical properties of cultured dorsal root ganglion neurons from normal and trisomy 21 human fetal tissue, *Brain Res.* **474**:246–254.

O'Hara, B. F., Fisher, S., Oster-Granite, M. L., Gearhart, J. D., and Reeves, R. H., 1989, Developmental expression of the amyloid precursor protein, growth associated protein 43 and somatostatin in normal and trisomy 16 mice, *Brain Res. Dev. Brain Res.* **49**:300–304.

OMIM (On-Line Mendelian Inheritance in Man), 1998, URL: http://www3.ncbi.nlm.nih.gov//-Omim.

Orozco, C. B., Smith, S. A., Epstein, C. J., and Rapoport, S. I., 1987, Electrophysiological properties of cultured dorsal root ganglion and spinal cord neurons of normal and trisomy 16 fetal mice, *Brain Res. Dev. Brain Res.* **32**:111–112.

Peng, S., Rapoport, S., Pearce, R., and Galdzicki, Z., 1996, Increased chloride conductance in cultured trisomy 16 mouse tongue muscle suggest a mechanism for the hypotonia in Down syndrome, *Soc. Neurosci. Abstr.* **22**:1100, Abstract.

Plioplys, A. V., and Bedford, H. M., 1989, Murine trisomy 16 model of Down's syndrome: Central nervous system electron microscopic observations, *Brain Res. Bull.* **22**:233–243.

Polani, P. E., and Adinolfi, M., 1980, Chromosome 21 of man, 22 of the great apes and 16 of the mouse, *Dev. Med. Child Neurol.* **22**:223–225.

Pueschel, S. M., 1990, Clinical aspects of Down syndrome from infancy to adulthood, *Am. J. Med. Genet.* **7**(Suppl):52–56.

Pueschel, S. M., Louis, S., and McKnight, P., 1991, Seizure disorders in Down syndrome, *Arch. Neurol.* **48**:318–320.

Ramirez-Solis, R., Liu, P., and Bradley, A., 1995, Chromosome engineering in mice, *Nature* **378**:720–724.

Rasore-Quartino, A., and Cominetti, M., 1995, Clinical follow-up of adolescents and adults with Down syndrome, *in* "Down Syndrome: Living and Learning in the Community" (L. Nadel and D. Rosenthal, eds.), pp. 238–245, Wiley-Liss, New York.

Raz, N., Torres, I. J., Briggs, S. D., Spencer, W. D., Thornton, A. E., Loken, W. J., Gunning, F. M., McQuain, J. D., Driesen, N. R., and Acker, J. D., 1995, Selective neuroanatomic abnormalities in Down's syndrome and their cognitive correlates: Evidence from MRI morphometry, *Neurology* **45**:356–366.

Reeves, R. H., Gallahan, D., O'Hara, B. F., Gallahan, R., and Gearhart, J. D., 1987, Genetic mapping of *Prm-1, Igl-1, Smst, Mtv-6, Sod-1,* and *Ets-2* and localization of the Down syndrome region on mouse chromosome 16, *Cytogenet. Cell Genet.* **44:**76–81.

Reeves, R. H., Yao, J., Crowley, M. R., Buck, S., Zhang, X., Yarowsky, P., Gearhart, J. D., and Hilt, D. C., 1994, Astrocytosis and axonal proliferation in the hippocampus of S100β transgenic mice, *Proc. Natl. Acad. Sci. USA* **91:**5359–5363.

Reeves, R. H., Irving, N. G., Moran, T. H., Wohn, A., Kitt, C., Sisodia, S. S., Schmidt, C., Bronson, R. T., and Davisson, M. T., 1995, A mouse model for Down syndrome exhibits learning and behaviour deficits, *Nat. Genet.* **11:**177–184.

Richards, S. J., Waters, J. J., Rogers, D. C., Martel, F. L., Sparkman, D. R., White, C. L., Beyreuther, K., Masters, C. L., and Dunnett, S. B., 1990, Hippocampal grafts derived from embryonic trisomy 16 mice exhibit amyloid (A4) and neurofibrillary pathology, *Prog. Brain Res.* **82:**215–223.

Rosen, D. R., Siddique, T., Patterson, D., Figlewicz, D. A., Sapp, P., Hentati, A., Donaldson, D., Goto, J., O'Regan, J. P., Deng, H. X., Rahmani, Z., Krizus, A., McKenna-Yasek, D., Cayabyab, A., Gaston, S. M., Berger, R., Tanzi, R. E., Halperin, J. J., Herzfeldt, B., Van den Bergh, R., Hung, W. Y., Bird, T., Deng, G., Mulder, D. W., Smyth, C., Laing, N. G., Soriano, E., Pericak-Vance, M. A., Haines, J., Rouleau, G. A., Gusella, J. S., Horvitz, H. R., and Brown, R. H., Jr., 1993, Mutations in Cu/Zn superoxide dismutase gene are associated with familial amyotrophic lateral sclerosis [published erratum appears in *Nature* 364:362], *Nature* **362:**59–62.

Rubin, E. M., 1996, Identification of genes affecting learning and memory in chromosome 21 YAC transgenic mice, presented at "Engineered Animal Models," May 7–8, Baltimore, Abstract.

Rumble, B., Retallack, R., Hilbich, C., Simms, G., Multhaup, G., Martins, R., Hockey, A., Montgomery, P., Beyreuther, K., and Masters, C. L., 1989, Amyloid A4 protein and its precursor in Down's syndrome and Alzheimer's disease, *N. Engl. J. Med.* **320:**1446–1452.

Sago, H., Carlson, E. J., Smith, D. J., Kilbridge, J., Rubin, E. M., Mobley, W. C., Epstein, C. J., and Huang, T. T., 1998, Ts1Cje, a partial trisomy 16 mouse model for Down syndrome, exhibits learning and behavioral abnormalities, *Proc. Natl. Acad. Sci. USA* **95:**6256–6261.

Sanberg, P. R., Moran, T. H., and Coyle, J. T., 1987, Microencephaly: Cortical hypoplasia induced by methylazoxymethanol, *in* "Animal Models of Dementia" (J. T. Coyle, ed.), pp. 253–278, Alan R. Liss, New York.

Sankaranarayanan, K., 1979, The role of non-disjunction in aneuploidy in man, an overview, *Mutat. Res.* **61:**1–28.

Shapiro, B. L., 1983, Down syndrome—a disruption of homeostatis, *Am. J. Med. Genet.* **14:**241–269.

Shetty, H. U., Holloway, H. W., Acevedo, L. D., and Galdzicki, Z., 1996, Brain accumulation of *myo*-inositol in the trisomy 16 mouse, an animal model of Down's syndrome, *Biochem. J.* **313:**31–33.

Shumway-Cook, A., and Woollacott, M. H., 1985, Dynamics of postural control in the child with Down syndrome, *Phys. Ther.* **65:**1315–1322.

Singer, H. S., Tiemeyer, M., Hedreen, J. C., Gearhart, J., and Coyle, J. T., 1984, Morphologic and neurochemical studies of embryonic brain development in murine trisomy 16, *Brain Res. Dev. Brain Res.* **15:**155–166.

Smith, D. J., Zhu, Y., Zhang, J., Cheng, J. F., and Rubin, E. M., 1995, Construction of a panel of transgenic mice containing a contiguous 2-Mb set of YAC/P1 clones from human chromosome 21q22.2, *Genomics* **27:**425–434.

Smith, D. J., Stevens, M. E., Patil, N., Cheng, J. F., and Ruin, E. M., 1996, Genotype/phenotype studies in YAC/P1 transgenic mice containing several megabases of contiguous DNA from human 21q22.2, *Am. J. Hum. Genet.* **59**(Suppl):A56, Abstract.

Sommer, B., Burnashev, N., Verdoorn, T. A., Keinanen, K. Sakmann, B., and Seeburg, P. H., 1992, A glutamate receptor channel with high affinity for domoate and kainate, *EMBO J.* **11:**1651–1656.

Stafstrom, C. E., Patxot, O. F., Gilmore, H. E., and Wisniewski, K. E., 1991, Seizures in children with Down syndrome: Etiology, characteristics and outcome, *Dev. Med. Child Neurol.* **33:**191–200.

Sumarsono, S. H., Wilson, T. J., Tymms, M. J., Venter, D. J., Corrick, C. M., Kola, R., Lahoud, M. H., Papas, T. S., Seth, A., and Kola, I., 1996, Down's syndrome-like skeletal abnormalities in *Ets2* transgenic mice, *Nature* **379:**534–537.

Sweeney, J. E., Höhmann, C. F., Oster-Granite, M. L., and Coyle, J. T., 1989, Neurogenesis of the basal forebrain in euploid and trisomy 16 mice: An animal model for developmental disorders in Down syndrome, *Neuroscience* **31:**413–425.

Trauner, D. A., Bellugi, U., and Chase, C., 1989, Neurologic features of Williams and Down syndromes, *Pediatr. Neurol.* **5:**166–168.

Webb, S., Anderson, R. H., and Brown, N. A., 1996, Endocardial cushion development and heart loop architecture in the trisomy 16 mouse, *Dev. Dyn.* **206:**301–309.

Wehner, J. M., and Silva, A., 1996, Importance of strain differences in evaluations of learning and memory processes in null mutants, *Ment. Retard. Dev. Dis.* **2:**243–248.

Wei, J., Dlouhy, S. R., Wang, Y., Zhu, J., Fitzpatrick, L., Ghetti, B., Hodes, M. E., 1996, Linkage mapping of microdissected clones from distal mouse chromosome 16, *Somat. Cell Mol. Genet.* **22:**227–232.

Wisden, W., and Seeburg, P. H., 1993, A complex mosaic of high-affinity kainate receptors in rat brain, *J. Neurosci.* **13:**3582–3598.

Wishart, J. G., 1995, Cognitive abilities in children with Down syndrome: Developmental instability and motivational deficits, *in* "Etiology and Pathogenesis of Down Syndrome. Progress in Clinical and Biological Research," Vol. 393 (C. J. Epstein, T. Hassold, L. T. Lott, L. Nadel, and D. Patterson, eds.), pp. 57–91, Wiley-Liss, New York.

Yates, C. M., Simpson, J., Gordon, A., Maloney, A. F., Allison, Y., Ritchie, I. M., Urquhart, A., 1983, Catecholamines and cholinergic enzymes in pre-senile and senile Alzheimer-type dementia and Down's syndrome, *Brain Res.* **280:**119–126.

Zheng, H., Jiang, M., Trumbauer, M. E., Sirinathsinghji, D. J. S., Hopkins, R., Smith, D. W., Heavens, R. P., Dawson, G. R., Boyce, S., Conner, M. W., Stevens, K. A., Slunt, H. H., Sisodia, S. S., Chen, H. Y., and Van der Ploeg, L. H. T., 1995, β-Amyloid precursor protein-deficient mice show reactive gliosis and decreased locomotor activity, *Cell* **81:**525–531.

CHAPTER 14

MODELING EPILEPTIC DISORDERS IN MICE

THOMAS N. SEYFRIED, MICHAEL PODERYCKI, and MARIANA TODOROVA

1. INTRODUCTION

Epilepsy can be defined generally as intermittent neuronal dysfunction resulting from sudden, disorderly discharges of central nervous system (CNS) neurons. With the exception of cerebrovascular disease, epilepsy is the most prevalent human neurological disorder. Seizures are a symptom of epilepsy and can be either convulsive (with overt motor manifestations) or nonconvulsive (without motor involvement). The classification of epilepsy based on seizure phenotype is complex and often ambiguous. A plethora of epilepsy phenotypes may arise from seizures originating in cortical or subcortical brain regions. The seizures are often described as generalized, involving the cortex of both hemispheres, or as partial, involving a specific cortical region. Partial seizures can be further categorized as simple (without loss of consciousness) or complex (consciousness impaired). Seizure phenotypes can also be modified significantly by age (Delgado-Escueta *et al.*, 1982). Because the seizure behavioral phenotype is removed from its

---

THOMAS N. SEYFRIED, MICHAEL PODERYCKI, and MARIANA TODOROVA • *Department of Biology, Boston College, Chestnut Hill, Massachusetts 02167.*

Mouse Models in the Study of Genetic Neurological Disorders, Volume 9 of *Advances in Neurochemistry,* edited by Popko, Kluwer Academic/Plenum Publishers, New York, 1999.

biochemical or physiological cause, epilepsy classification based on seizure phenotype is controversial.

Classification of epilepsy based on seizure etiology is less complex, but is also ambiguous. Generalized or partial seizures that occur without signs of organic brain disorder are often described as *idiopathic* epilepsy, whereas similar seizure types that occur from definable brain injury, disease, or neurostructural abnormality are often described as *symptomatic* or acquired epilepsy (Hauser, 1982; Delgado Escueta *et al.,* 1986; Wolf, 1994). Although inheritance is implicated in most idiopathic epilepsies, little is known about the genetic mechanisms involved. Genetic heterogeneity, small family size, and the general inaccessibility of human epileptic brain tissue has hindered the understanding of these mechanisms. Genetic factors can also influence symptomatic epilepsies by causing some brains to be more susceptible to damage-induced epilepsy than others. Furthermore, a wide spectrum of epileptic seizure phenotypes occur in association with many inherited human metabolic diseases. These epilepsies are obviously genetic, but are sometimes classified as symptomatic epilepsies because they occur together with metabolic or neurostructural abnormalities (Delgado Escueta *et al.,* 1986). Likewise, all inherited idiopathic epilepsies must involve some type of structural or metabolic defect in keeping with the one gene–one polypeptide hypothesis. These defects may be expressed as epilepsy at a critical age or following some environmental stress. It is possible that some idiopathic epilepsies are milder allelic variants of severe inborn errors of metabolism. Thus, classification of epilepsy based on idiopathic or symptomatic etiology can be ambiguous.

The classification of epilepsy based on genetic etiology may be less ambiguous than classification based on seizure type or neurostructural phenotype. Because human and murine nervous systems respond similarly to seizure-provoking stimuli (Krall *et al.,* 1978), and can display common neurostructural abnormalities (Ribak, 1991), it is likely that some mechanisms of naturally occurring spontaneous epilepsies are also similar in these species. Recent findings indicate that about 37% of the mouse autosomal genome shows linkage conservation with the human autosomal genome (Nadeau, 1989; Copeland *et al.,* 1993; Dietrich *et al.,* 1995). It is therefore possible that identification of epilepsy genes in mice may help identify genes in humans that might disrupt neural homeostasis and contribute to the pathogenesis of epilepsy. Our objective is to review some of these genetic convulsive disorders in mice that can serve as experimental models for human epilepsies.

2. MONOGENIC EPILEPSY MODELS IN MICE

Several monogenic convulsive disorders in mice can serve as models for idiopathic and symptomatic convulsive disorders in humans. Noebels previously reviewed a number of the monogenic epilepsy models (Noebels, 1986, 1991, 1995). Some of these models illustrate the influence of both allelic and nonallelic heterogeneity on the seizure phenotype.

2.1. Tottering and Leaner

Intralocus heterogeneity, arising from multiple alleles, occurs at the *tottering* (*tg*) locus on chromosome 8. Mice with different mutant alleles at this locus express markedly different seizure phenotypes. Homozygous *tottering* (*tg/tg*) mice express ataxia, simple partial seizures with motor symptoms (Jacksonian seizures), and abnormal spike-wave patterns resembling "absence" seizures, whereas homozygous *leaner* mice (tg^{la}/tg^{la}) mostly express severe immobilizing ataxia and early death, but do not express an overt convulsive phenotype (Kaplan *et al.,* 1979; Noebels and Sidman, 1979; Seyfried, 1982a; Noebels, 1986). On the other hand, mice carrying one *tg* allele and one tg^{la} allele, i.e., compound heterozygotes (tg/tg^{la}), express an intermediate level of ataxia and severe sustained motor seizures resembling a type of epilepsia partialis continua (Seyfried *et al.,* 1986). The *rolling* mouse Nagoya represents a third allele at this locus (tg^{rol}). These mice express the ataxic phenotype, but were not described as having focal motor seizures (Tamaki *et al.,* 1986). Cerebellar neuronal loss and selective overgrowth of axons in the locus coeruleus occur in both *tottering* and *leaner* mice (Seyfried *et al.,* 1981; Noebels, 1986; Seyfried *et al.,* 1986; Hess and Wilson, 1991; Noebels, 1991; Isaacs and Abbott, 1995). The cerebellum, a brain region not generally associated with paroxysmal events, is thought to be the critical link in the neural network involved with the seizures in *tottering* mice (Campbell and Hess, 1998).

Confirmation of genetic heterogeneity at the *tottering* locus comes from findings that *tg* is a missense mutation and tg^{la} a splice donor mutation in the gene for the α_{1A} voltage-sensitive calcium channel (Fletcher *et al.,* 1996). Mutations in a possible human homologue of the *tg* gene lead to familial hemiplegic migraine and episodic ataxia type-2 (Ophoff *et al.,* 1996). In contrast to mutations at the *tottering* locus, which are inherited as recessives, the human α_{1A} voltage-sensitive calcium channel mutations are inherited as dominants. It will be important to compare and contrast the pathogenesis of these mutations in the human and the mouse.

2.2. Lethargic and Stargazer

The abnormal spike-wave seizure patterns observed in *tottering* mice are not unique, but are also found in other independent autosomal recessive spontaneous epilepsy models including *lethargic* and *stargazer* (Noebels, 1986; Qiao and Noebels, 1991; Lin *et al.,* 1993; Noebels, 1994; Hosford *et al.,* 1995). Similar to the *tg* mutation, both the *lethargic* (*lh,* chromosome 2), and the *stargazer* (*stg,* chromosome 15) mutations cause ataxia and generalized "absense" seizures. The seizures in *stargazer* are more prolonged and frequent than those in *tottering* and *lethargic* mice. The ataxia and head-tossing are presumed to be pleiotropic consequences of the mutation in the cerebellum and inner ear, respectively (Letts *et al.,* 1998). The ataxia and absense seizures in the *lethargic* and *stargazer* mutants represent an example of interlocus genetic heterogeneity, i.e., independent nonallelic mutations that produce similar clinical phenotypes. The *lethargic* and *stargazer* mutations were recently defined. *Lethargic* arises from mutation of the β-subunit of the Cchb4 gene, which regulates voltage-dependent calcium currents through direct interaction with α 1 subunits. A four-nucleotide insertion into a splice donor site causes exon skipping, translational frameshift, and protein truncation with loss of the α 1-binding site (Burgess *et al.,* 1997). Letts and colleagues recently described a novel gene, *Cacng2,* whose expression is disrupted in two *stargazer* alleles (Letts *et al.,* 1998). The *stg* gene encodes a 36-kDa protein (stargazin) with structural similarity to the γ subunit of skeletal muscle voltage-gated calcium (Ca^{2+}) channels. The *stg* allele is associated with an intronic insertion of an ETn retrotransposon, greatly reducing the amount of normal transcript. Stargazin is brain-specific and, like other neuronal Ca^{2+} channel subunits, is enriched in synaptic plasma membranes. *In vitro,* stargazin increases steady-state inactivation of α 1 class A Ca^{2+} channels. The anticipated effect in *stargazer* mutants, inappropriate Ca^{2+} entry, may contribute to their more pronounced seizure phenotype compared to other mouse absence models with Ca^{2+} channel defects. The cloning of these single-locus epilepsy genes should provide interesting information on the shared biochemical and anatomical pathways of various epilepsy syndromes.

Genetic heterogeneity is a well documented phenomenon in human genetics and likely underlies several human epilepsy syndromes (Anderson *et al.,* 1986). Some of the monogenic epilepsy models in mice show that strikingly different epilepsy phenotypes can arise from defects at a single locus, i.e., *tottering/leaner,* whereas the same epilepsy phenotype (defined in terms of "absense" seizures) can arise from defects at several loci, e.g., *tg, lh, stg.* The expression of allelic and nonallelic heterogeneity in the epileptic mouse models may provide insight on the etiology of certain epilepsy conditions in man. Evidence for genetic heterogeneity in human epilepsy may come from finding that some patients with very different clinical epilepsies respond favorably to a particular anticonvulsant drug, whereas other patients with phenotypically similar epilepsies may show widely varying responses to the same drug. Allelic heterogeneity may be suspected in the former situation and nonallelic heterogeneity in the latter situation.

2.3. Weaver

Violent spontaneous generalized convulsions, that are occasionally lethal, occur in adult mice heterozygous (+/*wv*) for the weaver mutation (*wv,* chromosome 16) (Seyfried, 1982a; Eisenberg and Messer, 1989). The incidence of seizures in these mice can also be influenced by differences in background genome, suggesting an important role for nonallelic modifier loci (Dlouhy *et al.*, 1996). Spontaneous seizures are much less common, however, in the *wv/wv* homozygotes which express severe ataxia from extensive granule cell loss in the developing cerebellum. The *weaver* mutation involves a base pair substitution in the pore-forming domain of a G-protein-coupled inward rectifier potassium channel referred to as *Girk* (Patil *et al.*, 1995). It is not yet clear how one copy of this *Girk* mutation causes partial granule cell loss and spontaneous seizures in adult +/*wv* mice. In a more recent study, spontaneous seizures were reported in *GIRK2* knockout (−/−) mice that had normal cerebellar development (Signorini *et al.*, 1997). These and other findings indicate that the *weaver* mutation and the knockout have different effects on seizure susceptibility (Slesinger *et al.*, 1997). Catecholaminergic abnormalities in various brain stem regions have also been reported in a *wv/wv* mice (Schmidt *et al.*, 1982; Roffler Tarlov and Graybiel, 1987). Although these defects likely arise as pleiotropic effects of the potassium channel mutation, they could also contribute to the expression of generalized convulsions. The *weaver* mouse illustrates the importance of gene dosage and nonallelic interactions in the pathogenesis of convulsive behavior.

2.4. Lurcher

An unusual form of spontaneous convulsive behavior occurs in adult mice carrying the autosomal dominant gene *Lurcher* (*Lc*, chromosome 6). *Lurcher* mice also express severe ataxia from the loss of both cerebellar granule cells and Purkinje cells (Seyfried *et al.*, 1982). The *Lurcher* seizures, which can last as long as 10 min, involve back arching, tonic limb extension, and rolling. The seizures can also be induced in both young and adult mice by gentle tail twisting for about 30 sec. Seizure susceptibility in *Lurcher* mice, however, is not associated with an abnormal electrocorticogram (Noebels, 1986). It is important to mention that many murine convulsive disorders are likely of subcortical origin and may not present with cortical EEG abnormalities. The subcortical electrophysiology of murine epilepsy models has not been extensively studied (Hosford *et al.*, 1995). The *Lurcher* mutation was recently cloned and involves a G-to-A transition that changes a conserved alanine to a threonine residue in the transmembrane domain III of the mouse δ glutamate receptor gene (*GluR*δ*2*) (Zuo *et al.*, 1997).

2.5. Slow-Wave Epilepsy

Locomotor ataxia and frequent spike waves in the 3/sec range were described in the slow wave epilepsy (*swe*, chromosome 4) neurological mutant (Noebels *et al.*, 1995). Occasional spontaneous generalized seizures, which are sometimes lethal, occur in mutant mice that are 14–18 days old. Cox and coworkers recently found that *swe* represents a nonsense mutation in the gene for the Na^+H^+ exchanger, i.e., in the previously identified *Nhe1* gene (Morahan and Rakar, 1993; Cox *et al.,* 1997). The Na^+/H^+ exchanger may play a role in the ionic homeostasis within nerve terminals. It will be interesting to determine how a mutation in this ion exchanger can produce ataxia and slow spike-wave epileptogenesis.

2.6. Epilepsy Models Associated with Inherited Myelin Defects

Convulsive disorders have been associated with inherited defects in myelinogenesis in both man and mouse. Myelin abnormalities occur in a number of epilepsies of the West and Lennox–Gastaut syndromes (Gomez and Klass, 1983). Seizures are also associated with human X-linked adrenoleukodystrophy and other peroxisomal disorders. Because childhood seizure disorders associated with myelin defects are sometimes refractory to traditional anticonvulsant medications (Gomez and Klass, 1983), mouse epilepsy models involving myelin abnormalities may be useful for evaluating the efficacy of certain anticonvulsants.

There are a number of mouse epilepsy models involving myelin abnormalities. The *jimpy* mouse is an X-linked recessive that expresses generalized seizures, tremor, and ataxia from dysmyelination or demyelination due to a mutation in the gene for proteolipid protein (PLP). PLP is a major structural protein of the myelin sheath. Some jimpy mice die during the ictal period (Guenet, 1980). Seizures also occur in transgenic mice that overexpress the PLP gene (Kagawa *et al.,* 1994; Readhead *et al.,* 1994). The human counterpart of *jimpy* is Pelizaeus–Merzbacher disease, which is also associated with seizures in some patients (Gow and Lazzarini, 1996).

The quaking mouse is an autosomal recessive on chromosome 17 that expresses spontaneous tonic seizures in association with hypomyelination (Chauvel *et al.,* 1980; Valatx, 1980). Seizing quaking mice show tonic extension of the trunk, flexion of the neck and hindlimbs, and pedaling forepaw movements when the mouse lies on its side. Myoclonic seizures are also accompanied by an abnormal electrocorticogram (Chauvel *et al.,* 1980). The *quaking* seizures can be blocked by a variety of anticonvulsant drugs (Chermat *et al.,* 1979; Mitrovic *et al.,* 1990). The *quaking* gene was recently cloned and appears to influence signal transduction and RNA metabolism (Ebersole *et al.,* 1996). It is not clear how this gene defect produces myelin abnormalities or seizures.

The *shiverer* (*shi*) and myelin-deficient (shi^{mld}) mice arise from allelic mutations in the myelin basic protein gene on chromosome 18. Both mice express tremors and generalized seizures, but the seizure phenotype is more severe in *shiverer* mice than in myelin-deficient mice (Readhead and Hood, 1990; Shine *et al.*, 1992). Although the *shiverer* mice are behaviorally epileptic, the seizures are not associated with cortical EEG abnormalities (Noebels, personal communication). This suggests that the seizures in these mice might involve subcortical structures. Taken together, these findings establish an interesting association between mouse convulsive behavior and inherited defects in myelinogenesis. These models may therefore provide insight on the role of the myelin sheath or the oligodendrocyte in epilepsy.

2.7. Epilepsy Models Arising from Gene Disruptions

A number of new monogenic epilepsy models have been produced from disruption of specific genes through targeted homologous recombination or insertional mutagenesis (Noebels, 1996). The specific genes targeted were either knocked out (null mutants) or mutated to produce an abnormal protein (missense mutants). A number of these are discussed in the following subsections.

2.7.1. Ca^{2+}/Calmodulin-Dependent Kinase II α-Subunit

A model of limbic epilepsy was found in mice carrying a null mutation in the α subunit of Ca^{2+}/calmodulin-dependent kinase II (α-CaMK) (Butler *et al.*, 1995). This enzyme is thought to participate in neuronal excitability. The −/− mice exhibited both spontaneous seizures and electrically induced seizures. Remarkably, electrical stimulation caused status epilepticus in some of the −/− mice that involved repeated seizures, a running and bouncing seizure at 24 min, and a fatal seizure at 38 min (Butler *et al.*, 1995). These α CaMK null mutants will be useful for evaluating the role of intracellular calcium in neuronal excitability and convulsive behavior.

2.7.2. *Jerky*

Toth and colleagues described an insertional transgene that disrupts a novel cellular gene (*jerky*) on chromosome 15 that encodes a DNA binding protein (Toth *et al.*, 1995; Donovan *et al.*, 1996). Disruption of *jerky* produces generalized clonic seizures in both heterozygous and homozygous mice. In addition to seizures, homozygotes also show dwarfism. Although the *jerky* gene is expressed in many organs, it is heavily expressed in brain and especially in the hippocampus. Unknown environmental factors may also influence the expression of *jerky*, as incomplete penetrance of the seizure phenotype is observed in heterozygotes.

2.7.3. 5-HT^{2c} Serotonin Receptor

Spontaneous clonic–tonic seizures that are often lethal were observed in mice carrying an engineered nonsense mutation in exon five of the serotonin

5-HT^{2c} receptor gene (Tecott *et al.,* 1995; Brennan *et al.,* 1997). The mutation also caused enhanced susceptibility to seizures induced by a variety of convulsant agents including metrozol, kindling, and sound stimulation (audiogenic seizures) (Tecott *et al.,* 1995; Applegate and Tecott, 1996). These findings agree with numerous previous studies that defects in serotonergic mechanisms can influence convulsive behavior (Schlesinger *et al.,* 1965, 1968; Seyfried, 1982a; Yan *et al.,* 1995).

2.7.4. Calcium Channel Glutamate Receptor GluR-B

Recurrent seizures, phenotypically similar to kainic acid-induced limbic seizures, were observed in transgenic mice carrying one copy of an editing deficient GluR-B allele (Brusa *et al.,* 1995). This editing deficiency caused enhanced Ca^{2+} permeability through α-amino-3-hydroxy-5-methyl-4-isoxazolepropionate (AMPA)-sensitive glutamate receptors and selective neuronal destruction in the hippocampus. The expression of seizures and early postnatal lethality in heterozygous mice at P20 suggested that GluR-B pre-mRNA editing is essential for normal brain function.

2.7.5. Inositol 1,4,5-Triphosphate Calcium Release Channel

Similar to the γ-aminobutyric acid A ($GABA_A$) receptor subunit β3 knockout mice described later, most mice lacking the gene for the type 1 InsP3 (inositol 1,4,5-triphosphate, IP3R1) receptor die *in utero.* Those mice that survive, however, develop severe ataxia, tonic–clonic seizures, and die by the weaning period (Matsumoto *et al.,* 1996). These findings suggest that the IP3R1 receptor may be important for the regulation of neuronal excitability.

2.7.6. $GABA_A$ Receptor Subunit β3

Recent findings showed that targeted disruption of the gene encoding the β3 subunit of the $GABA_A$ receptor produced spontaneous seizures (Homanics *et al.,* 1997). Although most of the β3 knockout mice died (97%), those that survived developed spontaneous generalized seizures. The β3 subunit gene is located in a region of mouse chromosome 7 syntenic with human chromosome 15q11–13. These mice may serve as a model for the human Angelman syndrome, which is associated with mental retardation and epileptic seizures due to a maternally derived deletion on chromosome 15q11–13 (Williams *et al.,* 1990).

2.7.7. Synapsins

The synapsins are major proteins of synaptic vesicles and are thought to play a role in synaptogenesis and neurotransmitter release. A significant increase in seizure susceptibility, without a loss in viability, was found in mice with targeted disruptions (knockout) in the genes for synapsin I, synapsin II, or in double knockouts (Li *et al.*, 1995; Rosahl *et al.*, 1995). The loss of synapsins enhanced susceptibility to both electroshock seizures and to seizures induced by vestibular stimulation. Differences in genetic background could also influence the incidence of seizures (Rosahl *et al.*, 1995).

2.7.8. *Sphingolipid Disorders*

Seizures can sometimes occur as a pleiotropic effect of numerous lipid storage diseases (Menkes, 1974). Spontaneous generalized tonic seizures, that are lethal within approx. about 35 days, were reported recently in mice lacking the gene for prosaposin (sphingolipid activator precursor protein, sap) (Fujita *et al.*, 1996). The seizures were found in association with alterations in glycosphingolipid storage and distribution. Seizure-like activity was also found in mice lacking both the α and β subunits of β-hexosaminidase activity, but was not observed in mice with targeted disruptions in either subunit gene alone (Sango *et al.*, 1996). The double knockout mice differed from the human Tay–Sachs and Sandhoff diseases in accumulating mucopolysaccharides as well as glycolipids.

2.7.9. *Tissue Nonspecific Alkaline Phosphatase*

Reduced levels of brain GABA and defective metabolism of vitamin B_6 were found in mice lacking the gene for tissue nonspecific alkaline phosphatase (Waymire *et al.*, 1995). The homozygous mutant mice also developed lethal generalized seizures at about postnatal day 14. Pyridoxal treatment could prevent both the lethality and seizures, but the rescued mice developed abnormal incisors. These findings illustrate the interesting interrelationships between inborn metabolic errors and epilepsy.

2.7.10. *Amyloid Precursor Protein*

Spontaneous seizures were found in mice with targeted disruptions in the gene for amyloid precursor protein (APP) (Sisodia *et al.*, 1996). The seizures occur either spontaneously or from routine handling and are quite variable in severity from one mouse to the next (H. Zheng, personal communication). Besides serving as animal models for Alzheimer's disease, these knockout mice may also be useful as models for epilepsy.

2.7.11. *Huntington's Disease (HD) Gene*

Transgenic mice for the 5′ end of the human HD carrying 115–150 CAG repeat expansions display progressive neurological disorder and epileptic seizures (Mangiarini *et al.*, 1996). These mice express ataxia, tremor, and severe handling-induced seizures that can last for several minutes (Mangiarini *et al.*, 1996). Most of the transgenic mice die between 10 and 13 weeks of age and some may die during an epileptic seizure. Compared to normal littermates, the HD transgenic mice had smaller brains and neuronal loss in the caudate (Mangiarini *et al.*, 1996).

2.7.12. Brain Morphogenesis Gene, OTX1

The targeted disruption of the mouse *Otx1* gene, which is related to a homeobox-containing gene for *Drosophila* head development, caused brain malformations and epileptic seizures (Acampora *et al.*, 1996). Focal seizures, involving head and chewing automatisms, often progressed into generalized seizures that lasted about 60 sec. Generalized seizures occasionally progressed further into status epilepticus. Both the focal and generalized seizures were accompanied by electroencephalographic abnormalities measured in hippocampus and cortex (Acampora *et al.*, 1996). Severe abnormalities in brain development were found in cerebral cortex, hippocampus, cerebellum, and brain stem. About 30% of the B6/D2 derived *Otx1−/−* mice died in the first postnatal month. Although the *Otx1* gene product is not yet identified, the product is likely required for normal brain function. These knockout mice will be a useful model for epilepsies involving brain dysgenesis.

2.8. Epilepsy Models Arising from Abnormal Cortical Development

Mental retardation and seizures occur in several rare human disorders involving inherited defects in cortical development. These disorders include lissencephaly, double cortex, and periventricular heterotopia (Barinaga, 1996). The *OTX1* knockout mouse described previously may be considered a model for these types of human disorders. Although seizure activity was described in mutant rats with double cortex (Lee *et al.*, 1995), no seizure activity was reported in the *reeler, scrambler,* or *CDK5* mutant mice which also inherit defects in cortical development (Barinaga, 1996).

2.9. Analysis of Monogenic Epilepsy Models

Many of the monogenic epilepsy models, whether originating as spontaneous mutations or from targeted gene disruption, can be broadly classified as symptomatic epilepsies. Most of these models are associated with neurostructural abnormalities with varying degrees of reduced viability or lethality. Seizures associated with neurostructural abnormalities and reduced viability also occur in many human inborn errors of metabolism or congenital neurological disorders

(Menkes, 1974; Delgado Escueta *et al.,* 1986). The studies with knockout mice raise the possibility that epilepsy may arise as a pleiotropic effect from an alteration in any gene that is expressed in brain. Considering that about 30,000 genes are expressed in brain (Milner and Sutcliffe, 1983), it is possible that at least this number of epilepsy genes may be described eventually. The monogenic epilepsy models will be useful for tracing pathogenic pathways between altered gene expression at the molecular level and abnormalities at the biochemical, physiological, behavioral levels.

3. MULTIFACTORIAL EPILEPSY MODELS IN MICE

In addition to a number of single-locus epilepsy genes, some mouse epilepsy models involve multifactorial genetic disease, where the action of more than one gene together with environmental factors contribute to the epileptic phenotype. As a result of gene–gene and genotype–environmental interactions, these disorders do not usually follow simple Mendelian modes of inheritance (Falconer, 1960). This problem has hindered progress in identifying genes responsible for both human and mouse multifactorial disorders. In contrast to many simple Mendelian traits where the genotype largely determines the phenotype, the environment plays a greater role in the determination of the phenotype for multifactorial disorders. Although the genotype may render a predisposition or susceptibility to disease, it is the environment that mostly determines the degree of gene penetrance and expressivity.

Both external and internal environmental factors can influence the expression of multifactorial disorders (Strickberger, 1985). In the case of epilepsy, external environmental factors can include temperature, light, sound, nutrition, head trauma, infections, maternal effects, and stress. Internal factors, on the other hand, can include age, gender, circadian rhythms, hormones, and seizure history. It is important to mention that these categories of environmental factors are not firm, but serve only as general classifications. Besides environmental factors, epigenetic factors, e.g., genomic imprinting, may also influence the expression of multifactorial convulsive disorders. Because the most common forms of human idiopathic epilepsy are expressed as multifactorial disorders (Andermann, 1982), multifactorial mouse epilepsy models will be useful for characterizing the mechanisms by which genes and environmental factors interact to influence seizure susceptibility. The following mouse epilepsy models involve multifactorial etiology.

3.1. Epileptiform

Epileptiform mice express spontaneous motor seizures in association with a unique form of nonprogressive hydrocephalus (Flavin and Seyfried, 1990). Although the epileptiform mutation (*epf*) was described originally as a Mendelian autosomal recessive (Hare and Hare, 1979), we found that both the hydrocephalus and epilepsy are significantly influenced by environmental factors (Flavin and Seyfried, 1990). A high incidence of seizures and hydrocephalus occurred in the epileptiform mice (*epf/epf*) raised in Hare's laboratory in San Francisco, CA, but a low incidence of these phenotypes were expressed in these mice raised in the Boston College animal facility (Flavin and Seyfried, unpublished observation). It appears that differences in rearing environment may influence *epf* expression.

3.2. SWXL-4

SWXL-4 is a genetic epilepsy model derived as a recombinant inbred strain from the SWR/J and C57L/J inbred progenitor strains (Frankel *et al.,* 1994). During routine handling, the SWXL-4 mice expresses generalized seizures that are phenotypically similar to those seen in EL mice (described later). Despite a thorough genome sweep using more than 300 microsatellite DNA markers, no gene was found that had a major effect on seizure susceptibility in SWXL-4 mice (Frankel *et al.,* 1994). A gene that accounted for about 32% of the genetic variance (*Szf1*) was found on the proximal region of chromosome 7. The complexity of the genetic architecture in this epilepsy model was attributed to the overlapping function of multiple nonallelic seizure genes. It is also possible that this complexity arises in part from the procedure used to induce the seizures, i.e., repeated tossing in a mechanical shaker. This type of testing procedure significantly influences the manifestation of seizures in the EL epilepsy model (Poderycki *et al.,* 1998).

3.3. EL

The EL (epilepsy) mouse has been one of the most extensively studied mouse models of idiopathic epilepsy. The mouse was originally called Ep and was discovered by Imaizumi in 1954 in an outbred DDY (formally ddy) mouse colony (Naruse and Kurokawa, 1992). Both the seizure susceptible EL mice and nonsusceptible DDY mice are now maintained as inbred strains (Saito *et al.,* 1992; Frankel *et al.,* 1995a). The seizures in EL mice appear to originate in or near the parietal lobe and then spread quickly to the hippocampus and to other brain regions (Suzuki *et al.,* 1991; Kasamo *et al.,* 1992). The seizures are also accompanied by EEG abnormalities, vocalization (squeaking), incontinence, loss of postural equilibrium, excessive salivation, and head, limb, and chewing automatisms (Kurokawa *et al.,* 19966; Suzuki, 1976; Suzuki and Nakamoto, 1977; Naruse and Kurokawa, 1992; Seyfried *et al.,* 1992). Phenytoin and phenobarbital, the anticonvulsant drugs of choice for treatment of human partial epilepsies, inhibit the seizures in EL mice (Suzuki and Nakamoto, 1977; Matsumoto *et al.*, 1983). Based on these observations, the EL mouse is considered a genetic model for human complex partial seizures with secondary generalization.

The seizures in EL mice begin at about 80–100 days of age and usually occur during routine handling associated with cage changing. Seizure frequency, severity, and duration increase with age. Rhythmic vestibular stimulation, e.g., tossing, rotation, or rocking is often used to induce seizures in young EL mice (approx. 45–55 days) (Kurokawa *et al.*, 1966; Fueta *et al.*, 1983; Naruse and Kurokawa, 1992; Seyfried *et al.*, 1992; Frankel *et al.*, 1995a). Although vestibular stimulation was believed to trigger the EL seizures (Kurokawa *et al.*, 1966; Naruse and Kurokawa, 1992), EL mice older than 150 days will seize in the absence of vestibular stimulation, e.g., from removing the cage top. The trigger for seizure induction may be stress or fear, which can arise from numerous environmental stimuli including vestibular stimulation. Furthermore, the stress-related seizure threshold decreases as EL mice age. Of all the precipitating factors for seizures in man, fear or anxiety has long been recognized as one of the most potent (Gowers, 1901).

Besides an enhanced susceptibility to handling-induced seizures, susceptibility to metrazol-induced seizures is also greater in EL mice than in nonepileptic DDY mice (Sugaya *et al.*, 1986; Fueta and Mita, 1929). It is not yet known, however, whether a common genetic mechanism is responsible for the handling- and metrazol-induced seizures in EL mice. Although EL mice are also more susceptible to olfactory bulb kindling than DDY mice, they are less susceptible than the nonepileptic C57BL/6 mice (Green and Seyfried, 1991). These findings suggest that different genetic mechanisms contribute to these seizure susceptibilities (Green *et al.*, 1993).

The seizures in EL mice are not associated with gross neurostructural changes in the CNS (Suzuki *et al.*, 1983; Mizukawa and Mori, 1992; Naruse and Kurokawa, 1992). Nevertheless, an intense reactive gliosis occurs in the hippocampus of EL mice (Brigande *et al.*, 1992). Both the number of glial fibrillary acidic protein (GFAP)-positive cells and the relative GFAP concentration are significantly greater in EL mice than in age-matched seizure resistant C57BL/6J mice or in nonseizing young EL mice (Fig. 1 and Table 1). Because GFAP expression is a good marker for reactive glia in neurological disease, these findings indicate that the seizure activity in EL mice is associated with hippocampal gliosis. Although hippocampal gliosis or Ammon's horn sclerosis is a common finding in humans with complex partial seizures, there is considerable controversy as to whether the gliosis arises as a cause or an effect of the seizures (Babb and Brown, 1987; Ribak, 1991). Our results suggest that the gliosis in EL mice is associated with the effects rather than the cause of seizure activity (Brigande *et al.*, 1992). The EL mouse can serve as a useful natural model for examining the association between epilepsy and reactive gliosis.

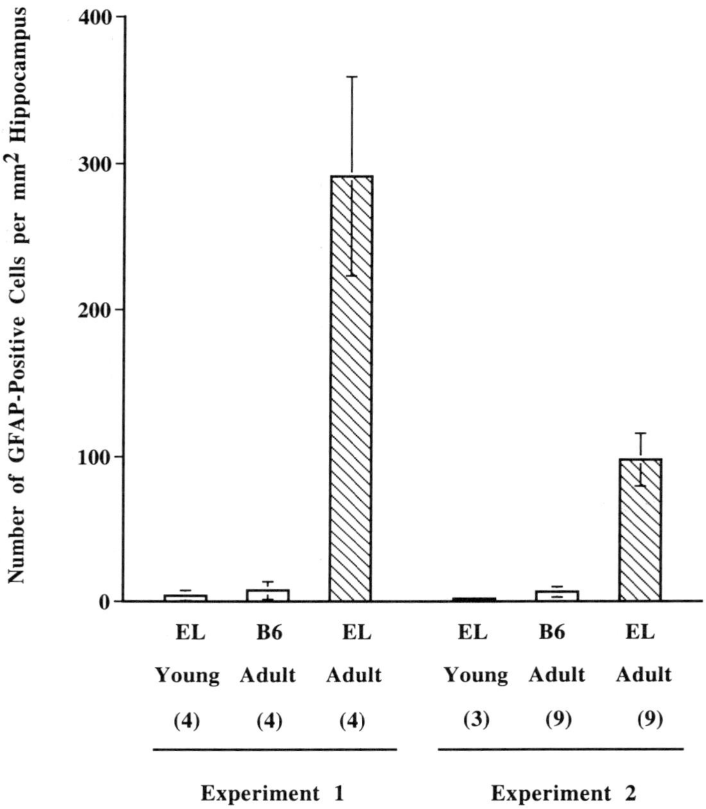

FIGURE 1. Number of GFAP-positive cells/mm^2 of hippocampus in adult seizing EL mice and in nonseizing control C57BL/6J (B6) mice and in young EL mice prior to seizure onset. The asterisk indicates that the value is significantly greater than those of the young EL and adult B6 mice at $p < 0.01$ (Mann-Whitney U test). The number of mice analyzed in each group is indicated in parentheses. The adult EL and B6 mice used in Experiment 1 did not receive vestibular stimulation and were analyzed between 240 and 370 days of age, whereas the adult EL and B6 mice used in Experiment 2 received vestibular stimulation (see Brigande *et al.,* 1992) and were analyzed between 130 and 160 days of age. The young EL mice in each experiment were analyzed at 43±2 days and did not receive vestibular stimulation. [Reprinted from Brigande *et al.* (1992) with permission.]

In contrast to human hippocampal sclerosis, where seizures and gliosis are associated with significant neuronal loss (Babb and Brown, 1987), the gliosis in EL mice is not associated with obvious neuronal loss (Suzuki *et al.,* 1983; Brigande *et al.,* 1992). Using the silver impregnation method, Mizukawa and Mori found a slight, but nonsignificant, reduction in the number of hippocampal pyramidal cells (Mizukawa and Mori, 1992). Furthermore, no reduction was

TABLE 1. Hippocampal and Cerebellar GFAP Content in Adult EL and Control B6 Mice[a]

Protein brain region (μg)	Strain	*n*	Relative GFAP content per milligram total tissue
Hippocampus	B6	3	3.5 ± 0.9
	EL	3	9.6 ± 1.4*
Cerebellum	B6	3	2.5 ± 0.4
	EL	3	1.8 ± 0.4

[a]The relative GFAP content was estimated by scanning densitometry of peroxidase stained western blots as we recently described (Brigande *et al.*, 1992). n, the number of separate samples analyzed.
Significantly different from B6 value at $p < 0.01$ (analyzed using two-tailed Student's *t* test).

found in EL mice for the concentration of hippocampal gangliosides (Brigande *et al.*, 1992). Gangliosides are complex sialoglycosphingolipids that are enriched in neural cell membranes and can serve as cell-surface markers for changes in neuronal cytoarchitecture that accompany brain development or disease (Brigande *et al.*, 1992). Neither the total ganglioside concentration, a sensitive index for neuronal content, nor the concentration of GD1a, a ganglioside enriched in dendritic and synaptic membranes, was reduced in EL hippocampus.

These findings are markedly different from those in human complex seizures, in which total hippocampal ganglioside and GD1a content were significantly reduced in associated with gliosis and neuronal loss (Yu *et al.*, 1987). The reason for the difference between man and mouse in seizure-associated neuronal loss is not clear, but may reflect a species difference in either neuronal or glial response to seizures. It is possible that mouse hippocampal neurons are more resistant to seizure-associated damage than human neurons or that mouse glia are more protective of neurons than human glia. The EL mouse may therefore be useful for examining factors that protect or buffer neurons from seizure-induced excitotoxicity.

Despite more than three decades of intensive neurochemical research on EL mice conducted in both Japan and the United States, the biochemical defects underlying the EL seizures are not yet clear (Kurokawa *et al.*, 1966; Mori, 1988; Mita *et al.*, 1991; Seyfried *et al.*, 1992). Difficulties in distinguishing innate neurochemical changes associated with seizure susceptibility from changes occurring as a consequence of repeated seizure activity have obscured underlying biochemical defects. Additional complicating factors also arise from the various methods used to induce the EL seizures. Some of these methods, e.g., vestibular stimulation or mechanical tossing, may produce neurochemical changes that are unrelated to either seizure susceptibility or to the effects of seizures. Furthermore, there is no ideal genetic control for the EL mouse. Although the DDY strain is more closely related to the EL strain than to other mouse strains, the EL and DDY mice differ in about 25–30% of their genes (Frankel *et al.,* 1995a). Consequently, it is necessary that several mouse groups be used to control for the effects of genetic background, age, number of prior seizures, and for the seizure testing procedure itself. Unfortunately, many of the neurochemical and physiological studies conducted on EL mice have not employed all of the control groups necessary to determine whether a particular neurochemical change is related to the cause of seizures or to the effects of seizures or the seizure test.

One neurochemical defect that may be causally related to the EL seizures involves the calcium-dependent release of aspartate. Aspartate is an important excitatory neurotransmitter in the mammalian CNS that may be involved in some epilepsies (Flavin *et al.,* 1991). Although the amount of releasable amino acid represents only a small portion of the total tissue amino acid content, this is the physiological pool most relevant to neuronal signal transmission and epilepsy. We found that the endogenous release of aspartate, but not of glutamate or GABA, was significantly higher in hippocampal slices of EL mice than in seizure resistant C57BL/6J or DDY control mice at both young and adult ages (Table 2). Enhanced aspartate release may be related to the cause rather than to the effects of seizure activity, as the enhanced release was apparent in young EL mice that neither had seizures nor were tested for seizures (Flavin *et al.,* 1991; Flavin and Seyfried, 1994). Hiramatsu *et al.* (1992) also showed that endogenous aspartate release from cortical slices was greater in seizing than in nonseizing EL mice. The aspartate findings are interesting in view of the hippocampal gliosis in EL mice and the postulated excitotoxicity hypothesis of epilepsy (Ribak, 1991).

In addition to the possible involvement of aspartate in EL seizures, GABAergic abnormalities may also play some role (Murashima *et al.,* 1992). Several other abnormalities have been reported in EL mice that may relate to global metabolic disturbance. Some of these abnormalities involve metals (Sutoo *et al.,* 1987; Fukahori *et al.,* 1988; Mori, 1988), biogenic amines (Sutoo *et al.,* 1992), RNA synthesis (Mui *et al.,* 1991), and the developmental properties of primary cultured EL neurons (Sugaya *et al.,* 1992). It will be interesting to determine how the neurochemical and physiological abnormalities in EL mice are related to the expression of specific seizure genes.

Early studies suggested that the inheritance of seizures in EL mice was dominant and complex (Fueta *et al.,* 1986; Naruse and Kurokawa, 1992). To better define the genetic basis of the EL seizures, we outcrossed EL mice to two seizure-resistant strains, ABP/Le (ABP) and DBA/2J (D2), and followed the inheritance of seizures and genetic markers in backcrosses to the seizure resistant ABP and D2 parental strains (Rise *et al.,* 1991). Although D2 mice are susceptible to audiogenic seizures, they do not experience handling seizures like EL mice. All of the animals used in these genetic studies were tested for seizure

TABLE 2. The Calcium-Dependent Potassium-Evoked Release of Aspartate, Glutamate, and GABA from Hippocampal Slices in Epileptic EL Mice and Nonepileptic B6 and DDY Mice[a]

		Released neurotransmitter[b]		
Strain	Age (days)	Aspartate	Glutamate	GABA
B6	320	3.3 ± 0.7 (5)	15.4 ± 3.6 (5)	12.7 ± 1.4 (5)
EL	320	6.6 ± 0.7 (5)*	18.2 ± 2.6 (5)	17.3 ± 2.9 (4)
B6	38	8.5 ± 1.4 (4)	23.0 ± 2.0 (4)	23.5 ± 0.7 (4)
DDY	38	8.4 ± 0.7 (4)	32.3 ± 5.8 (4)	19.5 ± 3.2 (4)
EL	38	15.8 ± 0.8 (3)*	25.9 ± 2.6 (3)	21.8 ± 3.2 (3)

[a]Data from 300 days are from Flavin *et al.* (1991) and from 38 days from Flavin and Seyfried (1994).
[b]Values are expressed as the mean pmol amino acid released/ml/min incubation/slice ± SEM. The number of independent samples analyzed is indicated in parentheses.
*Indicates that the EL aspartate release is significantly higher than that of B6 mice ($p < 0.02$; unpaired, two-tailed t-test).

susceptibility using a tossing procedure in a mechanical shaker (Seyfried *et al.*, 1992). The mice were tested twice per week beginning at 30 days of age and the mean seizure frequency per a total of 20 tests was calculated for each mouse. The results from these genetic studies indicated that seizure susceptibility in EL mice was inherited as an incompletely dominant multifactorial trait (Rise *et al.*, 1991).

Using several statistical tests to analyze the association between seizure susceptibility and the various genetic markers in the backcrosses, we mapped three genes that influenced seizure susceptibility in EL mice. One gene, designated *El1,* was found on the distal region of mouse chromosome 9 near the dilute/short ear locus and another, designated *El2,* was found on chromosome 2 near the microsatellite marker *D2Mit30* (Rise *et al.*, 1991; Frankel *et al.*, 1995a). In contrast to *El1* and *El2,* which were present in the EL genome, we also found a recessive "modifier" gene in the D2 genome that enhanced seizure susceptibility. This modifier mapped to the same region on chromosome 4 as *Asp2,* a previously identified enhancer of audiogenic seizure susceptibility. This seizure enhancer gene may modify either tossing-induced seizures in EL mice or audiogenic seizures in D2 mice.

Besides these seizure genes, several additional seizure frequency quantitative trait loci (QTL) were recently identified on chromosomes 10 (*El3*), 9 (*El4*), 14 (*El5*), and 11 (*El6*) (Frankel *et al.*, 1995a,b). The effects of these genes, however, were dependent upon the genetic background (the mouse strain to which EL is crossed) and upon the type of cross (backcrosses or F2 intercrosses). The authors concluded that no single *El* locus is essential for high seizure frequency and that the seizures in EL mice result from complex epistatic interactions between many nonallelic *El* seizure genes (Frankel *et al.,* 1995a,b).

The complexity of the genetic architecture associated with seizure susceptibility in EL mice may result in part from the methods used to induce the seizures, i.e., rhythmic vestibular stimulation. It is possible that different genotypes respond differently to the rhythmic vestibular stimulation. Such genotype–treatment interactions could produce phenocopies, i.e., false positives or false negatives, that would weaken associations with genes responsible for the spontaneous seizures. We recently found that the EL seizure phenotype is not independent of the seizure test, suggesting that gene–environment interactions contribute to the genetic complexity of the EL seizures (Poderycki *et al.,* 1998). Further studies are needed to determine if the previously mapped *El* "epilepsy" genes can influence the spontaneous or handling-induced seizures in these mice.

3.4. Audiogenic Seizures

Susceptibility to audiogenic seizures (AGS) is another widely studied multifactorial convulsive disorder. AGS are violent generalized convulsions that are induced in some mouse strains by loud high-frequency sound. The AGS phenotype manifests itself as a series of progressive phases (Seyfried *et al.,* 1980; Seyfried, 1982a; Schreiber, 1986). After initial sound stimulation, a susceptible mouse will burst into a wild, frenzied run. The wild run is then followed by the clonic phase in which the mouse falls on its side and kicks violently. If the seizure continues, the mouse enters a tonic phase, characterized by rigidly extended limbs, clenched forepaws, and splayed hind paws (Seyfried *et al.,* 1986). Respiratory arrest leading to asphyxiation and death may sometimes follow the tonic seizure, but some investigators do not consider death a more severe form of the seizure (Seyfried, 1982a). Cortical EEG abnormalities are not generally seen during the AGS due to the suspected brainstem origin of the seizures (Maxson and Cowen, 1976; Maxson and Sze, 1976). The audiogenic mouse is considered a genetic model for generalized brain stem or reflex epilepsies and also shares some characteristics with human "absence" and photic epilepsies (Seyfried *et al.,* 1986).

Marked interstrain variability occurs for the onset, incidence, and severity of AGS (Seyfried *et al.,* 1980; Seyfried, 1982a, 1983; Schreiber, 1986). The onset of AGS susceptibility occurs abruptly between 14 and 16 days of age in the DBA/2J (D2) strain. Susceptibility and severity are greatest around 21 days, but gradually weaken with age. By the time D2 mice reach adulthood, they are completely resistant to AGS (Seyfried, 1982b). It is important to mention that the gradual loss of AGS susceptibility in D2 mice is associated with age-dependent hearing loss (Ralls, 1967; Erway *et al.,* 1993; Willott *et al.,* 1995). In contrast to D2 mice, EP mice (derived from a line of BALB/c mice) are maximally susceptible at 30 days and remain susceptible into adult ages. Mice of the C57BL/6J (B6) are mostly resistant to AGS at all ages. Hence, AGS susceptibility is strongly dependent upon age and genetic background.

In general, AGS-susceptible mice are also more susceptible than AGS-resistant mice to seizures induced by other convulsant agents, e.g., electroshock, convulsant drugs, and kindling (Seyfried, 1982a; Engstrom and Woodbury, 1988; Green and Seyfried, 1991; Martin *et al.,* 1991; Ferraro *et al.,* 1995; Ferraro *et al.,* 1996). Different genetic mechanisms, however, may be responsible for the different forms of seizure susceptibilities (Seyfried, 1982a; Applegate *et al.,* 1988; Martin *et al.,* 1992; Ferraro *et al.,* 1996). A major gene for metrazol and kanic acid-induced seizures is located on chromosome 1, whereas a major gene for AGS susceptibility in D2 mice is located on chromosome 12 (Seyfried *et al.,* 1980; Seyfried and Glaser, 1981; Neumann and Seyfried, 1990; Ferraro *et al.,* 1996). Although AGS susceptibility dissipates with age in D2 mice, they generally maintain their high susceptibility to other convulsant agents (Seyfried, 1982a).

Wieraszko and Seyfried showed that AGS, in either D2 mice or in acoustically primed B6 mice, can enhance synaptic functions through a mechanism involving the NMDA receptor (Wieraszko and Seyfried, 1993). LTP (long-term potentiation) involves a permanent increase in synaptic efficiency following various types of electrical or chemical stimulation and has been extensively studied as a model for memory in the hippocampus. Some investigators suggest that LTP and epileptogenesis may share common molecular mechanisms (Wilkie Wilson, personal communication). Hence, AGS in mice may serve as a useful model system for identifying synaptic mechanisms common to memory and epilepsy.

Similar to the situation with EL mice, numerous neurochemical differences have been reported between mice susceptible and resistant to AGS (reviewed in Seyfried, 1982a). The multifactorial nature of AGS susceptibility together with the lack of adequate control groups makes it difficult to determine if observed neurochemical differences are related to differences in AGS susceptibility or to differences in genetic background (strain differences). For example, the deficiency of a brain Ca^{2+} or Mg^{2+}-stimulated ecto-ATPas was long considered a possible neurochemical defect associated with AGS susceptibility (Ginsburg, 1967; Rosenblatt *et al.,* 1976; Trams and Lauter, 1978; Seyfried, 1982a; Palayoor and Seyfried, 1984; Wieraszko and Seyfried, 1989). The gene for this ATPase, which regulates the hydrolysis of the four major nucleotide triphosphates (ATP, GTP, CTP, and UTP), was mapped to the proximal region of mouse chromosome 2 and was designated *Ntp.* No epilepsy or AGS genes, however, have been identified in the *Ntp* region. It is therefore unlikely that the ATPase deficiency in D2 mice is associated specifically with AGS susceptibility (Allen and Seyfried, 1994).

Previous studies showed that the gene for aryl hydrocarbon hydroxylase inducibility (*Ahr*, formally *Ah*) was closely linked to the major AGS susceptibility gene, *Asp1* (see later). Because the *Ahr* locus influences numerous biochemical differences between the B6 and D2 strains, some of the neuropharmacological differences reported between the B6 and D2 strains could arise from genetic differences at the *Ahr* locus (Nebert, 1980; Chang *et al.*, 1993). Hence, it can be difficult to determine if neurochemical differences between AGS-susceptible and -resistant mouse strains arise as pleiotropic effects of epilepsy genes or arise from tight linkage of the epilepsy gene to an unrelated biochemical variant. Recent studies indicate that the *Asp1* and *Ahr* loci are nonallelic (DiRocco *et al.*, 1998).

Several studies have suggested an association between AGS susceptibility and the content of brain serotonin (5-hydroxytryptamine, 5-HT). In general, increased 5-HT content reduces AGS susceptibility, whereas decreased 5-HT content enhances susceptibility (Schlesinger *et al.*, 1965, 1968; Seyfried, 1982a; Yan *et al.*, 1995). Further support for the involvement of serotonergic mechanisms in AGS susceptibility comes from the recent findings that mutant mice lacking the X-linked 5-HT_{2c} serotonin receptor subtype have enhanced susceptibility to spontaneous lethal seizures and to audiogenic seizures (Tecott *et al.*, 1995; Brennan *et al.*, 1997). It will be interesting to determine if genetic defects in other serotonin receptor genes might also influence AGS susceptibility.

Although the influence of genetic heterogeneity and various environmental factors have hindered progress in defining the genetic architecture of AGS susceptibility, several audiogenic seizure prone (*Asp*) genes have been identified. Two of these *Asp* genes were found in crosses between the seizure resistant B6 and susceptible D2 inbred strains. *Asp1* (formally *Ias*) was found through its tight linkage with the *Ahr* locus on chromosome 12, and *Asp2* was found through its tight linkage with the *b* coat color locus on chromosome 4 (Collins and Fuller, 1968; Collins, 1970; Seyfried *et al.*, 1980; Seyfried and Glaser, 1981; Neumann and Seyfried, 1990; Neumann and Collins, 1991). The $Asp1^b$ allele appears to block the spread rather than the initiation of AGS (Seyfried and Glaser, 1981; Neumann and Seyfried, 1990). A third *Asp* gene, *Asp-3,* was recently identified through its linkage with the *c* locus and several microsatellite DNA markers on chromosome 7 in crosses between D2 mice and A/J or EP mice (Neumann and Collins, 1991; Neumann and Collins, 1992; Banko *et al.*, 1997). The X-linked 5-HT_{2c} serotonin receptor gene can be considered another *Asp* gene (possibly *Asp4*), as a targeted nonsense mutation in this gene enhances AGS susceptibility (Brennan *et al.*, 1997).

Asp2 may represent a general modifier of neuronal excitability. This notion comes from finding that mice homozygous for *Asp2* have enhanced susceptibility to AGS as well as to tossing-induced seizures (see above for EL mice). In addition, recent findings from Ferraro and co-workers suggest that a modifier gene on chromosome 4 can enhance susceptibility to kainic acid and metrazol-induced seizures in D2 mice (Ferraro *et al.*, 1996; Ferraro *et al.*, 1997). It would be important to determine if *Asp2* is a common enhancer of seizure activity.

Although the AGS phenotype is similar in 30-day-old D2 and EP mice, we found that different genetic mechanisms are responsible for AGS susceptibility in these mice (Banko *et al.*, 1997). Support for this came from finding that EP and D2 mice possess different developmental profiles, and that the reciprocal EPD2F1 and D2EPF1 hybrids have low AGS susceptibility, unlike either the D2 or EP parents. These findings further suggest the influence of nonallelic genetic heterogeneity on the determination of AGS susceptibility in mice (Seyfried, 1983). It appears that different mechanisms can underlie phenotypically similar convulsive behaviors whether these behaviors occur at one age or at several ages. These original observations on AGS susceptible mice predicted that genetic heterogeneity would be eventually uncovered for human convulsive disorders (Seyfried, 1983).

In contrast to all previously identified epilepsy genes, *Asp-3* appears to be influenced by genomic imprinting (Banko *et al.*, 1997). This is an epigenetic phenomenon involving the differential expression of parental alleles in the offspring (Efstratiadis, 1994; Barlow, 1995). We found that the effects of *Asp3* on audiogenic seizure susceptibility in EP mice depend on whether the gene is transmitted through the male or female parent (Banko *et al.*, 1997). Although there are no reports for an influence of genomic imprinting on inherited cryptogenic human epilepsies, Ottman and co-workers reported a "maternally transmitted influence" on the risk of epilepsy in offspring of epileptics (Ottman *et al.*, 1988). Recent studies suggest that genomic imprinting might also influence other complex multifactorial disorders, e.g., manic depression in humans and alcohol preference in mice (McMahon *et al.*, 1995; Melo *et al.*, 1996). The EP mouse may therefore serve as a model for dissecting the effects of genomic imprinting on multifactorial behavioral traits including epilepsy.

In addition to genetic heterogeneity and genomic imprinting, a number of internal and external environmental factors can also influence AGS susceptibility. Hormones, circadian rhythms, and age can modify AGS susceptibility in some strains (Ginsburg, 1967; Seyfried, 1982a). An important external environmental factor is the audiogenic testing procedure itself. This is often referred to as the acoustic priming phenomenon and involves the transformation of normally AGS-resistant mice into susceptible mice through prior auditory exposure at a critical period of auditory development (Henry, 1967; Fuller and Collins, 1968; Iturrian and Fink, 1969). Hence, the acoustic priming phenomenon produces AGS phenocopies (false positives) that are phenotypically indistinguishable from inherited AGS (Seyfried, 1982a). Genetic factors may also influence the susceptibility of some to acoustic priming (Fuller, 1975). The acoustic priming phenomenon, like repeated tossing stimulation in EL mice, further illustrates the effects of the testing procedure on the manifestation of convulsive behavior.

4. GENETIC MODELS OF SPONTANEOUS LETHAL EPILEPSIES AND STATUS EPILEPTICUS

Most spontaneous idiopathic epilepsies in the mouse are not usually lethal. There are, however, a few exceptions. Adult mice (60–150 days of age) of the BXD-13 recombinant inbred strain express violent spontaneous generalized seizures that are often lethal (Seyfried, 1982a). These mice are also susceptible to AGS at adult ages, but no studies have been conducted to determine if a common genetic mechanism is responsible for the two types of seizures. Similar to the BXD-13 mice, weaver heterozygotes (*wv*/+) and mice with a targeted nonsense mutation in the X-linked 5-HT_{2c} serotonin receptor gene also express a high susceptibility to spontaneous lethal seizures and AGS (Seyfried, 1982a; Tecott *et al.*, 1995; Brennan *et al.*, 1997). Maxson and co-workers reported spontaneous lethal seizures between 40 and 70 days in C57BL/10Bg-*sps* mice that resembled psychomotor epilepsy (Maxson *et al.*, 1983). These mice also expressed enhanced AGS susceptibility. Fatal generalized convulsive status epilepticus following a single electroshock was reported in transgenic mice carrying a mutation in the Ca^{2+}/calmodulin-dependent kinase II α-subunit (Butler *et al.*, 1995). Continuous partial seizures resembling *epilepsia partialis continua* occur in mice that are compound heterozygotes at the *tottering* locus [containing one *tottering* (*tg*) and one *leaner* (tg^{la}) allele] (Seyfried, 1982a). These epilepsy models can be useful for addressing the pathophysiological consequences of acute severe seizures and chronic prolonged seizures.

5. MAPPING EPILEPSY GENES IN THE HUMAN AND THE MOUSE

Positional cloning strategies have been successful for the identification of many single-locus neurological mutations in man and mouse. Efforts are underway at present to isolate the genes for several of the monogenic epilepsy models. This genetic strategy, however, has had less success with complex multifactorial disorders. This comes largely from the uncertainty in phenotypic assessment owing to the influence of environmental factors and to the segregation of nonallelic modifier genes which can lead to the misclassification of some individuals as either false positives or false negatives. These gene–gene or gene–environmental interactions can distort map distances between genetic markers and the QTLs for multifactorial disorders.

Some epilepsy QTLs could map in regions with cloned genes which could then be screened for defects in probands and their relatives. The recent explosion of database information on mammalian chromosome maps makes positional candidate gene analysis the preferred strategy for disease gene identification

(Strachan and Read, 1996). Because about 37% of the mouse autosomal genome is syntenic with the human autosomal genome, mapping mouse genes can predict the chromosomal location of similar genes in man (Nadeau, 1989; Copeland *et al.,* 1993; Dietrich *et al.,* 1995). Indeed, the identification of the neuronal nicotinic acetylcholine receptor α4 subunit (*CHRNA4*) as a candidate for autosomal dominant nocturnal frontal lobe epilepsy was facilitated through comparative analysis of human and mouse homologous synteny (I. Scheffer, personal communication and (Steinlein *et al.,* 1995). Hence, the mapping of epilepsy genes in experimental mouse epilepsy models could increase the likelihood for identification of candidate genes for human monogenic and multifactorial epilepsies.

ACKNOWLEDGMENTS

This work was supported by NIH Grant (NS23355) and the Boston College Research Expense Fund.

REFERENCES

Acampora, D., Mazan, S., Avantaggiato, V., Barone, P., Tuorto, F., Lallemand, Y., Brulet, P., and Simeone, A., 1996, Epilepsy and brain abnormalities in mice lacking the *Otx1* gene, *Nat. Genet.* **14:**218–222.

Allen, K. M., and Seyfried, T. N., 1994, Genetic analysis of nucleotide triphosphatase activity in the mouse brain, *Genetics* **137:**257–265.

Andermann, E., 1982, Multifactorial inheritance of generalized and focal epilepsy, *in* "Genetic Basis of the Epilepsies" (V. E. Anderson, W. A. Hauser, J. K. Penry, and C. F. Sing, eds.), pp. 355–376, Raven Press, New York.

Anderson, V. E., Hauser, W. A., and Rich, S. S., 1986, Genetic heterogeneity in the epilepsies, *Adv. Neurol.* **44:**59–75.

Applegate, C. D., and Tecott, L., 1996, Global changes in seizure susceptibility in mice lacking 5-HT2c serotonin receptors, *Soc. Neurosci. (Abstr.)* **22:**2086.

Applegate, C., Burchfiel, J., and Neumann, P., 1988, Classical genetic and recombinant inbred strain analysis of seizure susceptibility of mice to pentylenetetrazol, *Mouse News Letter* **88:**61–62.

Babb, T. L., and Brown, W. J., 1987, Pathological findings in epilepsy, *in* "Surgical Treatment of the Epilepsies" (J. Engel, ed.), pp. 101–114, Raven Press, New York.

Banko, M. L., Allen, K., Dolina, S., Neumann, P. E., and Seyfried, T. N., 1997, Genomic imprinting and audiogenic seizures in mice, *Behav. Genet.* **27:**465–475.

Barinaga, M., 1996, Guiding neurons to the cortex, *Science* **274:**1100–1101.

Barlow, D. P., 1995, Gametic imprinting in mammals, *Science* **270:**1610–1613.

Brennan, T. J., Seeley, W. W., Kilgard, M., Schreiner, C. E., and Tecott, L. H., 1997, Sound-induced seizures in serotonin 5-HT2c receptor mutant mice, *Nat. Genet.* **16:**387–390.

Brigande, J. V., Wieraszko, A., Albert, M. D., Balkema, G. W., and Seyfried, T. N., 1992, Biochemical correlates of epilepsy in the El mouse: Analysis of glial fibrillary acidic protein and gangliosides, *J Neurochem.* **58:**752–760.

Brusa, R., Zimmermann, F., Koh, D. S., Feldmeyer, D., Gass, P., Seeburg, P. H., and Sprengel, R., 1995, Early-onset epilepsy and postnatal lethality associated with an editing-deficient GluR-B allele in mice, *Science* **270:**1677–1680.

Burgess, D. L., Jones, J. M., Meisler, M. H., and Noebels, J. L., 1997, Mutation of the Ca2+ channel beta subunit gene Cchb4 is associated with ataxia and seizures in the lethargic (1h) mouse, *Cell* **88:**385–392.

Butler, L. S., Silva, A. J., Abeliovich, A., Watanabe, Y., Tonegawa, S., and McNamara, J. O., 1995, Limbic epilepsy in transgenic mice carrying a Ca2+/calmodulin-dependent kinase II alpha-subunit mutation, *Proc. Natl. Acad. Sci. USA* **92:**6852–6855.

Campbell, D. B., and Hess, E. J., 1998, Cerebellar circuitry is activated during convulsive episodes in the tottering (*tg/tg*) mutant mouse, *Neuroscience* **85:**773–783.

Chang, C., Smith, D. R., Prasad, V. S., Sidman, C. L., Nebert, D. W., and Puga, A., 1993, Ten nucleotide differences, five of which cause amino acid changes, are associated with the Ah receptor locus polymorphism of C57BL/6 and DBA/2 mice, *Pharmacogenetics* **3:**312–321.

Chauvel, P., Louvel, J., Kurcewicz, I., and Debono, M., 1980, Epileptic seizures in the quaking mouse: Electroclinical correlations, *in* "Neurological Mutations Affecting Myelination" (N. Bauman, ed.), pp. 513–516, Elsevier, New York.

Chermat, R., Lachapelle, F., Baumann, N., and Simon, P., 1979, Anticonvulsant effect of yohimbine in quaking mice: Antagonism by clonidine and prazosine, *Life Sci.* **25:**1471–1476.

Collins, R. L., 1970, Unilateral inhibition of sound-induced convulsions in mice, *Science* **167:**1010–1011.

Collins, R. L., and Fuller, J. L., 1968, Audiogenic seizure prone (asp): A gene affecting behavior in linkage group 8 of the mouse, *Science* **162:**1137–1139.

Copeland, N. G., Jenkins, N. A., Gilbert, D. J., Eppig, J. T., Maltais, L. J., Miller, J. C., Dietrich, W. F., Weaver, A., Lincoln, S. E., Steen, R. G., Stein, L. O., Nadeau, J. H., and Lander, E., 1993, A genetic linkage map of the mouse: Current applications and future prospects, *Science* **262:**57–66.

Cox, G. A., Lutz, C. M., Yang, C.-L., Biemesderfer, D., Bronson, R. T., Fu, A., Aronson, P. S., Noebels, J. L., and Frankel, W. N., 1997, Sodium/hydrogen exchanger gene defect in slow-wave epilepsy mutant mice, *Cell* **91:**139–148.

Delgado-Escueta, A. V., Treiman, D. M., and Enrile-Bacsal, F., 1982, Phenotypic variations of seizures in adolescents and adults, *in* "Genetic Basis of the Epilepsies" (V. E. Anderson, W. A. Hauser, J. K. Penry, and C. F. Sing, eds.), pp. 49–82, Raven Press, New York.

Delgado-Escueta, A. V., White, R., Greenberg, D. A., and Treiman, L. J., 1986, Looking for epilepsy genes: Clinical and molecular genetic studies, *Adv Neurol.* **44:**77–95.

Dietrich, W. F., Copeland, N. G., Gilbert, D. J., Miller, J. C., Jenkins, N. A., and Lander, E. S., 1995, Mapping the mouse genome: Current status and future prospects, *Proc. Natl. Acad. Sci. USA* **92:**10849–108453.

DiRocco, L., Dalton, T., Liang, D., Nebert, D. W., and Seyfried, T. N., 1998, Nonallelism for the audiogenic seizure prone (*Asp1*) and the arylhydrocarbon receptor (*Ahr*) loci in mice, *J. Neurogenetics* in press.

Dlouhy, S. R., Richter, J., Stauss, C. J. W., Hodes, M. E., and Ghetti, B., 1996, Introduction of the weaver gene onto the BALB/cJ strain, *Soc. Neurosci. (Abstr.)* **22:**1649.

Donovan, G. P., Ho, L., and Toth, M., 1996, Molecular characterization of a novel epilepsy gene, *Soc. Neurosci. (Abstr.)* **22:**1417.

Ebersole, T. A., Chen, Q., Justice, M. J., and Artzt, K., 1996, The quaking gene product necessary in embryogenesis and myelination combines features of RNA binding and signal transduction proteins [see comments], *Nat. Genet.* **12:**260–265.

Efstratiadis, A., 1994, Parental imprinting of autosomal mammalian genes, *Curr. Opin. Genet. Dev.* **4:**265–280.

Eisenberg, B., and Messer, A., 1989, Tonic/clonic seizures in a mouse mutant carrying the weaver gene, *Neurosci. Lett.* **96:**168–172.

Engstrom, F. L., and Woodbury, D. M., 1988, Seizure susceptibility in DBA and C57 mice: The effects of various convulsants, *Epilepsia* **29:**389–395.

Erway, L. C., Willott, J. F., Archer, J. R., and Harrison, D. E., 1993, Genetics of age-related hearing loss in mice: I. Inbred and F1 hybrid strains, *Hear Res.* **65:**125–132.
Falconer, D. S., 1960, "Introduction to Quantitative Genetics," Ronald Press, New York.
Ferraro, T. N., Golden, G. T., Smith, G. G., and Berrettini, W. H., 1995, Differential susceptibility to seizures induced by systemic kainic acid treatment in mature DBA/2J and C57BL/6J mice, *Epilepsia* **36:**301–307.
Ferraro, T. N., Golden, G. T., Smith, G. S., Ballas, C., Mulholland, N., Choi, H., Hare, T. A., and Berrettini, W. H., 1996, Mapping murine loci for seizure response to pentylenetetrazole, *Soc. Neurosci. Abstr.* **22:**2081.
Ferraro, T. N., Golden, G. T., Smith, G. G., Schork, N. J., St. Jean, P., Ballas, C., Choi, H., and Berrettini, W. H., 1997, Mapping murine loci for seizure response to kainic acid, *Mamm. Genome* **8:**200–208.
Flavin, H. J., and Seyfried, T. N., 1990, Hydrocephalus in epileptiform mice, *Mouse Genome* **86:**221.
Flavin, H. J., and Seyfried, T. N., 1994, Enhanced aspartate release related to epilepsy in (EL) mice, *J. Neurochem.* **63:**592–595.
Flavin, H. J., Wieraszko, A., and Seyfried, T. N., 1991, Enhanced aspartate release from hippocampal slices of epileptic (EL) mice), *J. Neurochem.* **56:**1007–1011.
Fletcher, C. F., Lutz, C. M., O'Sullivan, T. N., Shaughnessy, J. D., Jr., Hawkes, R., Frankel, W. N., Copeland, N. G., and Jenkins, N. A., 1996, Absence epilepsy in tottering mutant mice is associated with calcium channel defects, *Cell* **87:**607–617.
Frankel, W. N., Taylor, B. A., Noebels, J. L., and Lutz, C. M., 1994, Genetic epilepsy model derived from common inbred mouse strains, *Genetics* **138:**481–489.
Frankel, W. N., Johnson, E. W., and Lutz, C. M., 1995b, Congenic strains reveal effects of the epilepsy quantitative locus, *El2,* separate from other *El* loci, *Mamm. Genome* **6:**839–843.
Frankel, W. N., Valenzuela, A., Lutz, C. M., Johnson, E. W., Dietrich, W. F., and Coffin, J. M., 1995a, New seizure frequency QTL and the complex genetics of epilepsy in EL mice, *Mamm. Genome* **6:**830–838.
Fueta, Y., and Mita, T., 1992, Susceptibility to pentylenetetrazole-induced convulsion of developing and adult EL mice, *Neurosciences* **18**(Suppl. 2):163–169.
Fueta, Y., Mita, T., and Matsuoka, S., 1983, Experimental animal epilepsy. A new device for the induction of epileptic seizures in the murine, EL mouse, *Sangyo Ika Daigaku Zasshi* **5:**359–364.
Fueta, Y., Matsuoka, S., and Mita, T., 1986, Crossbreeding analysis of the mouse epilepsy, *Sangyo Ika Daigaku Zasshi* **8:**417–424.
Fujita, N., Suzuki, K., Vanier, M. T., Popko, B., Maeda, N., Klein, A., Sandhoff, K., Nakayasu, H., and Suzuki, K., 1996, Genetic sphingolipid activator protein (SAP) deficiencies in human and a "knockout" mouse, *J. Neurochem.* **66**(Suppl. 1):S1.
Fukahori, M., Itoh, M., Oomagari, K., and Kawasaki, H., 1988, Zinc content in discrete hippocampal and amygdaloid areas of the epilepsy (EL) mouse and normal mice, *Brain Res.* **455:**381–384.
Fuller, J. L., 1975, Independence of inherited susceptibility to spontaneous and primed audiogenic seizures in mice, *Behav. Genet.* **5:**1–8.
Fuller, J. L., and Collins, R. L., 1968, Temporal parameters of sensitization for audiogenic seizures in SJL/J mice, *Dev. Psychobiol.* **1:**185–188.
Ginsburg, B., 1967, Genetic parameters in behavioral research, *in* "Behavior-Genetic Analysis" (J. Hirsch, ed.), pp. 135–153, McGraw-Hill, New York.
Gomez, M. R., and Klass, D. W., 1983, Epilepsies of infancy and childhood, *Ann. Neurol.* **13:**113–124.
Gow, A., and Lazzarini, R. A., 1996, A cellular mechanism governing the severity of Pelizaeus–Merzbacher disease, *Nat. Genet.* **13:**422–427.
Gowers, W. R., 1901, "Epilepsy and Other Chronic Convulsive Diseases: Their Causes, Symptoms and Treatment," Old Hickory Bookshop, Brinklow, MD.

Green, R. C., and Seyfried, T. N., 1991, Kindling susceptibility and genetic seizure predisposition in inbred mice, *Epilepsia* **32:**22–26.

Green, R. C., Rees, H. D., and Feller, D. J., 1993, Olfactory bulb kindling in mice susceptible and resistant to ethanol withdrawal, *Epilepsia* **34:**416–419.

Guenet, J.-L., 1980, Mutants of the mouse with abnormal myelination: A review for geneticists, *in* "Neurological Mutations Affecting Myelination" (N. Baumann, ed.), Elsevier, New York.

Hare, J. E., and Hare, A. S., 1979, Epileptiform mice, a new neurological mutant, *J. Hered.* **70:**417–420.

Hauser, W. A., 1982, Genetics and clinical characteristics of seizures, *in* "Genetic Basis of the Epilepsies" (V. E. Anderson, W. A. Hauser, J. K. Penry, and C. F. Sing, eds.), pp. 3–10, Raven Press, New York.

Henry, K. R., 1967, Audiogenic seizure susceptibility induced in C57BL-6J mice by prior auditory exposure, *Science* **158:**938–940.

Hess, E. J., and Wilson, M. C., 1991, Tottering and leaner mutations perturb transient developmental expression of tyrosine hydroxylase in embryologically distinct Purkinje cells, *Neuron* **6:**123–132.

Hiramatsu, M., Kinno, I., Kanakura, K., Sato, K., and Mori, A., 1992, Increased aspartate release from brain slices of epileptic experimental animals and effect of valproate on it, *Jpn. J. Psychiatry Neurol.* **46:**541–543.

Homanics, G. E., DeLorey, T. M., Firestone, L. L., Quinlan, J. J., Handforth, A., Harrison, N. L., Krasowski, M. D., Rick, C. E., Korpi, E. R., Makela, R., Brilliant, M. H., Hagiwara, N., Ferguson, C., Snyder, K., and Olsen, R. W., 1997, Mice devoid of gamma-aminobutyrate type A receptor beta3 subunit have epilepsy, cleft palate, and hypersensitive behavior, *Proc. Natl. Acad. Sci. USA* **94:**4143–4148.

Hosford, D. A., Lin, F. H., Kraemer, D. L., Cao, Z., Wang, Y., and Wilson, J. T., Jr., 1995, Neural network of structures in which GABAB receptors regulate absence seizures in the lethargic (lh/lh) mouse model, *J. Neurosci.* **15:**7367–7376.

Isaacs, K. R., and Abbott, L. C., 1995, Cerebellar volume decreases in the tottering mouse are specific to the molecular layer, *Brain Res. Bull.* **36:**309–314.

Iturrian, W. B., and Fink, G. B., 1969, Influence of age and brief auditory conditioning upon experimental seizures in mice, *Dev. Psychobiol.* **2:**10–18.

Kagawa, T., Ikenaka, K., Inoue, Y., Kuriyama, S., Tsujii, T., Nakao, J., Nakajima, K., Aruga, J., Okano, H., and Mikoshiba, K., 1994, Glial cell degeneration and hypomyelination caused by overexpression of myelin proteolipid protein gene, *Neuron* **13:**427–442.

Kaplan, B. J., Seyfried, T. N., and Glaser, G. H., 1979, Spontaneous polyspike discharges in an epileptic mutant mouse (tottering), *Exp. Neurol.* **66:**577–586.

Kasamo, K., Ishida, N., Murashima, Y. L., Ozawa, N., Nakamoto, Y., and Suzuki, J., 1992, The depth EEG and the multiunit activity in the hippocampal CA1 region during the epileptic seizure of an El mouse: Involvement of the hippocampal neurons in seizure manifestations, *Neuroscience* **18**(Suppl. 2):129–136.

Krall, R. L., Penry, J. K., Kupferberg, H. J., and Swinyard, E. A., 1978, Antiepileptic drug development: I. History and a program for progress, *Epilepsia* **19:**393–408.

Kurokawa, M., Naruse, H., and Kato, M., 1966, Metabolic studies on ep mouse, a special strain with convulsive predisposition, *Prog. Brain Res.* **21:**112–129.

Lee, K. S., Lanzino, G., Schottler, F., Collins, J., Kassell, N., Hiramatsu, K., Goto, Y., Hong, S., Berr, S., Caner, H., Yamamoto, H., Omary, R., Okonkwo, D., and Jane, J., 1995, A novel rat with a double cortex: I. General features, *Soc. Neurosci.* **21:**46.

Letts, V. A., Felix, R., Biddlecome, G. H., Arikkath, J., Mahaffey, C. L., Valenzuela, A., Bartlett II, F. S., Mori, Y., Campbell, K. P., and Frankel, W. N., 1998, The mouse stargazer gene encodes a neuronal Ca2+ channel gamma subunit, *Nat. Genet.* **19:**340–347.

Li, L., Chin, L. S., Shupliakov, O., Brodin, L., Sihra, T. S., Hvalby, O., Jensen, V., Zheng, D., McNamara, J. O., Greengard, P., and Andersen, P., 1995, Impairment of synaptic vesicle clustering and of synaptic transmission, and increased seizure propensity, in synapsin I-deficient mice, *Proc. Natl. Acad. Sci. USA* **92:**9235–9239.

Lin, F. H., Cao, Z., and Hosford, D. A., 1993, Increased number of GABAB receptors in the lethargic (lh/lh) mouse model of absence epilepsy, *Brain Res.* **608:**101–106.

Mangiarini, L., Sathasivam, K., Seller, M., Cozens, B., Harper, A., Hetherington, C., Lawton, M., Trottier, Y., Lehrach, H., Davies, S. W., and Bates, G. P., 1996, Exon 1 of the *HD* gene with an expanded CAG repeat is sufficient to cause a progressive neurological phenotype in transgenic mice, *Cell* **87:**493–506.

Martin, B., Desforges, C., and Chapouthier, G., 1991, Comparisons between patterns of convulsions induced by two beta-carbolines in 10 inbred strains of mice, *Neurosci. Lett.* **133:**73–76.

Martin, B., Chapouthier, G., and Motta, R., 1992, Analysis of B10.D2 recombinant congenic mouse strains shows that audiogenic and beta-CCM-induced seizures depend on different genetic mechanisms, *Epilepsia* **33:**11–13.

Matsumoto, M., Nakagawa, T., Inoue, T., Nagata, E., Tanaka, K., Takano, H., Minowa, O., Kuno, J., Sakakibara, S., Yamada, M., Yoneshima, H., Miyawaki, A., Fukuuchi, Y., Furuichi, T., Okano, H., Mikoshiba, K., and Noda, T., 1996, Ataxia and epileptic seizures in mice lacking type 1 inositol 1,4,5-triphosphate receptor, *Nature* **379:**168–171.

Matsumoto, Y., Hiramatsu, M., and Mori, A., 1983, Effects of phenytoin on convulsions and brain 5-hydroxytryptamine levels in El mice, *IRCS Med. Sci.* **11:**837.

Maxson, S. C., and Cowen, J. S., 1976, Electroencephalographic correlates of the audiogenic seizure response of inbred mice, *Physiol. Behav.* **16:**623–629.

Maxson, S. C., and Sze, P. Y., 1976, Electroencephalographic correlates of audiogenic seizures during ethanol withdrawal in mice, *Psychopharmacology Berl.* **47:**17–20.

Maxson, S. C., Fine, M. D., Ginsburg, B. E., and Koniecki, D. L., 1983, A mutant for spontaneous seizures in C57BL/10Bg mice, *Epilepsia* **24:**15–24.

McMahon, F. J., Stine, O. C., Meyers, D. A., Simpson, S. G., and DePaulo, J. R., 1995, Patterns of maternal transmission in bipolar affective disorder, *Am. J. Hum. Genet.* **56:**1277–1286.

Melo, J. A., Shendure, J., Pociask, K., and Silver, L. M., 1996, Identification of sex-specific quantitative trait loci controlling alcohol preference in C57BL/6 mice, *Nat. Genet.* **13:**147–153.

Menkes, J. H., 1974, "Textbook of Child Neurology," Lea & Febiger, Philadelphia.

Milner, R. J., and Sutcliffe, J. G., 1983, Gene expression in rat brain, *Nucleic Acids Res.* **11:**5497–5520.

Mita, T., Sashihara, S., Aramaki, I., Fueta, Y., and Hirano, H., 1991, Unusual biochemical development of genetically seizure-susceptible El mice, *Brain Res. Dev. Brain Res.* **64:**27–35.

Mitrovic, N., Besson, M. J., and Maurin, Y., 1990, Anticonvulsant effects of antagonists of the *N*-methyl-*D*-aspartate receptor complex in a genetic model of epilepsy: The quaking mouse, *Eur. J. Pharmacol.* **176:**357–361.

Mizukawa, K., and Mori, A., 1992, Cytoarchitecture of hippocampal formation of El mouse: Silver impregnation investigation and quantitative analysis, *Neurosciences* **18**(Suppl. 2):137–141.

Morahan, G., and Rakar, S., 1993, Localization of the mouse Na+/H+ exchanger gene on distal chromosome 4, *Genomics* **15:**231–232.

Mori, A., 1988, El Mice: Neurochemical approach to the seizure mechanism, *Neurosciences* **14:**275–285.

Mui, K., Yamagami, S., Kioka, T., Onishi, H., and Kawakita, Y., 1991, Sequence complexity of polyadenylated RNA from seizure-susceptible El mouse brain polysomes, *Epilepsy Res.* **10:**134–141.

Murashima, Y. L., Kasamo, K., and Suzuki, J., 1992, Developmental abnormalities of GABAergic system are involved in the formation of epileptogenesis in the EL, *Neurosciences* **18**(Suppl. 2):63–73.

Nadeau, J. H., 1989, Maps of linkage and synteny homologies between mouse and man, *Trends Genet.* **5**:82–86.

Naruse, H., and Kurokawa, M., 1992, The beginnings of studies on EL mice, *Neurosciences* **18**(Suppl. 2):1–3.

Nebert, D. W., 1980, Pharmacogenetics: An approach to understanding chemical and biologic aspects of cancer, *J. Natl. Cancer Inst.* **64**:1279–1290.

Neumann, P. E., and Collins, R. L., 1991, Genetic dissection of susceptibility to audiogenic seizures in inbred mice, *Proc. Natl. Acad. Sci. USA* **88**:5408–5412.

Neumann, P. E., and Collins, R. L., 1992, Confirmation of the influence of a chromosome 7 locus on susceptibility to audiogenic seizures, *Mamm. Genome.* **3**:250–253.

Neumann, P. E., and Seyfried, T. N., 1990, Mapping of two genes that influence susceptibility to audiogenic seizures in crosses of C57BL/6J and DBA/2J mice, *Behav. Genet.* **20**:307–323.

Noebels, J. L., 1986, Mutational analysis of inherited epilepsies, *Adv. Neurol.* **44**:97–113.

Noebels, J. L., 1991, Mutational analysis of spike-wave epilepsy phenotypes, *Epilepsy Res. Suppl.* **4**:201–212.

Noebels, J. L., 1994, Genetic and phenotypic heterogeneity of inherited spike-wave epilepsies, *in* "Idiopathic Generalized Epilepsies: Clinical, Experimental, and Genetic Aspects" (A. Malafosse, P. Genton, E. Hirschetal, eds.), pp.215–228, John Libbey & Co., London.

Noebels, J. L., 1995, Single locus mutations in mice expressing generalized spike-wave absence epilepsies, *Ital. J. Neurol. Sci.* **16**:107–111.

Noebels, J. L., 1996, Targeting epilepsy genes, *Neuron* **16**:241–244.

Noebels, J. L., and Sidman, R. L., 1979, Inherited epilepsy: Spike-wave and focal motor seizures in the mutant mouse tottering, *Science* **204**:1334–1336.

Noebels, J. L., Lutz, C. M., Fu, A., and Frankel, W. N., 1995, "Slow wave epilepsy," a mutant mouse with 3/sec spike wave seizures, *Soc. Neurosci. Abstr.* **21**:1965.

Ophoff, R. A., Terwindt, G. M., Vergouwe, M. N., van Eijk, R., Oefner, P. J., Hoffman, S. M. G., Lamerdin, J. E., Mohenweiser, H. W., Bulman, D. E., Ferrari, M., Hann, J., Lindhout, D., van Ommen, G.-J. B., Hofker, M. H., Ferrari, M. D., and Frants, R. R., 1996, Familial hemiplegic migraine and episodic ataxia type-2 are caused by mutations in the calcium channel gene CACNL1A4, *Cell* **87**:543–552.

Ottman, R., Annegers, J. F., Hauser, W. A., and Kurland, L. T., 1988, Higher risk of seizures in offspring of mothers than of fathers with epilepsy, *Am. J. Hum. Genet.* **43**:257–264.

Palayoor, S. T., and Seyfried, T. N., 1984, Genetic associations between Ca2+-ATPase activity and audiogenic seizures in mice, *J. Neurochem.* **42**:1771–1774.

Patil, N., Cox, D. R., Bhat, D., Faham, M., Myers, R. M., and Peterson, A. S., 1995, A potassium channel mutation in weaver mice implicates membrane excitability in granule cell differentiation [see comments], *Nat. Genet.* **11**:126–129.

Poderycki, M. J., Simoes, J. M., Todorova, M., Neumann, P. E., and Seyfried, T. N., 1998, Environmental influences on epilepsy gene mapping in EL mice, *J. Neurogenet.* **12**:67–86.

Qiao, X. X., and Noebels, J. L., 1991, Genetic and phenotypic heterogeneity of inherited spike-wave epilepsy: Two mutant gene loci with independent cerebral excitability defects, *Brain Res.* **555**:43–50.

Ralls, K., 1967, Auditory sensitivity in mice, Peromyscus and Mus musculus, *Anim. Behav.* **15**:123–128.

Readhead, C., and Hood, L., 1990, The dysmyelinating mouse mutations shiverer (shi) and myelin deficient (shimld), *Behav. Genet.* **20**:213–234.

Readhead, C., Schneider, A., Griffiths, I., and Nave, K. A., 1994, Premature arrest of myelin formation in transgenic mice with increased proteolipid protein gene dosage, *Neuron* **12**:583–595.

Ribak, C. E., 1991, Epilepsy and the cortex, *in* "Cerebral Cortex," Vol. 9 (A. Peters, ed.), pp. 427–483, Plenum Press, New York.

Rise, M. L., Frankel, W. N., Coffin, J. M., and Seyfried, T. N., 1991, Genes for epilepsy mapped in the mouse, *Science* **253**:669–673.

Roffler Tarlov, S., and Graybiel, A. M., 1987, The postnatal development of the dopamine-containing innervation of dorsal and ventral striatum: Effects of the weaver gene, *J. Neurosci.* **7**:2364–2372.

Rosahl, T. W., Spillane, D., Missler, M., Herz, J., Selig, D. K., Wolff, J. R., Hammer, R. E., Malenka, R. C., and Sudhof, T. C., 1995, Essential functions of synapsins I and II in synaptic vesicle regulation [see comments], *Nature* **375**:488–493.

Rosenblatt, D. E., Lauter, C. J., and Trams, E. G., 1976, Deficiency of a Ca2+-ATPase in brains of seizure prone mice, *J. Neurochem.* **27**:1299–1304.

Saito, M., Endo, S., Hioki, K., and Nomura, T., 1992, Rearing and maintenance of EL mice as laboratory animals, *Neurosciences* **18**(Suppl. 2):5–8.

Sango, K., McDonald, M. P., Crawley, J. N., Mack, M. L., Tifft, C. J., Skop, E., Starr, C. M., Hoffmann, A., Sandhoff, K., Suzuki, K., and Proia, R. L., 1996, Mice lacking both subunits of lysosomal β-hexosaminidase display gangliosidosis and mucopolysaccharidosis, *Nat. Genet.* **14**:348–352.

Schlesinger, K., Boggan, W., and Freedman, D. X., 1965, Genetics of audiogenic seizures: I. Relation to brain serotonin and norepinephrine in mice, *Life Sci.* **4**:2345–2351.

Schlesinger, K., Boggan, W., and Freedman, D. W., 1968, Genetics of audiogenic seizures: II. Effects of pharmacological manipulation of brain serotonin, norepinephrine and gamma-aminobutyric acid, *Life Sci.* **7**:437–447.

Schmidt, M. J., Sawyer, B. D., Perry, K. W., Fuller, R. W., Foreman, M. M., and Ghetti, B., 1982, Dopamine deficiency in the weaver mutant mouse, *J. Neurosci.* **2**:376–380.

Schreiber, R. A., 1986, Behavior genetics of audiogenic seizures in DBA/2J and Rb-1 mice, *Behav. Genet.* **16**:365–368.

Seyfried, T. N., 1982a, Convulsive Disorders, *in* "The Mouse in Biomedical Research," Vol. IV (H. L. Foster, J. D. Small, and J. G. Fox, eds.), pp. 97–124, Academic Press, New York.

Seyfried, T. N., 1982b, Developmental genetics of audiogenic seizure susceptibility in mice, *in* "Genetic Basis of the Epilepsies" (V. E. Anderson, W. A. Hauser, J. K. Penry, and C. F. Sing, eds.), pp. 198–210, Raven Press, New York.

Seyfried, T. N., 1993, Genetic heterogeneity for the development of audiogenic seizures in mice, *Brain Res.* **271**:325–329.

Seyfried, T. N., and Glaser, G. H., 1981, Genetic linkage between the AH locus and a major gene that inhibits susceptibility to audiogenic seizures in mice, *Genetics* **99**:117–126.

Seyfried, T. N., Yu, R. K., and Glaser, G. H., 1980, Genetic analysis of audiogenic seizure susceptibility in C57BL/6J X DBA/2J recombinant inbred strains of mice, *Genetics* **94**:701–718.

Seyfried, T. N., Itoh, T., Glaser, G. H., Miyazawa, N., and Yu, R. K., 1981, Cerebellar gangliosides and phospholipids in mutant mice with ataxia and epilepsy: The Tottering/Leaner syndrome, *Brain Res.* **216**:429–436.

Seyfried, T. N., Yu, R. K., and Miyazawa, N., 1982, Differential cellular enrichment of gangliosides in the mouse cerebellum: Analysis using neurological mutants, *J. Neurochem.* **38**:551–559.

Seyfried, T. N., Brigande, J. V., Flavin, H. J., Frankel, W. N., Rise, M. L., and Wieraszko, A., 1992, Genetic and biochemical correlates of epilepsy in the EL mouse, *Neurosciences* **18**(Suppl. 2):9–20.

Shine, H. D., Readhead, C., Popko, B., Hood, L., and Sidman, R. L., 1992, Morphometric analysis of normal, mutant, and transgenic CNS: Correlation of myelin basic protein expression to myelinogenesis, *J. Neurochem.* **58**:342–349.

Signorini, S., Liao, Y. J., Duncan, S. A., Jan, L. Y., and Stoffel, M., 1997, Normal cerebellar development but susceptibility to seizures in mice lacking G protein-coupled, inwardly rectifying K+ channel GIRK2, *Proc. Natl. Acad. Sci. USA* **94**:923–927.

Sisodia, S. S., Thinakaran, G., Slunt, H. H., Kitt, C. A., Von-Koch, C. S., Reed, R. R., Zheng, H., and Price, D. L., 1996, *In vivo* analysis of APP and APLP2 function, *J. Neurochem.* **66**(Suppl. 1):S53.

Slesinger, P. A., Stoffel, M., Jan, Y. N., and Jan, L. Y., 1997, Defective gamma-aminobutyric acid type B receptor-activated inwardly rectifying K+ currents in cerebellar granule cells isolated from weaver and Girk2 null mutant mice, *Proc. Natl. Acad. Sci. USA* **94:**12210–12217.

Steinlein, O. K., Mulley, J. C., Propping, P., Wallace, R. H., Phillips, H. A., Sutherland, G. R., Scheffer, I. E., and Berkovic, S. F., 1995, A missense mutation in the neuronal nicotinic acetylcholine receptor alpha 4 subunit is associated with autosomal dominant nocturnal frontal lobe epilepsy, *Nat. Genet.* **11:**201–203.

Strachan, T., and Read, A. P., 1996, "Human Molecular Genetics," John Wiley & Sons, New York.

Strickberger, M. W., 1985, "Genetics," Macmillan, New York.

Sugaya, E., Ishige, A., Sekiguchi, K., Iizuka, S., Ito, K., Sugimoto, A., Aburada, M., and Hosoya, E., 1986, Pentylenetetrazol-induced convulsion and effect of anticonvulsants in mutant inbred strain El mice, *Epilepsia* **27:**354–358.

Sugaya, E., Sugaya, A., Yuyama, N., Tsuda, T., Kajiwara, K., Kubota, K., Katoh, K., Hosoya, S., Takagi, T., and Motoki, M., 1992, Developmental defects of primary cultured neurons from the cerebral cortex of the El mouse and their amelioration with a herbal mixture formulation, TJ-960, *Neurosciences* **18**(Suppl. 2):111–128.

Sutoo, D., Akiyama, K., and Takita, H., 1987, The relationship between metal ion levels and biogenic amine levels in epileptic mice, *Brain Res.* **418:**205–213.

Sutoo, D., Akiyama, K., and Takita, H., 1992, Abnormal behavior in epileptic El mice related to a calcium-dependent-catacholamine synthesis disorder of the central nervous system, *Neurosciences* **18**(Suppl. 2):55–62.

Suzuki, J., 1976, Paroxysmal discharges in the electroencephalogram of the El mouse, *Experientia* **32:**336–338.

Suzuki, J., and Nakamoto, Y., 1977, Seizure patterns and electroencephalograms of El mouse, *Electroencephalogr. Clin. Neurophysiol.* **43:**299–311.

Suzuki, J., Matsushita, M., and Nakamoto, Y., 1983, Histopathological alterations in the hippocampus of an El mouse, *Folia Psychiatr. Jpn.* **37:**362–363.

Suzuki, J., Kasamo, K., Ishida, N., and Murashima, Y. L., 1991, Initiation, propagation and generalization of paroxysmal discharges in an epileptic mutant animal, *Jpn. J. Psychiatry Neurol.* **45:**271–274.

Tamaki, Y., Oda, S., and Kameyama, Y., 1986, Postnatal locomotion development in a neurological mutant of rolling mouse Nagoya, *Dev. Psychobiol.* **19:**67–77.

Tecott, L. H., Sun, L. M., Akana, S. F., Strack, A. M., Lowenstein, D. H., Dallman, M. F., and Julius, D., 1995, Eating disorder and epilepsy in mice lacking 5-HT2c serotonin receptors, *Nature* **374:**542–546.

Toth, M., Grimsby, J., Buzsaki, G., and Donovan, G. P., 1995, Epileptic seizures caused by inactivation of a novel gene, jerky, related to centromere binding protein-B in transgenic mice, *Nat. Genet.* **11:**71–75.

Trams, E. G., and Lauter, C. J., 1978, Ecto-ATPase deficiency in glia of seizure-prone mice, *Nature* **271:**270–271.

Valatx, J.-L., 1980, Quaking mutation: Relationship between sleep and seizures, *in* "Neurological Mutations Affecting Myelination" (N. Baumann, ed.), pp. 507–512, Elsevier, New York.

Waymire, K. G., Mahuren, J. D., Jaje, J. M., Guilarte, T. R., Coburn, S. P., and MacGregor, G. R., 1995, Mice lacking tissue non-specific alkaline phosphatase die from seizures due to defective metabolism of vitamin B-6, *Nat. Genet.* **11:**45–51.

Wieraszko, A., and Seyfried, T. N., 1989, Increased amount of extracellular ATP in stimulated hippocampal slices of seizure prone mice, *Neurosci. Lett.* **106:**287–293.

Wieraszko, A., and Seyfried, T. N., 1993, Influence of audiogenic seizures on synaptic facilitation in mouse hippocampal slices is mediated by *N*-methyl-*D*-aspartate receptor, *Epilepsia* **34:**979–984.

Williams, C. A., Zori, R. T., Stone, J. W., Gray, B. A., Cantu, E. S., and Ostrer, H., 1990, Maternal origin of 15q11-13 deletions in Angelman syndrome suggests a role for genomic imprinting, *Am. J. Med. Genet.* **35:**350–353.

Willott, J. F., Erway, L. C., Archer, J. R., and Harrison, D. E., 1995, Genetics of age-related hearing loss in mice. II. Strain differences and effects of caloric restriction on cochlear pathology and evoked response thresholds, *Hear. Res.* **88:**143–155.

Wolf, P., 1994, Historical aspects: the concept of idiopathy, *in* "Idiopathic Generalized Epilepsies: Clinical, Experimental, and Genetic Aspects" (A. Malafosse, P. Genton, and E. Hirschetal, eds.), pp. 3–6, John Libbey & Co., London.

Yan, Q. S., Jobe, P. C., and Dailey, J. W., 1995, Further evidence of anticonvulsant role for 5-hydroxytryptamine in genetically epilepsy-prone rats, *Br. J. Pharmacol.* **115:**1314–1318.

Yu, R. K., Holley, J. A., Macala, L. J., and Spencer, D. D., 1987, Ganglioside changes associated with temporal lobe epilepsy in the human hippocampus, *Yale J. Biol. Med.* **60:**107–117.

Zuo, J., De Jager, P. L., Takahashi, K. A., Jiang, W., Linden, D. J., and Heintz, N., 1997, Neurodegeneration in Lurcher mice caused by mutation in delta2 glutamate receptor gene, *Nature* **388:**769–773.

INDEX